Biology For
Medical, Dental and Pharmaceutical
Admission Tests
(IMAT, MCAT, BMAT, PCAT, UKCAT, TOLC-F, UCAS, MFP, A-Level, …)

Dr. Fatemeh Hojjati, Biology PhD.

Biology For
Medical, Dental and Pharmaceutical Admission Tests
(IMAT, MCAT, BMAT, PCAT, UKCAT, TOLC-F, UCAS, MFP, A-Level, ...)

Volume 1: Chemistry And Cell

Dr. Fatemeh Hojjati, Biology PhD.

Preface

I am honored to present the "Biology For Medical, Dental and Pharmaceutical Admission Tests (IMAT, MCAT, BMAT, PCAT, UKCAT, TOLC-F, UCAS, MFP, A-Level, …)" book. I am PhD in biology and a researcher of biology in Baylor university, Texas, USA. As a biology teacher in high schools and universities, in recent years, I realized the serious need for a perfect resource of biology for candidates of medical, dental and pharmaceutical admission tests to study.

The available biology resources are mainly academic text books that contain a lot of topics that have explained them to high levels, while what medical, dental and pharmaceutical admission tests candidates should study in biology are often specific topics at specific levels. Therefore, I decided to provide a perfect resource of biology for such candidates so that by focusing on it instead of reading various resources they can save their time and energy and this book is the result. The topics that are mentioned for medical, dental and pharmaceutical admission tests are mainly divided into five general categories which include: Biochemistry, Cell, Genetics, Evolution and Human body. This book contains all these topics that are explained as much as needed.

Such a book has been compiled for the first time in the world. I have used 16 biology text books of the best ones in the world to write it and you can find them in references section at the end of this book. This book is written in three volumes: volume 1 is about chemistry and cell, volume 2 contains genetics and evolution and volume 3 is about anatomy and physiology of human organ systems.

By using 16 biology text books and by working hard to write it, I guarantee this book at the highest level. By reading this book, you will not need to study any other text book and you will reach the scientific level required for medicine, dentistry and pharmacy entrance exams.

Finally, I will be glad to receive comments about this book from candidates, biologists and everyone else who reads it.

Good luck!

Dr. Fatemeh Hojjati,

Biology PhD, 2024

Contents

Chapter 1: Life, Chemistry, and Water

Matter, elements and compounds **3**

Subatomic particles **4**

Isotopes **5**

The energy levels of electrons **7**

Electron distribution and chemical properties **8**

Chemical bonds **11**

Chemical reactions make and break chemical bonds **17**

Types of chemical reactions **18**

Energy flow in chemical reactions **20**

Factors influencing the rate of chemical reactions **20**

Functional groups **21**

Biochemistry **22**

Water and life **22**

Hydrophilic and hydrophobic substances **29**

Salts **30**

Acidic and basic conditions affect living organisms **31**

Buffers **33**

Key concepts **34**

Chapter 2: Biological Molecules

Living organisms are composed of about 25 key elements **39**

Life is based on carbon compounds including **40**

Macromolecules are polymers, built from monomers **41**

The synthesis and breakdown of polymers **43**

Carbohydrates include sugars and polymers of sugars **44**

Lipids are a diverse group of hydrophobic molecules **50**

Proteins **58**

Nucleic acids **72**

Key concepts **80**

Chapter 3: Cell and Microscopy

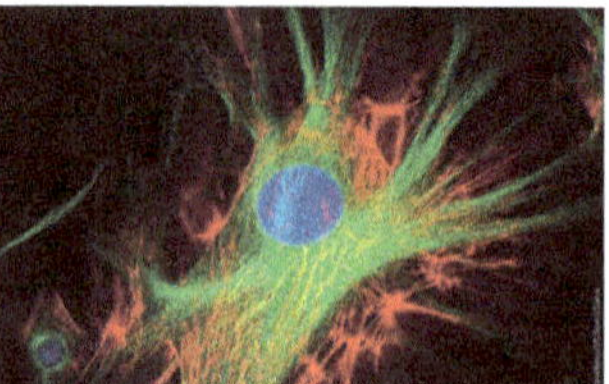

Microscopy **86**

Preparation of samples for study **89**

Light microscopes are used to study stained or living cells **91**

Electron microscopes **93**

Cell theory is the unifying foundation of cell biology **95**

Cell size is limited **97**

Comparing prokaryotic and eukaryotic cells **99**

Prokaryotic cells have relatively simple organization **100**

Eukaryotic cells **104**

Plasma membrane **106**

The cytoplasm **108**

Nucleus, the information center **109**

Ribosomes: the cell's protein synthesis machinery **113**

The endoplasmic reticulum: biosynthetic factory **116**

The Golgi apparatus: shipping and receiving center **119**

Lysosomes: digestive compartments **122**

Vacuoles: diverse maintenance compartments **124**

Chloroplasts: capture of light energy **125**

Mitochondrion: chemical energy conversion **126**

Peroxisomes: oxidation **129**

The cytoskeleton **131**

Cell surface specializations (cell coverings and cell junctions) **142**

Key concepts **156**

Chapter 4: Membrane Transport and Cell Signaling

Cellular membranes consist of three component groups **161**

The fluidity of membranes **164**

Membrane proteins and their functions **166**

The role of membrane carbohydrates in cell-cell recognition **168**

Membrane structure results in selective permeability **169**

Passive transport **170**

Active transport **178**

Bulk transport **183**

Cell communication **187**

Signal reception **191**

Signal transduction **199**

Cellular response **211**

Regulation of the response **214**

Apoptosis **218**

Key concepts **223**

Chapter 5: Metabolism, Energy and Enzyme

An organism's metabolism transforms matter and energy **227**

Forms of energy **228**

The laws of energy transformation **229**

ATP powers cellular work **232**

Redox reactions **236**

Enzymes are organized into teams in metabolic pathways **241**

Catalysis in the enzyme's active site **243**

Effect of temperature on enzyme activity **244**

Effect of PH on enzyme activity **244**

Effect of substrate concentration or enzyme concentration on enzyme activity **245**

Regulation of enzyme activity helps control metabolism **246**

Enzyme inhibitors **248**

Comparing enzyme affinities **249**

RNA-based biological catalysts: ribozymes **251**

Metabolic pathways are regulated in three general ways **252**

Key concepts **253**

Chapter 6: Cellular Respiration and Fermentation

Catabolic pathways yield energy by oxidizing organic fuels **257**

Catabolic pathways and production of ATP **258**

Oxidation of organic fuel molecules during cellular respiration **260**

The stages of cellular respiration **262**

Glycolysis harvests chemical energy by oxidizing glucose to pyruvate **264**

Oxidation of pyruvate to Acetyl CoA **265**

The citric acid cycle **266**

The pathway of electron transport **268**

Chemiosmosis: the energy-coupling mechanism **271**

An accounting of ATP production by cellular respiration **275**

Fermentation and anaerobic respiration **277**

Types of fermentation **278**

Comparing fermentation with anaerobic and aerobic respiration **280**

Glycolysis and the citric acid cycle connect to many other metabolic pathways **281**

Biosynthesis (anabolic pathways) **283**

Glycolysis and citric acid cycle are regulated by feedback mechanisms **285**

Photosynthesis and respiration are ancient pathways **287**

Key concepts **288**

Chapter 7: Photosynthesis

Photosynthesis feeds the biosphere **293**

Chloroplasts: the sites of photosynthesis in plants **294**

The two stages of photosynthesis **296**

The light reactions convert solar energy to the chemical energy **297**

Photosynthetic pigments: the light receptors **298**

Excitation of chlorophyll by light **302**

A photosystem **303**

Linear electron flow **305**

Cyclic electron flow **307**

A comparison of chemiosmosis in chloroplasts and mitochondria **308**

The Calvin cycle **310**

Alternative mechanisms of carbon fixation have evolved in hot, arid climates **314**

C_4 plants have evolved to minimize photorespiration **315**

CAM plants **317**

Limiting factors in photosynthesis **319**

Key concepts **320**

Chapter 8: The Cell Cycle

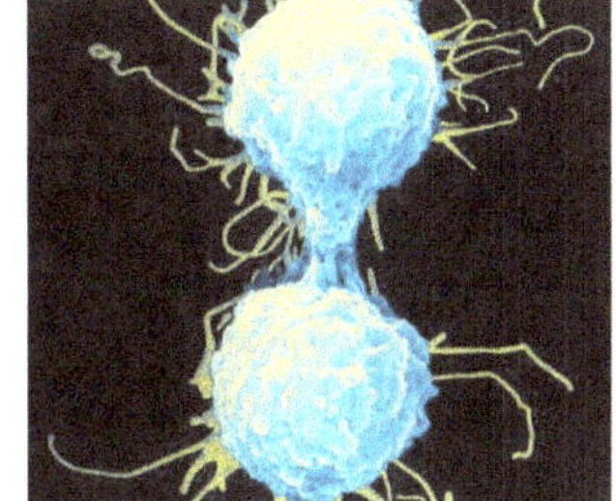

Cell division **325**

Genetic material **326**

A chromosome consists of a DNA molecule packed together with proteins **327**

Distribution of chromosomes during eukaryotic cell division **336**

Phases of the cell cycle **338**

The cell cycle is divided into five phases **339**

Mitosis **340**

Cytokinesis **346**

Binary fission in bacteria **347**

The cell cycle control system **349**

Internal and external signals at the checkpoints **356**

Cancer **359**

Cell death is part of life **372**

Cell aging **374**

Cellular diversity **375**

Stem cells **377**

Key concepts **380**

References **383**

CHAPTER 1

Life, Chemistry and Water

Chapter Contents:

- Matter, elements and compounds
- Subatomic Particles
- Atomic Number and Atomic Mass
- Isotopes
- The Energy Levels of Electrons
- Electron Distribution and Chemical Properties
- Chemical bonds
- Chemical reactions make and break chemical bonds
- Functional groups
- Biochemistry
- Water and Life
- Emergent properties of water
- Hydrophilic and Hydrophobic Substances
- Salts
- Acidic and basic conditions affect living organisms

Matter, elements and compounds

Matter is the "stuff" of the universe. More precisely, matter is anything that occupies space and has mass. With some exceptions, it can be seen, smelled, and felt. Matter exists in solid, liquid, and gaseous states. Examples of each state are found in the human body. Solids, like bones and teeth, have a definite shape and volume. Liquids such as blood plasma have a definite volume, but they conform to the shape of their container. Gases have neither a definite shape nor a definite volume. The air we breathe is a gas.

Matter is made up of **elements**. An element is a substance that cannot be broken down to other substances by chemical reactions. At present, 118 elements are recognized. Of these, 92 occur in nature. The rest are made artificially in particle accelerator devices. Each element has a symbol, usually the first letter or two of its name. Some symbols are derived from Latin or German; for instance, the symbol for sodium is Na, from the Latin word natrium. A **compound** is a substance consisting of two or more different elements combined in a fixed ratio. A compound has characteristics different from those of its elements. Water (H_2O), as a compound, consists of the elements hydrogen (H) and oxygen (O) in a 2:1 ratio.

▶ **Figure 1.1** The emergent properties of a compound. The metal sodium combines with the poisonous gas chlorine, forming the edible compound sodium chloride, or table salt.

Of the 92 natural elements, about 20–25% are essential elements that an organism needs to live a healthy life and reproduce. The essential elements are similar among organisms, but there is some variation—for example, humans need 25 elements, but plants need only 17.

Just four elements (**major elements**), oxygen (O), carbon (C), hydrogen (H), and nitrogen (N), make up approximately 96% of living matter. Calcium (Ca), phosphorus (P), potassium (K), sulfur (S), and a few other elements account for most of the remaining 4% or so of an organism's mass (**lesser elements**).

Trace elements are required by an organism in only minute quantities. Some trace elements, such as iron (Fe), are needed by all forms of life; others are required only by certain species. For example, in vertebrates (animals with backbones), the element iodine (I) is an essential ingredient of a hormone produced by the thyroid gland. Some naturally occurring elements are toxic to organisms. In humans, for instance, the element arsenic has been linked to numerous diseases and can be lethal. In some areas of the world, arsenic occurs naturally and can make its way into the groundwater.

Element	Symbol	% Human body mass	% All atoms in human body
Most abundant in living organisms (approximately 95% of total mass)			
Oxygen	O	65	25.5
Carbon	C	18	9.5
Hydrogen	H	9	63.0
Nitrogen	N	3	1.4

Element	Symbol	Element	Symbol	Element	Symbol
Mineral elements (less than 1% of total mass)		**Trace elements (less than 0.01% of total mass)**			
Calcium	Ca	Boron	B	Manganese	Mn
Chlorine	Cl	Chromium	Cr	Molybdenum	Mo
Magnesium	Mg	Cobalt	Co	Selenium	Se
Phosphorus	P	Copper	Cu	Silicon	Si
Potassium	K	Fluorine	F	Tin	Sn
Sodium	Na	Iodine	I	Vanadium	V
Sulfur	S	Iron	Fe	Zinc	Zn

▲ **Table 1.1** Chemical Elements Essential for Life in Most Organisms

Subatomic Particles

Each element is composed of more or less identical particles or building blocks, called **atoms**. Although the atom is the smallest unit having the properties of an element, these tiny bits of matter are composed of even smaller parts, called subatomic particles. Using high-energy collisions, physicists have produced more than 100 types of particles from the atom, but only three kinds of particles are relevant here: **neutrons**, **protons**, and **electrons**. Protons and electrons are electrically charged. Each proton has one unit of positive charge, and each electron has one unit of negative charge. A neutron, as its name implies, is electrically neutral.

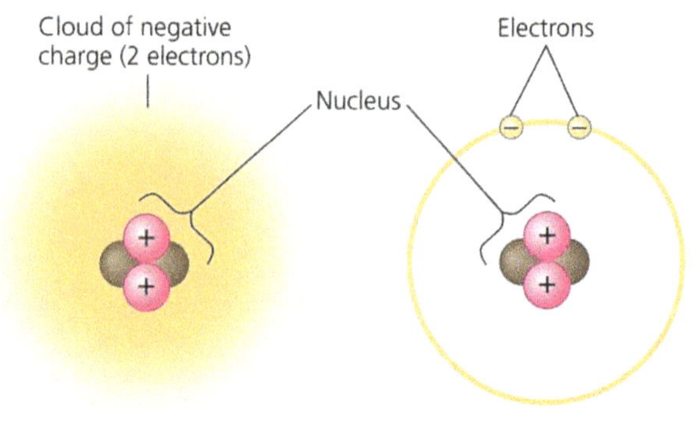

▶ **Figure 1.2** Simplified models of a helium (He) atom. The helium nucleus consists of 2 neutrons (brown) and 2 protons (pink). Two electrons (yellow) exist outside the nucleus. These models are not to scale; they greatly overestimate the size of the nucleus in relation to the electron.

Center of an atom; protons give the nucleus a positive charge. The rapidly moving electrons form a "cloud" of negative charge around the nucleus, and it is the attraction between opposite charges that keeps the electrons in the vicinity of the nucleus. Almost all of the volume of an atom is empty space. This is because the electrons are usually far away from the nucleus, relative to its size. All atoms of a particular element have the same number of protons in their nuclei. This number of protons, which is unique to that element, is called the **atomic number** and is written as a subscript to the left of the symbol for the element.

The atomic number tells us the number of protons and also the number of electrons in an electrically neutral atom. We can deduce the number of neutrons from a second quantity, the **mass number**, which is the total number of protons and neutrons in the nucleus of an atom. The mass number is written as a superscript to the left of an element's symbol. The simplest atom is hydrogen which has no neutrons; it consists of a single proton with a single electron.

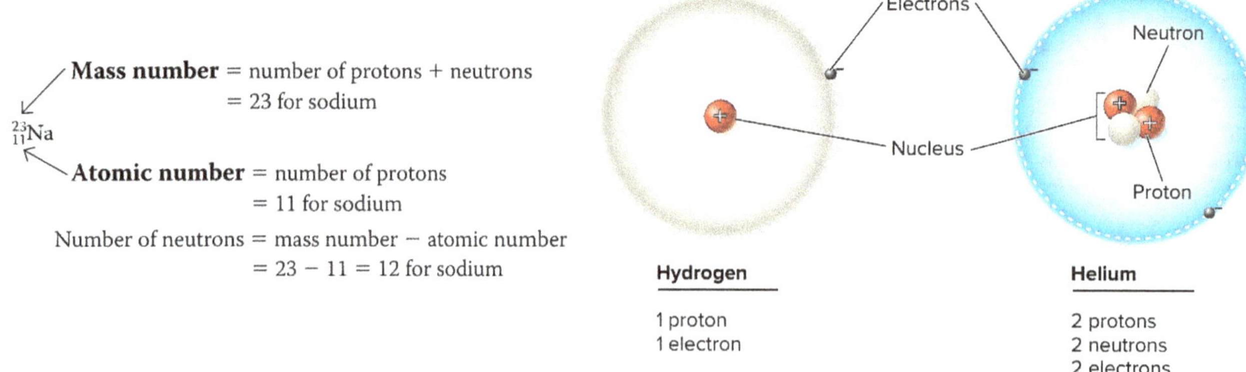

▲ **Figure 1.3** Diagrams of two simple atoms. These are models of the two simplest atoms, hydrogen and helium. The nucleus of most atoms (except hydrogen) consists of protons and neutrons, whereas electrons are found outside the nucleus.

Isotopes

All atoms of a given element have the same number of protons, but some atoms have more neutrons than other atoms of the same element and therefore have greater mass. These different atomic forms of the same element are called isotopes of the element. In nature, an element may occur as a mixture of its isotopes. Although the isotopes of an element have slightly different masses, they behave identically in chemical reactions. It is for this reason that an atom's electrons, not its protons or neutrons, determine its chemical behavior. Carbon has several isotopes. The most abundant of these are ^{12}C, ^{13}C, and ^{14}C. Each of the carbon isotopes has six protons (otherwise it would not be carbon), but ^{12}C has six neutrons, ^{13}C has seven, and ^{14}C has eight.

The three isotopes of hydrogen, ^{1}H (ordinary hydrogen), ^{2}H (deuterium), and ^{3}H (tritium), contain 0, 1, and 2 neutrons, respectively. In medicine, radioisotopes are used for both diagnosis and treatment. The location and/or metabolism of a substance such as a hormone or drug can be followed in the body by labeling the substance with a radioisotope such as carbon-14 or tritium. Radioisotopes are used to test thyroid gland function, to provide images of blood flow in the arteries supplying the cardiac muscle, and to study many other aspects of body function and chemistry. Because radiation can interfere with cell division, radioisotopes have been used therapeutically in treating cancer, a disease often characterized by rapidly dividing cells.

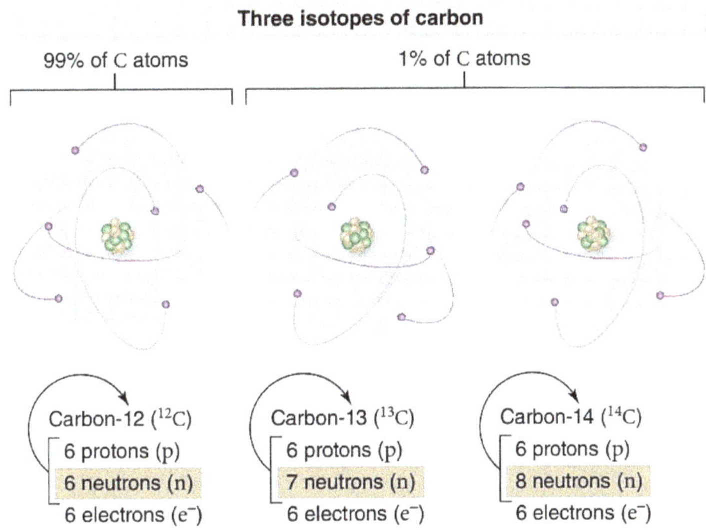

▲ **Figure 1.4** Three Isotopes of Carbon. Carbon's atomic number is 6, so its nucleus always contains six protons. These three carbon isotopes, however, have different numbers of neutrons.

A **radioactive isotope** is one in which the nucleus decays spontaneously, giving off particles and energy (subatomic alpha (α) particles, beta (β) particles or gamma (γ) rays). When the radioactive decay leads to a change in the number of protons, it transforms the atom to an atom of a different element. Radioactive isotopes have many useful applications in biology. For example, when a carbon-14 (^{14}C) atom decays, a neutron decays into a proton, transforming the atom into a nitrogen (^{14}N) atom. A "parent" isotope decays into its "daughter" isotope at a fixed rate, expressed as the **half-life** of the isotope—the time it takes for 50% of the parent isotope to decay.

Term	Definition
Element	A fundamental type of substance
Atom	The smallest unit of an element that retains the characteristics of that element
Atomic number	The number of protons in an atom's nucleus
Mass number	The number of protons plus the number of neutrons in an atom's nucleus
Isotope	Any of the different forms of the same element, distinguished from one another by the number of neutrons in the nucleus
Atomic weight	The average mass of all isotopes of an element

▲ **Table 1.2** A Mini glossary of Matter

A technique called **radiometric dating** uses the clocklike decay of unstable isotopes to estimate the age of organic material, rocks, or fossils that contain them. With respect to evolution, radiometric dating is a particularly important technique for tracing evolutionary lineages through analysis of fossils. The radioactive isotopes are incorporated into biologically active molecules, which are then used as **tracers** to track atoms during metabolism, the chemical processes of an organism.

The Energy Levels of Electrons

An atom's electrons vary in the amount of energy they possess. **Energy** is defined as the capacity to cause change, for instance, by doing work. **Potential energy** is the energy that matter possesses because of its location or structure. The electrons of an atom have potential energy due to their distance from the nucleus. It takes work to move a given electron farther away from the nucleus, so the more distant an electron is from the nucleus, the greater its potential energy. An electron's potential energy is determined by its energy level. An electron can exist only at certain energy levels, not between them.

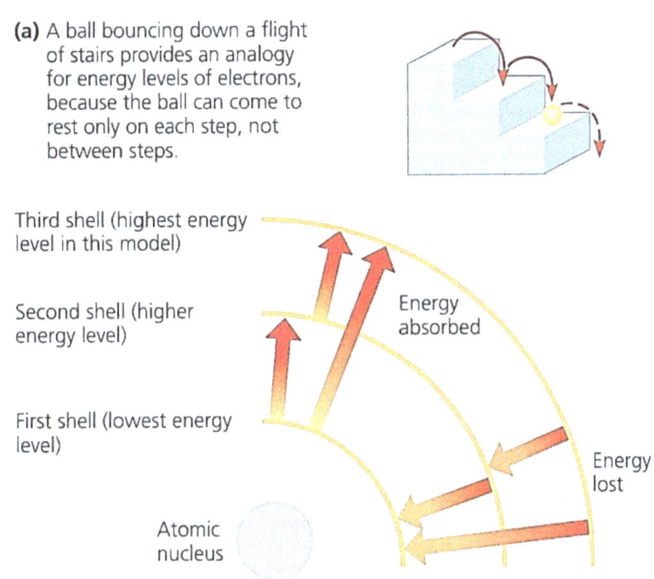

▲ **Figure 1.5** Energy levels of an atom's electrons.

An electron's energy level is correlated with its average distance from the nucleus. Electrons are found in different **electron shells**, each with a characteristic average distance and energy level. In diagrams, shells can be represented by concentric circles. The first shell is closest to the nucleus, and electrons in this shell have the lowest potential energy. Electrons in the second shell have more energy, and electrons in the third shell even more energy. An electron can move from one shell to another, but only by absorbing or losing an amount of energy equal to the difference in potential energy between its position in the old shell and that in the new shell. When an electron absorbs energy, it moves to a shell farther out from the nucleus.

Electron Distribution and Chemical Properties

The chemical behavior of an atom is determined by the distribution of electrons in the atom's electron shells. Beginning with hydrogen, the simplest atom, we can imagine building the atoms of the other elements by adding 1 proton and 1 electron at a time (along with an appropriate number of neutrons).

The chemical behavior of an atom depends mostly on the number of electrons in its outermost shell. We call those outer electrons **valence electrons** and the outermost electron shell the **valence shell**. Atoms with the same number of electrons in their valence shells exhibit similar chemical behavior. An atom with a completed valence shell is unreactive; that is, it will not interact readily with other atoms. Such elements are said to be inert, meaning chemically unreactive like helium, neon, and argon.

Atom name	Hydrogen	Oxygen	Nitrogen	Carbon
Electron number needed to complete outer shell (typical number of covalent bonds)	1	2	3	4

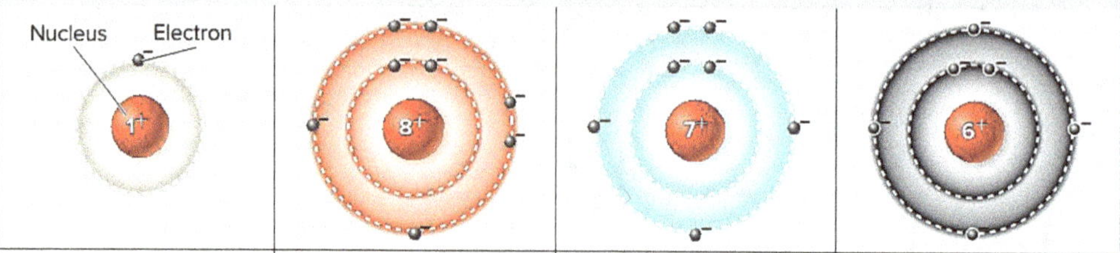

▲ **Figure 1.6** The number of covalent bonds formed by common elements found in living organisms.

In the real world, electrons really like to be in pairs when they occupy atoms. Solitary atoms that have unpaired electrons are called free radicals. With some exceptions, free radicals have a very strong tendency to interact with other atoms, and such interactions make them dangerous to life. A free radical sodium atom can easily evict its one unpaired electron, so that its second shell, which is full of electrons, becomes its outermost, and no vacancies remain. This is the atom's most stable state, and in fact the vast majority of sodium atoms on Earth are like this one, with 11 protons and 10 electrons.

A free radical is an ion or molecule that has an unpaired electron in its outermost shell. (Most of an atom's electrons associate in pairs.) A common example of a free radical is superoxide, which is formed by the addition of an electron to an oxygen molecule. Having an unpaired electron makes a free radical unstable and destructive to nearby molecules. Free radicals break apart important body molecules by either giving up their unpaired electron to or taking on an electron from another molecule.

The three-dimensional space where an electron is found 90% of the time is called an **orbital**. Some electron orbitals near the nucleus are spherical (s orbitals), whereas others are dumbbell shaped (p orbitals). Still other orbitals, farther away from the nucleus, may have different shapes.

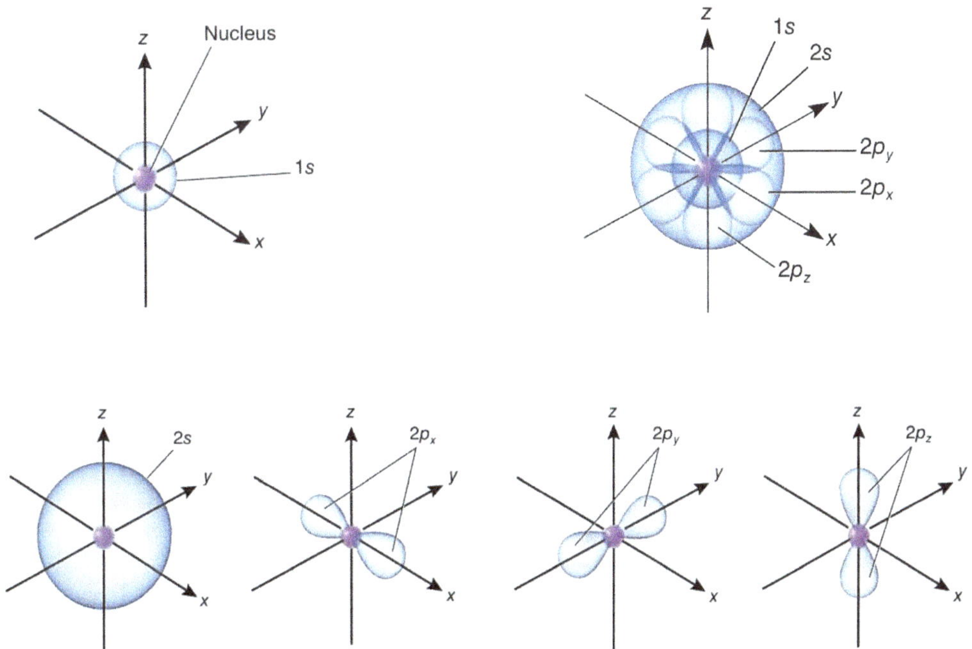

▲ **Figure 1.7** Atomic orbitals. Electrons occupy orbitals corresponding to energy levels. Each orbital is represented as an "electron cloud." The arrows labeled x, y, and z establish the imaginary axes of the atom.

Each electron shell contains electrons at a particular energy level, distributed among a specific number of orbitals of distinctive shapes and orientations. No more than 2 electrons can occupy a single orbital. The reactivity of an atom arises from the presence of unpaired electrons in one or more orbitals of the atom's valence shell. In fact atoms interact in a way that completes their valence shells. When they do so, it is the unpaired electrons that are involved.

Ninety elements occur naturally, each with a different number of protons and a different arrangement of electrons. When the 19th century Russian chemist Dmitri Mendeleev arranged the known elements in a table according to their atomic number, he discovered one of the great generalizations of science: The elements exhibit a pattern of chemical properties that repeats itself in groups of eight. This periodically repeating pattern lent the table its name: the periodic table of elements.

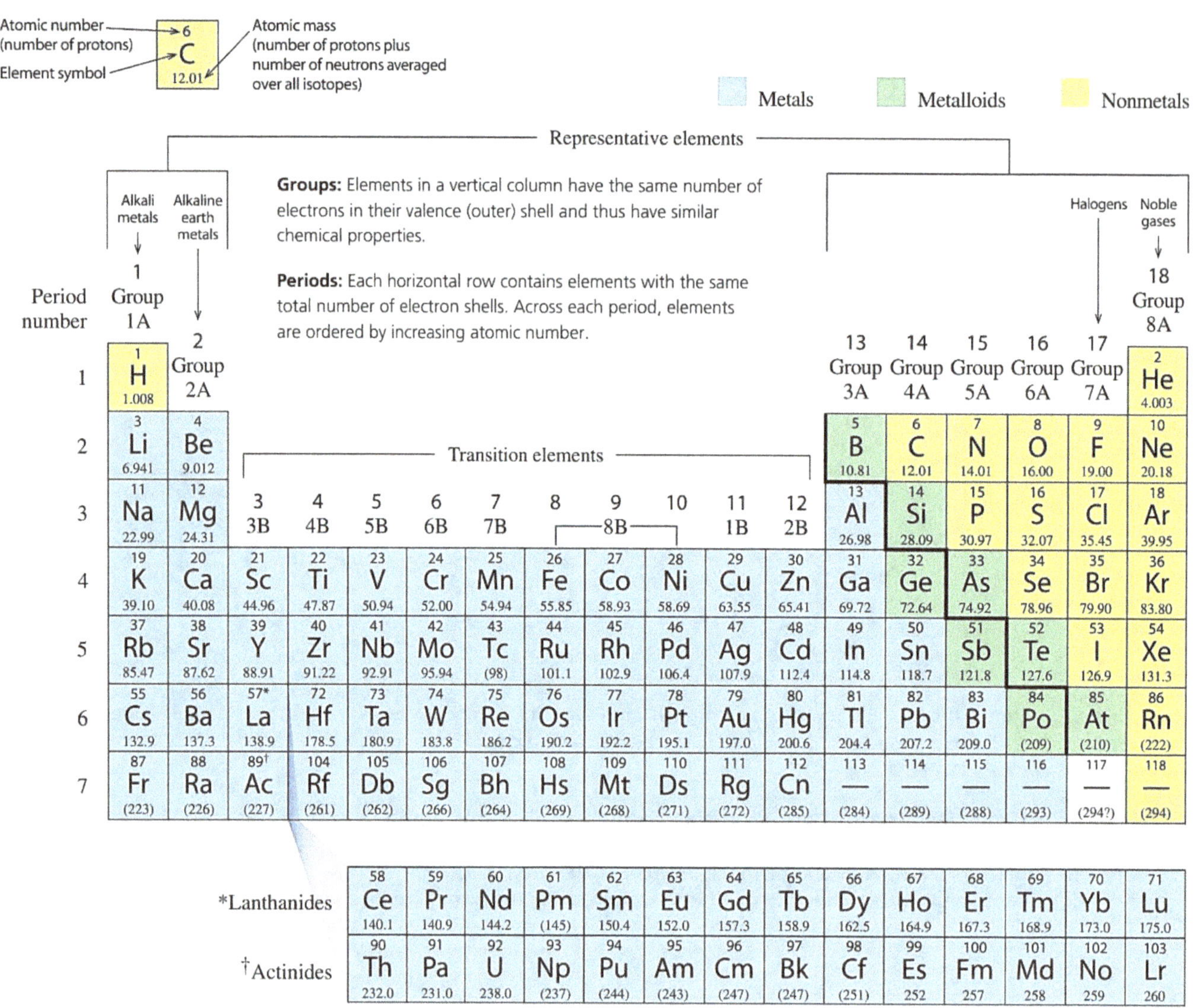

▲ **Figure 1.8** Periodic Table of the Elements

Chemical bonds

Most atoms do not exist in the free state, but instead are chemically combined with other atoms. Such a combination of two or more atoms held together by chemical bonds is called a **molecule**. If two or more atoms of the same element combine, the resulting substance is called a molecule of that **element**. When two hydrogen atoms bond, the product is a molecule of hydrogen gas and is written as H_2. Similarly, when two oxygen atoms combine, a molecule of oxygen gas (O_2) is formed. Sulfur atoms commonly combine to form sulfur molecules containing eight sulfur atoms (S_8).

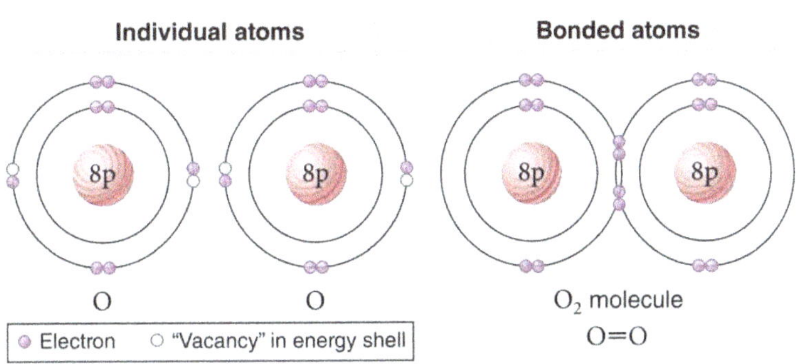

▲ **Figure 1.9** Double Bond.

When two or more different kinds of atoms bind, they form molecules of a **compound**. Two hydrogen atoms combine with one oxygen atom to form the compound water (H_2O). Four hydrogen atoms combine with one carbon atom to form the compound methane (CH_4). Compounds are chemically pure, and all of their molecules are identical. So, just as an atom is the smallest particle of an element that still has the properties of the element, a molecule is the smallest particle of a compound that still has the specific characteristics of the compound. This concept is important because the properties of compounds are usually very different from those of the atoms they contain. Water, for example, is very different from the elements hydrogen and oxygen.

Atoms with incomplete valence shells can interact with certain other atoms in such a way that each partner atom completes its valence shell: The atoms either share or transfer valence electrons. These interactions usually result in atoms staying close together, held by attractions called **chemical bonds**. The strongest kinds of chemical bonds are covalent bonds in molecules and ionic bonds in dry ionic compounds. (Ionic bonds in aqueous, or water-based, solutions are weak interactions).

In some cases, two atoms are so unequal in their attraction for valence electrons that the more electronegative atom strips an electron completely away from its partner. The two resulting oppositely charged atoms (or molecules) are called **ions**, in fact Atoms with an unequal number of protons and electron are called ions. A positively charged ion is called a **cation**, while a negatively charged ion is called an **anion**.

Because of their opposite charges, cations and anions attract each other; this attraction is called an **ionic bond**. Thus, one "end" of an ionically bonded molecule has a positive charge, and the other "end" has a negative charge. Any such separation of charge into distinct positive and negative regions is called **polarity.** Compounds formed by ionic bonds are called **ionic compounds**, or **salts**. An ionic compound does not consist of molecules. The term **ion** also applies to entire molecules that are electrically charged.

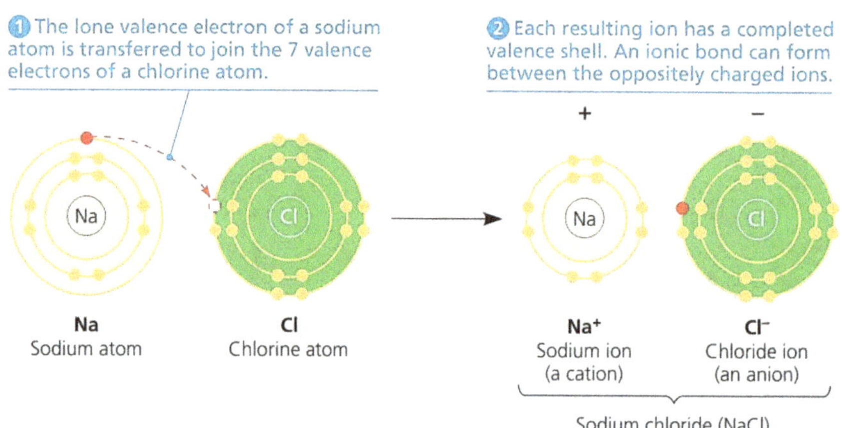

▲ **Figure 1.10** Electron transfer and ionic bonding. The attraction between oppositely charged atoms, or ions, is an ionic bond. An ionic bond can form between any two oppositely charged ions, even if they have not been formed by transfer of an electron from one to the other electrons, forming a double covalent bond.

A **covalent bond** is the sharing of a pair of valence electrons by two atoms. Two or more atoms held together by covalent bonds constitute a **molecule**, in this case a hydrogen molecule. **Molecular formula**, simply indicates that the molecule consists of which atoms. We can also use a **structural formula**, where the line represents a single bond, a pair of shared electrons.

Each atom that can share valence electrons has a bonding capacity corresponding to the number of covalent bonds the atom can form. When the bonds form, they give the atom a full complement of electrons in the valence shell. This bonding capacity is called the atom's valence and usually equals the number of electrons required to complete the atom's outermost (valence) shell.

Atoms in a molecule attract shared bonding electrons to varying degrees, depending on the element. The attraction of a particular atom for the electrons of a covalent bond is called its **electronegativity**. The more electronegative an atom is, the more strongly it pulls shared electrons toward itself. In general, electronegativity increases left to right across a row of the periodic table and decreases down the column. Thus the elements in the upper-right corner have the highest electronegativity.

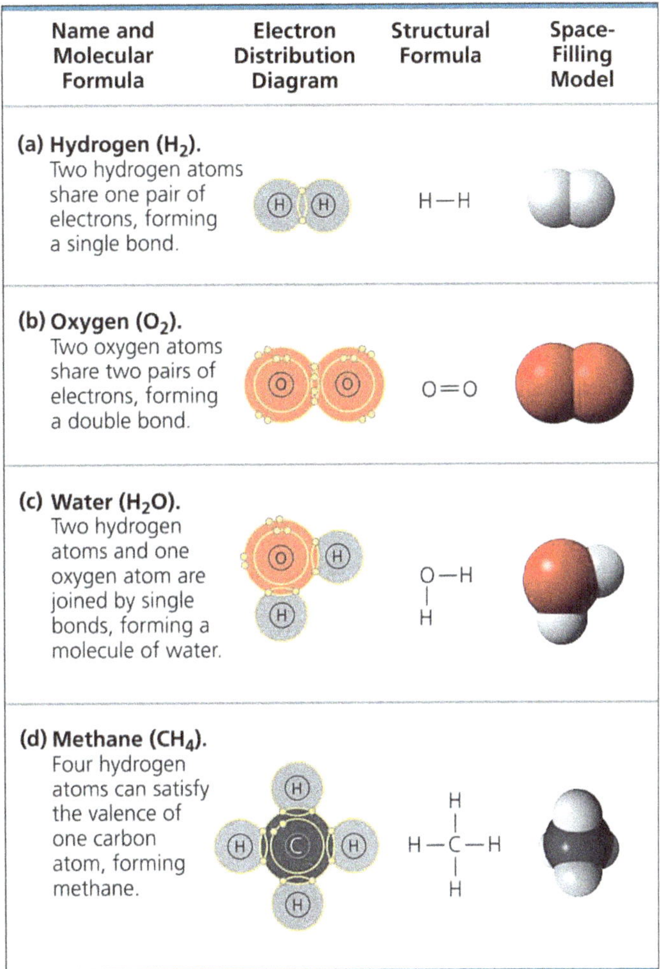

▲ **Figure 1.11** Covalent bonding in four molecules. The number of electrons required to complete an atom's valence shell generally determines how many covalent bonds that atom will form. This figure shows several ways of indicating covalent bonds.

In a covalent bond between two atoms of the same element, the electrons are shared equally because the two atoms have the same electronegativity. Such a bond is called a **nonpolar covalent bond**. For example, the single bond of H_2 is nonpolar, as is the double bond of O_2.

However, when an atom is bonded to a more electronegative atom, the electrons of the bond are not shared equally. This type of bond is called a **polar covalent bond**. Such bonds vary in their polarity, depending on the relative electronegativity of the two atoms. For example, the bonds between the oxygen and hydrogen atoms of a water molecule are quite polar.

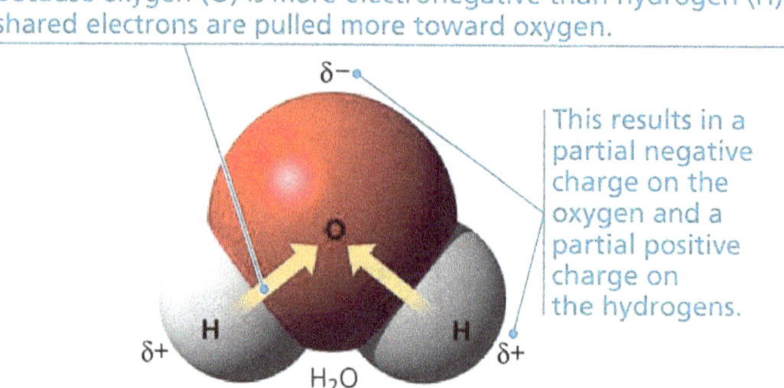

▲ **Figure 1.12** Polar covalent bonds in a water molecule.

Oxygen is one of the most electronegative elements, attracting shared electrons much more strongly than hydrogen does. In a covalent bond between oxygen and hydrogen, the electrons spend more time near the oxygen nucleus than near the hydrogen nucleus. Because electrons have a negative charge and are pulled toward oxygen in a water molecule, the oxygen atom has partial negative charges (indicated by the Greek letter δ with a minus sign, δ-, or "delta minus"), and the hydrogen atoms have partial positive charges (δ+, or "delta plus"). In contrast, the individual bonds of methane (CH_4) are much less polar because the electronegativities of carbon and hydrogen are quite similar.

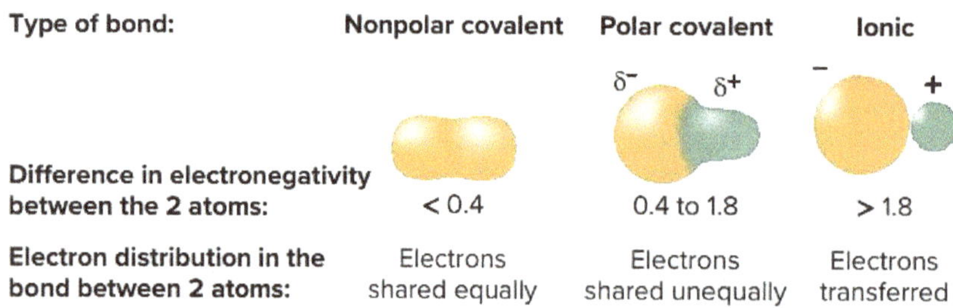

▲ **Figure 1.13** Types of chemical bonds and their relationship to electronegativity. The three drawings schematically represent the sharing of electrons between two atoms. In a nonpolar covalent bond, the two atoms share a pair of electrons equally. In the polar covalent bond shown here, the atom on the left has a higher electronegativity, and the pair of electrons are closer to this atom giving it a partial negative charge, designated δ−. The atom with lower electronegativity has a partial positive charge, δ+. In an ionic bond, one or more electrons are actually transferred to the atom on the left.

The strength of a covalent bond depends on the number of shared electrons. Thus double bonds, which satisfy the octet rule by allowing two atoms to share two pairs of electrons, are stronger than single bonds in which only one electron pair is shared. In practical terms, more energy is required to break a double bond than a single bond. The strongest covalent bonds are triple bonds, such as those that link the two nitrogen atoms of nitrogen gas molecules (N_2).

Atom	Electronegativity
O	3.5
N	3.0
C	2.5
H	2.1

▲ **Figure 1.14** Relative Electronegativities of Some Important Atoms

Weak Chemical Interactions

In organisms, most of the strongest chemical bonds are covalent bonds, which link atoms to form a cell's molecules. But weaker interactions within and between molecules are also indispensable, contributing greatly to the emergent properties of life. Many large biological molecules are held in their functional form by weak interactions. In addition, when two molecules in the cell make contact, they may adhere temporarily by weak interactions. The reversibility of weak interactions can be an advantage: Two molecules can come together, affect one another in some way, and then separate.

Several types of weak chemical interactions are important in organisms. One is the **ionic bond** as it exists between ions dissociated in water. **Hydrogen bonds** and **van der Waals interactions** are also crucial to life. Van der Waals interactions, hydrogen bonds, ionic bonds in water, and other weak interactions may form not only between molecules but also between parts of a large molecule, such as a protein or nucleic acid. The cumulative effect of weak interactions is to reinforce the three-dimensional shape of the molecule.

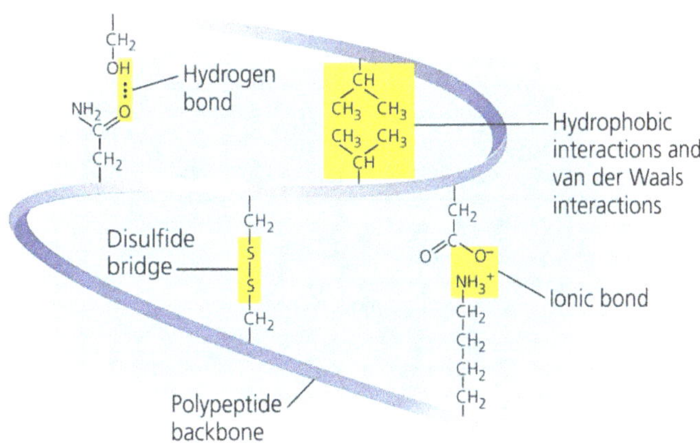

Hydrogen Bonds

Among weak chemical interactions, hydrogen bonds are so central to the chemistry of life that they deserve special attention. When a hydrogen atom is covalently bonded to an electronegative atom, the hydrogen atom has a partial positive charge that allows it to be attracted to a different electronegative atom with a partial negative charge nearby. This noncovalent attraction between a hydrogen and an electronegative atom is called a hydrogen bond. In living cells, the electronegative partners are usually oxygen or nitrogen atoms. Hydrogen bonds are important intramolecular bonds (literally, bonds within molecules), which hold different parts of a single large molecule in a specific three-dimensional shape. Some large biological molecules, such as proteins and DNA, have numerous hydrogen bonds that help maintain and stabilize their structures.

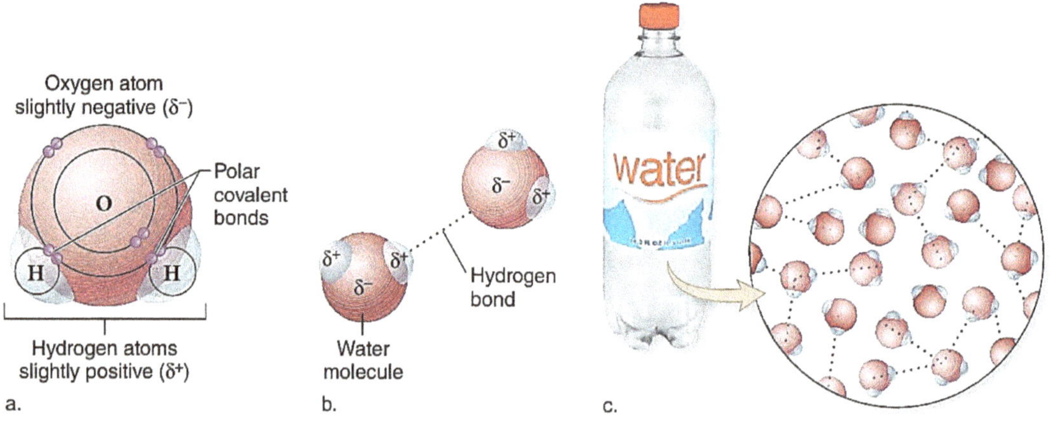

▲ **Figure 1.15** Hydrogen Bonds in Water. (a) An oxygen atom attracts electrons more strongly than do the hydrogen atoms in water. The O atom therefore bears a partial negative charge (δ−), and the H atoms carry partial positive charges (δ+). (b) The hydrogen bond is the attraction between partial charges on adjacent molecules. (c) In liquid water, many molecules stick to one another with hydrogen bonds.

Van der Waals Interactions

Even electrically neutral and nonpolar molecules may have positively and negatively charged regions. Electrons are not always evenly distributed; at any instant, they may accumulate by chance in one part of a molecule or another. The results are everchanging regions of positive and negative charge that enable all atoms and molecules to stick to one another. These van der Waals interactions are individually weak and occur only when atoms and molecules are very close together. When many such interactions occur simultaneously, however, they can be powerful.

Type	Chemical Basis	Strength	Example
Ionic bond	One atom donates one or more electrons to another atom; electronegativity difference between atoms is very large (>1.7). The resulting oppositely charged ions attract each other.	Strong but breaks easily in water	Sodium chloride (NaCl)
Covalent bond	Two atoms share pairs of electrons.	Strong	
Nonpolar	Electronegativity difference between atoms is small (<0.4)		H—H bond in H_2 molecule
Polar	Electronegativity difference between atoms is moderate or large (0.4 to 1.7)		O—H bond within water molecule
Hydrogen bond	An atom with a partial negative charge attracts a hydrogen atom with a partial positive charge in an adjacent molecule or in a different part of a large molecule.	Weak	Attraction between adjacent water molecules

▲ **Table 1.3** Chemical Bonds: A Summary

Chemical reactions make and break chemical bonds

The making and breaking of chemical bonds, leading to changes in the composition of matter, are called **chemical reactions**. When we write the equation for a chemical reaction, we use an arrow to indicate the conversion of the starting materials, called the reactants, to the resulting materials, or products. The coefficients indicate the number of molecules involved. Notice that all atoms of the reactants must be accounted for in the products. Matter is conserved in a chemical reaction: Reactions cannot create or destroy atoms but can only rearrange (redistribute) the electrons among them.

Notice that there are as many atoms of each element to the left of the arrow as there are to the right, even though the products are different from the reactants. This balance reflects the fact that in such reactions, atoms may be rearranged but not created or destroyed. Chemical reactions written in balanced form are known as **chemical equations**.

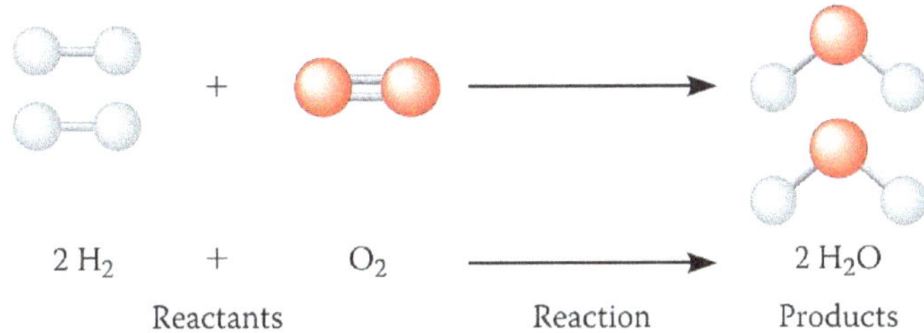

▲ **Figure 1.16** Polar covalent bonds in a water molecule.

Types of Chemical reactions

Most chemical reactions can be categorized as one of three types: **synthesis, decomposition**, or **exchange** reactions.

- When atoms or molecules combine to form a larger, more complex molecule, the process is a **synthesis**, or **combination** reaction. A synthesis reaction always involves bond formation. Synthesis reactions are the basis of constructive, or **anabolic** activities in body cells, such as joining small molecules called amino acids into large protein molecules It can be represented as:

$$A + B \rightarrow AB$$

- A **decomposition** reaction occurs when a molecule is broken down into smaller molecules or its constituent atoms. Essentially, decomposition reactions are reverse synthesis reactions: Bonds are broken. Decomposition reactions underlie all degradative, or **catabolic**, processes in body cells. For example, the bonds of glycogen molecules are broken to release simpler molecules of glucose:

$$AB \rightarrow A + B$$

- **Exchange**, or displacement reactions involve both synthesis and decomposition. Bonds are both made and broken. An exchange reaction occurs when ATP reacts with glucose and transfers its end phosphate group to glucose, forming glucose-phosphate. At the same time, the ATP becomes ADP. This important reaction occurs whenever glucose enters a body cell, and it effectively traps the glucose fuel molecule inside the cell.

$$AB + C \rightarrow AC + B \quad \text{and} \quad AB + CD \rightarrow AD + CB$$

The sum of all the chemical reactions in the body is called **metabolism**, all of the synthesis reactions that occur in your body are collectively referred to as **anabolism** and the decomposition reactions that occur in your body are collectively referred to as **catabolism**.

Some chemical reactions may be reversible. Reversible reactions can go in either direction under different conditions and are indicated by two half arrows pointing in opposite directions:

$$CH_4 + 2\,O_2 \quad \rightleftharpoons \quad CO_2 + 2\,H_2O$$

$$\text{(methane) (oxygen)} \quad \text{(carbon dioxide) (water)}$$

A group of important chemical reactions in living systems is **oxidation-reduction** reactions, called **redox** reactions for short. Oxidation-reduction reactions are decomposition reactions in that they are the basis of all reactions in which food fuels are broken down for energy (that is, in which ATP is produced). They are also a special type of exchange reaction because electrons are exchanged between the reactants.

The reactant losing the electrons is the electron donor and is said to be oxidized. The reactant taking up the transferred electrons is the electron acceptor and is said to become reduced. Electrons are not easily removed from covalent compounds unless an entire atom is removed. In cells, oxidation often involves the removal of a hydrogen atom from a covalent compound; reduction often involves the addition of the equivalent of a hydrogen atom.

Redox reactions occur simultaneously because one substance must accept the electrons that are removed from the other. In a redox reaction, one component, the **oxidizing agent**, accepts 1 or more electrons and becomes reduced. Another reaction component, the **reducing agent**, gives up 1 or more electrons and becomes oxidized.

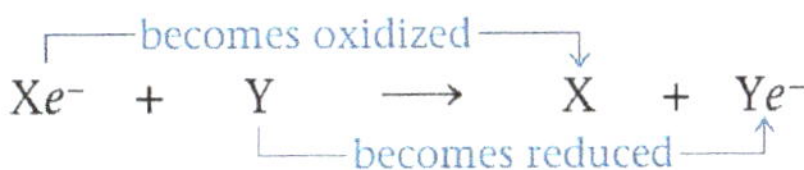

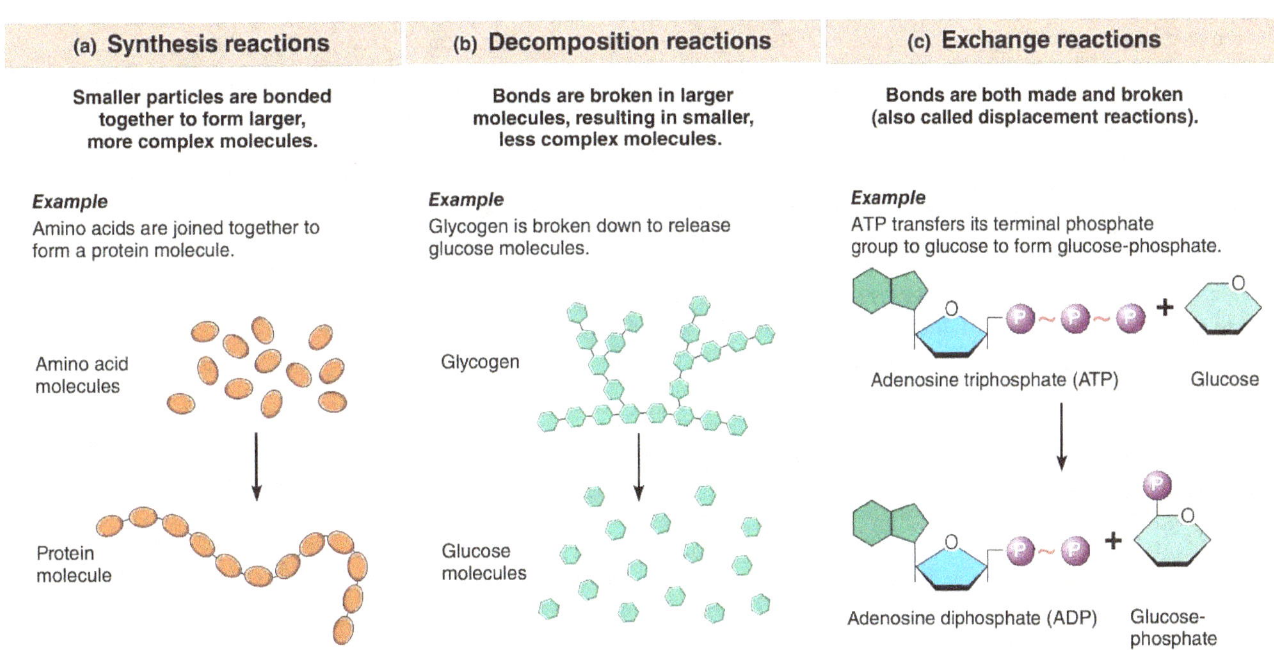

▲ **Figure 1.17** Types of chemical reactions.

Energy Flow in Chemical reactions

Energy is the capacity to do work. The two main forms of energy are **potential energy**, energy stored by matter due to its position, and **kinetic energy**, the energy of matter in motion. **Chemical energy** is a form of potential energy that is stored in the bonds of molecules. In chemical reactions, breaking old bonds requires an input of energy and forming new bonds releases energy. Because most chemical reactions involve both breaking old bonds and forming new bonds, the overall reaction may either release energy or require energy.

Reactions that release energy are **exergonic** or **exothermic** reactions. These reactions yield products with less energy than the initial reactants, along with energy that can be harvested for other uses. With a few exceptions, catabolic and oxidative reactions are exergonic. In contrast, the products of energy-absorbing or **endergonic** or **endothermic** reactions contain more potential energy in their chemical bonds than did the reactants. **Anabolic** reactions are typically endergonic reactions.

Factors influencing the rate of Chemical reactions

- **Temperature** Increasing the temperature of a substance increases the kinetic energy of its particles and the force of their collisions. For this reason, chemical reactions proceed more quickly at higher temperatures.

- **Concentration** Chemical reactions progress most rapidly when the reacting particles are present in high numbers, because the chance of successful collisions is greater. As the concentration of the reactants declines, the reaction slows.

- **Particle Size** Smaller particles move faster than larger ones (at the same temperature) and tend to collide more frequently and more forcefully. Hence, the smaller the reacting particles, the faster a chemical reaction goes at a given temperature and concentration.

- **Catalysts** Many chemical reactions in nonliving systems can be speeded up simply by heating, but drastic increases in body temperature are life threatening because important biological molecules are destroyed. Still, at normal body temperatures, most chemical reactions would proceed far too slowly to maintain life were it not for the presence of catalysts. Catalysts are substances that increase the rate of chemical reactions without themselves becoming chemically changed or part of the product. Biological catalysts are called enzymes.

Functional groups

A functional group is a group of atoms that help determine the types of chemical reactions and associations in which the compound participates. Most functional groups readily form associations, such as ionic and hydrogen bonds, with other molecules. Polar and ionic functional groups are hydrophilic because they associate strongly with polar water molecules. When we know what kinds of functional groups are present in an organic compound, we can predict its chemical behavior. Note that the symbol R is used to represent the remainder of the molecule of which each functional group is a part.

The seven chemical groups most important in biological processes are the hydroxyl, carbonyl, carboxyl, amino, sulfhydryl, phosphate, and methyl groups. The first six groups can be chemically reactive; of these six, all except the sulfhydryl group are also hydrophilic and thus increase the solubility of organic compounds in water. The methyl group is not reactive, but instead often serves as a recognizable tag on biological molecules.

Functional Group and Description	Structural Formula	Class of Compound Characterized by Group
HYDROXYL Polar because electronegative oxygen attracts covalent electrons	$R{-}OH$	Alcohols 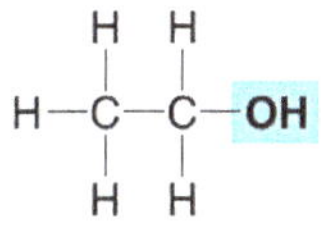 Example, ethanol
CARBONYL **Aldehydes:** Carbonyl group carbon is bonded to at least one H atom; polar because electronegative oxygen attracts covalent electrons **Ketones:** Carbonyl group carbon is bonded to two other carbons; polar because electronegative oxygen attracts covalent electrons	$R{-}C(=O){-}H$ $R{-}C(=O){-}R$	Aldehydes Example, formaldehyde Ketones Example, acetone
CARBOXYL Weakly acidic; can release an H^+	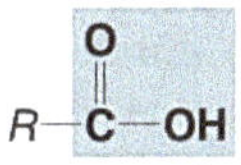 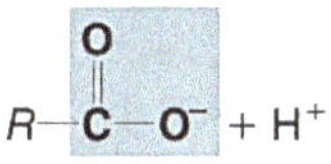 $R{-}C(=O){-}OH$ (Non-ionized) $R{-}C(=O){-}O^- + H^+$ (Ionized)	Carboxylic acids (organic acids) 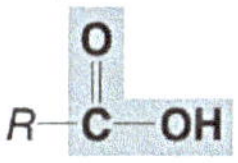 Example, amino acid
AMINO Weakly basic; can accept an H^+	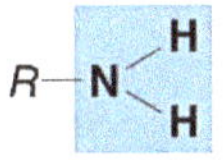 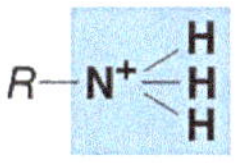 $R{-}NH_2$ (Non-ionized) $R{-}N^+H_3$ (Ionized)	Amines 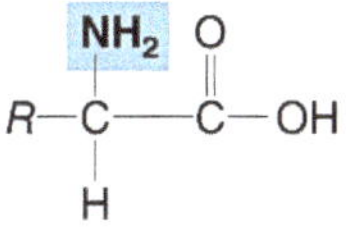 Example, amino acid
PHOSPHATE Weakly acidic; one or two H^+ can be released	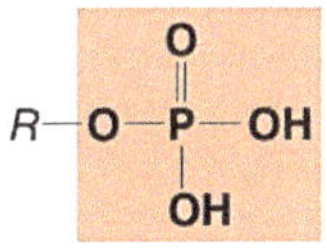 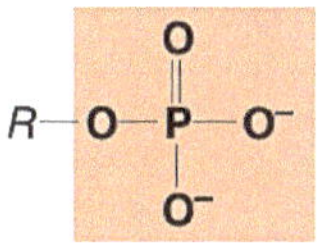 $R{-}O{-}P(=O)(OH){-}OH$ (Non-ionized) $R{-}O{-}P(=O)(O^-){-}O^-$ (Ionized)	Organic phosphates 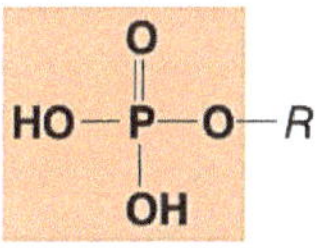 Example, phosphate ester (as found in ATP)
SULFHYDRYL Helps stabilize internal structure of proteins	$R{-}SH$	Thiols 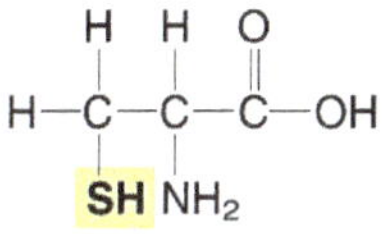 Example, cysteine

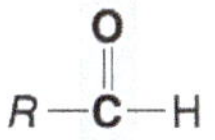

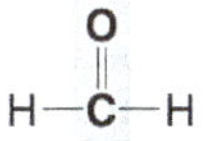

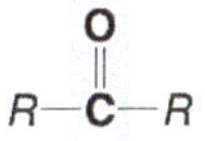

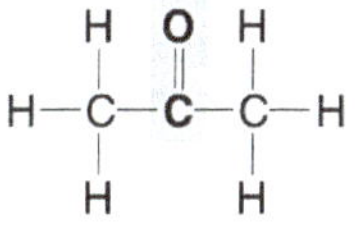

▲ **Table 1.4** Some Biologically Important Functional Groups

Biochemistry

Biochemistry is the study of the chemical composition and reactions of living matter. All chemicals in the body fall into one of two major classes: **organic** or **inorganic** compounds that are equally essential for life. Organic compounds are molecules unique to living systems, contain carbon, usually contain hydrogen, always have covalent bonds and many are large. Examples include carbohydrates, lipids, proteins and nucleic acids. Large organic molecules called **macromolecules** are formed by covalent bonding of many identical or similar building-block subunits termed **monomers**. All other chemicals in the body are considered inorganic compounds. These include water, salts, and many acids and bases. Acids and bases react with one another to form salts:

$$\text{HCl} + \text{KOH} \longrightarrow \text{KCl} + \text{H}_2\text{O}$$

Acid   Base   Salt   Water

Water and Life

Three-fourths of the Earth is covered by liquid water. Without water, life would not exist on this planet. Water is the most important and most abundant inorganic compound in all living systems. It is important for two reasons. First, it is a major component of cells, typically forming between 70% and 95% of the mass of the cell. You are about 60% water. Second, it provides an environment for those organisms that live in water. Three-quarters of the planet is covered in water. Polar covalent bonds in water molecules result in hydrogen bonding:

The water molecule is deceptively simple. It is shaped like a wide V, with its two hydrogen atoms joined to the oxygen atom by single covalent bonds. Oxygen is more electronegative than hydrogen, so the electrons of the covalent bonds spend more time closer to oxygen than to hydrogen; these are polar covalent bonds.

This unequal sharing of electrons and water's V-like shape (tetrahedron: a pyramid with a triangle as its base) make it a polar molecule, meaning that its overall charge is unevenly distributed. In water, the oxygen of the molecule has partial negative charges (δ-),and the hydrogens have partial positive charges (δ+).

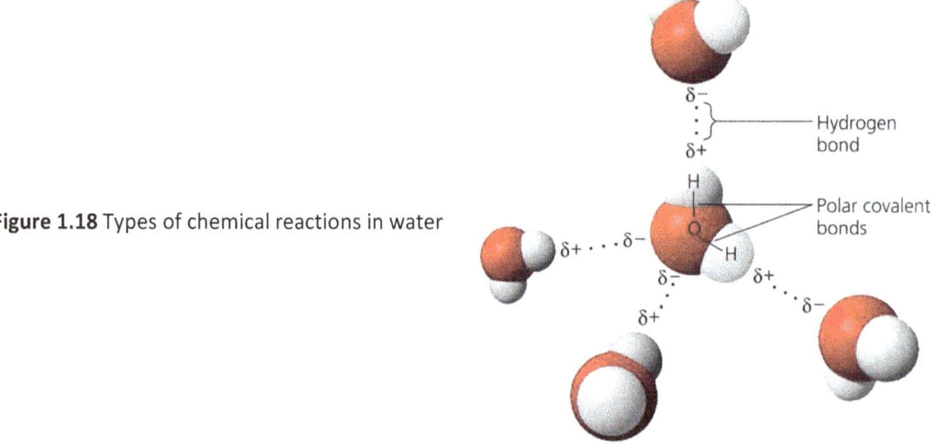

▶ **Figure 1.18** Types of chemical reactions in water

The properties of water arise from attractions between oppositely charged atoms of different water molecules: The partially positive hydrogen of one molecule is attracted to the partially negative oxygen of a nearby molecule. The two molecules are thus held together by a hydrogen bond. Each water molecule can therefore form hydrogen bonds with a maximum of four neighboring water molecules.

When water is in its liquid form, its hydrogen bonds are very fragile, each only about 1/20 as strong as a covalent bond. The hydrogen bonds form, break, and re-form with great frequency. Each lasts only a few trillionths of a second, but the molecules are constantly forming new hydrogen bonds with a succession of partners. Therefore, at any instant, most of the water molecules are hydrogen-bonded to their neighbors.

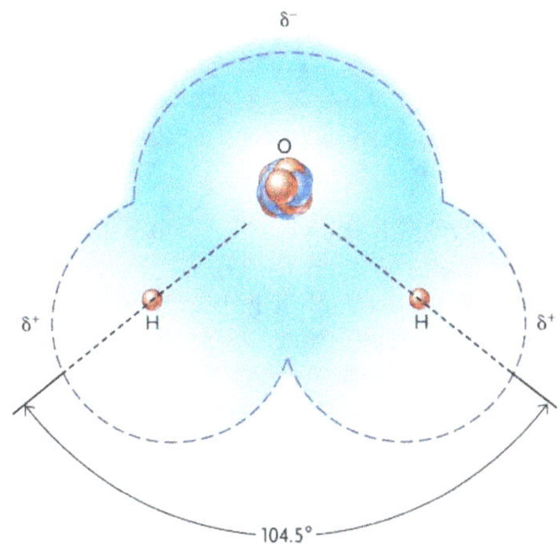

Four emergent properties of water contribute to Earth's suitability for life

- **Cohesion**

Water molecules stay close to each other as a result of hydrogen bonding. Although the arrangement of molecules in a sample of liquid water is constantly changing, at any given moment many of the molecules are linked by multiple hydrogen bonds. These linkages make water more structured than most other liquids. Collectively, the hydrogen bonds hold the substance together, a phenomenon called **cohesion**.

Water has a high degree of **surface tension** (a measure of how difficult it is to stretch or break the surface of a liquid) because of the cohesion of its molecules, which have a much greater attraction for one another than for molecules in the air. The spider takes advantage of the surface tension of water to walk across a pond without breaking the surface, and some plants can float on water as well.

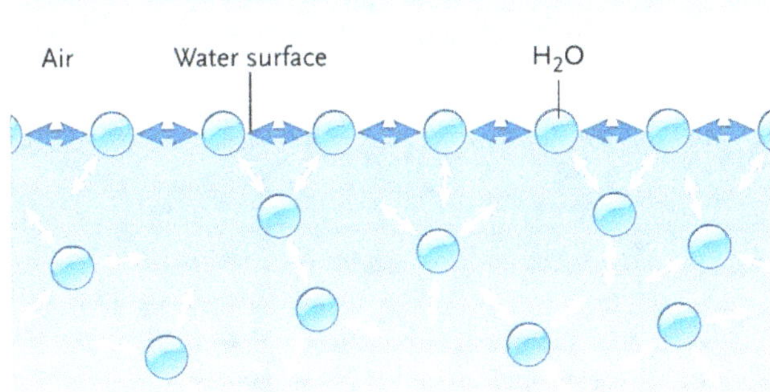

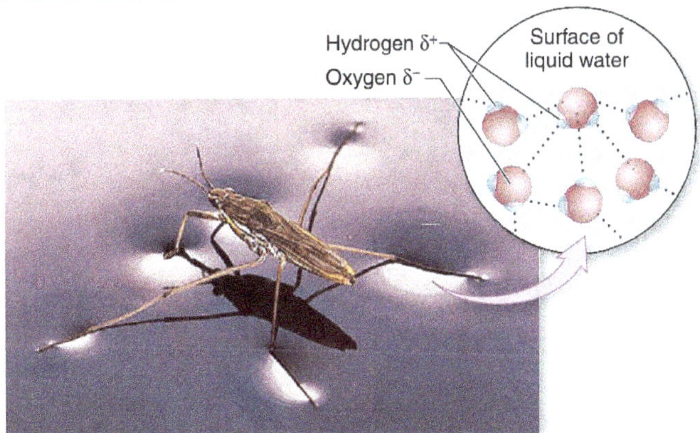

▶ **Figure 1.19** running on Water. A lightweight body and water-repellent legs allow this water strider to "skate" across a pond without breaking the water's surface tension.

- **Adhesion**, the clinging of one substance to another, also plays a role. Adhesion refers to the ability of water molecules to cling to a surface. These adhesive forces explain how water makes things wet. Because water is polar, it is attracted to other polar surfaces. Many animals, including humans, contain internal vessels in which water assists in the transport of nutrients and wastes, because the cohesion and adhesion of water allows blood to fill the tubular vessels of the cardiovascular system. For example, the liquid portion of our blood, which transports dissolved and suspended substances about the body, is 92% water. The water in our blood assists in the transport of nutrients and oxygen to our cells and in the removal of waste material from the cells.

A combination of adhesive and cohesive forces accounts for **capillary action**, which is the tendency of water to move in narrow tubes, even against the force of gravity as it moves from a plant's roots to its highest leaves. Water from the roots reaches the leaves through a network of water-conducting cells. As water evaporates from a leaf, hydrogen bonds cause water molecules leaving the veins to tug on molecules farther down, and the upward pull is transmitted through the water-conducting cells all the way to the roots. Adhesion to the walls of the conducting tubes also helps lift water to the topmost leaves of trees.

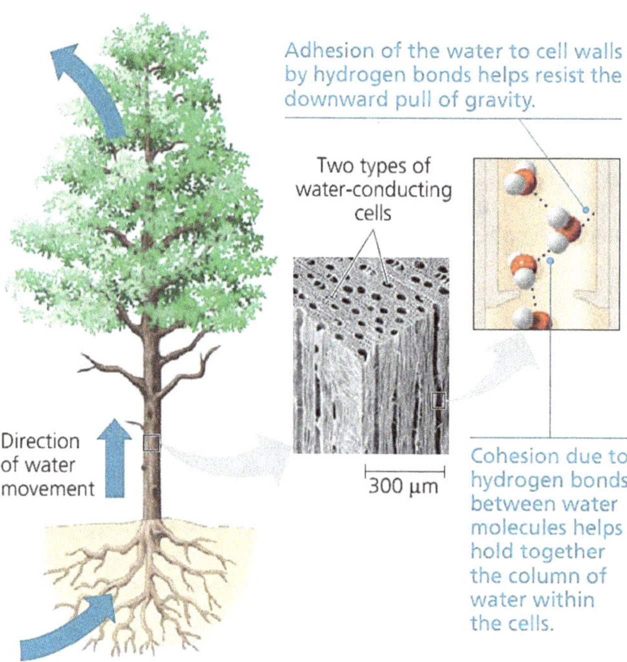

▲ **Figure 1.20** Water transport in plants. Evaporation from leaves pulls water upward from the roots through water-conducting cells. Because of the properties of cohesion and adhesion, the tallest trees can transport water more than 100 m upward.

- **Water's High Specific Heat:**

The ability of water to stabilize temperature stems from its relatively high specific heat. The specific heat of a substance is defined as the amount of heat that must be absorbed or lost for 1 g of that substance to change its temperature by 1°C. Raising the temperature of a substance involves adding heat energy to make its molecules move faster, that is, to increase the energy of motion, kinetic energy, of the molecules. The term heat refers to the total amount of kinetic energy in a sample of a substance; temperature is a measure of the average kinetic energy of the particles. Because of the high specific heat of water relative to other materials, water will change its temperature less than other liquids when it absorbs or loses a given amount of heat.

Water resists changing its temperature; when it does change its temperature, it absorbs or loses a relatively large quantity of heat for each degree of change. This means that the temperature within cells and within the bodies of organisms (which have a high proportion of water) tends to be more constant than that of the air around them. As a result, biochemical reactions operate at relatively constant rates and are less likely to be adversely affected by extremes of temperature.

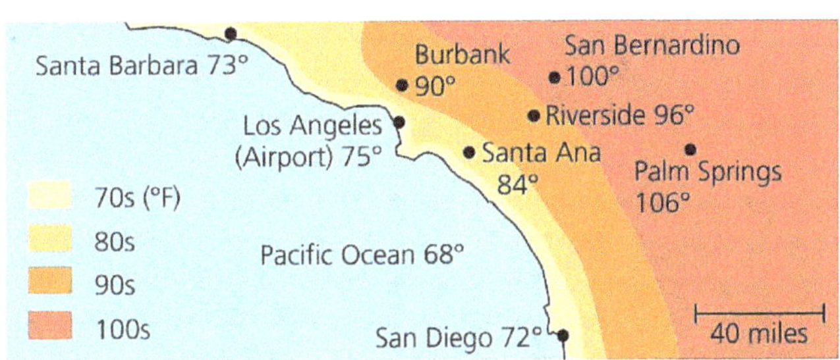

▲ **Figure 1.21** Effect of a large body of water on climate. By absorbing or releasing heat, oceans moderate coastal climates. In this example from an August day in Southern California, the relatively cool ocean reduces coastal air temperatures by absorbing heat. (The temperatures are in degrees Fahrenheit.)

A large body of water can absorb and store a huge amount of heat from the sun in the daytime and during summer while warming up only a few degrees. At night and during winter, the gradually cooling water can warm the air. This capability of water serves to moderate air temperatures in coastal areas. The high specific heat of water also tends to stabilize ocean temperatures, creating a favorable environment for marine life. Thus, because of its high specific heat, the water that covers most of Earth keeps temperature fluctuations on land and in water within limits that permit life. Also, because organisms are made primarily of water, they are better able to resist changes in their own temperature than if they were made of a liquid with a lower specific heat.

- **Evaporative Cooling**

Transformation from a liquid to a gas is called **vaporization**, or evaporation. **Heat of vaporization** is the quantity of heat a liquid must absorb for 1 g of it to be converted from the liquid to the gaseous state. For the same reason that water has a high specific heat, it also has a high heat of vaporization relative to most other liquids. Water's high heat of vaporization is another emergent property resulting from the strength of its hydrogen bonds, which must be broken before the molecules can exit from the liquid in the form of water vapor.

The high amount of energy required to vaporize water has a wide range of effects. On a global scale, for example, it helps moderate Earth's climate. As a liquid evaporates, the surface of the liquid that remains behind cools down (**evaporative cooling**). Evaporation of water from the leaves of a plant helps keep the tissues in the leaves from becoming too warm in the sunlight. Evaporation of sweat from human skin dissipates body heat and helps prevent overheating on a hot day. Animals without sweat glands, such as elephants, may spray water on themselves to cool down.

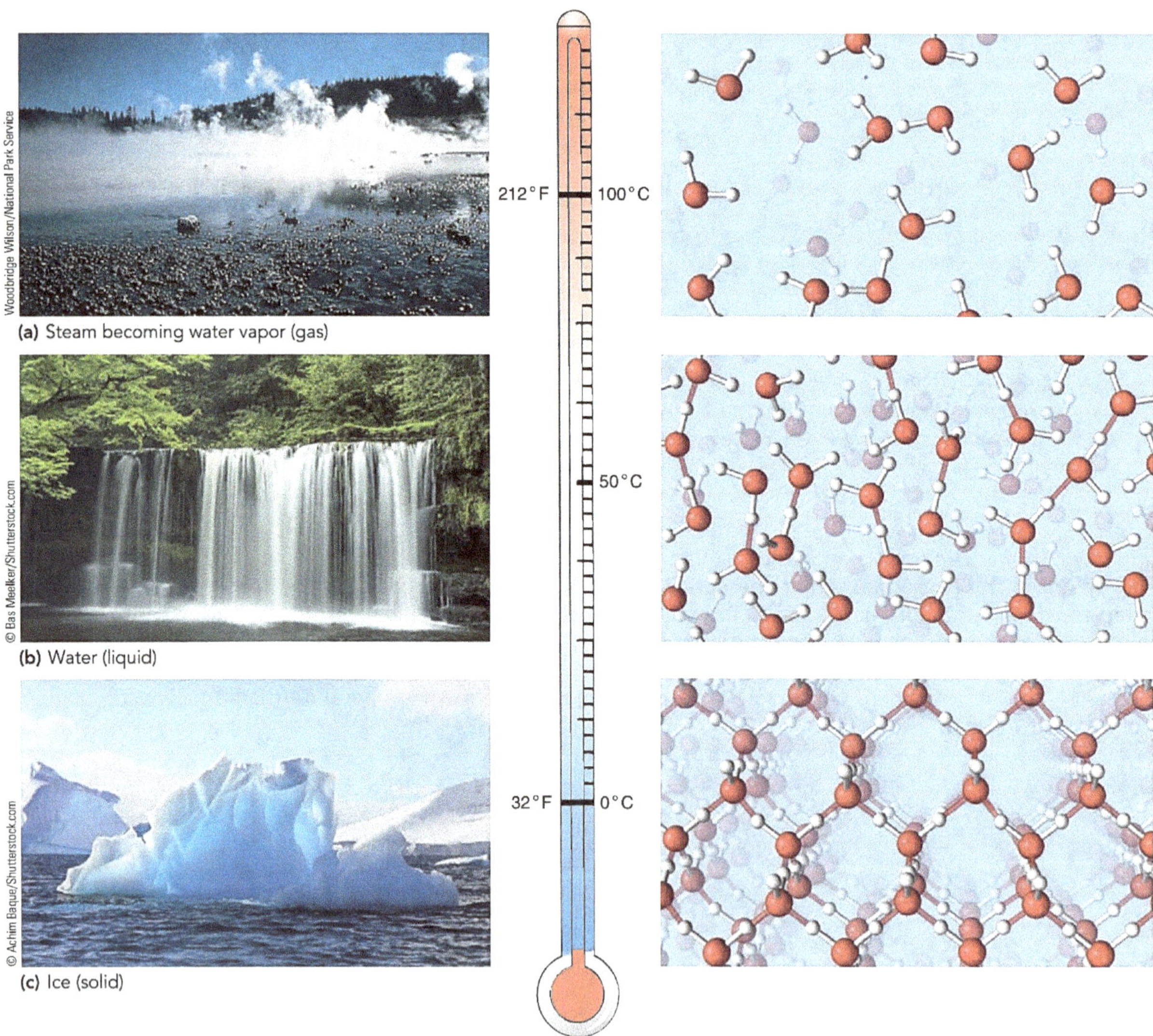

▲ **Figure 1.22** Three forms of water. (a) When water boils, as in this hot spring at Yellowstone National Park, many hydrogen bonds are broken, causing steam, which consists of minuscule water droplets, to form. If most of the remaining hydrogen bonds break, the molecules move more freely as water vapor (a gas). (b) Water molecules in a liquid state continually form, break, and re-form hydrogen bonds with one another. (c) In ice, each water molecule participates in four hydrogen bonds with adjacent molecules, resulting in a regular, evenly distanced crystalline lattice structure.

- **Floating of Ice on Liquid Water:**

Water is one of the few substances that are less dense as a solid than as a liquid. As a result, ice floats on liquid water. While other materials contract and become denser when they solidify, water expands. The cause of this exotic behavior is, once again, hydrogen bonding. At temperatures above 4°C, water behaves like other liquids, expanding as it warms and contracting as it cools. Water is most dense at 4°C. As the temperature falls from 4°C to 0°C, water begins to freeze because more and more of its molecules are moving too slowly to break hydrogen bonds. At 0°C, the molecules become locked into a crystalline lattice, each water molecule hydrogen bonded to four partners.

The hydrogen bonds keep the molecules at "arm's length," far enough apart to make ice about 10% less dense (10% fewer molecules in the same volume) than liquid water at 4°C. The ability of ice to float due to its lower density is an important factor in the suitability of the environment for life. If ice sank, then eventually ponds, lakes, and even oceans could freeze solid, making life as we know it impossible on Earth.

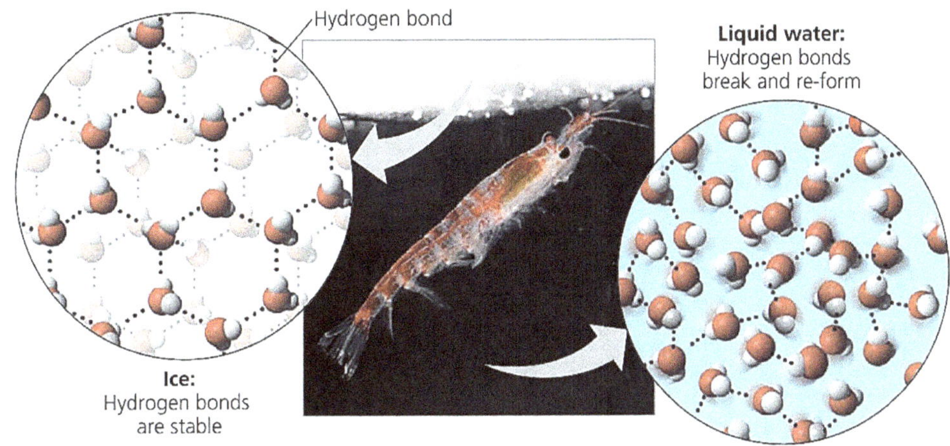

▲ **Figure 1.23** Ice: crystalline structure and floating barrier. In ice, each molecule is hydrogen-bonded to four neighbors in a three-dimensional crystal. Because the crystal is spacious, ice has fewer molecules than an equal volume of liquid water. In other words, ice is less dense than liquid water. Floating ice becomes a barrier that protects the liquid water below from the colder air.

- **Water, The Solvent of Life:**

The small size, polarity and its ability to form hydrogen bonds make water an excellent solvent, which means that many other substances easily dissolve in it. A sugar cube placed in a glass of water will dissolve. In time, the glass will contain a uniform mixture of sugar and water; the concentration of dissolved sugar will be the same everywhere in the mixture.

A liquid that is a completely homogeneous mixture of two or more substances is called a **solution**. The dissolving agent of a solution is the **solvent**, and the substance that is dissolved is the **solute**. In this case, water is the solvent and sugar is the solute.

An **aqueous solution** is one in which the solute is dissolved in water; water is the solvent. Water is a very versatile solvent, a quality we can trace to the polarity of the water molecule. Water molecules are attracted to the ions and polar molecules and therefore collect around them and separate them.

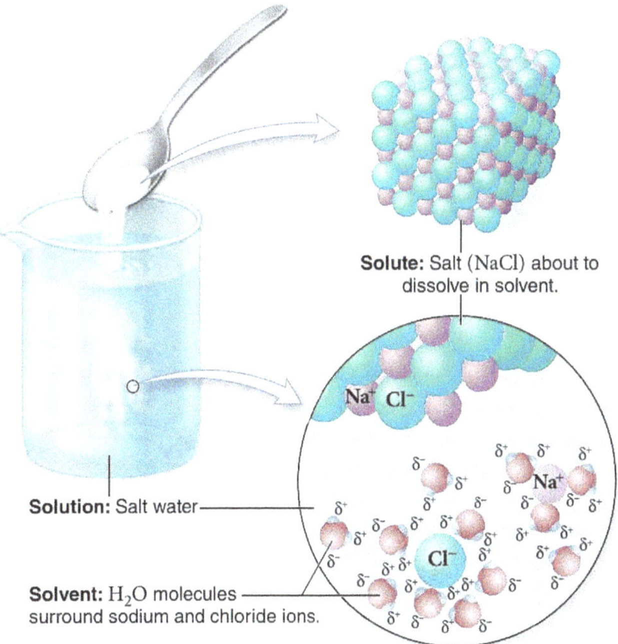

▲ **Figure 1.24** Dissolving salt. As salt crystals dissolve, polar water molecules surround each sodium and chloride ion.

In solution each cation and anion of the ionic compound is surrounded by oppositely charged ends of the water molecules. This process is known as **hydration** and the sphere of water molecules around each dissolved ion is called a **hydration shell.**

A compound does not need to be ionic to dissolve in water; many compounds made up of nonionic polar molecules, such as the sugar. Such compounds dissolve when water molecules surround each of the solute molecules, forming hydrogen bonds with them. Even molecules as large as proteins can dissolve in water if they have ionic and polar regions on their surface. Polar molecules and ions are soluble in water.

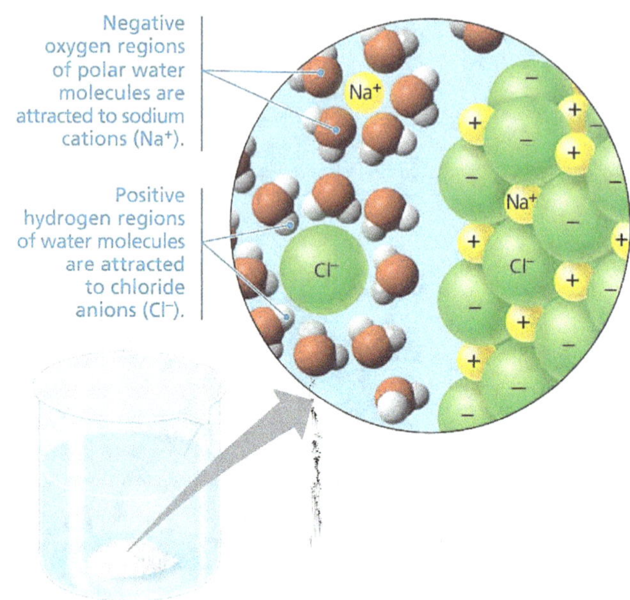

▲ **Figure 1.25** Table salt dissolving in water. A sphere of water molecules, called a hydration shell, surrounds each solute ion.

Many different kinds of polar compounds are dissolved (along with ions) in the water of such biological fluids as blood, the sap of plants, and the liquid within all cells. Water is the solvent of life. Water is the body's major transport medium because it is such an excellent solvent. Nutrients, respiratory gases, and metabolic wastes carried throughout the body are dissolved in blood plasma, and many metabolic wastes are excreted from the body in urine, another watery fluid. Lubricants (e.g. mucus) also use water as their dissolving medium.

Hydrophilic and Hydrophobic Substances

Any substance that has an affinity for water is said to be **hydrophilic**. In some cases, substances can be hydrophilic without actually dissolving. For example, some molecules in cells are so large that they do not dissolve. Substances that are nonionic and nonpolar (or otherwise cannot form hydrogen bonds) actually seem to repel water; these substances are said to be **hydrophobic.** Many hydrophobic ("water-fearing") substances found in living things are especially important because of their ability to form associations or structures that are not disrupted. Hydrophobic interactions occur between groups of nonpolar molecules. Such molecules are insoluble in water and tend to cluster together. This tendency is not due to formation of bonds between the nonpolar molecules but rather to the hydrogen-bonded water molecules excluding them and in a sense "driving them together." **Hydrophobic interactions** explain why oil tends to form globules when added to water.

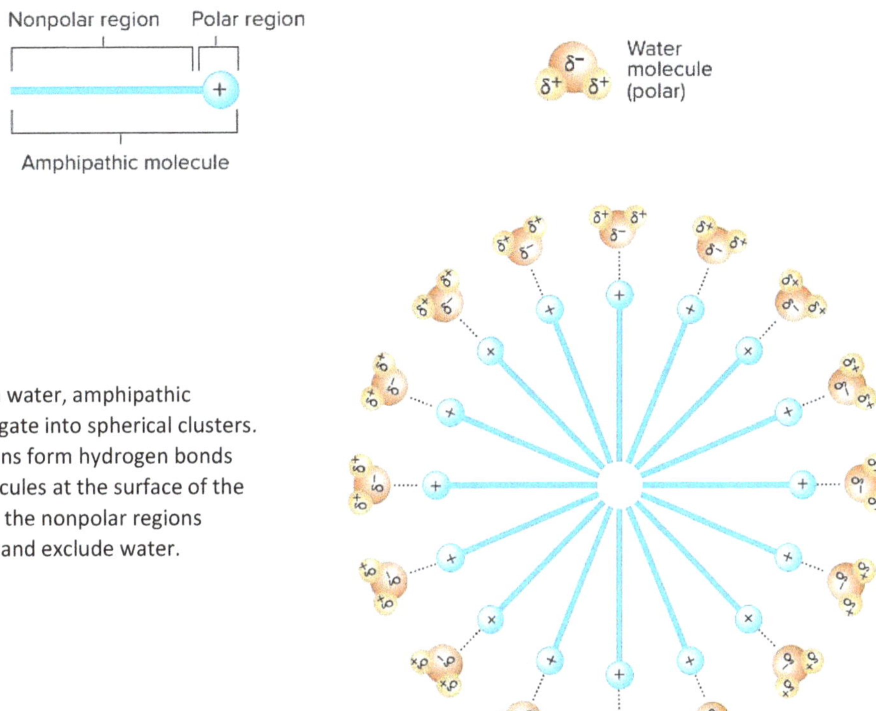

▶ **Figure 1.26** In water, amphipathic molecules aggregate into spherical clusters. Their polar regions form hydrogen bonds with water molecules at the surface of the cluster, whereas the nonpolar regions cluster together and exclude water.

By forcing the hydrophobic portions of molecules together, water causes these molecules to assume particular shapes. This property can also affect the structure of proteins, DNA, and biological membranes. In carbon dioxide (CO_2), carbon shares four electron pairs with two oxygen atoms (two pairs are shared with each oxygen). Oxygen is very electronegative and so attracts the shared electrons much more strongly than does carbon. However, because the carbon dioxide molecule is linear and symmetrical, the electron-pulling ability of one oxygen atom offsets that of the other, like a standoff between equally strong teams in a game of tug-of-war. As a result, the shared electrons orbit the entire molecule and carbon dioxide is a nonpolar compound.

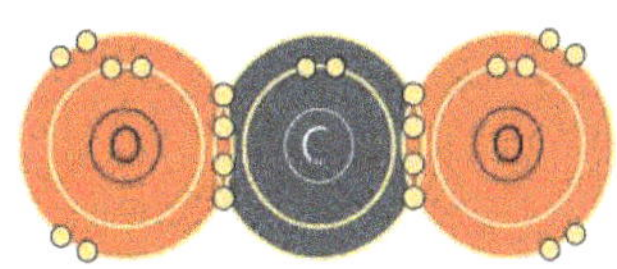

Salts

A salt is an ionic compound containing cations other than H^+ and anions other than the hydroxyl ion (OH^-). When salts are dissolved in water, they dissociate into their component ions. For example, sodium sulfate (Na_2SO_4) dissociates into two Na^+ ions and one SO_4^{2-} ion. It dissociates easily because the ions are already formed. All that remains is for water to overcome the attraction between the oppositely charged ions.

All ions are **electrolytes**, substances that conduct an electrical current in solution. Salts commonly found in the body include NaCl, $CaCO_3$ (calcium carbonate), and KCl (potassium chloride). However, the most plentiful salts are the calcium phosphates that make bones and teeth hard. In their ionized form, salts play vital roles in body function. For instance, the electrolyte properties of sodium and potassium ions are essential for nerve impulse transmission and muscle contraction. Ionic iron forms part of the hemoglobin molecules that transport oxygen within red blood cells, and zinc and copper ions are important to the activity of some enzymes.

Main Chemical Elements in the Body

CHEMICAL ELEMENT (SYMBOL)	% OF TOTAL BODY MASS	SIGNIFICANCE
MAJOR ELEMENTS	about 96%	
Oxygen (O)	65.0	Part of water and many organic (carbon-containing) molecules; used to generate ATP, a molecule used by cells to temporarily store chemical energy
Carbon (C)	18.5	Forms backbone chains and rings of all organic molecules: carbohydrates, lipids (fats), proteins, and nucleic acids (DNA and RNA)
Hydrogen (H)	9.5	Constituent of water and most organic molecules; ionized form (H^+) makes body fluids more acidic.
Nitrogen (N)	3.2	Component of all proteins and nucleic acids
LESSER ELEMENTS	about 3.6%	
Calcium (Ca)	1.5	Contributes to hardness of bones and teeth; ionized form (Ca^{2+}) needed for blood clotting, release of hormones, contraction of muscle, and many other processes
Phosphorus (P)	1.0	Component of nucleic acids and ATP; required for normal bone and tooth structure
Potassium (K)	0.35	Ionized form (K^+) most plentiful cation (positively charged particle) in intracellular fluid; needed to generate action potentials
Sulfur (S)	0.25	Component of some vitamins and many proteins
Sodium (Na)	0.2	Ionized form (Na^+) most plentiful cation in extracellular fluid; essential for maintaining water balance; needed to generate action potentials
Chlorine (Cl)	0.2	Ionized form (Cl^-) most plentiful anion (negatively charged particle) in extracellular fluid; essential for maintaining water balance
Magnesium (Mg)	0.1	Ionized form (Mg^{2+}) needed for action of many enzymes, molecules that increase the rate of chemical reactions in organisms
Iron (Fe)	0.005	Ionized forms (Fe^{2+} and Fe^{3+}) part of hemoglobin (oxygen-carrying protein in red blood cells) and some enzymes
TRACE ELEMENTS	about 0.4%	Aluminum (Al), boron (B), chromium (Cr), cobalt (Co), copper (Cu), fluorine (F), iodine (I), manganese (Mn), molybdenum (Mo), selenium (Se), silicon (Si), tin (Sn), vanadium (V), and zinc (Zn)

Acidic and basic conditions affect living organisms

Like salts, acids and bases are electrolytes. They ionize and dissociate in water and can then conduct an electrical current. Occasionally, a hydrogen atom participating in a hydrogen bond between two water molecules shifts from one molecule to the other. When this happens, the hydrogen atom leaves its electron behind, and what is actually transferred is a hydrogen ion (H^+), a single proton with a charge of 1+. The water molecule that lost a proton is now a hydroxide ion (OH^-), which has a charge of 1-. The proton binds to the other water molecule, making that molecule a hydronium ion ($H3O^+$).

 In pure water, only one water molecule in every 554 million is dissociated; the concentration (the amount of a particular solute that is dissolved in a given volume of fluid) of H^+ and of OH^- in pure water is therefore 10^{-7} M (at 25°C). So the concentrations of H^+ and OH^- are equal in pure water, but adding certain kinds of solutes, called acids and bases, disrupts this balance. Biologists use something called the pH scale to describe how acidic or basic (the opposite of acidic) a solution is.

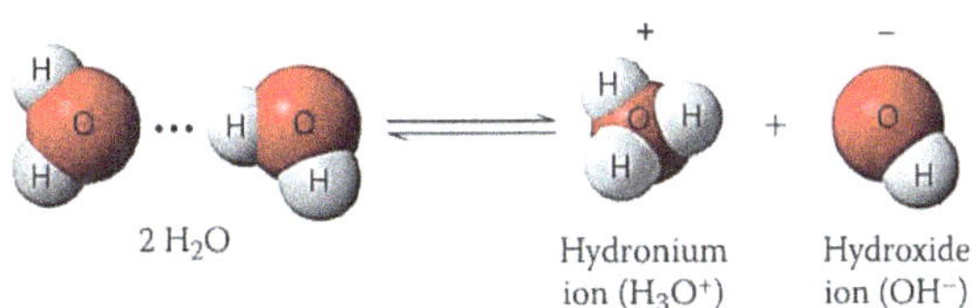

An **acid** is a substance that increases the hydrogen ion concentration of a solution. For example, when hydrochloric acid (HCl) is added to water, hydrogen ions dissociate from chloride ions. This source of H^+ results in an acidic solution one having more H^+ than OH^-:

$$HCl \rightarrow H^+ + Cl^-$$

A substance that reduces the hydrogen ion concentration of a solution is called a base. Some bases reduce the H^+ concentration directly by accepting hydrogen ions. Ammonia ($NH3$), for instance, acts as a base when the unshared electron pair in nitrogen's valence shell attracts a hydrogen ion from the solution, resulting in an ammonium ion ($NH4^+$):

$$NH_3 + H^+ \rightleftharpoons NH_4^{\,+}$$

Other bases reduce the H^+ concentration indirectly by dissociating to form hydroxide ions, which combine with hydrogen ions and form water. One such base is sodium hydroxide (NaOH), which in water dissociates into its ions.

$$NaOH \rightarrow Na^+ + OH^-$$

Solutions with a higher concentration of OH^- than H^+ are known as basic solutions. A solution in which the H^+ and OH^- concentrations are equal is said to be neutral.

The pH Scale

In any aqueous solution at 25°C, the product of the H^+ and OH^- concentrations is constant at 10^{-14}. This can be written:

$$[H^+][OH^-] = 10^{-14}$$

The pH scale is a simple numerical method for expressing the range of H^+ concentrations. The pH of a solution is defined as the negative logarithm (base 10) of the H^+ concentration:

$$pH = -\log [H^+]$$

For a neutral aqueous solution, $[H^+]$ is 10^{-7} M, giving us: $\qquad -\log 10^{-7} = -(-7) = 7$

Notice that pH decreases as H^+ concentration increases. The pH scale is used to indicate the acidity or basicity (alkalinity) of a solution. The pH scale ranges from 0 to 14. As we move down the pH scale from pH 14 to pH 0, each unit is 10 times more acidic than the previous unit and as we move up the scale from 0 to 14, each unit is 10 times more basic than the previous unit. The pH of a neutral aqueous solution at 25°C is 7, the midpoint of the pH scale. A pH value less than 7 denotes an acidic solution; the lower the number, the more acidic the solution. The pH for basic solutions is above 7. Most biological fluids, such as blood and saliva, are within the range of pH 6–8. There are a few exceptions, however, including the strongly acidic digestive juice of the human stomach (gastric juice), which has a pH of about 2.

All species have optimal pH requirements. Some organisms, such as the bacteria that cause ulcers in human stomachs, are adapted to low-pH environments. In contrast, the normal pH of human blood is 7.35 to 7.45. Extremely shallow breathing or kidney failure can cause the blood's pH to drop below 7 because carbon dioxide gas that accumulates in tissues ends up as excess carbonic acid in blood. The resulting decline in blood pH may cause the person to enter a dangerous level of unconsciousness called a coma.

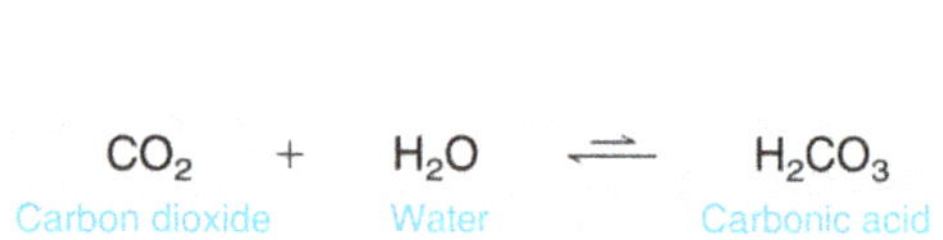

▶ **Figure 1.27** The pH scale and pH values of some aqueous solutions.

Vomiting, hyperventilating, or taking some types of alkaloid drugs, on the other hand, can raise the blood's pH above 7.8. Straying too far from the normal pH in either direction can cause death by destroying the shapes of critical proteins.

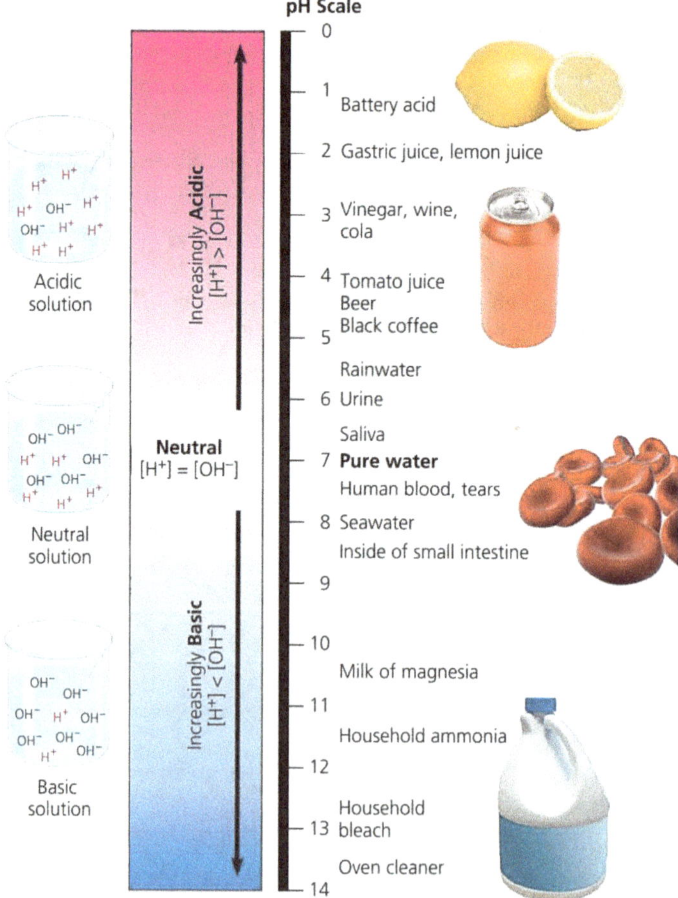

Buffers

The internal pH of most living cells is close to 7. Even a slight change in pH can be harmful because the chemical processes of the cell are very sensitive to the concentrations of hydrogen and hydroxide ions. The pH of human blood is very close to 7.4, which is slightly basic. A person cannot survive for more than a few minutes if the blood pH drops to 7 or rises to 7.8, and a chemical system exists in the blood that maintains a stable pH.

The presence of substances called buffers allows biological fluids to maintain a relatively constant pH despite the addition of acids or bases. A **buffer** is a substance that minimizes changes in the concentrations of H^+ and OH^- in a solution. It does so by accepting hydrogen ions from the solution when they are in excess and donating hydrogen ions to the solution when they have been depleted. Most buffer solutions contain a weak acid and its corresponding base, which combine reversibly with hydrogen ions.

Several buffers contribute to pH stability in human blood and many other biological solutions. One of these is carbonic acid (H_2CO_3), which is formed when CO_2 reacts with water in blood plasma. As mentioned earlier, carbonic acid dissociates to yield a bicarbonate ion (HCO_3^-) and a hydrogen ion (H^+).

The chemical equilibrium between carbonic acid and bicarbonate acts as a pH regulator, the reaction shifting left or right as other processes in the solution add or remove hydrogen ions. If the H^+ concentration in blood begins to fall (that is, if pH rises), the reaction proceeds to the right and more carbonic acid dissociates, replenishing hydrogen ions. But when the H^+ concentration in blood begins to rise (when pH drops), the reaction proceeds to the left, with HCO_3^- (the base) removing the hydrogen ions from the solution and forming H_2CO_3. Thus, the carbonic acid-bicarbonate buffering system consists of an acid and a base in equilibrium with each other. Most other buffers are also acid-base pairs.

$$CO_2 + H_2O \rightleftharpoons H_2CO_3 \rightleftharpoons H^+ + HCO_3^-$$

KEY CONCEPTS

- Matter is anything that occupies space and has mass. Matter is composed of elements, each consisting of atoms of the same kind.

- Atoms combine chemically in fixed numbers and ratios to form the molecules of living and nonliving matter. Compounds are molecules in which the component atoms are different.

- Atoms consist of an atomic nucleus that contains protons and neutrons surrounded by one or more electrons traveling in orbitals. Each orbital can hold a maximum of two electrons.

- All atoms of an element have the same number of protons, but the number of neutrons is variable. The number of protons in an atom is designated by its atomic number; the number of protons plus neutrons is designated by the mass number.

- Isotopes are atoms of an element with differing numbers of neutrons. The isotopes of an atom differ in physical but not chemical properties.

- Electrons surround an atomic nucleus in orbitals occupying energy levels that increase in discrete steps.

- The chemical activities of atoms are determined largely by the number of electrons in the outermost energy level. Atoms that have the outermost level filled with electrons are nonreactive, whereas atoms in which that level is not completely filled with electrons are reactive. Atoms tend to lose, gain, or share electrons to fill the outermost energy level.

- An ionic bond forms between atoms that gain or lose electrons in the outermost energy level completely, that is, between a positively charged cation and a negatively charged anion.

- A covalent bond is established by a pair of electrons shared between two atoms. If the electrons are shared equally, the covalent bond is nonpolar.

- If electrons are shared unequally in a covalent bond, the atoms carry partial positive and negative charges and the bond is polar.

- Polar molecules tend to associate with other polar molecules and to exclude nonpolar molecules. Polar molecules that associate readily with water are hydrophilic; nonpolar molecules excluded by water are hydrophobic.

- A hydrogen bond is a weak attraction between a hydrogen atom made partially positive by unequal electron sharing and another atom—usually oxygen, nitrogen, or sulfur—made partially negative by unequal electron sharing.

- Van der Waals forces, bonds even weaker than hydrogen bonds, can form when natural changes in the electron density of molecules produce regions of positive and negative charge, which cause the molecules to stick together briefly.

- The three-dimensional arrangement of the atoms in a molecule— its molecular geometry or shape—is characteristic of the molecule and determines the function of the molecule.

- Chemical reactions occur when molecules form or break chemical bonds. The atoms or molecules entering into a chemical reaction are the reactants, and those leaving a reaction are the products.

- The hydrogen-bond lattice gives water unusual properties that are vital to living organisms, including high specific heat, boiling point, cohesion, and surface tension.

- The polarity of the water molecules in the hydrogen-bond lattice makes it difficult for nonpolar substances to penetrate the lattice. The distinct polar and nonpolar environments created by water are critical to the organization of cells.

- The polar properties of water allow it to form a hydration layer over the surfaces of polar and charged biological molecules, particularly proteins. Many chemical reactions depend on the special molecular conditions created by the hydration layer.

- The polarity of water allows ions and polar molecules to dissolve readily in water, making it a good solvent.

- Acids are substances that increase the H^+ concentration by releasing additional H^+ as they dissolve in water; bases are substances that decrease the H^+ concentration by gathering H^+ or releasing OH^- as they dissolve.

- The relative concentrations of H^+ and OH^- in a water solution determine the acidity of the solution, which is expressed quantitatively as pH on a number scale ranging from 0 to 14. Neutral solutions, in which the concentrations of H^+ and OH^- are equal, have a pH of 7. Solutions with pH less than 7 have H^+ in excess and are acidic; solutions with pH greater than 7 have OH^- in excess and are basic or alkaline.

- The pH of living cells is regulated by buffers, which absorb or release H^+ to compensate for changes in H^+ concentration.

Biological Molecules

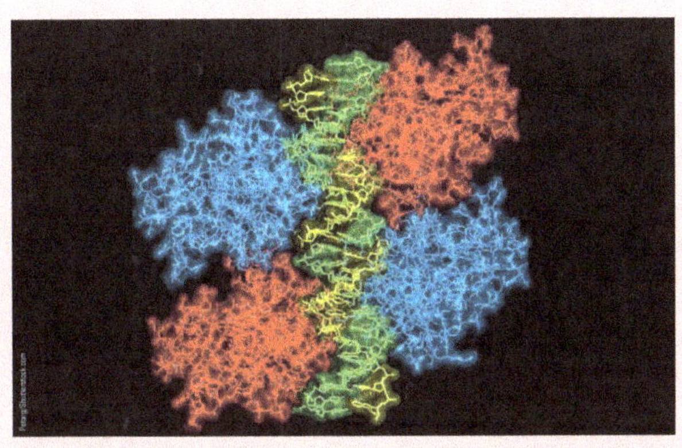

Chapter Contents:

- Living organisms are composed of about 25 key elements
- Life is based on carbon compounds
- Macromolecules are polymers
- The Synthesis and Breakdown of Polymers
- Carbohydrates include sugars and polymers of sugars
- Structure of the glucose molecule
- Isomers and stereoisomers
- Lipids are a diverse group of hydrophobic molecules
- Proteins
- Amino Acids (Monomers)
- Four Levels of Protein Structure
- Denaturation inactivates proteins
- Nucleic acids
- Other nucleotides
- KEY CONCEPTS

Living Organisms Are Composed of about 25 Key Elements

The four most common elements in living organisms are, in order of abundance, **hydrogen**, **carbon**, **oxygen** and **nitrogen**. They make up more than 96% of the weight of living organisms. Seven other elements—calcium, phosphorus, potassium, sulfur, sodium, chlorine, and magnesium—contribute most of the remaining 4%. Several other elements occur in organisms in quantities so small (less than 0.01%) that they are known as trace elements for example iodine that makes up only about 0.0004% of a human's weight. However, a lack of iodine in the human diet severely impairs the function of the thyroid gland, which produces hormones that regulate metabolism and growth.

▶ **Table 2.1** Valences of the major elements of organic molecules. Valence is the number of covalent bonds an atom can form. It is generally equal to the number of electrons required to complete the valence (outermost) shell. Note that carbon can form four bonds.

	Hydrogen	Oxygen	Nitrogen	Carbon
Lewis dot structure showing existing valence electrons	H·	·Ö:	·N̈·	·C̈·
Electron distribution diagram with red circles showing electrons needed to fill the valence shell	(H)	(O)	(N)	(C)
Number of electrons needed to fill the valence shell	1	2	3	4
Valence: Number of bonds the element can form	1	2	3	4

Carbon is particularly important because carbon atoms can join together to form long chains or ring structures. They can be thought of as the basic skeletons of organic molecules to which groups of other atoms are attached. Carbon has 6 electrons, with 2 in the first electron shell and 4 in the second shell; thus, it has 4 valence electrons in a shell that can hold up to 8 electrons.

A carbon atom usually completes its valence shell by sharing its 4 electrons with other atoms so that 8 electrons are present. Each pair of shared electrons constitutes a covalent bond. In organic molecules, carbon usually forms single or double covalent bonds. Each carbon atom acts as an intersection point from which a molecule can branch off in as many as four directions. This enables carbon to form large, complex molecules. Organic molecules always contain carbon and hydrogen.

(a) Length

Ethane

Propane

Carbon skeletons vary in length.

(b) Branching

Butane

2-Methylpropane
(commonly called isobutane)

Skeletons may be unbranched or branched.

(c) Double bond position

1-Butene

2-Butene

The skeleton may have double bonds, which can vary in location.

(d) Presence of rings

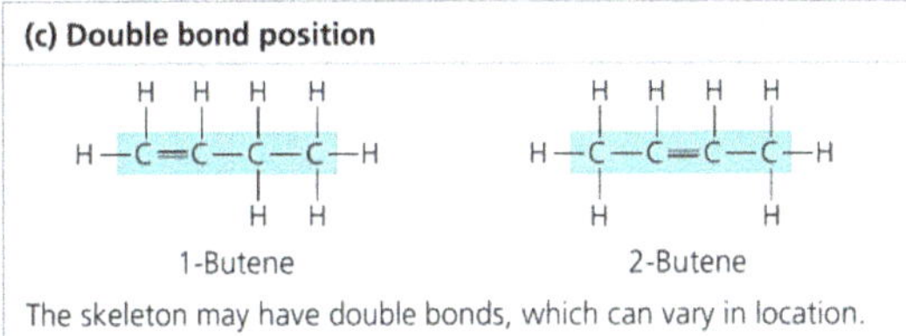

Cyclohexane

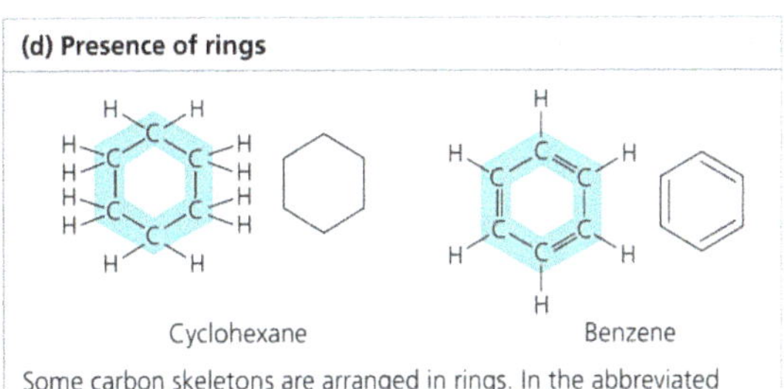

Benzene

Some carbon skeletons are arranged in rings. In the abbreviated structural formula for each compound (to its right), each corner represents a carbon and its attached hydrogens.

▶ **Figure 2.1** Four ways that carbon skeletons can vary

Life is based on carbon compounds including

- **Hydrocarbons** organic molecules consisting of only carbon and hydrogen. Atoms of hydrogen are attached to the carbon skeleton wherever electrons are available for covalent bonding. Hydrocarbons are the major components of petroleum, which is called a fossil fuel because it consists of the partially decomposed remains of organisms that lived millions of years ago.

- **Carbohydrates** are characterized by their composition. They are composed of carbon, hydrogen and oxygen, with hydrogen and oxygen in the ratio of two hydrogen atoms to one oxygen, hence the name carbohydrate

- **Lipids** are a broad class of molecules that are insoluble in water , including steroids, waxes, fatty acids and triglycerides. In common language, triglycerides are fats if they are solid at room temperature or oils if they are liquid at room temperature.

- **Proteins** are composed of one or more chains of amino acids. All of the amino acids in these chains contain the elements carbon, hydrogen, oxygen and nitrogen, but two of the twenty amino acids also contain sulphur (S).

- **Nucleic acids** are chains of subunits called nucleotides, which contain carbon, hydrogen, oxygen, nitrogen and phosphorus. There are two types of nucleic acid: ribonucleic acid (RNA) and deoxyribonucleic acid (DNA).

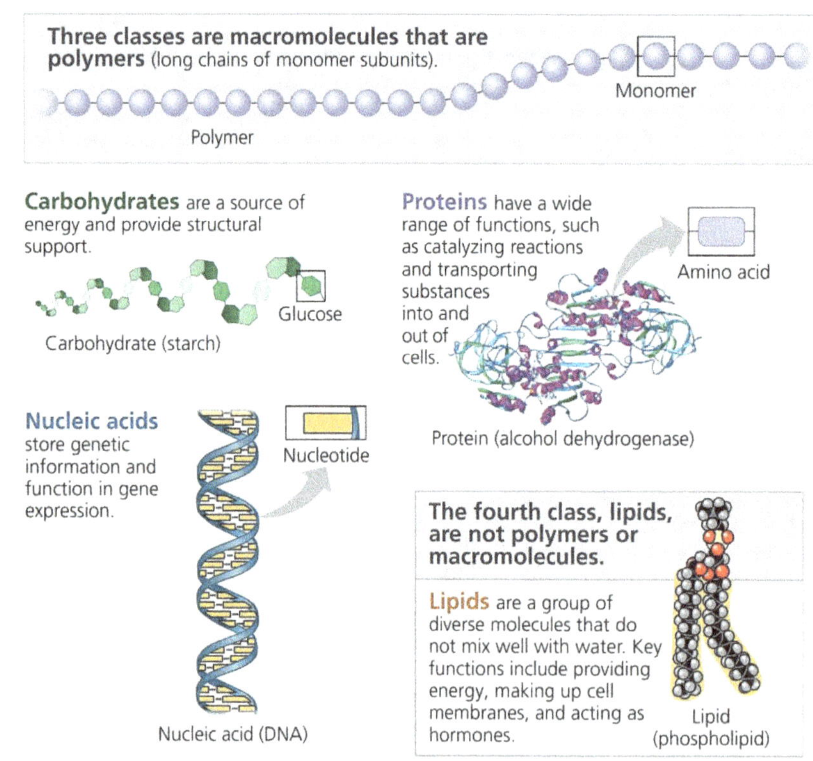

Macromolecules are polymers, built from monomers

Macromolecules are polymers, built from monomers. Large carbohydrates, proteins, and nucleic acids, also known as **macromolecules** for their huge size, are chain-like molecules called **polymers**. A polymer is a long molecule consisting of many similar or identical building blocks linked by covalent bonds. The repeating units that serve as the building blocks of a polymer are smaller molecules called **monomers.**

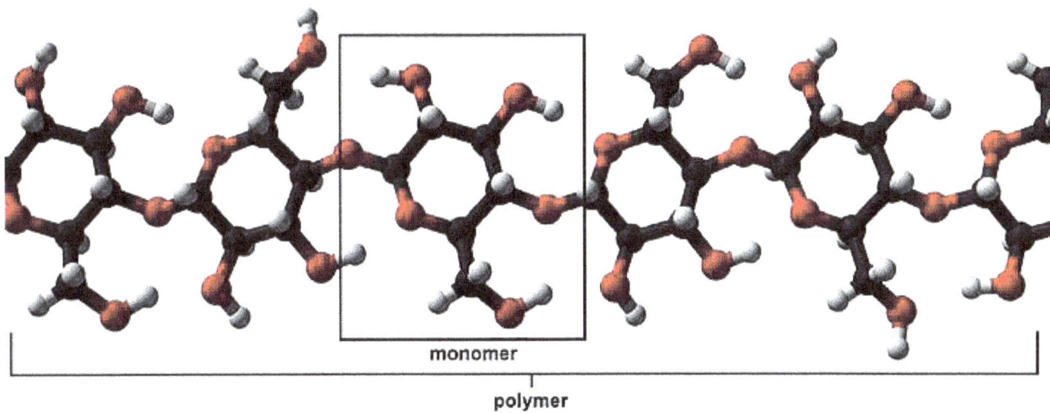

Name and Comment	Molecular Formula	Structural Formula	Ball-and-Stick Model (molecular shape in pink)	Space-Filling Model
(a) Methane. When a carbon atom has four single bonds to other atoms, the molecule is tetrahedral.	CH_4			
(b) Ethane. A molecule may have more than one tetrahedral group of single-bonded atoms. (Ethane consists of two such groups.)	C_2H_6			
(c) Ethene (ethylene). When two carbon atoms are joined by a double bond, all atoms attached to those carbons are in the same plane; the molecule is flat.	C_2H_4			

▲ **Figure 2.2** The shapes of three simple organic molecules

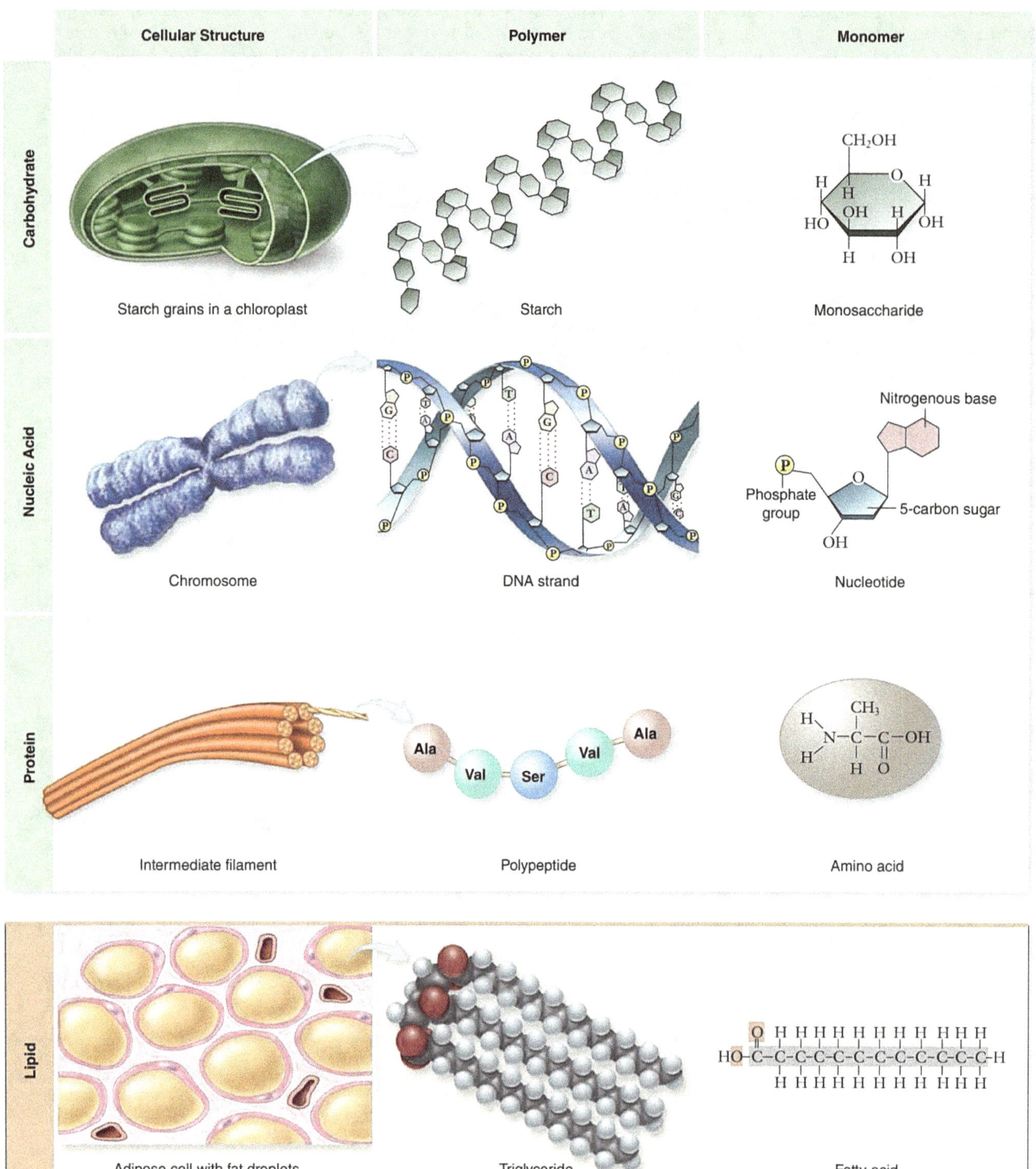

▲ **Figure 2.3** Polymer macromolecules. The four major biological macromolecules are shown. Carbohydrates, nucleic acids, and proteins all form polymers and are shown with the monomers used to make them.
Lipids do not fit this simple monomer–polymer relationship.

The Synthesis and Breakdown of Polymers

The reaction that connects a monomer to another monomer or a polymer is a **condensation** reaction, a reaction in which two molecules are covalently bonded to each other with the loss of a small molecule. If a water molecule is lost, it is known as a **dehydration reaction**.

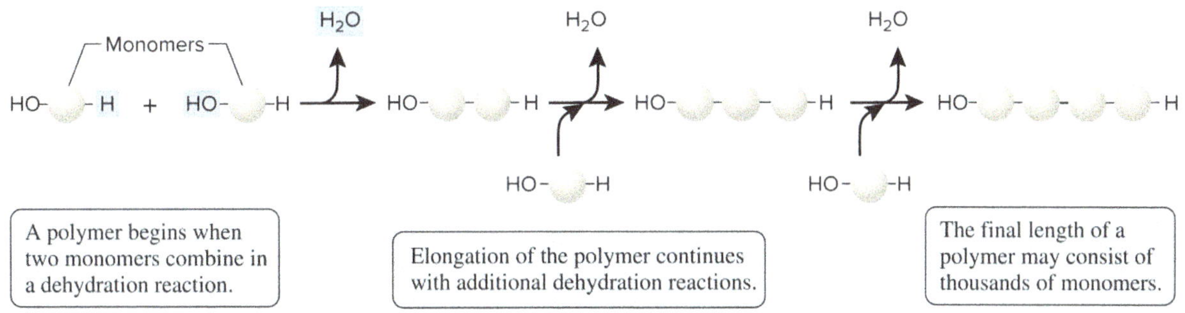

(a) Polymer formation by dehydration reactions

Polymers are disassembled to monomers by **hydrolysis**, a process that is essentially the reverse of the dehydration reaction. The bond between monomers is broken by the addition of a water molecule, with a hydrogen from water attaching to one monomer and the hydroxyl group attaching to the other. An example of hydrolysis within our bodies is the process of digestion.

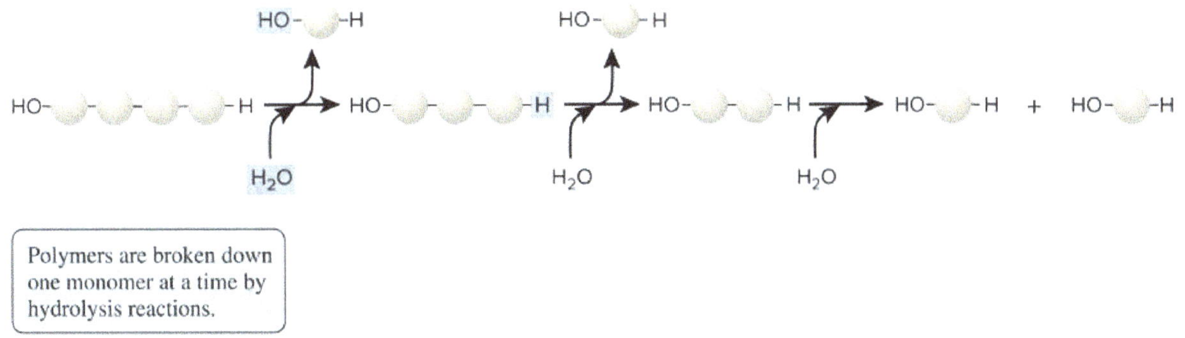

(b) Breakdown of a polymer by hydrolysis reactions

▲ **Figure 2.4** Formation and breakdown of polymers.
(a) Monomers combine to form polymers in living organisms by dehydration reactions, in which a molecule of H2O is removed each time a new monomer is added to the growing polymer.
(b) Polymers can be broken down into their constituent monomers by hydrolysis reactions, in which a molecule of H2O is added each time a monomer is released.

Carbohydrates include sugars and polymers of sugars

Monosaccharides (from the Greek monos, single, and sacchar, sugar) generally have molecular formulas that are some multiple of the unit CH2O. Glucose (C6H12O6), the most common monosaccharide, is of central importance in the chemistry of life. Depending on the location of the carbonyl group, a monosaccharide is either an aldose (aldehyde sugar) or a ketose (ketone sugar). Glucose, for example, is an aldose; fructose, an isomer of glucose, is a ketose.

Glucose, fructose, and other sugars that have six carbons are called hexoses. Trioses (three-carbon sugars) and pentoses (five-carbon sugars) are also common.

Monosaccharides, particularly glucose, are major nutrients for cells. In the process known as cellular respiration, cells extract energy from glucose molecules by breaking them down in a series of reactions. Not only are monosaccharides a major fuel for cellular work, but their carbon skeletons also serve as raw material for the synthesis of other types of small organic molecules, such as amino acids and fatty acids.

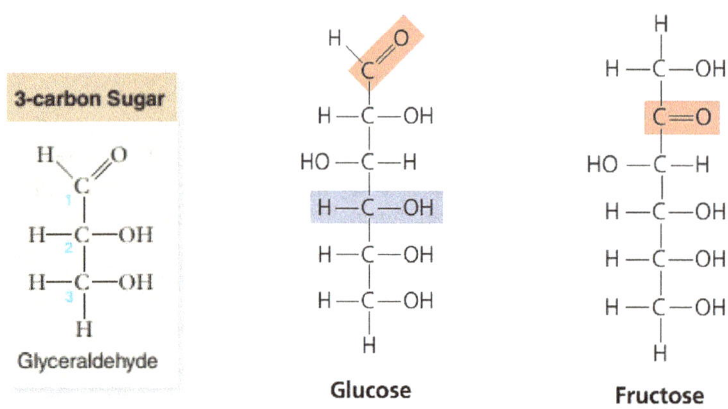

Monomers of carbohydrates

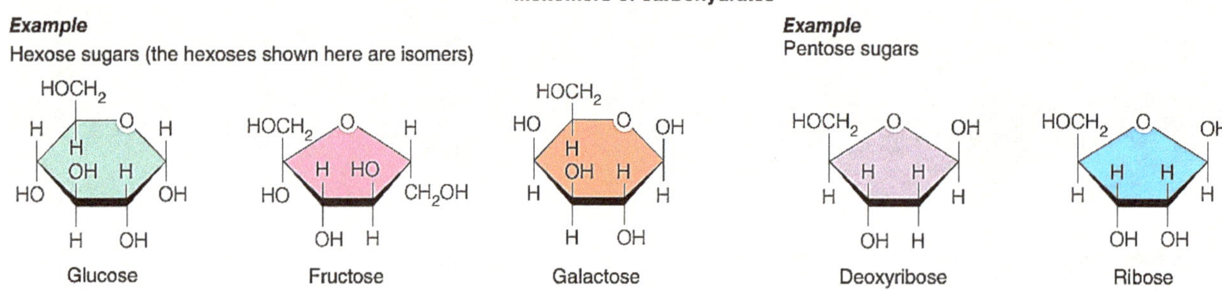

▲ **Figure 2.5** Monosaccharides.
Monosaccharides, or simple sugars, can contain as few as three carbon atoms and are often used as building blocks to form larger molecules. The 5-carbon sugars ribose and deoxyribose are components of nucleic acids

Structure of the glucose molecule

Glucose is a linear, 6-carbon molecule that forms a six-membered ring in solution. Ring closure occurs such that two forms can result: **α-glucose** and **β-glucose**. These structures differ only in the position of the - OH bound to carbon 1.

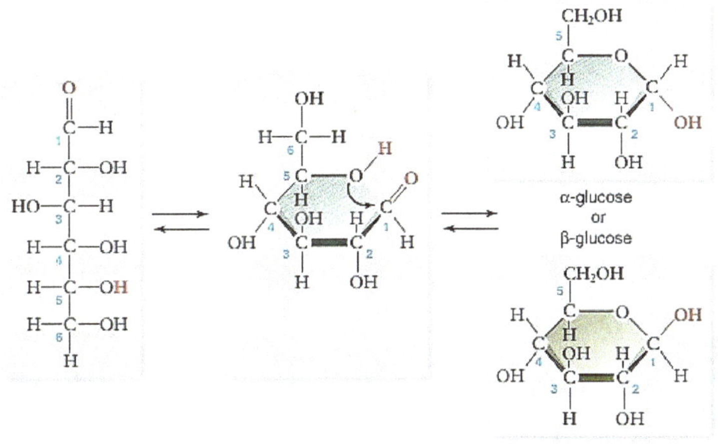

▲ **Figure 2.6** Structure of the glucose molecule. Glucose is a linear, 6-carbon molecule that forms a six-membered ring in solution. Ring closure occurs such that two forms can result: α-glucose and β-glucose. These structures differ only in the position of the -OH bound to carbon 1.

Isomers and stereoisomers

Glucose, fructose, and **galactose** are isomers with the empirical formula $C_6H_{12}O_6$. A structural isomer of glucose, such as fructose, has identical chemical groups bonded to different carbon atoms that this results in a five-membered ring in solution.

A stereoisomer of glucose, such as galactose, has identical chemical groups bonded to the same carbon atoms but in different orientations (the -OH at carbon 4).

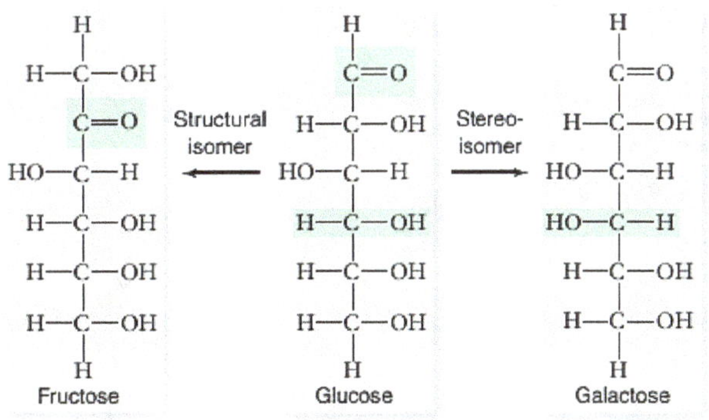

▲ **Figure 2.7** Isomers and stereoisomers. Glucose, fructose, and galactose are isomers with the empirical formula $C_6H_{12}O_6$. A structural isomer of glucose, such as fructose, has identical chemical groups bonded to different carbon atoms. A stereoisomer of glucose, such as galactose, has identical chemical groups bonded to the same carbon atoms but in different orientations (the -OH at carbon 4).

Disaccharide consists of two monosaccharides joined by a glycosidic linkage, a covalent bond formed between two monosaccharides by a dehydration reaction (glycorefers to carbohydrate). For example, maltose is a disaccharide formed by the linking of two molecules of glucose.

The most prevalent disaccharide is sucrose, or table sugar. Its two monomers are glucose and fructose. Plants generally transport carbohydrates from leaves to roots and other nonphotosynthetic organs in the form of sucrose. Lactose, the sugar present in milk, is another disaccharide, in this case a glucose molecule joined to a galactose molecule. Disaccharides must be broken down into monosaccharides to be used for energy by organisms.

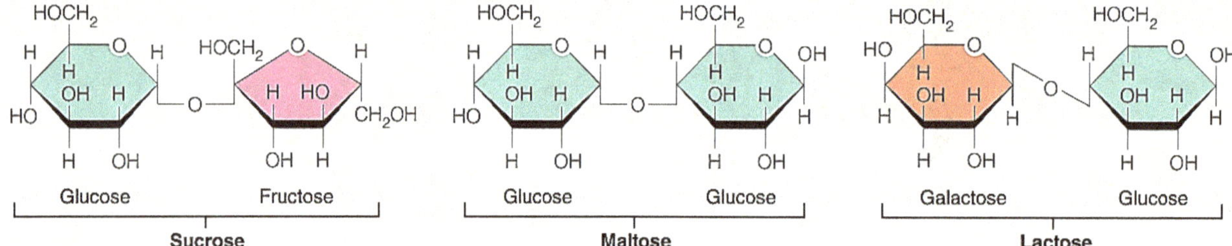

▲ **Figure 2.8** Disaccharides.

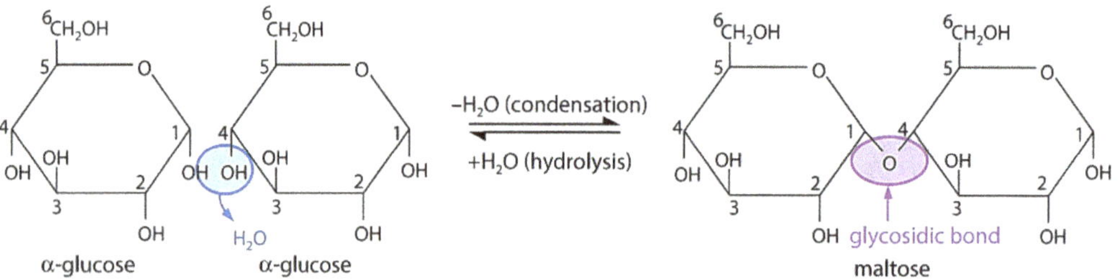

B. Sucrose

Sucrose is assembled from glucose and fructose.

C. Lactose

Lactose is assembled from galactose and glucose.

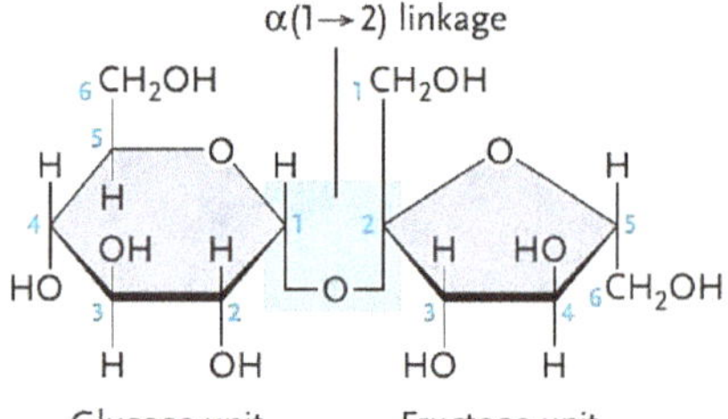

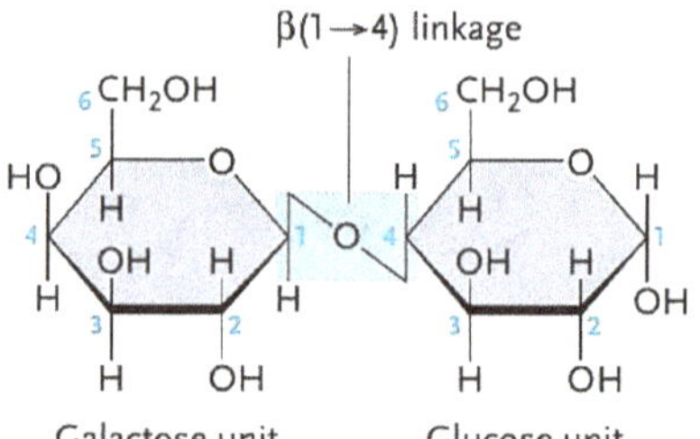

Polysaccharides are macromolecules, polymers with a few hundred to a few thousand monosaccharides joined by glycosidic linkages. Some polysaccharides serve as storage material, hydrolyzed as needed to provide monosaccharides for cells. Both plants and animals store sugars for later use in the form of storage polysaccharides.

Plants store **starch**, a polymer of glucose monomers, as granules within cellular structures known as plastids. Potato tubers and grains—the fruits of wheat, maize (corn), rice, and other grasses—are the major sources of starch in the human diet. Most of the glucose monomers in starch are joined by 1–4 linkages (number 1 carbon to number 4 carbon), like the glucose units in maltose. The simplest form of starch, **amylose**, is unbranched. **Amylopectin**, a more complex starch, is a branched polymer with 1–6 linkages at the branch points. Mixtures of amylose and amylopectin molecules build up into relatively large starch grains. Starch grains are easily seen with a light microscope, especially if stained.

False-colour scanning electron micrograph of a slice through a raw potato shows cells containing starch grains or starch-containing organelles as red coloured (staining with iodine–potassium iodide solution).

A. Amylose, a plant starch
Amylose, formed from α-glucose units joined end to end in α(1→4) linkages. The coiled structures are induced by the bond angles in the α-linkages.

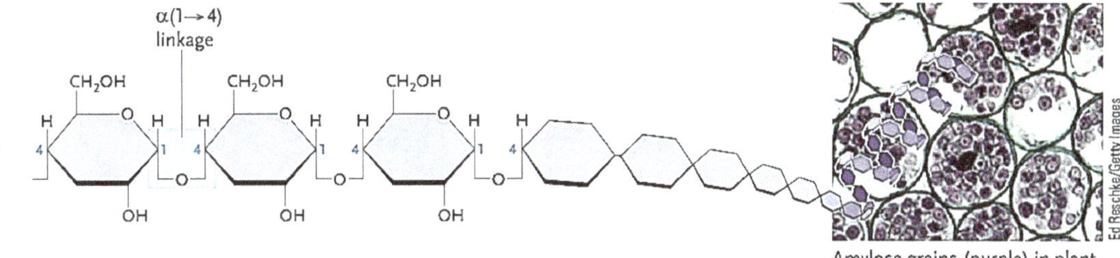

Amylose grains (purple) in plant root tissue

Animals store a polysaccharide called **glycogen**, a polymer of glucose that is like amylopectin but more extensively branched. Vertebrates store glycogen mainly in liver and muscle cells. Breakdown of glycogen in these cells releases glucose when the demand for energy increases.

B. Glycogen, found in animal tissues
Glycogen, formed from glucose units joined in chains by α(1→4) linkages; side branches are linked to the chains by α(1→6) linkages (boxed in blue).

Glycogen particles (magenta) in liver cell

Organisms build strong materials from structural polysaccharides. For example, the polysaccharide called **cellulose** is a major component of the tough walls that enclose plant cells. It is the most abundant organic molecule on the planet due to its presence in plant cell walls and its slow rate of breakdown in nature. Like starch, cellulose is a polymer of glucose with 1–4 glycosidic linkages, but the linkages in these two polymers differ.

In starch, all the glucose monomers are in the α configuration (the hydroxyl group attached to the number 1 carbon is positioned below the plane of the ring). In contrast, the glucose monomers of cellulose are all in the β configuration (the hydroxyl group attached to the number 1 carbon is positioned above the plane of the ring), making every glucose monomer "upside down" with respect to its neighbors.

c. Cellulose, the primary fiber in plant cell walls

Cellulose, formed from glucose units joined end to end by $\beta(1\rightarrow4)$ linkages. Hundreds to thousands of cellulose chains line up side by side, in an arrangement reinforced by hydrogen bonds between the chains, to form cellulose microfibrils in plant cells.

Biophoto Associates/Science Source

Cellulose is never branched, and some hydroxyl groups on its glucose monomers are free to hydrogen-bond with the hydroxyls of other cellulose molecules lying parallel to it. Between 60 and 70 cellulose molecules become tightly cross-linked by hydrogen bonding to form bundles called microfibrils.

Microfibrils are in turn held together by hydrogen bonding in bundles called fibres. A plant cell wall typically has several layers of fibres, running in different directions to increase strength. Without the wall, the cell would burst when in a dilute solution. These pressures help provide support for the plant by making tissues rigid, and are responsible for cell expansion during growth. The arrangement of fibres around the cell helps to determine the shape of the cell as it grows.

The unbranched structure of cellulose fits its function: imparting strength to parts of the plant. Few organisms possess enzymes that can digest cellulose. Almost all animals, including humans, do not. Some microorganisms can digest cellulose, breaking it down into glucose monomers. A cow harbors cellulose digesting prokaryotes and protists in its gut. Some fungi can also digest cellulose in soil and elsewhere.

Another important structural polysaccharide is **chitin**, the carbohydrate used by arthropods (insects, spiders, crustaceans, and related animals) to build their exoskeletons—hard cases that surround the soft parts of an animal. Chitin is also found in fungi, which use this polysaccharide rather than cellulose as the building material for their cell walls. Chitin is similar to cellulose, with β linkages, except that the glucose monomer of chitin has a nitrogen-containing attachment.

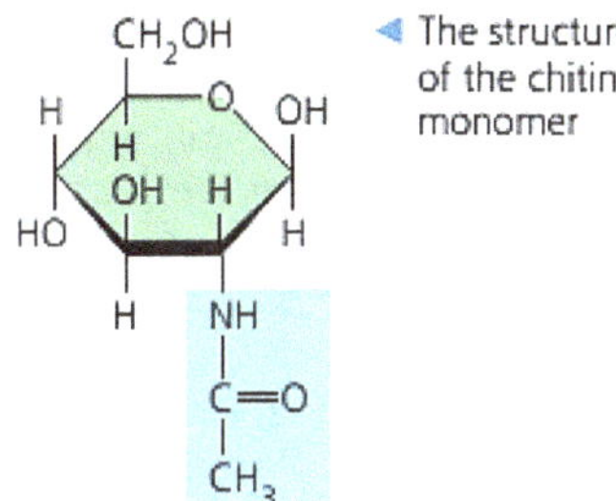

D. Chitin, a reinforcing fiber in the external skeleton of arthropods and the cell walls of some fungi

Chitin, formed from β(1→4) linkages joining glucose units modified by the addition of nitrogen-containing groups. The external body armor of the tick is reinforced by chitin fibers.

Carolina K. Smith, M.D/Shutterstock.com

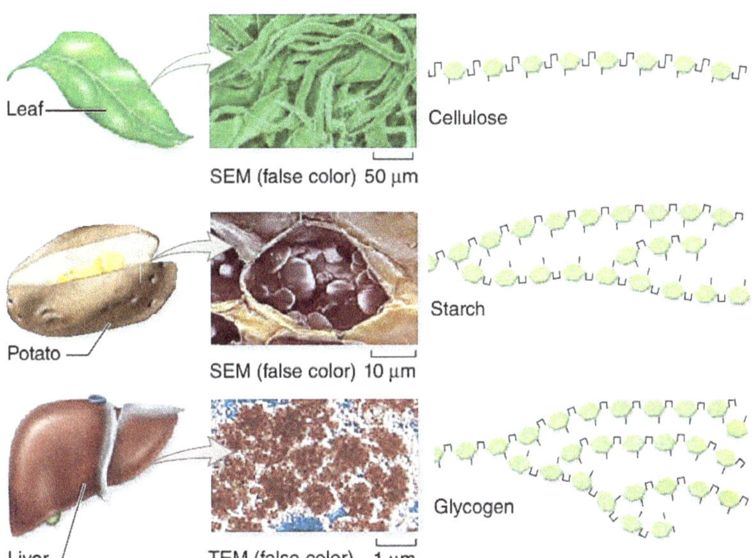

▲ **Figure 2.9** Carbohydrates—simple and Complex. (a) Monosaccharides are composed of single sugar molecules, such as glucose or fructose. (b) Disaccharides form by dehydration synthesis. In this example, glucose and fructose bond to form sucrose. (c) Polysaccharides are long chains of monosaccharides such as glucose. Different orientations of covalent bonds produce different characteristics in the polymers.

Lipids are a diverse group of hydrophobic molecules

Lipids are the one class of large biological molecules that does not include true polymers, and they are generally not big enough to be considered macromolecules. The compounds called lipids are grouped with each other because they share one important trait: They are hydrophobic: They mix poorly, if at all, with water. This behavior of lipids is based on their molecular structure. Although they may have some polar bonds associated with oxygen, lipids consist mostly of hydrocarbon regions with relatively non-polar C—H bonds. Lipids are varied in form and function. They include fats, phospholipids, steroids, waxes and certain pigments.

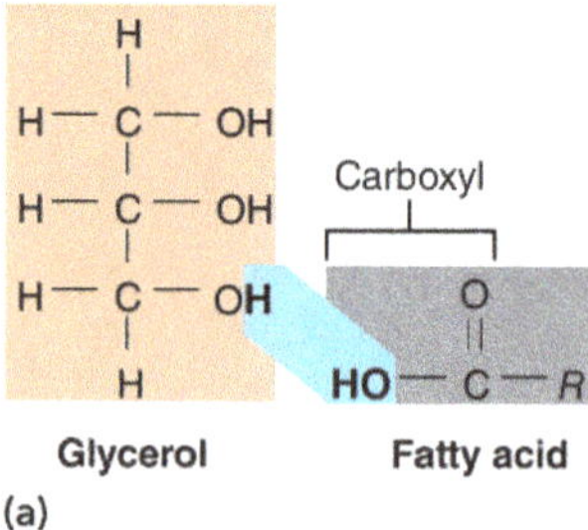

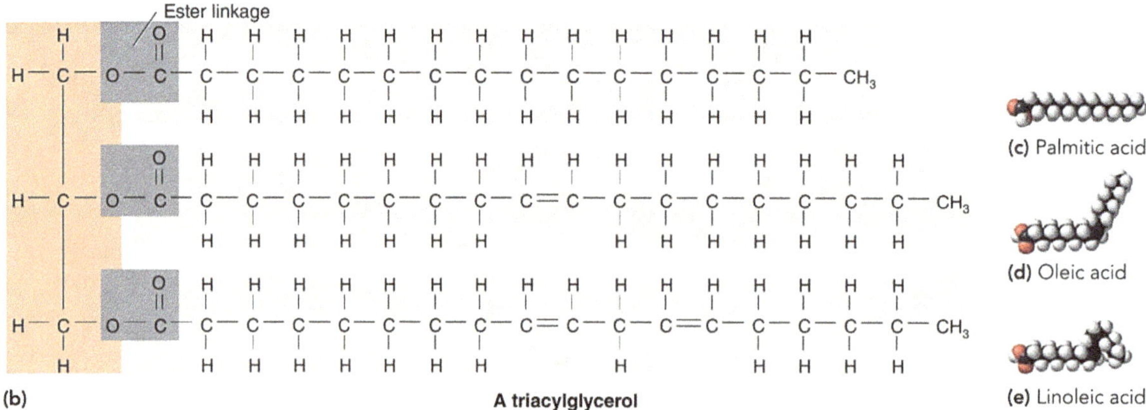

▲ **Figure 2.10** Triacylglycerol, the main storage lipid.
a) Glycerol and fatty acids are the components of fats.
b) Glycerol is attached to fatty acids by ester linkages (in gray). The space-filling models show the actual shapes of the fatty acids.
c) Palmitic acid, a saturated fatty acid, is a straight chain.
d) Oleic acid (monounsaturated)
e) linoleic acid (polyunsaturated) are bent or kinked wherever a carbon-to-carbon double bond appears.

- **Triglyceride or Triacylglycerol or Fat:** A fat consists of a glycerol molecule joined to three fatty acids. Glycerol is an alcohol; each of its three carbons bears a hydroxyl group. A fatty acid has a long carbon skeleton, usually 16 or 18 carbon atoms in length. The carbon at one end of the skeleton is part of a carboxyl group, the functional group that gives these molecules the name fatty acid. The rest of the skeleton consists of a hydrocarbon chain. The relatively nonpolar C—H bonds in the hydrocarbon chains of fatty acids are the reason fats are hydrophobic so triglycerides are insoluble in water but are soluble in certain organic solvents such as ethanol. In making a fat, each fatty acid molecule is joined to glycerol by a dehydration reaction. This results in an ester linkage, a bond between a hydroxyl group and a carboxyl group. The completed fat consists of three fatty acids linked to one glycerol molecule.

Triglycerides make excellent energy stores because they are even richer in carbon–hydrogen bonds than carbohydrates. A given mass of triglyceride will therefore yield more energy on oxidation than the same mass of carbohydrate (it has a higher calorific value), an important advantage for a storage product. Triglycerides are stored in a number of places in the human body, particularly just below the skin and around the kidneys. Below the skin they also act as an insulator against loss of heat. An unusual role for triglycerides is as a metabolic source of water. When oxidised in respiration, triglycerides are converted to carbon dioxide and water. The water may be of importance in very dry habitats.

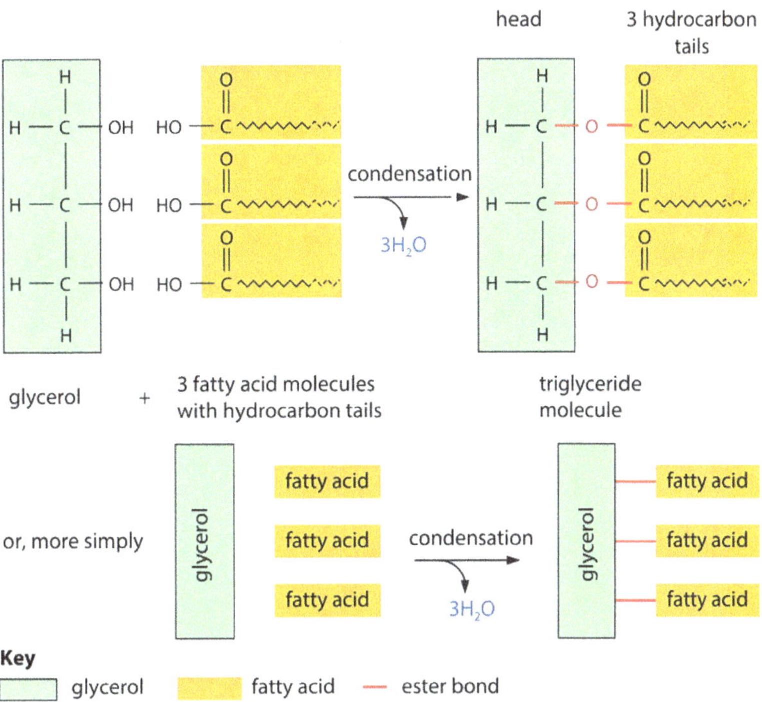

(a) A triglyceride consists of three fatty acids bonded to glycerol. In saturated fats such as butter, the fatty acid chains contain only single carbon–carbon bonds. (b) In unsaturated fats, one or more double bonds bend the fatty acid tails, making the lipid more fluid. Vegetable oil is an unsaturated fat.

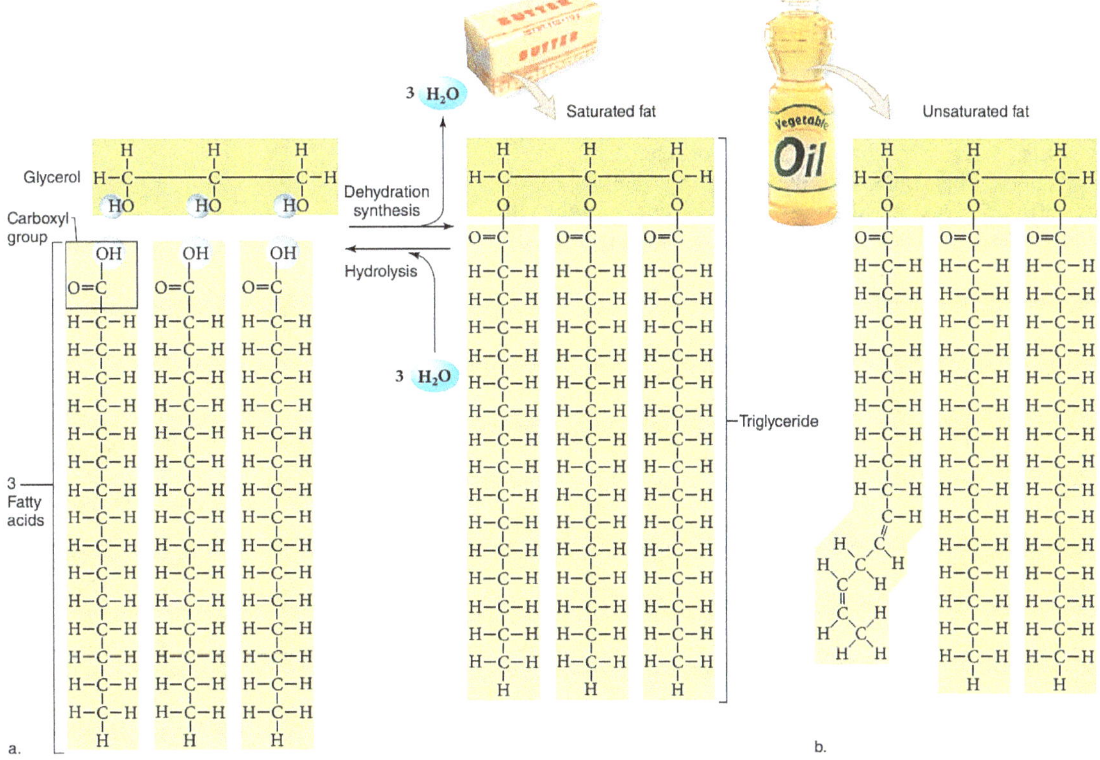

The terms saturated fats and unsaturated fats are commonly used in the context of nutrition, These terms refer to the structure of the hydrocarbon chains of the fatty acids. If there are no double bonds between carbon atoms composing a chain, then as many hydrogen atoms as possible are bonded to the carbon skeleton. Such a structure is said to be saturated with hydrogen, and the resulting fatty acid is therefore called a **saturated fatty acid**. An **unsaturated fatty acid** has one or more double bonds, with one fewer hydrogen atom on each double-bonded carbon. Most animal fats are saturated and are solid at room temperature. In contrast, the fats of plants and fishes are generally unsaturated, meaning that they are composed of one or more types of unsaturated fatty acids. Usually liquid at room temperature, plant and fish fats are referred to as oils. Where the **cis** double bonds are located prevent the molecules from packing together closely enough to solidify at room temperature.

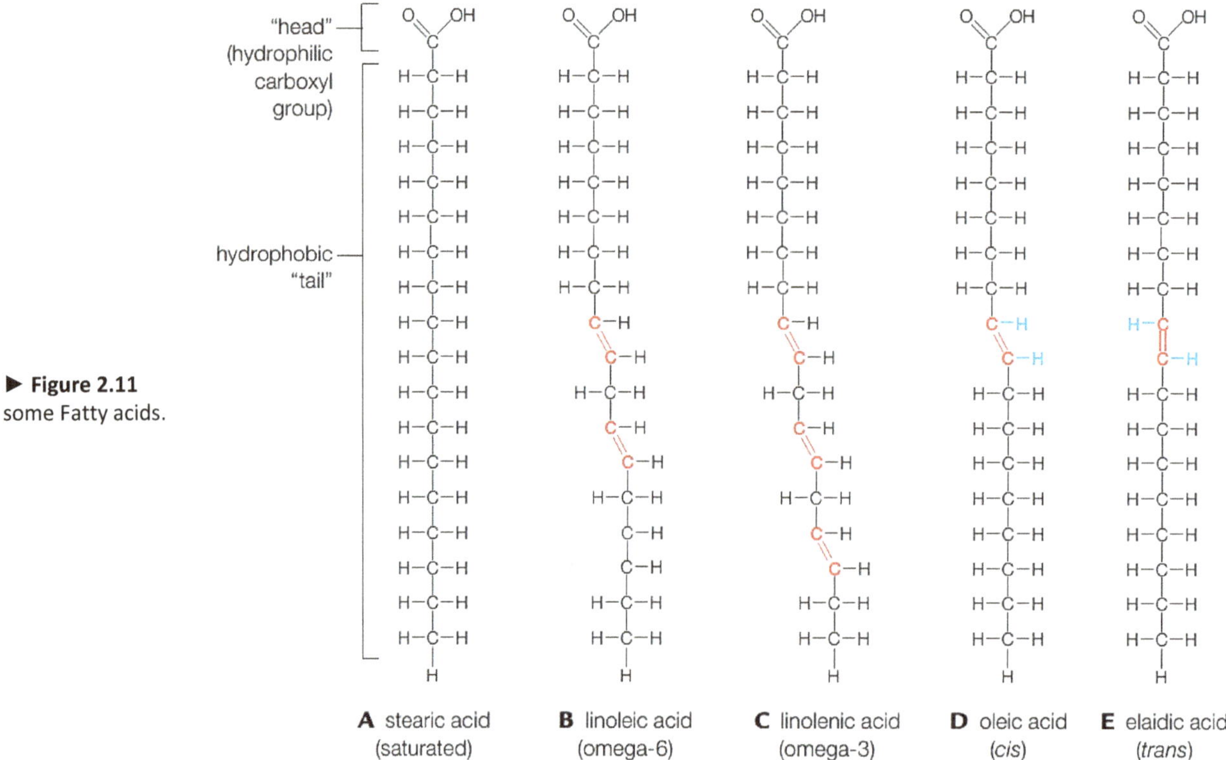

► **Figure 2.11** some Fatty acids.

In **cis-fatty acids**, there is a bend in the hydrocarbon chain at the double bond. This makes triglycerides containing cis-unsaturated fatty acids less good at packing together in regular arrays than saturated fatty acids, so it lowers the melting point. Triglycerides with cis-unsaturated fatty acids are therefore usually liquid at room temperature – they are oils.

Trans-fatty acids do not have a bend in the hydrocarbon chain at the double bond, so they have a higher melting point and are solid at room temperature. Trans-fatty acids are produced artificially by partial hydrogenation of vegetable or fish oils. This is done to produce solid fats for use in margarine and some other processed foods.

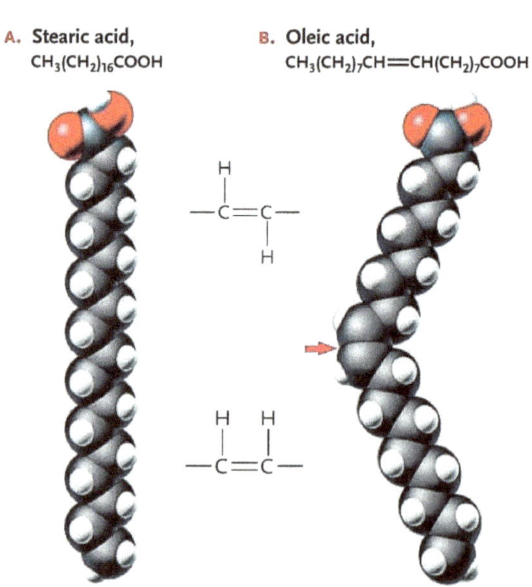

A diet rich in saturated fats is one of several factors that may contribute to the cardiovascular disease known as atherosclerosis. In this condition, deposits called plaques develop within the walls of blood vessels, causing inward bulges that impede blood flow and reduce the resilience of the vessels.

The major function of fats is energy storage. The hydrocarbon chains of fats are similar to gasoline molecules and just as rich in energy. A gram of fat stores more than twice as much energy as a gram of a polysaccharide, such as starch. In addition to storing energy, adipose tissue also cushions such vital organs as the kidneys, and a layer of fat beneath the skin insulates the body. This subcutaneous layer is especially thick in whales, seals, and most other marine mammals, insulating their bodies in cold ocean water.

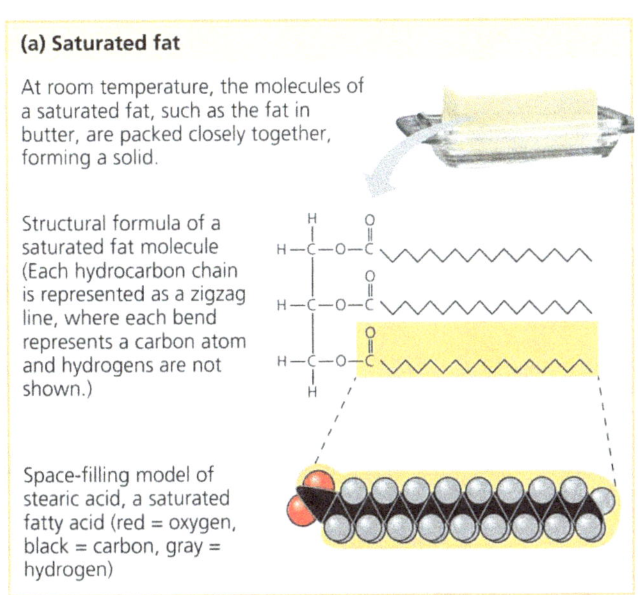

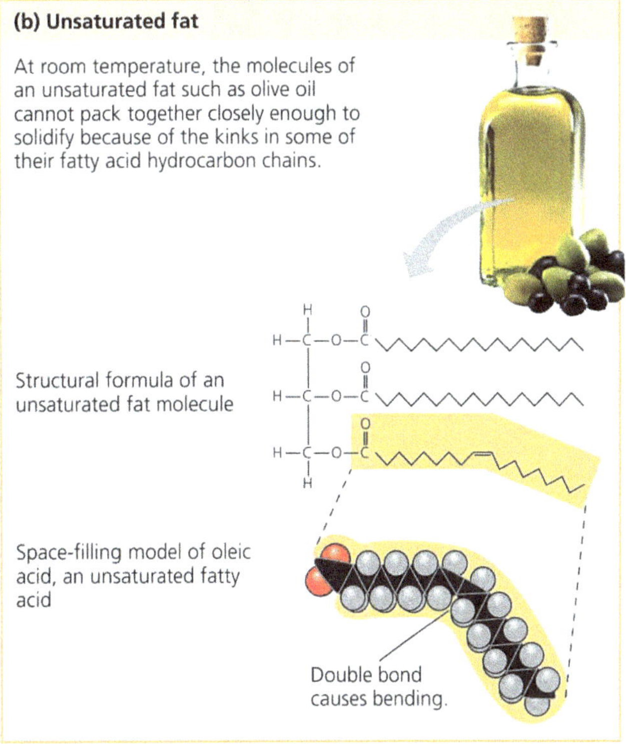

▲ **Figure 2.12** Saturated and unsaturated fats and fatty acids

- **Phospholipids**

Phospholipids are essential for cells because they are major constituents of cell membranes. A phospholipid is similar to a fat molecule but has only two fatty acids attached to glycerol rather than three. The third hydroxyl group of glycerol is joined to a phosphate group, which has a negative electrical charge in the cell. Typically, an additional small charged or polar molecule is also linked to the phosphate group. Choline is one such molecule.

The two ends of phospholipids show different behaviors with respect to water. The hydrocarbon tails are hydrophobic and are excluded from water. However, the phosphate group and its attachments form a hydrophilic head that has an affinity for water. When phospholipids are added to water, they self-assemble into a double-layered sheet called a "bilayer" that shields their hydrophobic fatty acid tails from water.

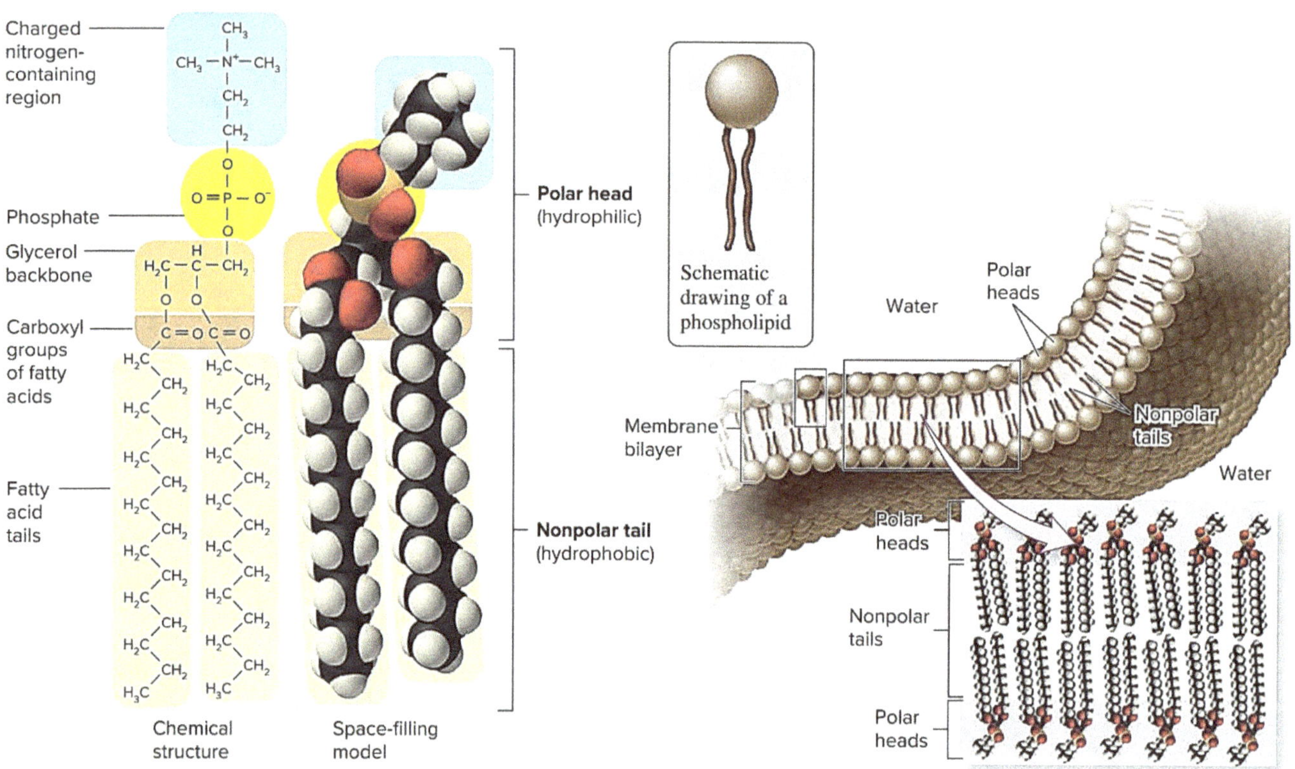

▲ **Figure 2.13** Structure of phospholipids.
(a) Chemical structure and space-filling model of phosphatidylcholine, a common phospholipid found in living organisms. Phospholipids contain both polar and nonpolar regions, making them amphipathic. The fatty acid tails are nonpolar. The rest of the molecule is polar.
(b) Arrangement of phospholipids in a biological membrane, such as the plasma membrane that encloses cells. The polar region of the phospholipid faces the watery environment, whereas the nonpolar regions associate with each other in the interior of the membrane, forming a bilayer.

At the surface of a cell, phospholipids are arranged in a similar bilayer. The hydrophilic heads of the molecules are on the outside of the bilayer, in contact with the aqueous solutions inside and outside of the cell. The hydrophobic tails point toward the interior of the bilayer, away from the water. The phospholipid bilayer forms a boundary between the cell and its external environment and establishes separate compartments within eukaryotic cells; in fact, the existence of cells depends on the properties of phospholipids.

In an aqueous environment, lipid molecules orient so that their polar (hydrophilic) heads are in the polar medium, water, and their nonpolar (hydrophobic) tails are held away from the water. a. Droplets called **micelles** can form, or (b) phospholipid molecules can arrange themselves into two layers that called **phospholipid bilayer**; in both structures, the hydrophilic heads extend outward and the hydrophobic tails inward.

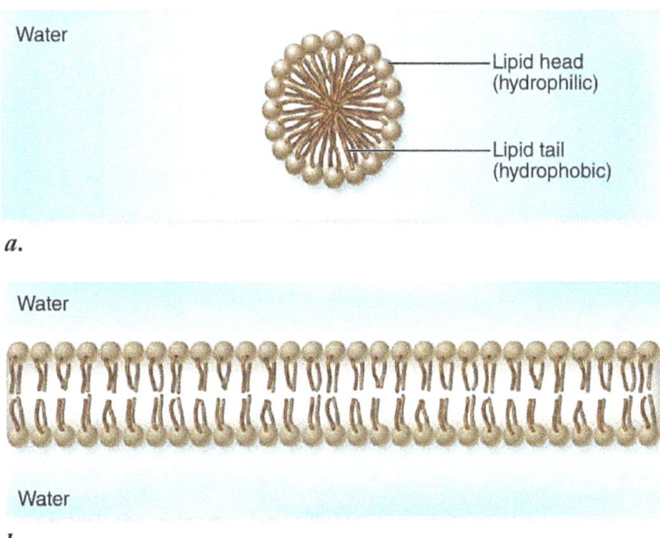

▶ **Figure 2.14** Lipids spontaneously form micelles or lipid bilayers in water. In an aqueous environment, lipid molecules orient so that their polar (hydrophilic) heads are in the polar medium, water, and their nonpolar (hydrophobic) tails are held away from the water.
a. Droplets called micelles can form, or b. phospholipid molecules can arrange themselves into two layers; in both structures, the hydrophilic heads extend outward and the hydrophobic tails inward. This second example is called a phospholipid bilayer.

- **Steroids**

Steroids are lipids characterized by a carbon skeleton consisting of four fused rings; three of the rings contain six carbon atoms, and the fourth contains five. Different steroids are distinguished by the particular chemical groups attached to this ensemble of rings. Cholesterol, a type of steroid, is a crucial molecule in animals. It is a common component of animal cell membranes and is also the precursor from which other steroids, such as the vertebrate reproductive hormones (testosterone and estrogen) and cortisol as well as other hormones secreted by the adrenal cortex, are synthesized. In vertebrates, cholesterol is synthesized in the liver and is also obtained from the diet. A high level of cholesterol in the blood may contribute to atherosclerosis, although some researchers are questioning the roles of cholesterol and saturated fats in the development of this condition.

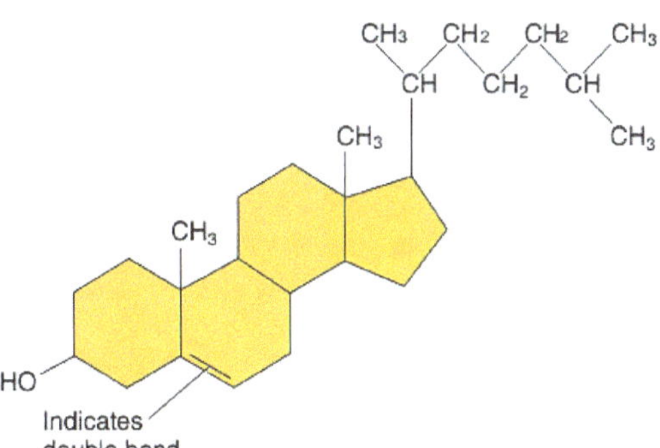

▶ **Figure 2.15** Steroids. Four attached rings—three 6-carbon rings and one with 5 carbons—make up the fundamental structure of a steroid.

- **Terpenes**: Terpenes are long-chain lipids that are components of many biologically important pigments, such as chlorophyll and the visual pigment retinal. Rubber is also a terpene.

- **Prostaglandins:** are a group of about 20 lipids that are modified fatty acids, with two nonpolar "tails" attached to a 5-carbon ring. Prostaglandins which have varied roles, including promoting inflammation, smooth muscle contraction and as local chemical messengers in many vertebrate tissues.

- Certain hormones, such as the **juvenile hormone** of insects, are also fatty acid derivatives.

- **Eicosanoids:** The eicosanoids are diverse lipids chiefly derived from a 20-carbon fatty acid (arachidonic acid) found in all cell membranes. Most important of these are the prostaglandins and their relatives, which play roles in various body processes including blood clotting, regulation of blood pressure, inflammation, and labor contractions. Their synthesis and inflammatory actions are blocked by NSAIDs (nonsteroidal anti-inflammatory drugs).

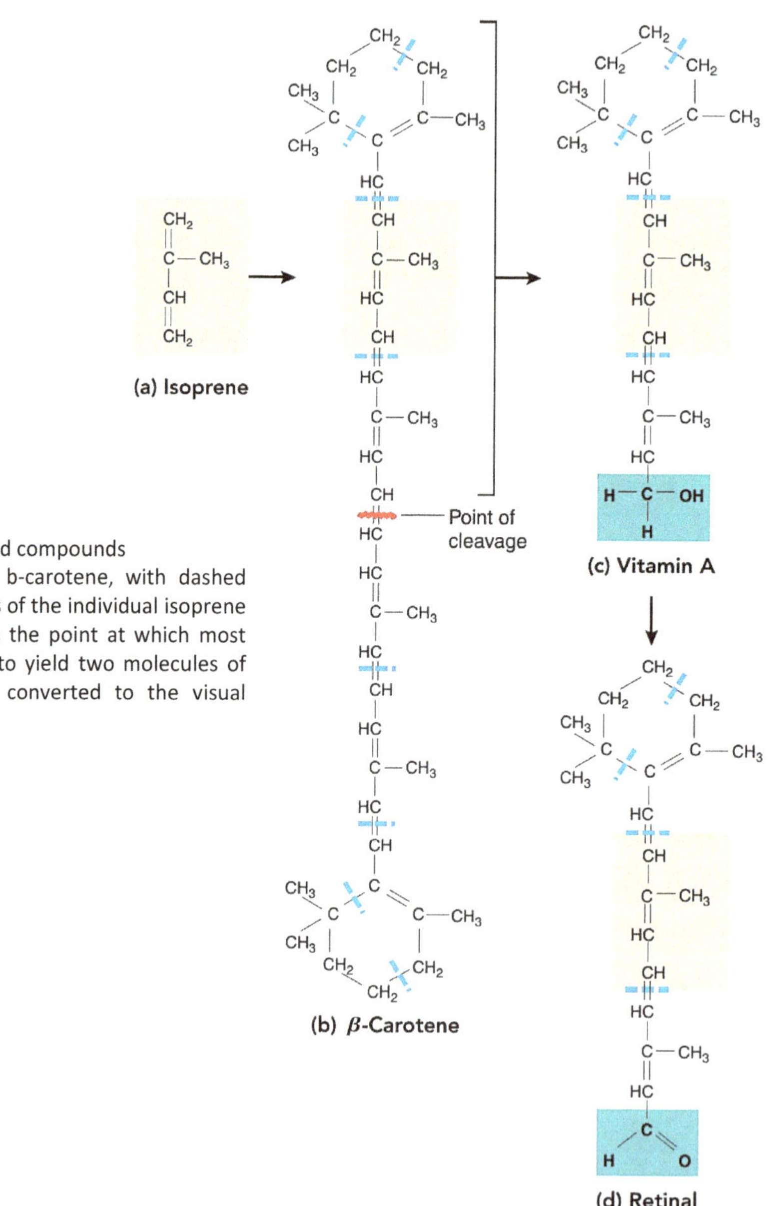

► **Figure 2.16** Isoprene-derived compounds
(a) An isoprene subunit. (b) b-carotene, with dashed lines indicating the boundaries of the individual isoprene units within. The wavy line is the point at which most animals cleave the molecule to yield two molecules of (c) vitamin A. Vitamin A is converted to the visual pigment (d) retinal.

LIPID TYPE	LOCATION/FUNCTION
Triglycerides (Neutral Fats)	
	Fat deposits (in subcutaneous tissue and around organs) protect and insulate body organs, and are the major source of *stored* energy in the body.
Phospholipids (phosphatidylcholine; cephalin; others)	
	Chief components of cell membranes. Participate in the transport of lipids in plasma. Prevalent in nervous tissue.
Steroids	
Cholesterol	The structural basis for manufacture of all body steroids. A component of cell membranes.
Bile salts	These breakdown products of cholesterol are released by the liver into the digestive tract, where they aid fat digestion and absorption.
Vitamin D	Fat-soluble vitamin produced in the skin on exposure to UV radiation. Necessary for normal bone growth and function.
Sex hormones	Estrogen and progesterone (female hormones) and testosterone (a male hormone) are produced in the gonads. Necessary for normal reproductive function.
Adrenocortical hormones	Cortisol, a glucocorticoid, is a metabolic hormone necessary for maintaining normal blood glucose levels. Aldosterone helps to regulate salt and water balance of the body by targeting the kidneys.
Other Lipoid Substances	
Fat-soluble vitamins:	
A	Ingested in orange-pigmented vegetables and fruits. Converted in the retina to retinal, a part of the photoreceptor pigment involved in vision.
E	Ingested in plant products such as wheat germ and green leafy vegetables. Claims have been made (but not proved in humans) that it promotes wound healing, contributes to fertility, and may help to neutralize highly reactive particles called free radicals believed to be involved in triggering some types of cancer.
K	Prevalent in a wide variety of ingested foods; also made available to humans by the action of intestinal bacteria. Necessary for proper clotting of blood.
Eicosanoids (prostaglandins; leukotrienes; thromboxanes)	Group of molecules derived from fatty acids found in all cell membranes. The potent prostaglandins have diverse effects, including stimulation of uterine contractions, regulation of blood pressure, control of gastrointestinal tract motility, and secretory activity. Both prostaglandins and leukotrienes are involved in inflammation. Thromboxanes are powerful vasoconstrictors.
Lipoproteins	Lipoid and protein-based substances that transport fatty acids and cholesterol in the bloodstream. Major varieties are high-density lipoproteins (HDLs) and low-density lipoproteins (LDLs).
Glycolipids	Components of cell membranes. Lipids associated with carbohydrate molecules determine blood type, play a role in cell recognition, and in recognition of foreign substances by immune cells.

▲ **Table 2.2** Representative lipids Found in the body

Proteins

→ **Living organisms synthesize many different proteins with a wide range of functions.**

Proteins are the most diverse group of biological macromolecules, both chemically and functionally.

- Catalysis: there are thousands of different enzymes to catalyse specific chemical reactions within the cell or outside it.

- Muscle contraction: actin and myosin together cause the muscle contractions used in locomotion and transport around the body .

- Cytoskeletons: tubulin is the subunit of microtubules that give animals cells their shape and pull on chromosomes during mitosis.

- Tensile strengthening: fibrous proteins give tensile strength needed in skin, tendons, ligaments and blood vessel walls.

- Blood clotting: plasma proteins act as clotting factors that cause blood to turn from a liquid to a gel in wounds.

- Transport of nutrients and gases: proteins in blood help transport oxygen, carbon dioxide, iron and lipids.

- Cell adhesion: membrane proteins cause adjacent animal cells to stick to each other within tissues.

- Membrane transport: membrane proteins are used for facilitated diffusion and active transport, and also for electron transport during cell respiration and photosynthesis.

- Hormones: some such as insulin, FSH and LH are proteins, but hormones are chemically very diverse.

- Receptors: binding sites in membranes and cytoplasm for hormones, neurotransmitters, tastes and smells, and also receptors for light in the eye and in plants.

- Packing of DNA: histones are associated with DNA in eukaryotes and help chromosomes to condense during mitosis.

- Immunity: this is the most diverse group of proteins, as cells can make huge numbers of different antibodies

▶ **Figure 2.17** An overview of protein functions.

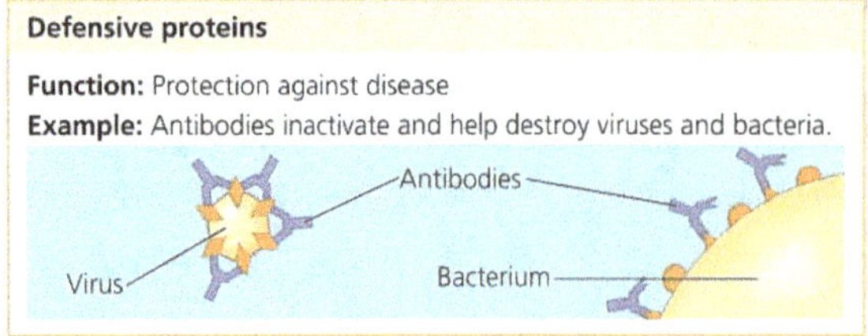

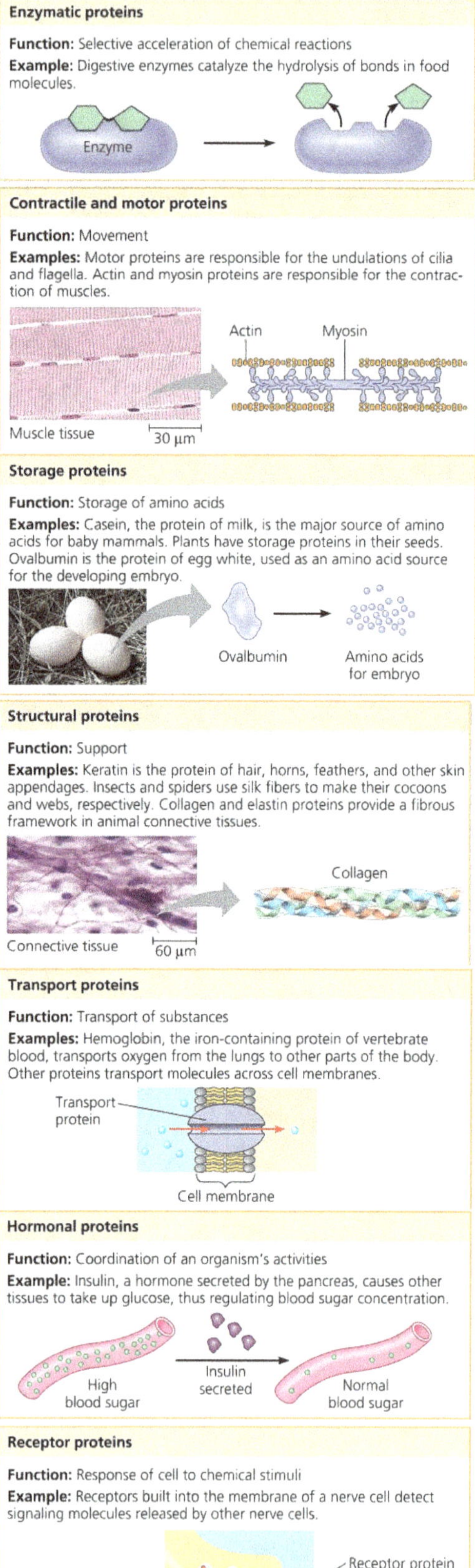

Amino Acids (Monomers)

→ There are twenty different amino acids in polypeptides synthesized on ribosomes.

→ All amino acids share a common structure. An amino acid is an organic molecule with both an amino group and a carboxyl group

→ At the center of the amino acid is an asymmetric carbon atom called the alpha(a) carbon. Its four different partners are an amino group, a carboxyl group, a hydrogen atom, and a variable group symbolized by R. The R group, also called the side chain, differs with each amino acid.

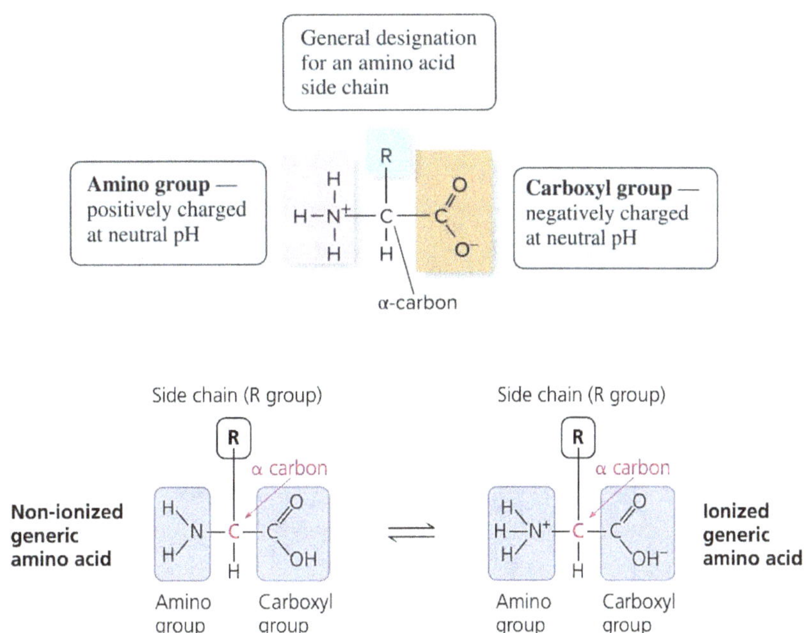

▲ **Figure 2.18** The general structure of an amino acid.

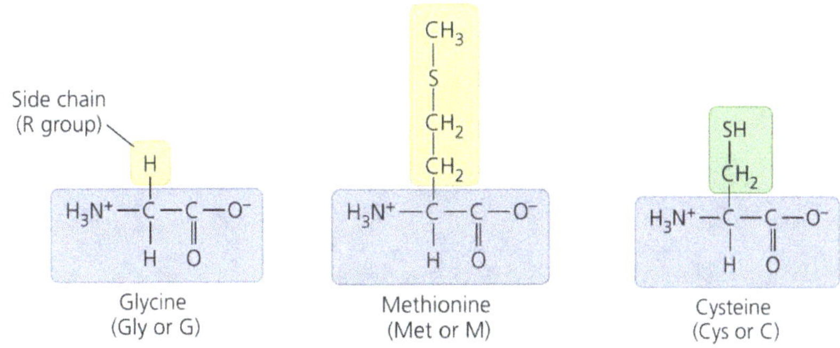

▲ **Figure 2.19** Some amino acids.

→ Amino acids are linked together by condensation to form polypeptides. When two amino acids are positioned so that the carboxyl group of one is adjacent to the amino group of the other, they can become joined by a dehydration reaction, with the removal of a water molecule. The resulting covalent bond is called a peptide bond. Repeated over and over, this process yields a polypeptide, a polymer of many amino acids linked by peptide bonds. Note that one end of the polypeptide chain has a free amino group (the N-terminus of the polypeptide), while the opposite end has a free carboxyl group (the C-terminus).

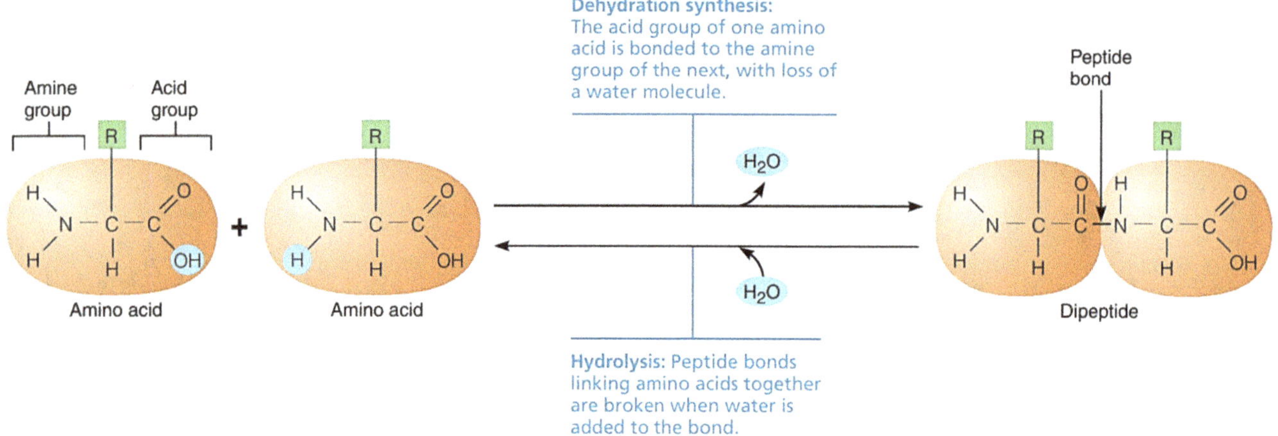

▲ **Figure 2.20** Amino acids are linked together by peptide bonds. Peptide bonds are formed by dehydration synthesis and broken by hydrolysis reactions.

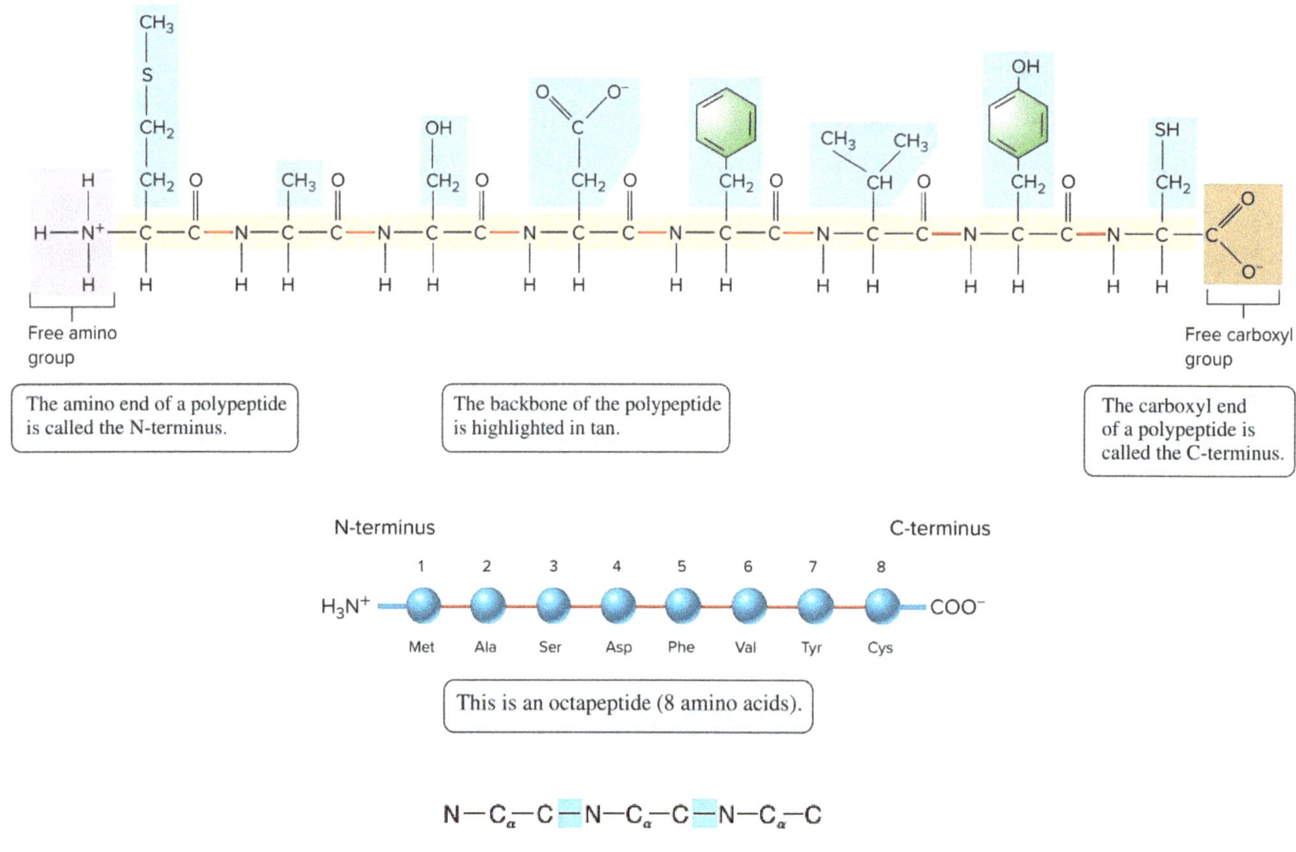

▲ **Figure 2.21** Polypeptide—a linear chain of amino acids

Four Levels of Protein Structure

In spite of their great diversity, proteins share three superimposed levels of structure, known as primary, secondary, and tertiary structure. A fourth level, quaternary structure, arises when a protein consists of two or more polypeptide chains.

→ **Primary (1°) structure** (Linear chain of amino acids): The primary structure of a protein is its sequence of amino acids. The primary structure is like the order of letters in a very long word. If left to chance, there would be 20^{127} different ways of making a polypeptide chain 127 amino acids long. However, the precise primary structure of a protein is determined not by the random linking of amino acids, but by inherited genetic information. The primary structure in turn dictates secondary structure (α helices and β pleated sheets) and tertiary structure, due to the chemical nature of the backbone and the side chains (R groups) of the amino acids along the polypeptide.

▲ **Figure 2.22** Primary structure: The linear sequence of amino acids is the primary structure.

Most proteins have segments of their polypeptide chains repeatedly coiled or folded in patterns that contribute to the protein's overall shape. These coils and folds, collectively referred to as secondary structure, are the result of hydrogen bonds between the repeating constituents of the polypeptide backbone (not the amino acid side chains). Within the backbone, the oxygen atoms have a partial negative charge, and the hydrogen atoms attached to the nitrogens have a partial positive charge; therefore, hydrogen bonds can form between these atoms. Individually, these hydrogen bonds are weak, but because they are repeated many times over a relatively long region of the polypeptide chain, they can support a particular shape for that part of the protein.

One such secondary structure is the **α helix**, a delicate coil held together by hydrogen bonding between every fourth amino acid. The other secondary structure is the **β pleated sheet**. Two or more segments of the polypeptide chain lying side by side (called β strands) are connected by hydrogen bonds between parts of the two parallel segments.

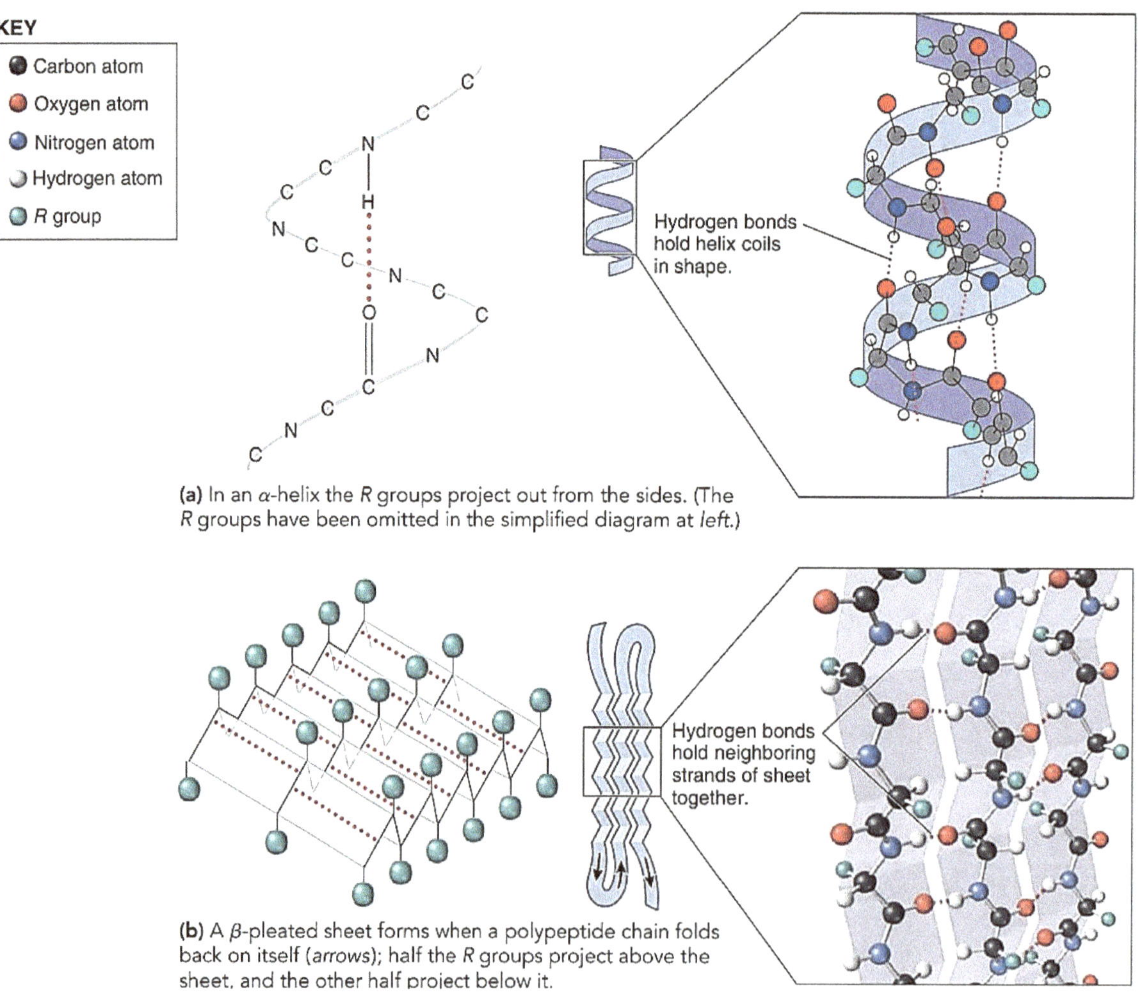

(a) In an α-helix the *R* groups project out from the sides. (The *R* groups have been omitted in the simplified diagram at *left*.)

(b) A β-pleated sheet forms when a polypeptide chain folds back on itself (*arrows*); half the *R* groups project above the sheet, and the other half project below it.

▲ **Figure 2.23** Secondary structure of a protein. Hydrogen bonding of the backbone can produce two types of secondary structure: the α-helix and the β-pleated sheet.

Motif

As biologists discovered the three-dimensional structure of proteins, they noticed similarities between otherwise dissimilar proteins. These similar structures are called motifs, or sometimes "supersecondary structure."

One very common protein motif is the β-α-βmotif, which creates a fold or crease; the so-called "Rossmann fold" at the core of nucleotide-binding sites in a wide variety of proteins. A second motif that occurs in many proteins is the βbarrel, which is a βsheet folded around to form a tube. A third type of motif, the helix-turn-helix, consists of two αhelices separated by a bend. This motif is important because many proteins use it to bind to the DNA double helix.

Motifs have been useful in determining the function of unknown proteins. Databases of protein motifs are used to search new unknown proteins. Finding motifs with known functions may allow an investigator to infer the function of a new protein.

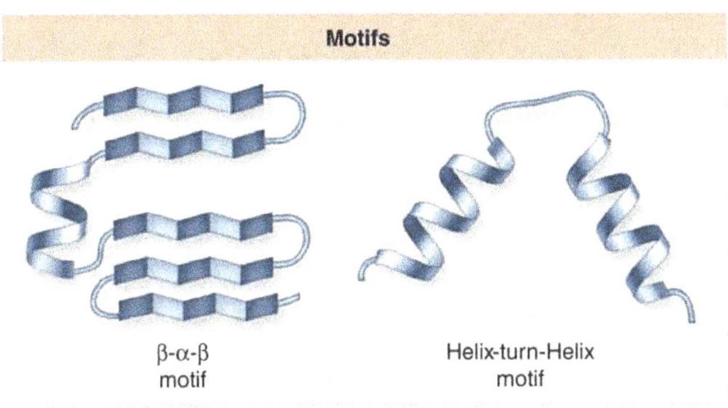

→ **Tertiary (3°) structure:** Three-dimensional shape stabilized by interactions between side chains. While secondary structure involves interactions between backbone constituents, tertiary structure is the overall shape of a polypeptide resulting from interactions between the side chains (R groups) of the various amino acids.

One type of interaction that contributes to tertiary structure is called—somewhat misleadingly—a hydrophobic interaction. As a polypeptide folds into its functional shape, amino acids with hydrophobic (nonpolar) side chains usually end up in clusters at the core of the protein, out of contact with water. Thus, a "hydrophobic interaction" is actually caused by the exclusion of nonpolar substances by water molecules. Once nonpolar amino acid side chains are close together, van der Waals interactions help hold them together. Meanwhile, hydrogen bonds between polar side chains and ionic bonds between positively and negatively charged side chains also help stabilize tertiary structure. These are all weak interactions in the aqueous cellular environment, but their cumulative effect helps give the protein a unique shape.

Tertiary structure depends on side chain interactions.

▶ **Figure 2.24** Primary structure: The linear sequence of amino acids is the primary structure.

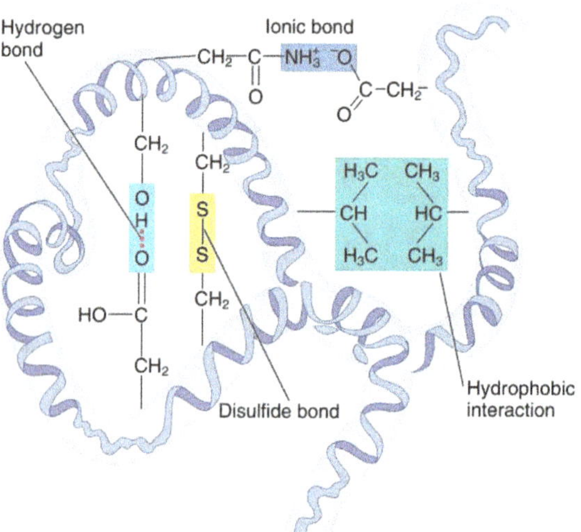

(a) Hydrogen bonds, ionic bonds, hydrophobic interactions, and disulfide bridges between *R* groups hold the parts of the molecule in the designated shape.

Covalent bonds called disulfide bridges may further reinforce the shape of a protein. Disulfide bridges form where two cysteine monomers, which have sulfhydryl groups (—SH) on their side chains are brought close together by the folding of the protein.

The sulfur of one cysteine bonds to the sulfur of the second, and the disulfide bridge (—S—S—) rivets parts of the protein together. All of these different kinds of interactions can contribute to the tertiary structure of a protein.

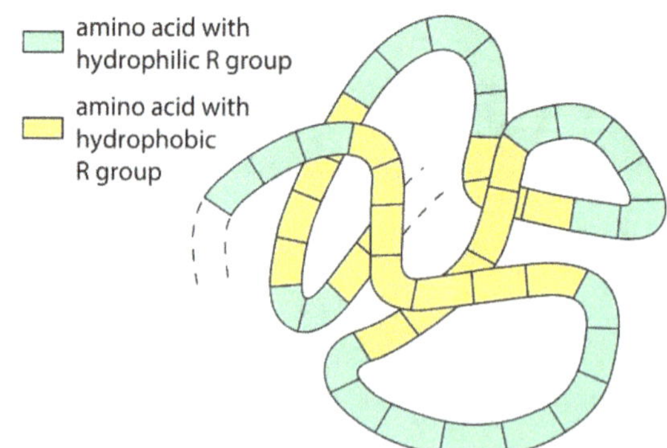

Domain

Domains of proteins are functionally distinct parts of a protein. They can be thought of as substructure within the tertiary structure of a protein. Most proteins are made up of multiple domains that perform different parts of the protein's function. In many cases, these domains can be physically separated.

For example, transcription factors are proteins that bind to DNA and initiate its transcription. If the DNA-binding region is exchanged with a different transcription factor, then the specificity of the factor for DNA can be changed without changing its ability to stimulate transcription, it indicates that the DNA-binding and activation domains are functionally separate. These functional domains of proteins may also help the protein to fold into its proper shape.

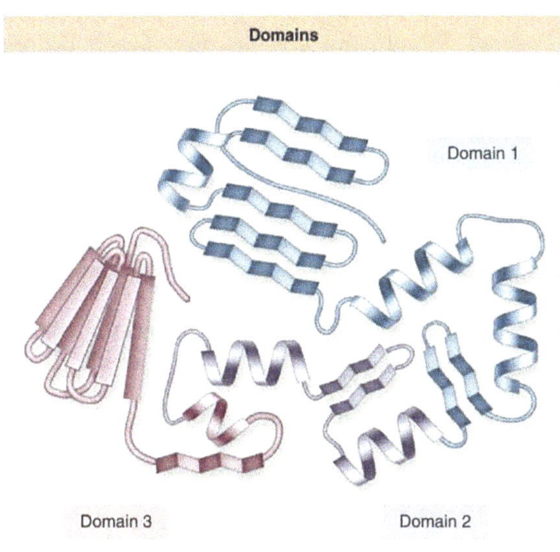

→ Quaternary (4°) structure

Association of two or more polypeptides (some proteins only). Some proteins consist of two or more polypeptide chains aggregated into one functional macromolecule like **Hemoglobin** and **collagen**. Quaternary structure is the overall protein structure that results from the aggregation of polypeptide subunits.

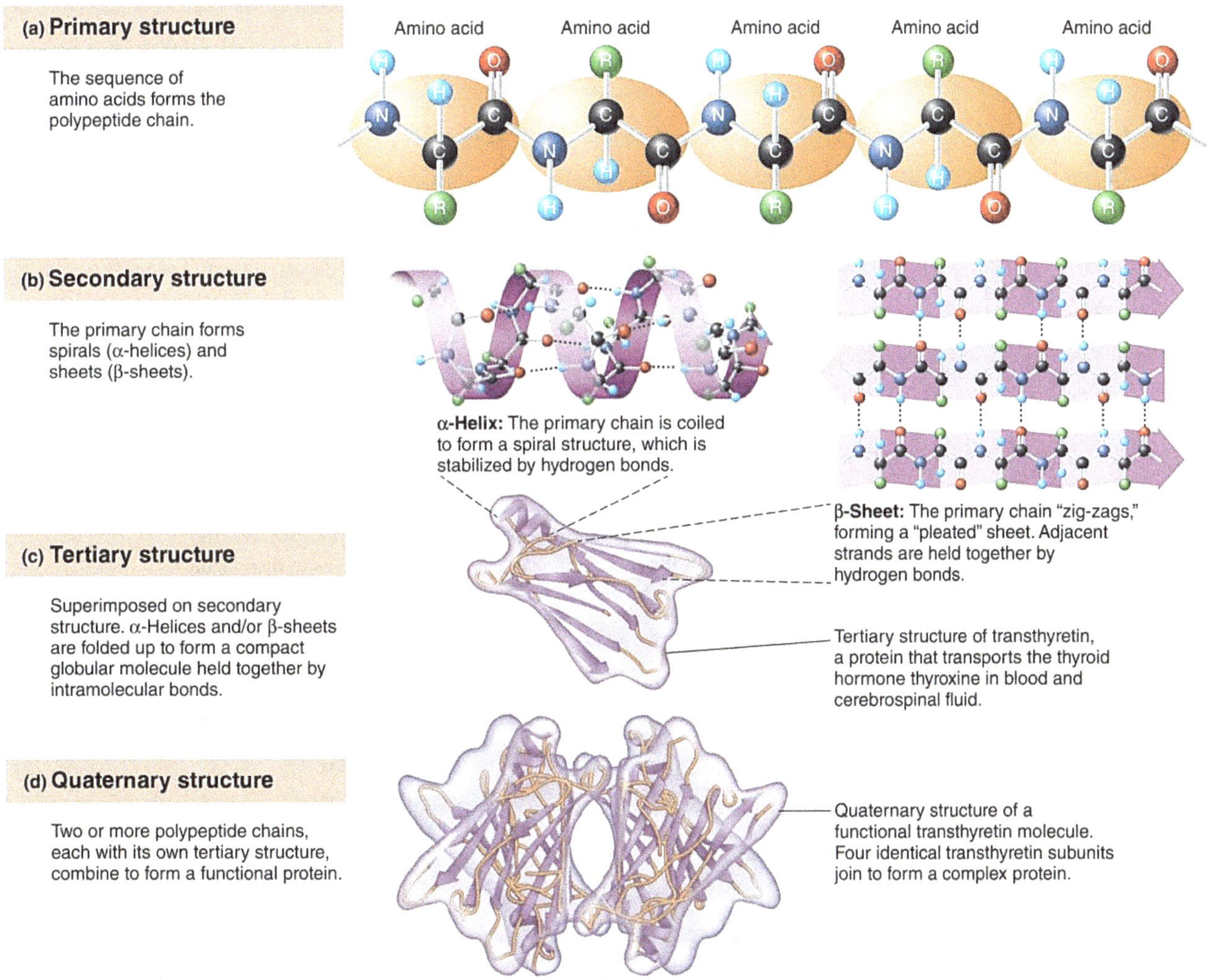

(a) Primary structure

The sequence of amino acids forms the polypeptide chain.

(b) Secondary structure

The primary chain forms spirals (α-helices) and sheets (β-sheets).

(c) Tertiary structure

Superimposed on secondary structure. α-Helices and/or β-sheets are folded up to form a compact globular molecule held together by intramolecular bonds.

(d) Quaternary structure

Two or more polypeptide chains, each with its own tertiary structure, combine to form a functional protein.

Several factors determine the way proteins adopt their secondary, tertiary, and quaternary structures. The following factors are critical for protein folding and stability:

- Hydrogen bonds. The large number of weak hydrogen bonds within a polypeptide and between polypeptides collectively produces a strong force that promotes protein folding and stability

- Ionic bonds and other polar interactions. Some amino acid side chains are positively or negatively charged. Positively charged side chains may bind to negatively charged side chains via ionic bonds

- Hydrophobic effect. Some amino acid side chains are nonpolar. As a protein folds, the hydrophobic amino acids are likely to be found in the center of the protein, minimizing their contact with water. Some proteins have stretches of nonpolar amino acids that anchor the proteins in the hydrophobic portion of membranes.

- van der Waals dispersion forces. Atoms within molecules have temporary attractions for each other if they are an optimal distance apart. The van der Waals dispersion forces contribute to tertiary structure and quaternary structure.

- Disulfide bridges. The side chain of the amino acid cysteine contains a sulfhydryl group (—SH), which can react with an —SH group in another cysteine side chain and form a covalent bond. The resulting bond is a disulfide bridge, which links the two amino acid side chains together (—S—S—). Disulfide bridges can occur within a polypeptide or between different polypeptides.

The first four factors are also important in the ability of different proteins to interact with each other. Many cellular processes involve steps in which two or more different proteins interact with each other. For such an interaction to occur, proteins must bind to each other. Such binding is usually very specific. The shape of one protein may permit it to fit precisely into a region of another protein. Such protein-protein interactions are critically important in allowing cellular processes to occur in a series of defined steps. In addition, protein-protein interactions are important in building complicated cellular structures and in processes that provide organization and function to cells.

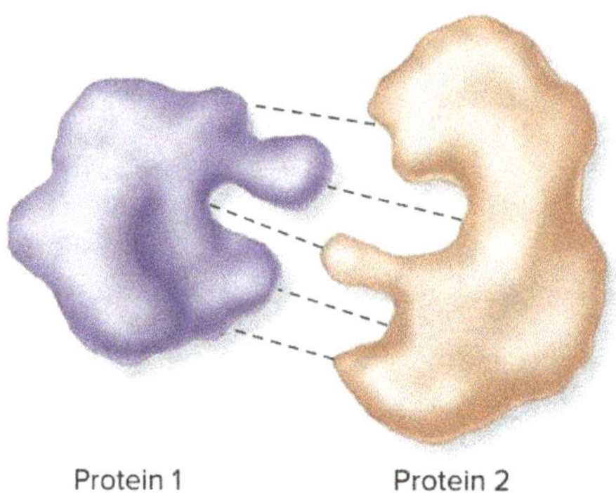

▲ **Figure 2.25** Protein-protein interaction. Two different proteins may interact with each other due to hydrogen bonding, ionic bonding, the hydrophobic effect, and van der Waals dispersion forces. Interaction is also facilitated by their respective three-dimensional shapes.

Globular proteins

Globular proteins, also called functional proteins, are compact, spherical proteins that have at least tertiary structure. Some also exhibit quaternary structure. The globular proteins are water-soluble, chemically active molecules, and they play crucial roles in virtually all biological processes. Some (antibodies) help to provide immunity, others (protein-based hormones) regulate growth and development, and still others (enzymes) are catalysts that oversee just about every chemical reaction in the body.

Hemoglobin, the oxygen-binding protein of red blood cells, is another example of a globular protein with quaternary structure. It consists of four polypeptide subunits, two of one kind (α) and two of another kind (β). Both α and β subunits consist primarily of α-helical secondary structure. Each subunit has a nonpolypeptide component, called heme, with an iron atom that binds oxygen.

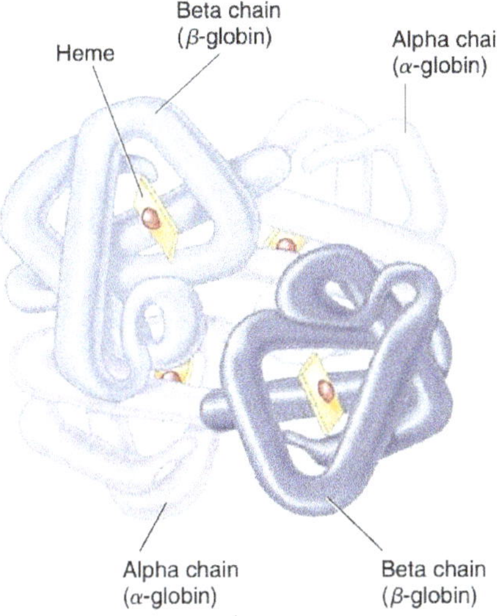

(a) Hemoglobin, a globular protein, consists of four polypeptide chains, each joined to an iron-containing molecule, a heme.

Fibrous Proteins

The overall structure of a protein determines its biological function. In general, proteins are classified according to their overall appearance and shape as either fibrous or globular.

Fibrous proteins, also known as structural proteins, are extended and strandlike. Some exhibit only secondary structure, but most have tertiary or even quaternary structure as well. For example, **collagen** in connective tissue in skin, bone, tendons, ligaments, and other body parts. (Collagen accounts for 40% of the protein in a human body).

A collagen molecule (tropocollagen) consists of three polypeptide chains, each in the shape of a helix. These three helical polypeptides are wound around each other, forming a three-stranded 'rope' or 'triple helix'. The three strands are held together by hydrogen bonds and some covalent bonds. Almost every third amino acid in each polypeptide is glycine, the smallest amino acid.

Glycine is found on the insides of the strands and its small size allows the three strands to lie close together and so form a tight coil. Any other amino acid would be too large. Each complete, three-stranded molecule of collagen interacts with other collagen molecules running parallel to it. Covalent bonds form between the R groups of amino acids lying next to each other. These cross-links hold many collagen molecules side by side, forming fibrils. Finally, many fibrils lie alongside each other, forming strong bundles called fibres. Note that many collagen molecules make up one collagen fibre.

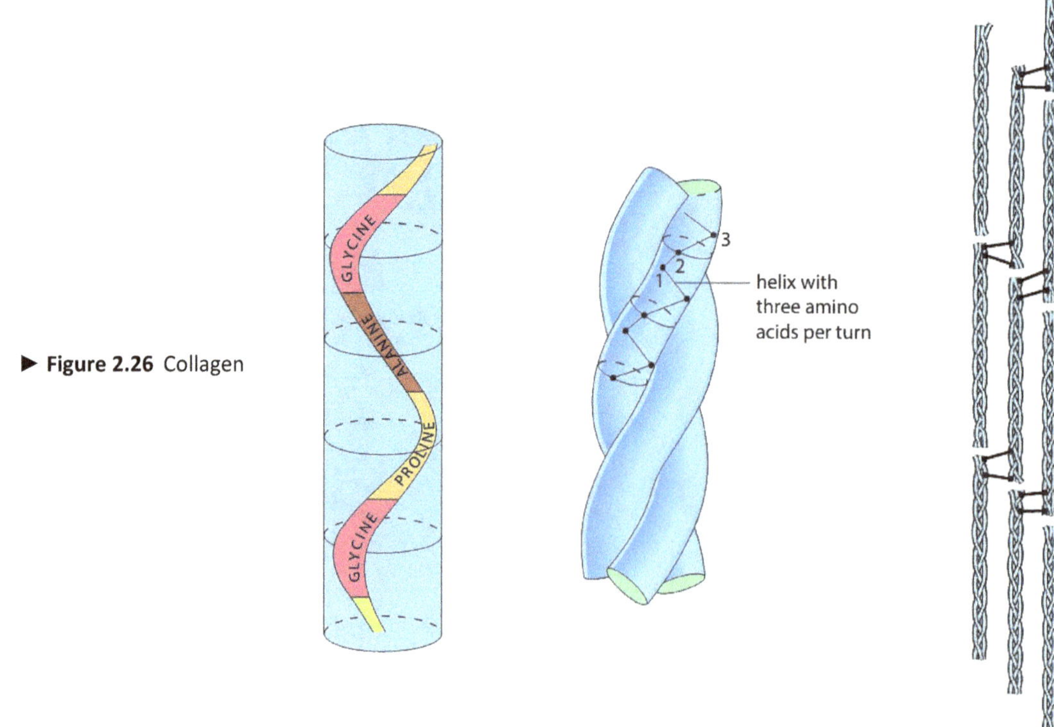

▶ **Figure 2.26** Collagen

The advantage of collagen is that it is flexible but at the same time has tremendous tensile strength. High tensile strength means it can withstand large pulling forces without stretching or breaking. The human Achilles tendon, which is almost pure collagen fibres, can withstand a pulling force of 300 N per mm^2 of cross-sectional area, about one-quarter the tensile strength of mild steel. In tendons, they line up in parallel bundles along the length of the tendon, in the direction of tension. In skin, they may form layers, with the fibres running in different directions in the different layers.

Fibrous proteins are not usually soluble in water and most have structural roles. For example, keratin forms hair, nails and the outer layers of skin, making these structures waterproof. Besides collagen, which is the single most abundant protein in the body, the fibrous proteins include keratin, elastin, and certain contractile proteins of muscle.

CLASSIFICATION ACCORDING TO		
OVERALL STRUCTURE	**GENERAL FUNCTION**	**EXAMPLES FROM THE BODY**
Fibrous		
	Structural framework/ mechanical support	*Collagen*, found in all connective tissues, is the single most abundant protein in the body. It is responsible for the tensile strength of bones, tendons, and ligaments.
		Keratin is the structural protein of hair and nails and a water-resistant material of skin.
		Elastin is found, along with collagen, where durability and flexibility are needed, such as in the ligaments that bind bones together.
		Spectrin internally reinforces and stabilizes the plasma membrane of some cells, particularly red blood cells. *Dystrophin* reinforces and stabilizes the plasma membrane of muscle cells. *Titin* helps organize the intracellular structure of muscle cells and accounts for the elasticity of skeletal muscles.
	Movement	*Actin* and *myosin*, contractile proteins, are found in substantial amounts in muscle cells, where they cause muscle cell shortening (contraction); they also function in cell division in all cell types. Actin is important in intracellular transport, particularly in nerve cells.
Globular		
	Catalysis	Protein enzymes are essential to virtually every biochemical reaction in the body; they increase the rates of chemical reactions by at least a millionfold. Examples include *salivary amylase* (in saliva), which catalyzes the breakdown of starch, and *oxidase enzymes*, which act to oxidize food fuels.
	Transport	*Hemoglobin* transports oxygen in blood, and *lipoproteins* transport lipids and cholesterol. Other transport proteins in the blood carry iron, hormones, or other substances. Some globular proteins in plasma membranes are involved in membrane transport (as carriers or channels).
	Regulation of pH	Many plasma proteins, such as *albumin,* function reversibly as acids or bases, thus acting as buffers to prevent wide swings in blood pH.
	Regulation of metabolism	*Peptide* and *protein hormones* help to regulate metabolic activity, growth, and development. For example, *growth hormone* is an anabolic hormone necessary for optimal growth; *insulin* helps regulate blood sugar levels.
	Body defense	*Antibodies* (immunoglobulins) are specialized proteins released by immune cells that recognize and inactivate foreign substances (bacteria, toxins, some viruses).
		Complement proteins, which circulate in blood, enhance both immune and inflammatory responses.
	Protein management	*Molecular chaperones*, originally called "heat shock proteins," aid folding of new proteins in both healthy and damaged cells and transport of metal ions into and within the cell. They also promote breakdown of damaged proteins.

▲ **Table 2.3** Representative Types of Proteins in the Body

Misfolding of polypeptides in cells is a serious problem that has come under increasing scrutiny by medical researchers. Many diseases—such as cystic fibrosis, Alzheimer's, Parkinson's, and mad cow disease—are associated with an accumulation of misfolded proteins.

Normal cells contain **chaperone proteins**, which help other proteins to fold correctly. Molecular biologists have now identified many proteins that act as molecular chaperones. This large class of proteins can be divided into subclasses, one class of these proteins, called chaperonins. Chaperonins associate to form a large   macromolecular complex that resembles a cylindrical container. Proteins can move into the container, and the container itself can change its shape considerably. Experiments have shown that an improperly folded protein can enter the chaperonin and be refolded.

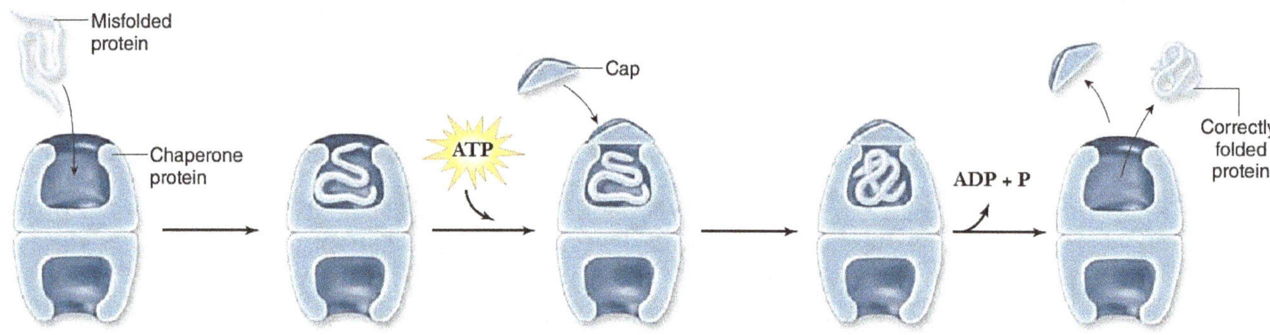

▲ **Figure 2.27** How one type of chaperone protein works. This barrel-shaped chaperonin is from the GroE family of chaperone proteins. It is composed of two identical rings each with seven identical subunits, each of which has three distinct domains. An incorrectly folded protein enters one chamber of the barrel, and a cap seals the chamber. Energy from the hydrolysis of ATP fuels structural alterations to the chamber, changing it from hydrophobic to hydrophilic. This change allows the protein to refold. After a short time, the protein is ejected, either folded or unfolded, and the cycle can repeat itself.

At least some of these proteins have been shown to be necessary for viability, illustrating their fundamental importance. Many are so-called heat shock proteins, produced in large amounts in response to elevated temperature. High temperatures cause proteins to unfold, and heat shock chaperone proteins help the cell's proteins to refold properly.

Chaperone protein deficiencies may be implicated in certain diseases in which key proteins are improperly folded. Cystic fibrosis is a hereditary disorder in which a mutation disables a vital protein that moves ions across cell membranes. As a result, people with cystic fibrosis have thicker than normal mucus. This results in breathing problems, lung disease, and digestive difficulties, among other things.

One interesting feature of the molecular analysis of this disease has been the number of different mutations found in human populations. One diverse class of mutations all result in problems with protein folding. The number of different mutations that can result in improperly folded proteins may be related to the fact that the native protein often fails to fold properly.

Denaturation inactivates proteins

If a protein's environment is altered, the protein may change its shape or even unfold completely. This process is called **denaturation**. Proteins can be denatured when the pH, temperature, or ionic concentration of the surrounding solution changes.

Denatured proteins are usually biologically inactive. This action is particularly significant in the case of enzymes. Because practically every chemical reaction in a living organism is catalyzed by a specific enzyme, it is vital that a cell's enzymes work properly.

The traditional methods of food preservation, salt curing and pickling, involve denaturation of proteins. Prior to the general availability of refrigerators and freezers, the only practical way to keep microorganisms from growing in food was to keep the food in a solution containing a high concentration of salt or vinegar, which denatured the enzymes of most microorganisms and prevented them from growing on the food.

Most enzymes function within a very narrow range of environmental conditions. Blood-borne enzymes that course through a human body at a pH of about 7.4 would rapidly become denatured in the highly acidic environment of the stomach. Conversely, the protein-degrading enzymes that function at a pH of 2 or less in the stomach would be denatured in the relatively basic pH of the blood.

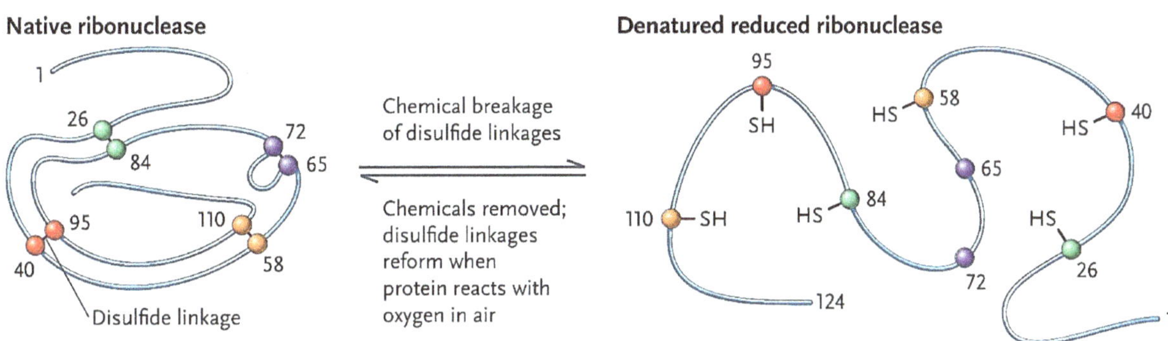

When a protein's normal environment is reestablished after denaturation, a small protein may spontaneously refold into its natural shape, driven by the interactions between its nonpolar amino acids and water. This process is termed **renaturation**. Although this process requires temperature and ion concentration shifts, it indicates an amazing degree of self-assembly. That complex structures can arise by self-assembly is a key idea in the study of modern biology.

Protein denaturation. Environmental changes, such as variation in temperature or pH, can cause a protein to unfold and lose its shape. This loss of structure is called denaturation. Denatured proteins are biologically inactive.

Nucleic acids

The amino acid sequence of a polypeptide is programmed by a discrete unit of inheritance known as a gene. Genes consist of DNA. The information that programs all the cell's activities is encoded in the structure of the DNA which belongs to the class of compounds called nucleic acids. Nucleic acids are polymers made of monomers called **nucleotides**. The two types of nucleic acids, **deoxyribonucleic acid (DNA)** and **ribonucleic acid (RNA),** enable living organisms to reproduce their complex components from one generation to the next.

▶ **Figure 2.28** A nucleotide consists of a sugar, one or more phosphate groups, and one of several nitrogenous bases. In DNA, the sugar is deoxyribose, whereas RNA nucleotides contain ribose. In addition, the base thymine appears only in DNA; uracil is only in RNA.

Dehydration synthesis joins two nucleotides together.

Nucleic Acids: DNA and RNA. DNA consists of two strands of nucleotides entwined to form a double-helix shape held together by hydrogen bonds. RNA is usually single-stranded.

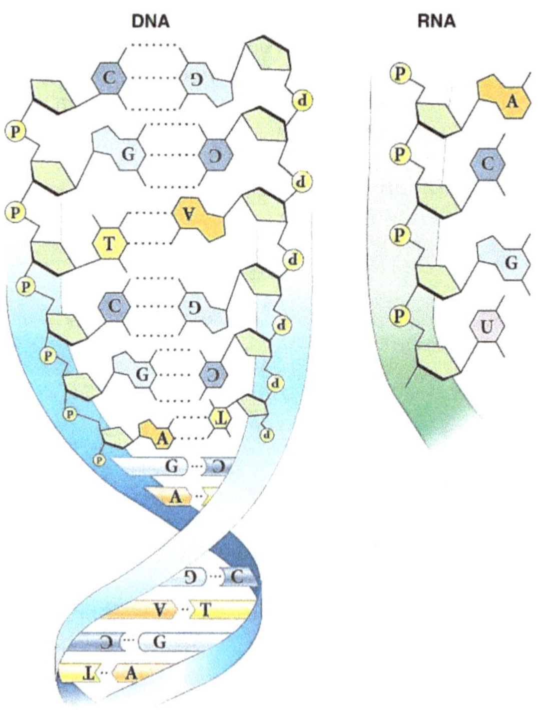

Nucleic acids are macromolecules that exist as polymers called polynucleotides. As indicated by the name, each polynucleotide consists of monomers called nucleotides. A nucleotide, in general, is composed of three parts: a five-carbon sugar (a pentose), a nitrogen-containing (nitrogenous) base, and one to three phosphate groups.

A structure containing only a nitrogenous base and a five carbon sugar is called a nucleoside. Thus, nucleotides are nucleoside phosphates. For example, the nucleoside containing adenine and ribose is called adenosine. Adding one phosphate group to this structure produces adenosine monophosphate (AMP), adding two phosphate groups produces adenosine diphosphate (ADP), and adding three produces adenosine triphosphate (ATP).

The corresponding adeninedeoxyribose complexes are named deoxyadenosine monophosphate (dAMP), deoxyadenosine diphosphate (dADP), and deoxyadenosine triphosphate (dATP). The lowercase d in the abbreviations indicates that the nucleoside contains the deoxyribose form of the sugar. Equivalent names and abbreviations are used for the other nucleotides

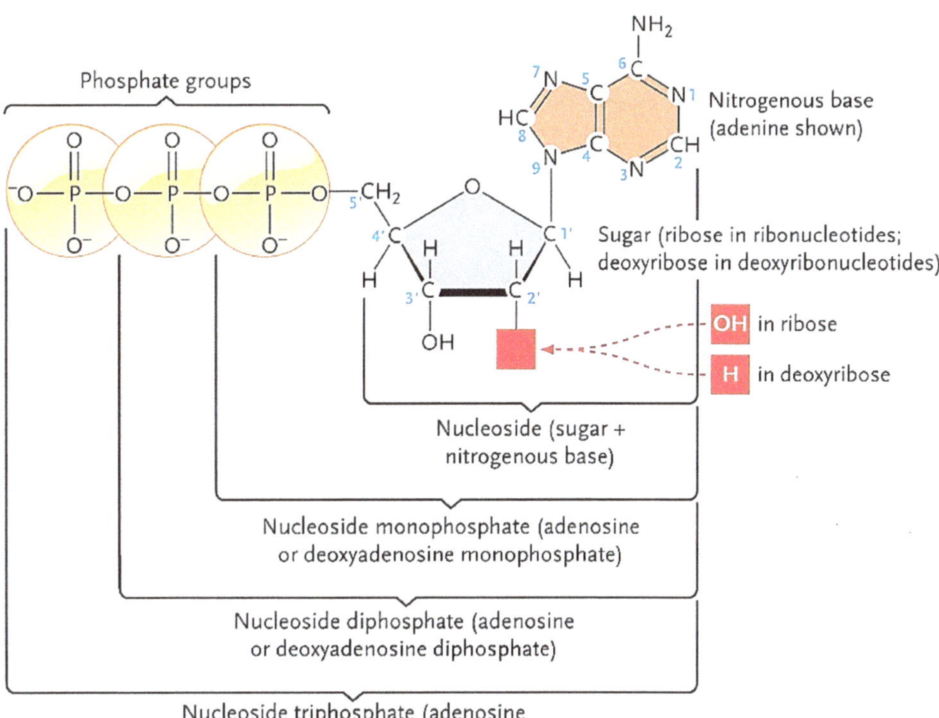

▲ **Figure 2.29** Nucleotide structure.

Each nitrogenous base has one or two rings that include nitrogen atoms. There are two families of nitrogenous bases: **pyrimidines** and **purines**. A pyrimidine has one six-membered ring of carbon and nitrogen atoms. The members of the pyrimidine family are **cytosine (C)**, **thymine (T)**, and **uracil (U)**.

Purines are larger, with a six-membered ring fused to a five membered ring. The purines are **adenine (A)** and **guanine (G)**. The specific pyrimidines and purines differ in the chemical groups attached to the rings. Adenine, guanine, and cytosine are found in both DNA and RNA; thymine is found only in DNA and uracil only in RNA.

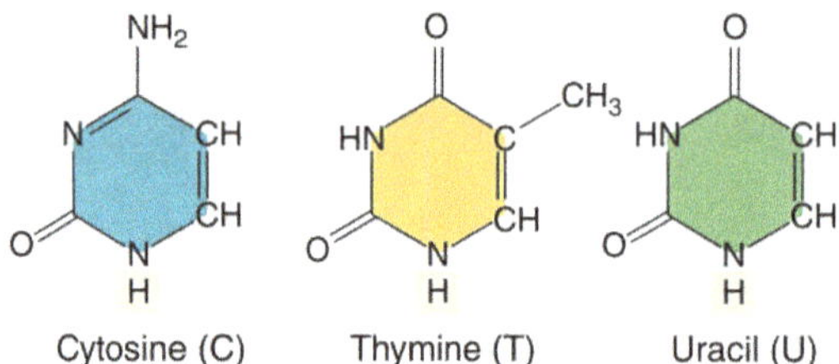

(a) **Pyrimidines.** The three major pyrimidine bases found in nucleotides are cytosine, thymine (in DNA only), and uracil (in RNA only).

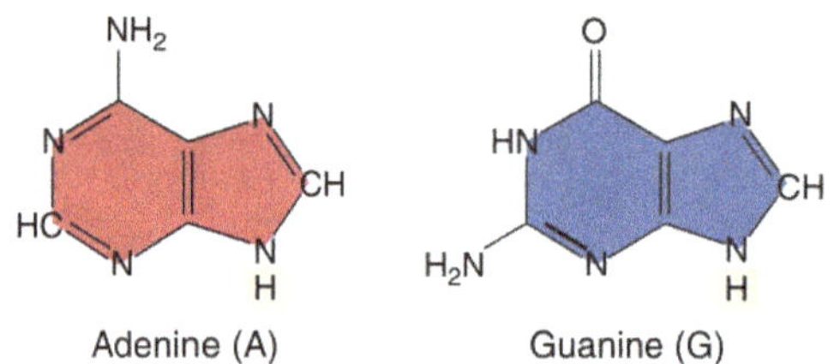

(b) **Purines.** The two major purine bases found in nucleotides are adenine and guanine.

In DNA the sugar is **deoxyribose**; in RNA it is **ribose**. The only difference between these two sugars is that deoxyribose lacks an oxygen atom on the second carbon in the ring, hence the name deoxyribose. To complete the construction of a nucleotide, we attach one to three phosphate groups to the 5' carbon of the sugar (the carbon numbers in the sugar include '). With one phosphate, this is a nucleoside monophosphate, more often called a nucleotide.

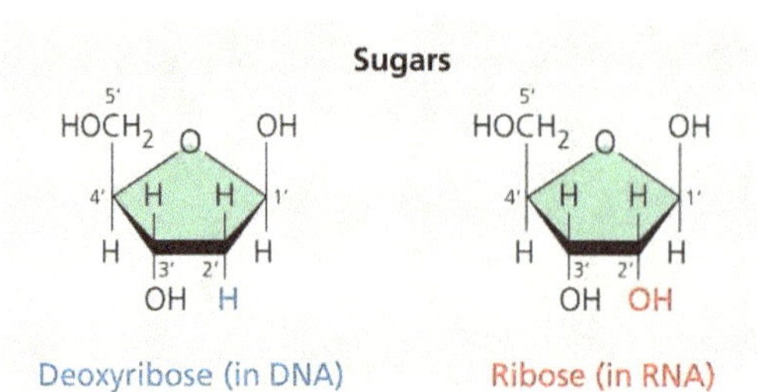

The linkage of nucleotides into a polynucleotide involves a condensation reaction. In the polynucleotide, adjacent nucleotides are joined by a **phosphodiester** linkage, which consists of a phosphate group that covalently links the sugars of two nucleotides. This bonding results in a repeating pattern of sugar-phosphate units called the **sugar-phosphate backbone**. The two free ends of the polymer are distinctly different from each other. One end has a phosphate attached to a 5' carbon, and the other end has a hydroxyl group on a 3' carbon.

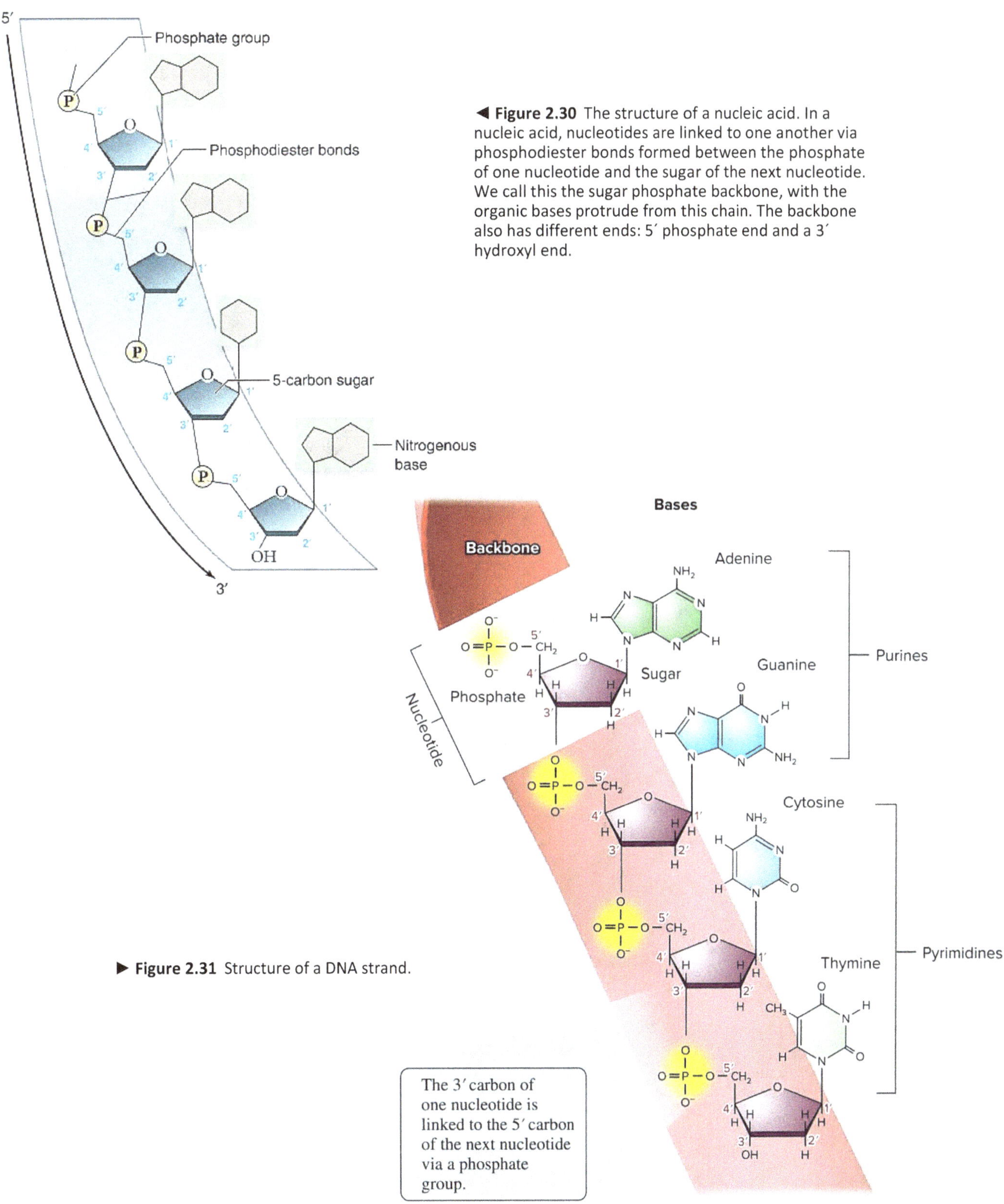

◄ **Figure 2.30** The structure of a nucleic acid. In a nucleic acid, nucleotides are linked to one another via phosphodiester bonds formed between the phosphate of one nucleotide and the sugar of the next nucleotide. We call this the sugar phosphate backbone, with the organic bases protrude from this chain. The backbone also has different ends: 5' phosphate end and a 3' hydroxyl end.

► **Figure 2.31** Structure of a DNA strand.

➜ The Structures of DNA Molecule

A DNA molecule has two polynucleotides, or "strands," that wind around an imaginary axis, forming a **double helix**. The two sugar-phosphate backbones run in opposite 5' → 3' directions from each other; this arrangement is referred to as **antiparallel**, somewhat like a divided highway. The sugar-phosphate backbones are on the outside of the helix, and the nitrogenous bases are paired in the interior of the helix. The two strands are held together by hydrogen bonds between the paired bases.

Most DNA molecules are very long, with thousands or even millions of base pairs. The one long DNA double helix in a eukaryotic chromosome includes many genes, each one a particular segment of the double-stranded molecule.

In base pairing, only certain bases in the double helix are compatible with each other. The bases that participate in base-pairing are said to be **complementary** to each other. Adenine (A) in one strand always pairs with thymine (T) in the other, and guanine (G) always pairs with cytosine (C). According to these base-pairing rules, ATGA on one DNA nucleotide strand would necessarily be bonded to TACT (a complementary base sequence) on the other strand.

In eukaryotic organisms, the DNA is further complex with protein to form structures we call chromosomes. This actually forms a higher order structure that affects the function of DNA as it is involved in the control of gene expression.

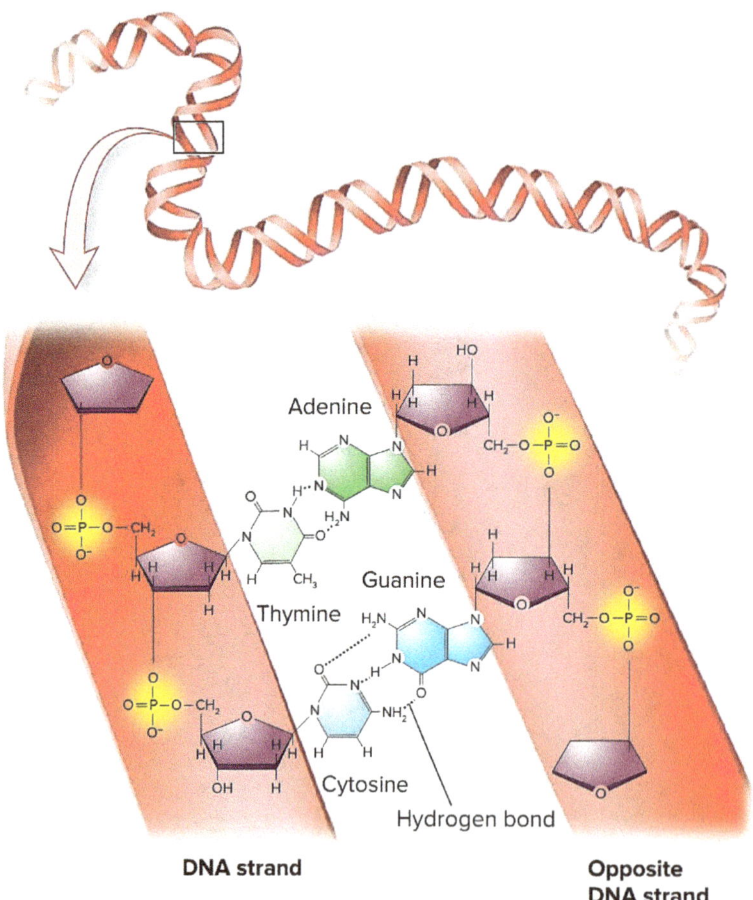

▲ **Figure 2.32** The double-stranded structure of DNA. DNA consists of two strands coiled together into a double helix. The bases form hydrogen bonds (dashed lines) in which A pairs with T, and G pairs with C.

RNA is similar to DNA, but with two major chemical differences. First, RNA molecules contain ribose sugars, in which the C-2 is bonded to a hydroxyl group. (In DNA, a hydrogen atom replaces this hydroxyl group). Second, RNA molecules use uracil in place of thymine. Uracil has a similar structure to thymine, except that one of its carbons lacks a methyl (—CH3) group.

Another difference between RNA and DNA is that DNA almost always exists as a double helix, whereas RNA molecules are more variable in shape. Complementary base pairing can occur, however, between regions of two RNA molecules or even between two stretches of nucleotides in the same RNA molecule. In fact, base pairing within an RNA molecule allows it to take on the particular three-dimensional shape necessary for its function for example, the type of RNA called transfer RNA (tRNA), which brings amino acids to the ribosome during the synthesis of a polypeptide.

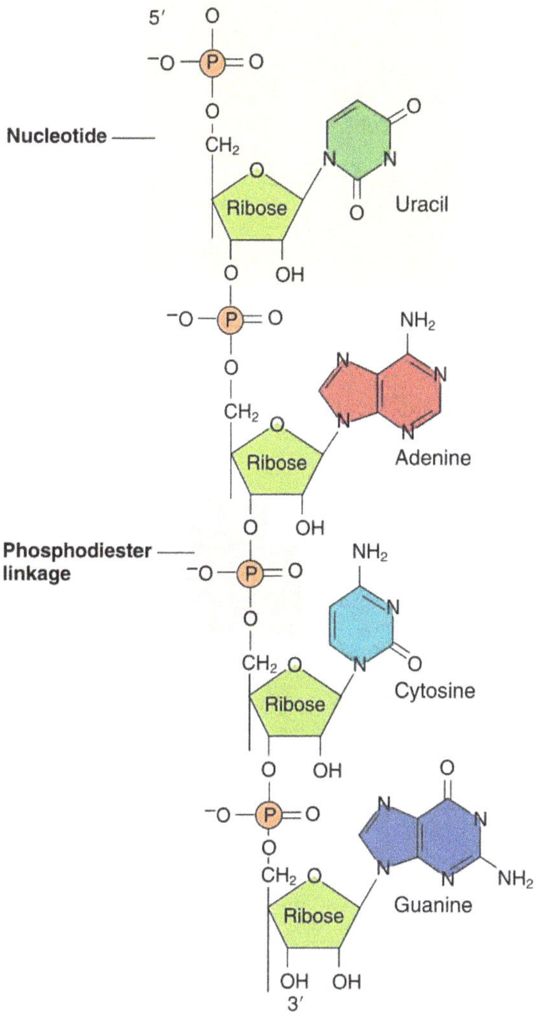

▶ **Figure 2.33** RNA, a nucleic acid

RNA is produced by transcription (copying) from DNA, and is usually single-stranded. The role of RNA in cells is quite varied: it carries information in the form of mRNA, it is part of the ribosome, in the form of ribosomal RNA (rRNA), and it carries amino acids in the form of transfer RNA (tRNA). Transfer RNA and rRNA fold up into complex structures, but mRNA remains as an unfolded strand. RNA has been found to function as an enzyme, and other forms of RNA are involved in regulating gene expression so that several types of small RNA molecules, including microRNAs appear to control genetic expression by shutting down genes or altering their expression.

CHARACTERISTIC	DNA	RNA
Major cellular site	Nucleus	Cytoplasm (cell area outside the nucleus)
Major functions	Is the genetic material; directs protein synthesis; replicates itself before cell division	Carries out the genetic instructions for protein synthesis
Sugar	Deoxyribose	Ribose
Bases	Adenine, guanine, cytosine, thymine	Adenine, guanine, cytosine, uracil
Structure	Double strand coiled into a double helix	Single strand, straight or folded

Other nucleotides are vital components of energy reactions

Nucleotides perform many functions in cells in addition to serving as the building blocks of nucleic acids. Two ribose-containing nucleotides in particular, adenosine triphosphate (ATP) and guanosine triphosphate (GTP), are the primary molecules that transport chemical energy from one reaction system to another. Cells use ATP as energy in a variety of transactions, the way we use money in society. ATP is used for example to drive transport across membranes, to power the movement of cells, for muscle contraction and chemical works.

ATP is a high-energy molecule, because the last two phosphate bonds are unstable and easily broken. Usually in cells, the last phosphate bond is hydrolyzed, leaving the molecule ADP (adenosine diphosphate) and a molecule of inorganic phosphate P_i.

A nucleotide may be converted to an alternative form with specific cell functions. ATP, for example, is converted to cyclic adenosine monophosphate (cyclic AMP, or cAMP) by the enzyme adenylyl cyclase. Cyclic AMP regulates certain cell functions, such as cell signaling, and is important in the mechanism by which some hormones act. A related molecule, cyclic guanosine monophosphate (cGMP), also plays a role in certain cell signaling processes.

Two other important nucleotide-containing molecules are nicotinamide adenine dinucleotide (**NAD⁺**) and flavin adenine dinucleotide (**FAD**). These molecules function as electron carriers in a variety of cellular processes.

▶ **Figure 2.34** Structure of ATP (adenosine triphosphate). ATP is an adenine nucleotide to which two additional phosphate groups have been attached during breakdown of food fuels. When the terminal phosphate group is cleaved off, energy is released to do useful work and ADP (adenosine diphosphate) is formed. When the terminal phosphate group is cleaved off ADP, a similar amount of energy is released and AMP (adenosine monophosphate) is formed.

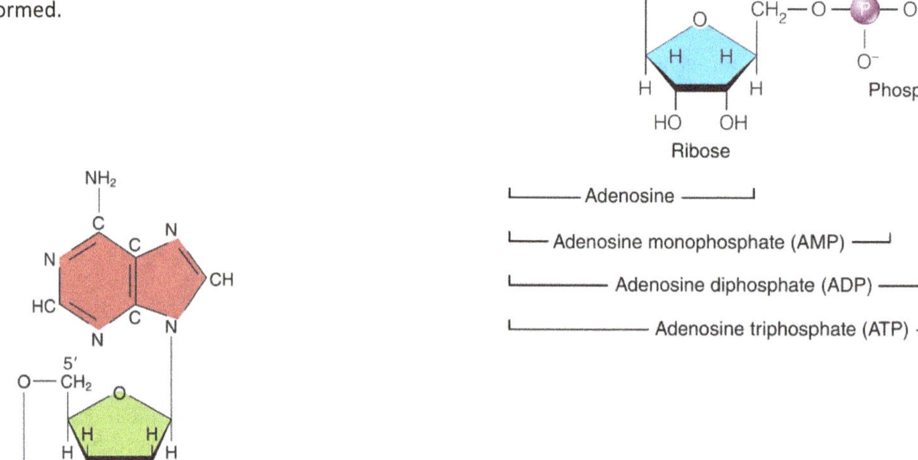

◀ **Figure 2.35** Cyclic adenosine monophosphate (cAMP). The single phosphate is part of a ring connecting two regions of the ribose.

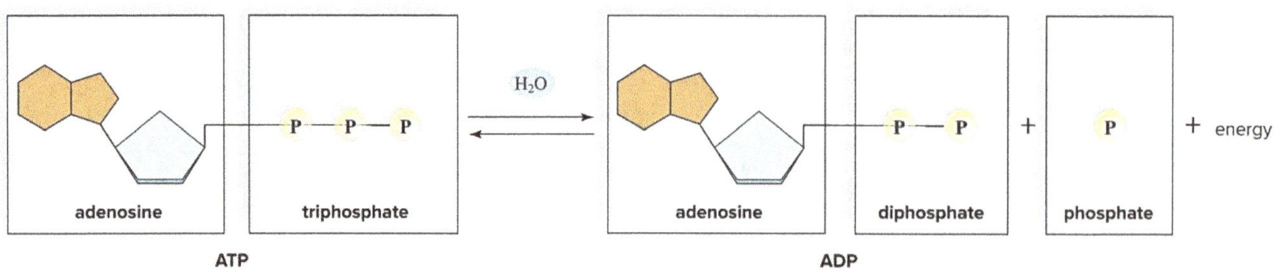

Components	Examples	Functions
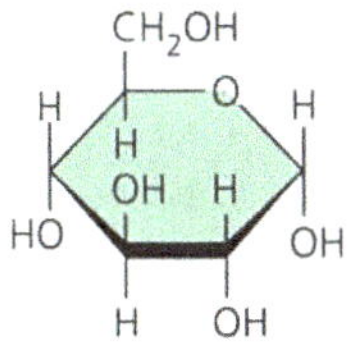 Monosaccharide monomer	**Monosaccharides:** glucose, fructose	Fuel; carbon sources that can be converted to other molecules or combined into polymers
	Disaccharides: lactose, sucrose	
	Polysaccharides: • Cellulose (plants) • Starch (plants) • Glycogen (animals) • Chitin (animals and fungi)	• Strengthens plant cell walls • Stores glucose for energy in plants • Stores glucose for energy in animals • Strengthens exoskeletons and fungal cell walls
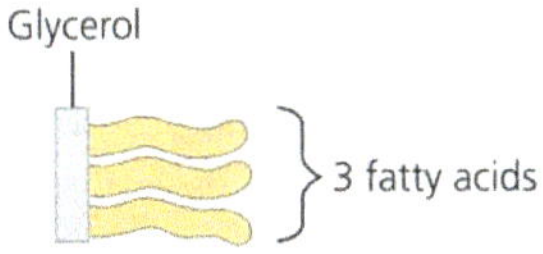 Glycerol / 3 fatty acids	**Triacylglycerols** (fats or oils): glycerol + 3 fatty acids	Important energy source
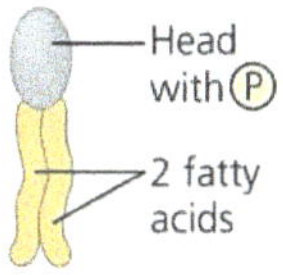 Head with P / 2 fatty acids	**Phospholipids:** phosphate group + glycerol + 2 fatty acids	Lipid bilayers of membranes 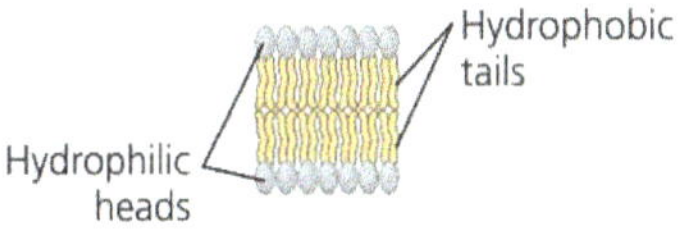 Hydrophobic tails / Hydrophilic heads
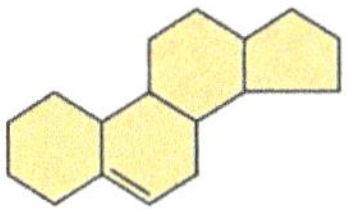 Steroid backbone	**Steroids:** four fused rings with attached chemical groups	• Component of cell membranes (cholesterol) • Signaling molecules that travel through the body (hormones)
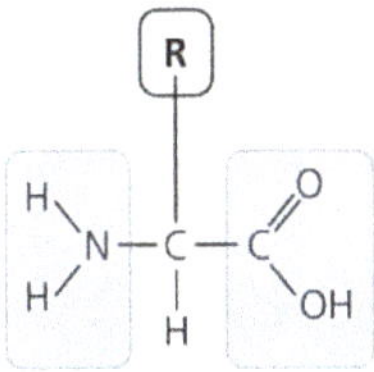 Amino acid monomer (20 types)	• Enzymes • Structural proteins • Storage proteins • Transport proteins • Hormones • Receptor proteins • Motor proteins • Defensive proteins	• Catalyze chemical reactions • Provide structural support • Store amino acids • Transport substances • Coordinate organismal responses • Receive signals from outside cell • Function in cell movement • Protect against disease
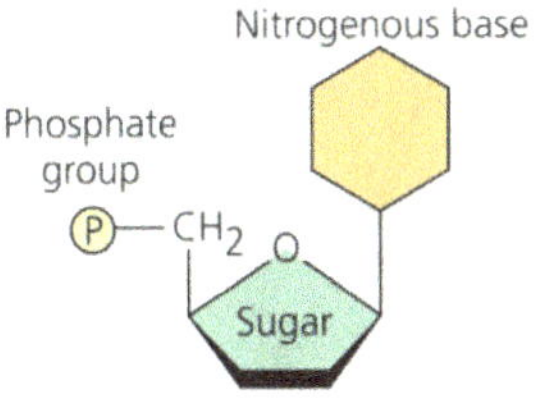 Nitrogenous base / Phosphate group / Sugar / Nucleotide monomer	**DNA:**  • Sugar = deoxyribose • Nitrogenous bases = C, G, A, T • Usually double-stranded	Stores hereditary information
	RNA:  • Sugar = ribose • Nitrogenous bases = C, G, A, U • Usually single-stranded	Various functions in gene expression, including carrying instructions from DNA to ribosomes

KEY CONCEPTS

- The empirical formula of a carbohydrate is $(CH_2O)_n$. Carbohydrates are used for energy storage and as structural molecules.
- Simple sugars contain three to six or more carbon atoms. Examples are glyceraldehyde (3 carbons), deoxyribose (5 carbons), and glucose (6 carbons).
- The general formula for 6-carbon sugars is $C_6H_{12}O_6$, and many isomeric forms are possible. Living systems often have enzymes for converting isomers from one to the other.
- Plants convert glucose into the disaccharide sucrose for transport within their bodies. Female mammals produce the disaccharide lactose to nourish their young.
- Glucose is used to make three important polymers: glycogen (in animals), and starch and cellulose (in plants). Chitin is a related structural material found in arthropods and many fungi.
- Most enzymes are proteins. Proteins also provide defense, transport, motion, and regulation, among many other roles.
- Amino acids are joined by peptide bonds to make polypeptides. The 20 common amino acids are characterized by R groups that determine their properties.
- Protein structure is defined by the following hierarchy: primary (amino acid sequence), secondary (hydrogen bonding patterns), tertiary (three dimensional folding), and quaternary (associations between two or more polypeptides).
- Motifs are similar structural elements found in dissimilar proteins. They can create folds, creases, or barrel shapes. Domains are functional subunits or sites within a tertiary structure.
- Chaperone proteins assist in the folding of proteins. Heat shock proteins are an example of chaperone proteins.
- Some forms of cystic fibrosis and Alzheimer disease are associated with misfolded proteins.
- Denaturation refers to an unfolding of tertiary structure, which usually destroys function. Some denatured proteins may recover function when conditions are returned to normal. This implies that primary structure strongly influences tertiary structure.
- Dissociation refers to separation of quaternary subunits with no changes to their tertiary structure.
- Lipids are insoluble in water because they have a high proportion of nonpolar C—H bonds.
- Many lipids exist as triglycerides, three fatty acids connected to a glycerol molecule. Saturated fatty acids contain the maximum number of hydrogen atoms. Unsaturated fatty acids contain one or more double bonds between carbon atoms.
- The energy stored in the C—H bonds of fats is more than twice that of carbohydrates: 9 kcal/g compared with 4 kcal/g. For this reason, excess carbohydrate is converted to fat for storage.
- Phospholipids contain two fatty acids and one phosphate attached to glycerol. In phospholipid-bilayer membranes, the phosphate heads are hydrophilic and cluster on the two faces of the membrane, and the hydrophobic tails are in the center.

- Deoxyribonucleic acid (DNA) and ribonucleic acid (RNA) are polymers composed of nucleotide monomers. Cells use nucleic acids for information storage and transfer.
- Nucleic acids contain four different nucleotide bases. In DNA these are adenine, guanine, cytosine, and thymine. In RNA, thymine is replaced by uracil.
- DNA exists as a double helix held together by specific base pairs: adenine with thymine and guanine with cytosine. The nucleic acid sequence constitutes the genetic code.
- RNA is made by copying DNA. RNA carries information from DNA and forms part of the ribosome. RNA can also be an enzyme and affect gene expression.
- Adenosine triphosphate (ATP) provides energy in cells; NAD^+ and FAD transport electrons in cellular processes.

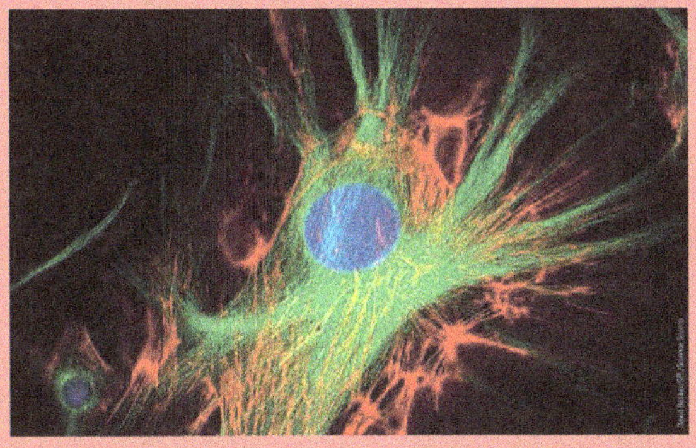

CHAPTER 3

Cell and Microscopy

Chapter Contents:

- Microscopy
- Cell theory is the unifying foundation of cell biology
- Cell size is limited
- Comparing Prokaryotic and Eukaryotic Cells
- Prokaryotic cells have relatively simple organization
- Eukaryotic cells have internal membranes
- Plasma membrane
- The Cytoplasm
- Nucleus: The information center
- Ribosome: The cell's protein synthesis machinery
- The Endoplasmic Reticulum: Biosynthetic Factory
- The Golgi Apparatus: Shipping and Receiving Center
- Lysosomes: Digestive Compartments
- Vacuoles: Diverse Maintenance Compartments
- Chloroplasts: Capture of Light Energy
- Mitochondrion
- Peroxisomes: Oxidation
- The cytoskeleton
- Cell coverings
- Cell junctions
- Key concepts

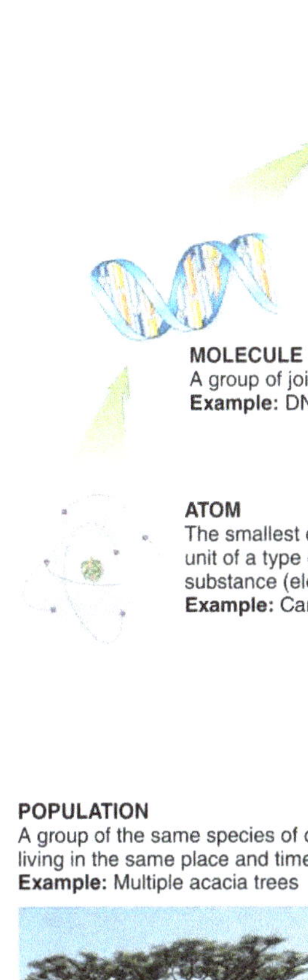
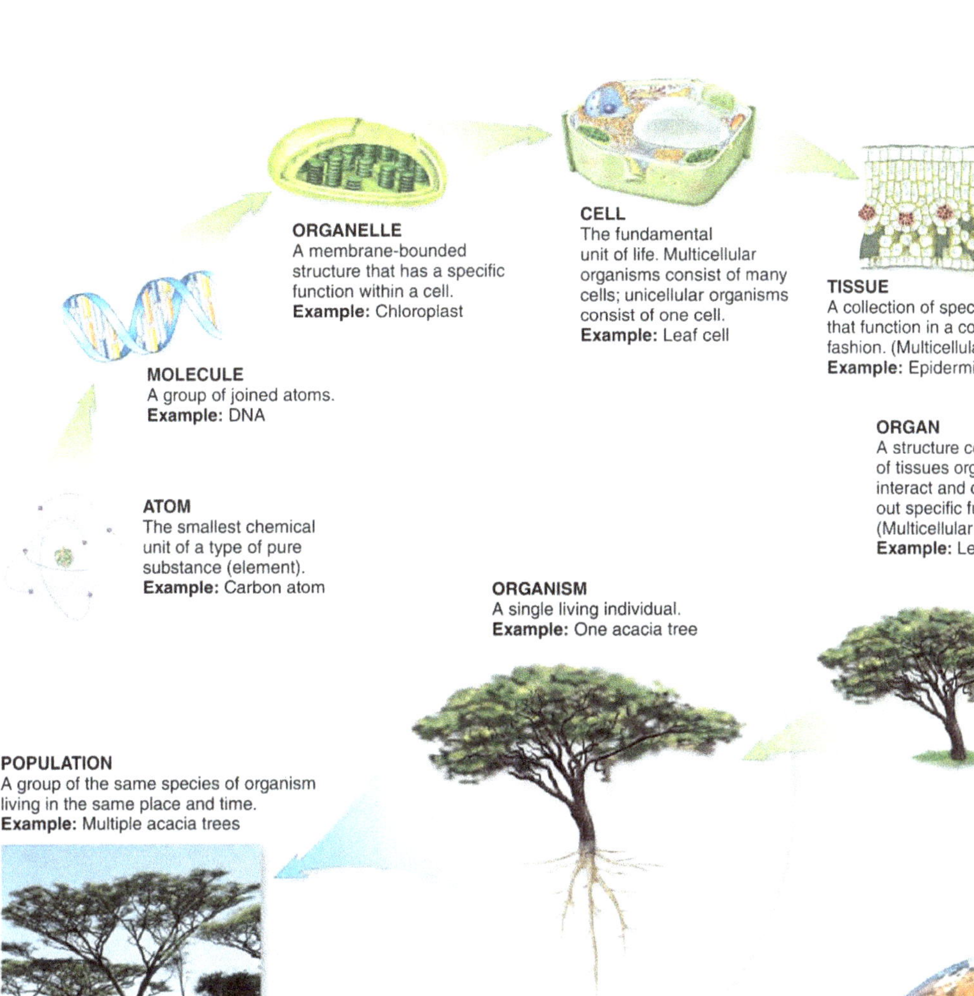

Life's Organizational Hierarchy. This diagram applies life's organizational hierarchy to a multicellular organism (an acacia tree). At the smallest level, atoms are arranged into molecules, which form organelles in the plant's cells. Multiple cells are organized into tissues, which make up organs and, in turn, organ systems. A population consists of individuals of the same species, and communities are multiple populations sharing the same space. Communities interact with the nonliving environment to form ecosystems, and the biosphere consists of all places on Earth where life occurs.

Microscopy

A cell is the basic unit of life. **Cells** are the structural units of all living things, from one celled "generalists" like amoebas to complex multicellular organisms such as humans, dogs, and trees. The human body has 50 to 100 trillion of these tiny building blocks include over 250 different cell types that vary greatly in shape, size, and function. The study of cells is called **cytology**, and scientists who study cells are called **cytologists**.

The unaided eye can see objects that are larger than about 0.2 mm. The development of instruments that extend the human senses allowed the discovery and early study of cells. Cell walls were first seen on dead cells of oak bark by Robert Hooke in 1665 and living cells by Antoni van Leeuwenhoek a few years later. Leeuwenhoek was highly skilled at fabricating lenses and was able to magnify images more than 200 times. Among his important discoveries were bacteria, protists, blood cells, and sperm cells. Leeuwenhoek was among the first scientists to report cells in animals.

▲ **Figure 3.1** The cork cells drawn by Robert Hooke and the compound microscope he used to examine them.

Humans cannot see objects smaller than about 0.1 mm in diameter. The smallest bacteria have diameters of about 0.5 mm (a micrometer is 1,000 times smaller than a millimeter). Most animal and plant cells range from 10 to 100 mm in diameter. The human body contains trillions of cells representing a few hundred different kinds, and virtually all but one type is invisible without a microscope. Egg, the only human cell visible to the naked eye, is approximately as big as the period at the end of this sentence (a human egg cell is around 100 µm in size, placing it right at the limit of what can be viewed by our eyes).

To see cells and the structures within them biologists use **microscopy**, a technique for producing visible images of objects, biological or otherwise, that are too small to be seen by the human eye. The instrument of microscopy is the **microscope**. The two common types of microscopes are **light microscopes**, which use light to illuminate the specimen (the object being viewed), and **electron microscopes**, which use electrons to illuminate the specimen.

Microscopes are the most important tools of cytology, the study of cell structure. Understanding the function of each structure, however, required the integration of cytology and biochemistry, the study of the chemical processes (metabolism) of cells.

Three important parameters in microscopy are magnification, resolution, and contrast.

Magnification is the ratio of an object's image size to its real size. Light microscopes can magnify effectively to about 1,000 times the actual size of the specimen; at greater magnifications, additional details cannot be seen clearly. Cells and organelles can be measured with a microscope by means of an **eyepiece graticule**. This is a transparent scale. It usually has 100 divisions. The eyepiece graticule is placed in the microscope eyepiece so that it can be seen at the same time as the object to be measured. To calibrate the eyepiece graticule, a miniature transparent ruler called a **stage micrometer** is placed on the microscope stage and is brought into focus. This scale may be etched onto a glass slide or printed on a transparent fjlm. It commonly has subdivisions of 0.1 and 0.01 mm. The images of the stage micrometer and the eyepiece graticule can then be superimposed (placed on top of one another)

$$M = \frac{I}{A}$$

$$\text{magnification} = \frac{\text{observed size of the image}}{\text{actual size}} \quad \text{or}$$

M = magnification

I = observed size of the image (what you can measure with a ruler)

A = actual size (the real size – for example, the size of a cell before it is magnified).

Resolution is a measure of the clarity of the image; it is the minimum distance two points can be separated and still be distinguished as separate points. The light microscope cannot resolve detail finer than about 0.2 micrometer (µm), or 200 nanometers (nm), regardless of the magnification. As the wavelength decreases, the resolution increases. The visible light used by ordinary light microscopes has wavelengths ranging from about 400 nm (violet) to 700 nm (red); this limits the resolution of the light microscope to details no smaller than the diameter of a small bacterial cell (about 0.2 µm or 200 nm).

The general rule when viewing specimens is that the limit of resolution is about one half the wavelength of the radiation used to view the specimen. In other words, if an object is any smaller than half the wavelength of the radiation used to view it, it cannot be seen separately from nearby objects. This means that the best resolution that can be obtained using a microscope that uses visible light (a light microscope) is 200 nm, since the shortest wavelength of visible light is 400 nm (violet light).

Contrast, is the difference in brightness between the light and dark areas of an image. Methods for enhancing contrast include staining or labeling cell components to stand out visually.

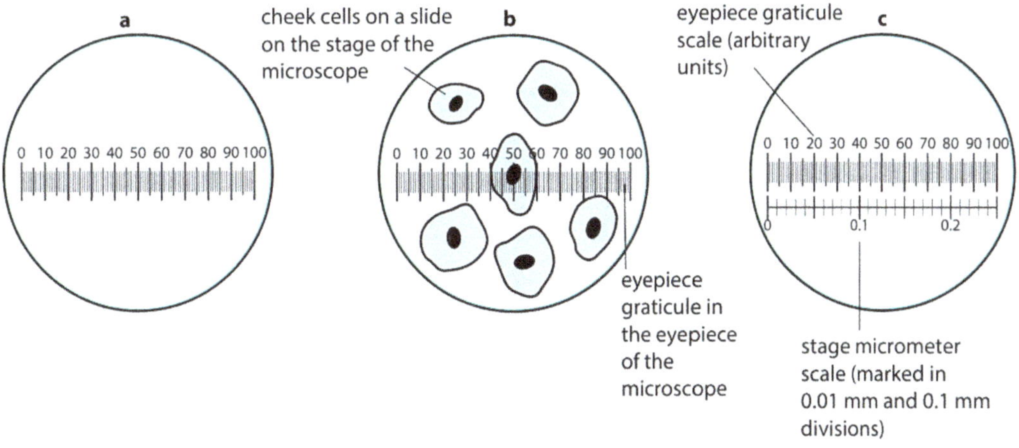

▲ **Figure 3.2** Microscopical measurement.

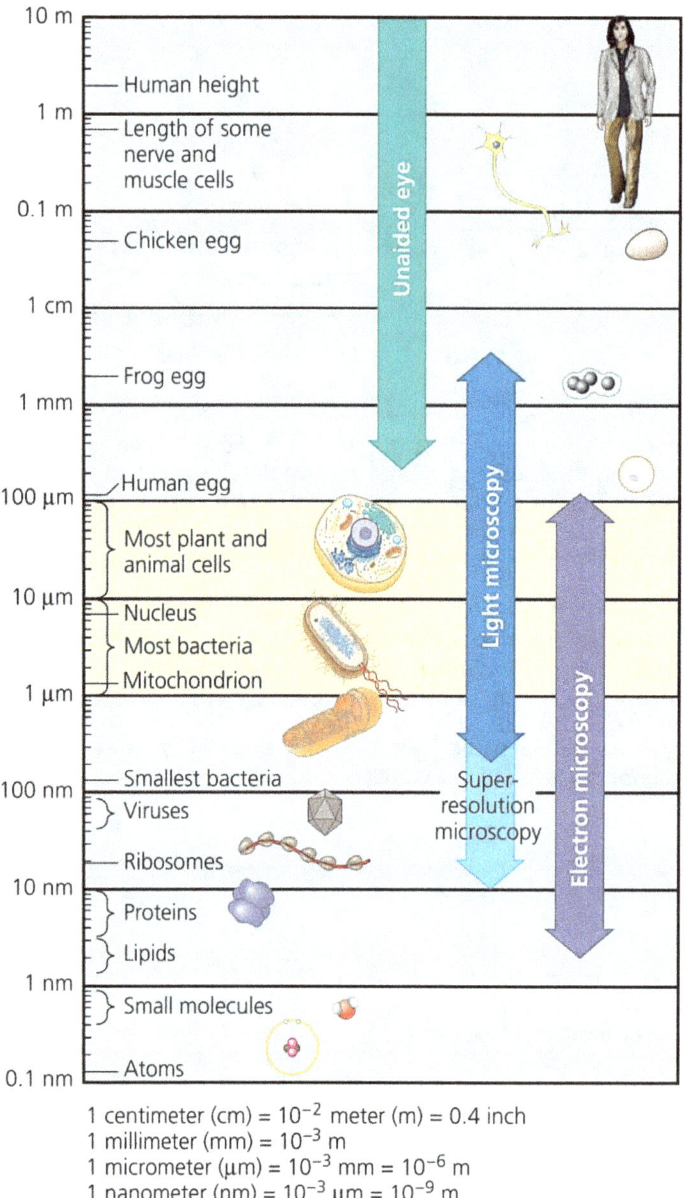

1 centimeter (cm) = 10^{-2} meter (m) = 0.4 inch
1 millimeter (mm) = 10^{-3} m
1 micrometer (µm) = 10^{-3} mm = 10^{-6} m
1 nanometer (nm) = 10^{-3} µm = 10^{-9} m

▲ **Figure 3.3** The size range of cells and how we view them. Most cells are between 1 and 100 µm in diameter and their components are even smaller, as are viruses.

Preparation of samples for study

- **Cell fractionation**

Is a technique for separating (fractionating) different parts of cells so that they can be studied by physical and chemical methods. Generally, cells are broken apart in a blender. The resulting mixture, called the cell homogenate, is subjected to centrifugal force by spinning in a **centrifuge**.

Differential centrifugation involves the separation of cell components through a series of centrifugation stages run at increasingly higher speeds. This allows various cell components to be separated on the basis of their different sizes and densities. At each step centrifugal force separates the extract into two fractions: a pellet and a supernatant. The pellet that forms at the bottom of the tube contains heavier materials packed together. (In the first low-speed step, this pellet is typically composed of nuclei.) The supernatant, the liquid above the pellet, contains lighter organelles, dissolved molecules, and ions.

After the pellet is removed, the supernatant is centrifuged again at a higher speed to obtain a pellet that contains the next-heaviest cell components, for example, mitochondria and chloroplasts. To separate smaller, less dense components, the supernatant is then centrifuged in the powerful ultracentrifuge.

Pellets can be resuspended and their components further purified by **density gradient centrifugation**. In this procedure, the ultracentrifuge tube is filled with a series of solutions of decreasing density. For example, sucrose solutions can be used. The concentration of sucrose is highest at the bottom of the tube and decreases gradually so that it is lowest at the top. The resuspended pellet is placed in a layer on top of the density gradient. Because the densities of organelles differ, each migrates during centrifugation to a position in the sucrose gradient that corresponds to its own density. These purified organelles can then be studied to determine what kinds of proteins and other molecules they might contain, or what types of biochemical reactions take place within them.

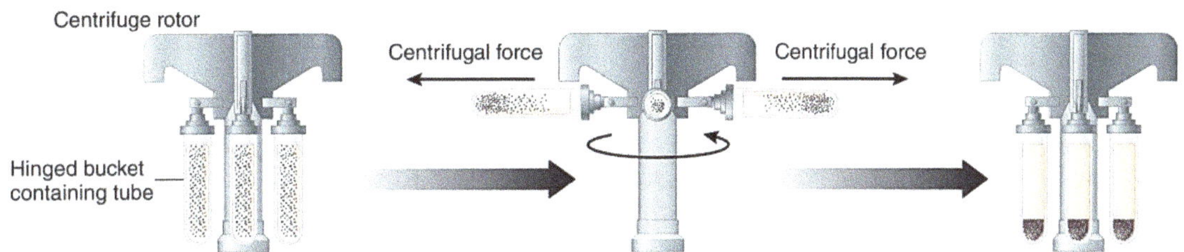

(a) Centrifugation. Due to centrifugal force, large or very dense particles move toward the bottom of a tube and form a pellet.

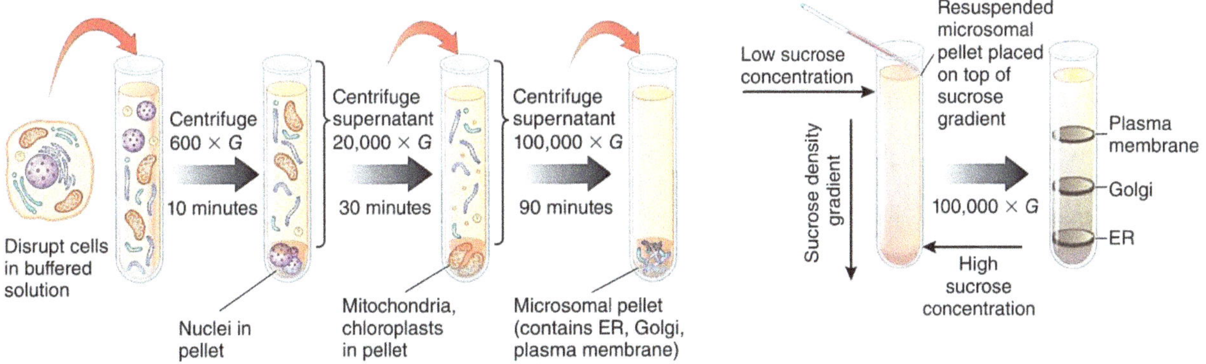

(b) Differential centrifugation.

(c) Density gradient centrifugation.

- **The basic steps used in tissue preparation for light microscopy**

Fixation: Small pieces of tissue are placed in solutions of chemicals that cross-link proteins and inactivate degradative enzymes, which preserve cell and tissue structure.

Dehydration: The tissue is transferred through a series of increasingly concentrated alcohol solutions, ending in 100%, which removes all water.

Clearing: Alcohol is removed in organic solvents in which both alcohol and paraffin are miscible.

Infiltration: The tissue is then placed in melted paraffin until it becomes completely infiltrated with this substance.

Embedding: The paraffin-infiltrated tissue is placed in a small mold with melted paraffin and allowed to harden.

Trimming: The resulting paraffin block is trimmed to expose the tissue for sectioning (slicing) on a microtome.

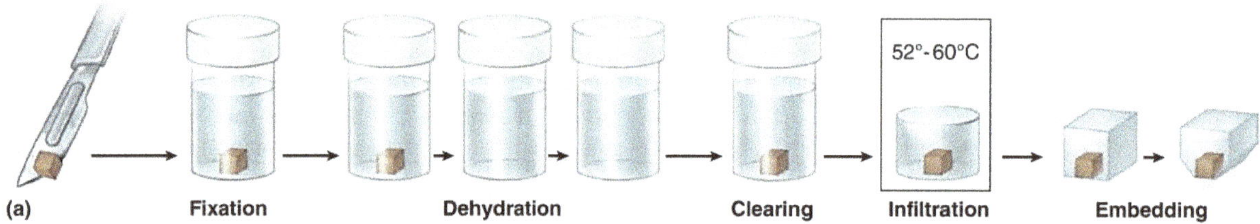

- **Using stains to view cell structure**

Although resolution remains a physical limit, we can improve the images we see by altering the sample. Most cells and extracellular material are completely colorless. If an object is transparent, it will allow light waves to pass through it and therefore will still not be visible. This is why many biological structures have to be stained (dyed) before they can be seen. Certain chemical stains increase the contrast between different cellular components. Structures within the cell absorb or exclude the stain differentially, producing contrast that aids resolution. Staining has enabled biologists to discover the many different internal cell structures, the organelles. Unfortunately, most methods used to prepare and stain cells for observation also kill them in the process.

Dyes stain material more or less selectively, often behaving like acidic or basic compounds and forming electrostatic (salt) linkages with ionizable radicals of macromolecules in tissues. Cell components, such as nucleic acids with a net negative charge (anionic), have an affinity for basic dyes and are termed **basophilic**; cationic components, such as proteins with many ionized amino groups, stain more readily with acidic dyes and are termed **acidophilic.** Examples of basic dyes include toluidine blue, alcian blue, and methylene blue. Hematoxylin behaves like a basic dye, staining basophilic tissue components. The main tissue components that ionize and react with basic dyes do so because of acids in their composition (DNA, RNA, and glycosaminoglycans). Acid dyes (eg, eosin, orange G, and acid fuchsin) stain the acidophilic components of tissues such as mitochondria, secretory granules, and collagen.

Stains that bind to specific types of molecules have made these techniques even more powerful. This method uses antibodies that bind, for example, to a particular protein. This process, called **immunohistochemistry**, uses antibodies generated in animals such as rabbits or mice.

Light microscopes are used to study stained or living cells

The first microscopists used glass lenses to magnify small cells and cause them to appear larger than the 100-μm limit imposed by the human eye. The glass lens increases focusing power. Because the glass lens makes the object appear closer, the image on the back of the eye is bigger than it would be without the lens.

Modern light microscopes, which operate with visible light, use two magnifying lenses (and a variety of correcting lenses) to achieve very high magnification and clarity. The first lens focuses the image of the object on the second lens, which magnifies it again and focuses it on the back of the eye. Microscopes that magnify in stages using several lenses are called **compound microscopes**.

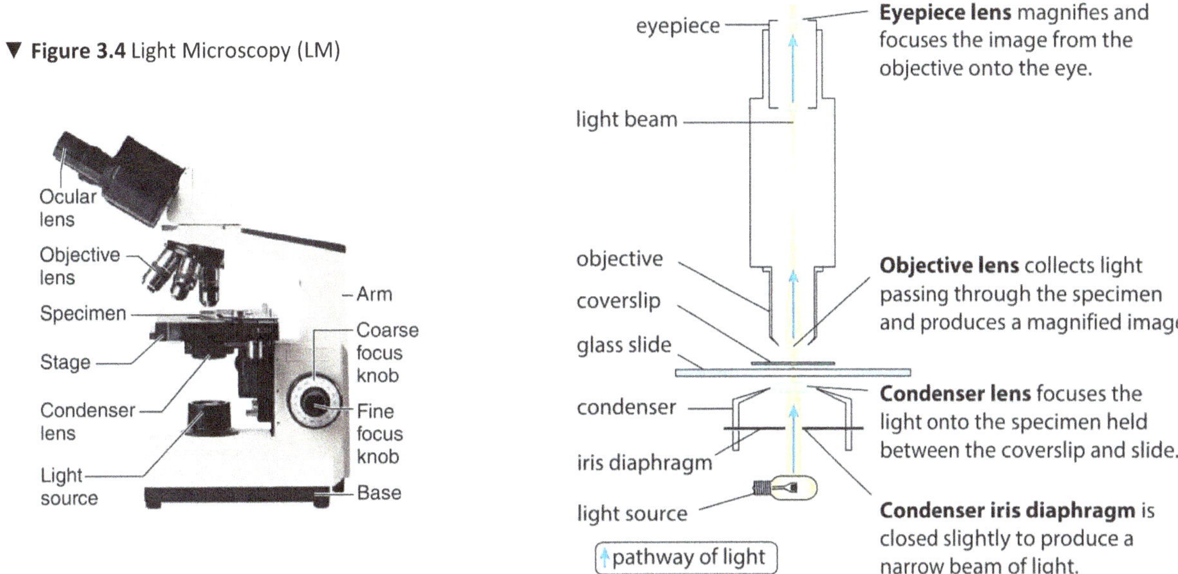

▼ **Figure 3.4** Light Microscopy (LM)

In the past several decades, light microscopy has been revitalized by major technical advances:

- Labeling individual cellular molecules or structures with fluorescent markers has made it possible to see such structures with increasing detail.
- In addition, both confocal and deconvolution microscopy have produced sharper images of three-dimensional tissues and cells.
- Finally, a group of new techniques and labeling molecules developed in recent years, called super-resolution microscopy, has allowed researchers to "break" the resolution barrier and distinguish subcellular structures as small as 10–20 nm across.

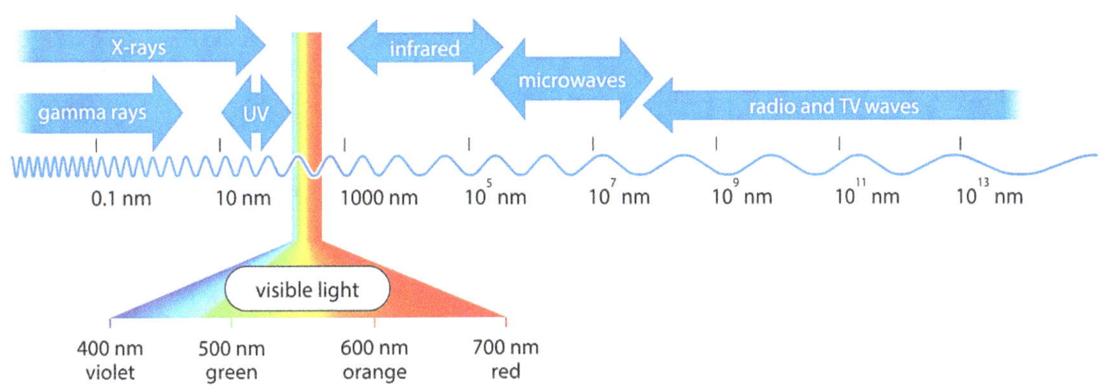

▲ **Figure 3.5** Diagram of the electromagnetic spectrum.

Micrograph is a picture taken with the aid of a microscope; photomicrograph (or light micrograph) is taken using a light microscope; an electron micrograph is taken using an electron microscope.

- In **bright-field microscopy**, an image is formed by transmitting light through a cell (or other specimen). Because there is little contrast, the details of cell structure are not visible. Unstained image has little contrast. Staining with dyes enhances contrast. Most stains require cells to be preserved, which kills them.

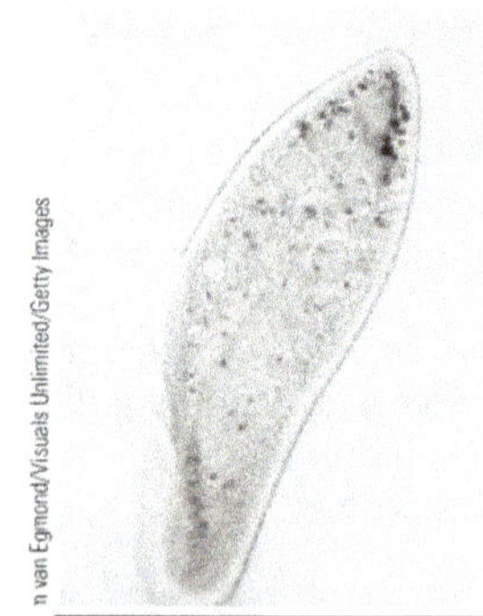

- In **dark-field microscopy**, rays of light are directed from the side, and only light scattered by the specimen enters the lenses. The cell is seen as a bright image against a dark background. The specimen does not need to be stained.

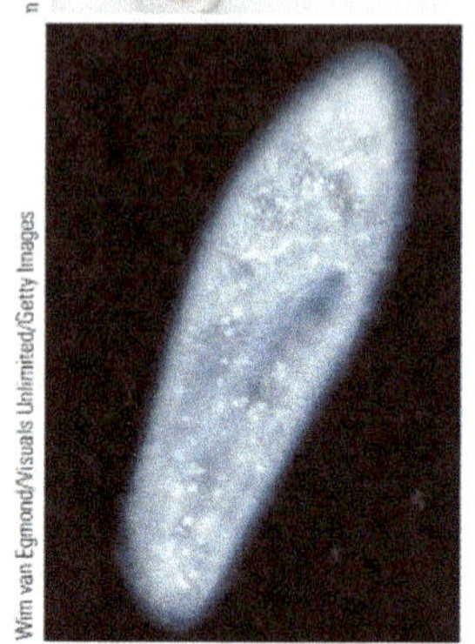

- **Phase contrast microscopy** and **Nomarski differential interference-contrast microscopy** take advantage of variations in density within the cell. Differences in refraction (the way light is bent) caused by variations in the density of the specimen are visualized as differences in contrast. Using these microscopes, scientists can observe living cells in action and can view numerous internal structures that are constantly changing shape and location. Phase-contrast is especially useful for examining living, unstained cells. Nomarski image appears almost 3-D.

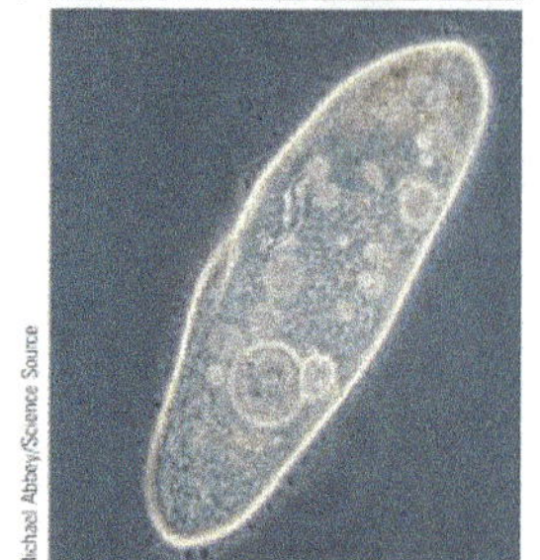

- **Fluorescence microscopy** identifies the intracellular locations of specific molecules. In the fluorescence microscope, filters transmit light that is emitted by fluorescent molecules, or fluorophores. Fluorophores are molecules that absorb light energy of one wavelength and then release some of that energy as light of a longer wavelength. The stained structures or molecules fluoresce when the microscope illuminates them with ultraviolet light, and their locations are seen by viewing the emitted visible light.

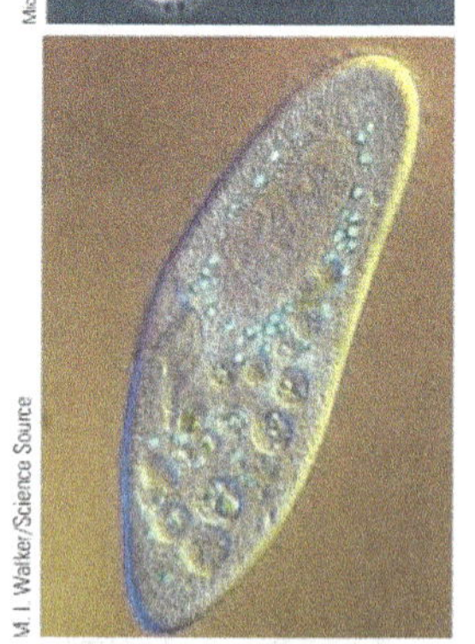

- **Confocal microscopy** has led to significant advances in our understanding of intracellular structural dynamics. These microscopes produce a sharper image than standard fluorescence microscopy. A confocal microscope uses a laser to excite fluorophores in just a thin "slice" through a cell, enabling an investigator to visualize objects in a single plane of sharp focus. By capturing sharp images at many different planes, a 3-D reconstruction can be created.

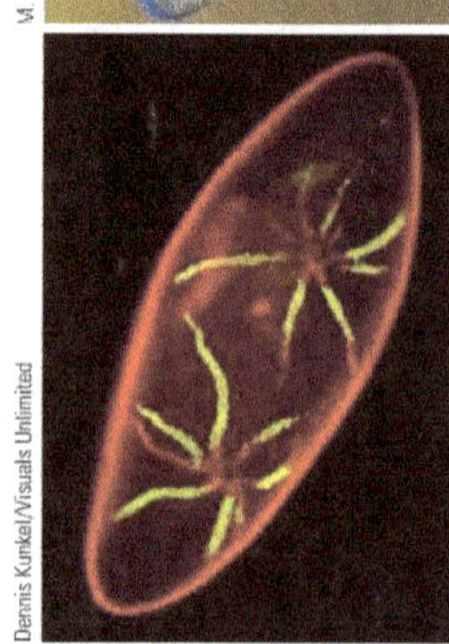

▲ **Figure 3.6** Light Microscopy (LM)

Electron microscopes

In the 1950s, the electron microscope was introduced to biology. Rather than focusing light, the electron microscope (EM) focuses a beam of electrons through the specimen or onto its surface. Electrons are negatively charged particles which orbit the nucleus of an atom. When a metal becomes very hot, some of its electrons gain so much energy that they escape from their orbits, similar to a rocket escaping from Earth's gravity. Free electrons behave like electromagnetic radiation. They have a very short wavelength.

Resolution is inversely related to the wavelength of the light (or electrons) a microscope uses for imaging. The wavelength in an electron beam is much shorter than that of light, allowing 400 times increase in resolution.

The **scanning electron microscope (SEM)** is especially useful for detailed study of the topography of a specimen. The electron beam scans the surface of the sample, usually coated with a thin film of gold. The beam excites electrons on the surface, and these secondary electrons are detected by a device that translates the pattern of electrons into an electronic signal sent to a video screen. The result is an image of the specimen's surface that appears three-dimensional. The maximum magnification of SEM is about 250,000 times, and the resolution is between 3 nm and 20 nm. A disadvantage of the SEM is that it cannot achieve the same resolution as a TEM.

The **transmission electron microscope (TEM)** is used to study the internal structure of cells. The TEM aims an electron beam through a very thin section of the specimen, much as a light microscope aims light through a sample on a slide. For the TEM, the specimen has been stained with atoms of heavy metals, which attach to certain cellular structures, thus enhancing the electron density of some parts of the cell more than others.

The electrons passing through the specimen are scattered more in the denser regions, so fewer are transmitted. The image displays the pattern of transmitted electrons. TEMs permits resolution around 0.5 nm to 3 nm. This high resolution allows isolated particles magnified as much as 400,000 times to be viewed in detail.

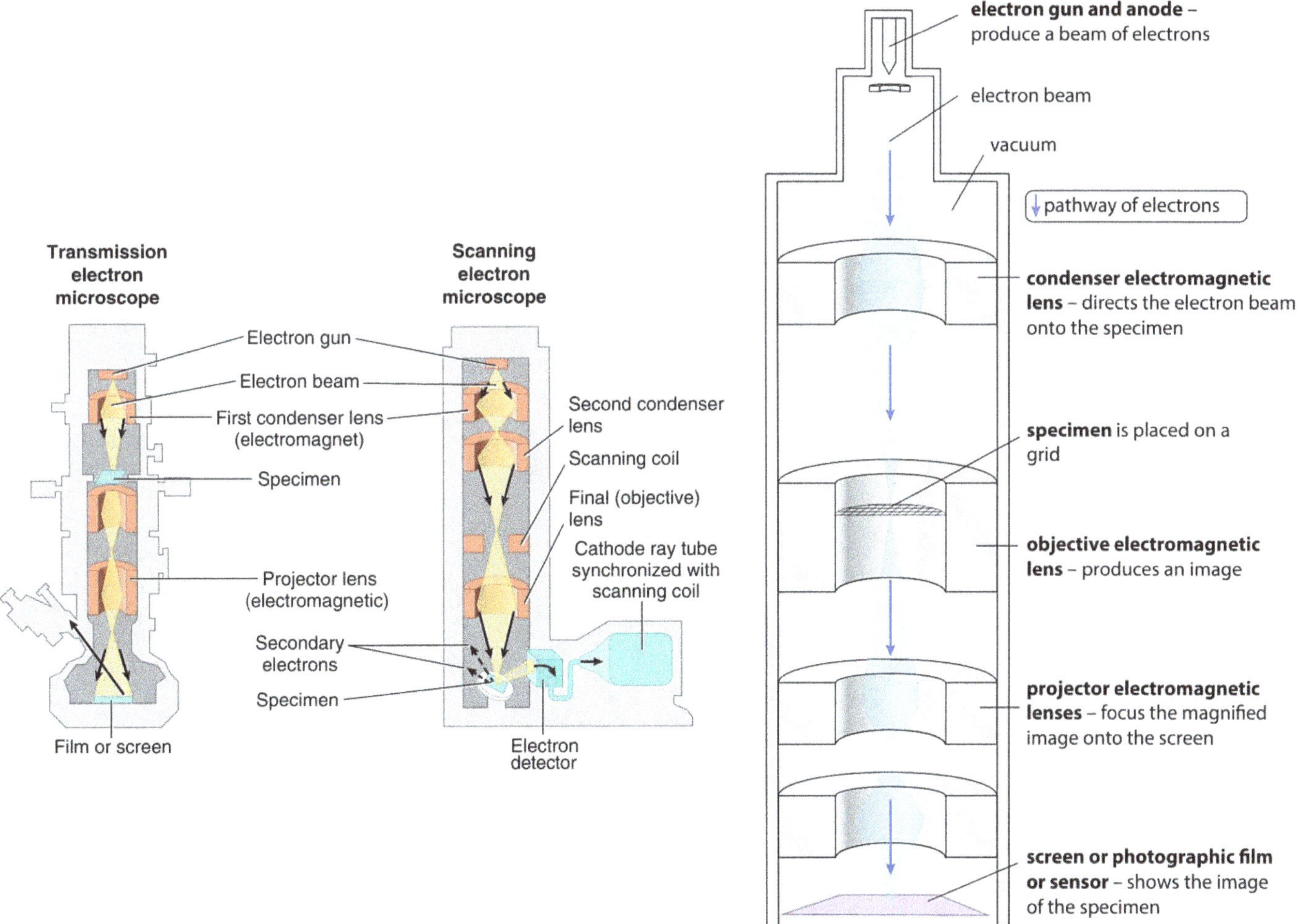

A recently developed new type of TEM called cryo-electron microscopy (cryo-EM) allows specimens to be preserved at extremely low temperatures. This avoids the use of preservatives, allowing visualization of structures in their cellular environment. This method is increasingly used to complement X-ray crystallography in revealing protein complexes and subcellular structures like ribosomes. Cryo-EM has even been used to resolve some individual proteins. The Nobel Prize for Chemistry was awarded in 2017 to the developers of this valuable technique.

Cryo-electron microscopy (cryo-EM). Specimens of tissue or aqueous solutions of proteins are frozen rapidly at temperatures less than −160°C, locking the molecules into a rigid state. A beam of electrons is passed through the sample to visualize the molecules by electron microscopy, and software is used to merge a series of such micrographs, creating a 3-D image like the one below.

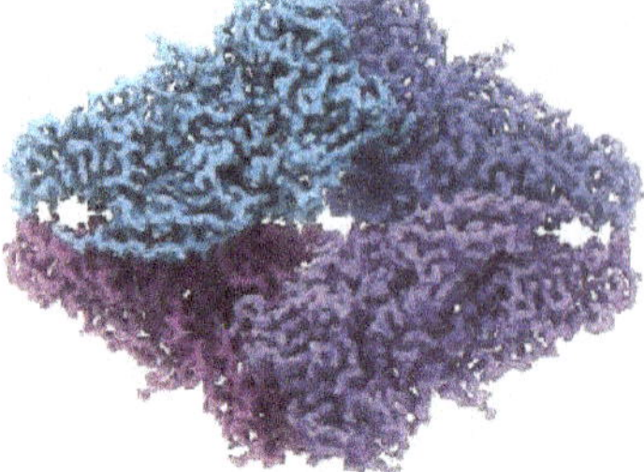

Computer-generated image of the bacterial enzyme β-galactosidase, which breaks down lactose. This image was compiled from more than 90,000 cryo-EM images.

Instead of using glass lenses, both the SEM and TEM use electromagnets as lenses to bend the paths of the electrons, ultimately focusing the image onto a monitor for viewing. It is not possible to see an electron beam, so to make the image visible the electron beam has to be projected onto a fluorescent screen. The areas hit by electrons shine brightly, giving overall a black and white picture.

The electron beam, and therefore the specimen and the fluorescent screen, must be in a vacuum. If the electrons collided with air molecules, they would scatter, making it impossible to achieve a sharp picture. Also, water boils at room temperature in a vacuum, so all specimens must be dehydrated before being placed in the microscope. This means that only dead material or non-living can be examined. So a disadvantage of electron microscopy is that the methods customarily used to prepare the specimen kill the cells and can introduce artifacts, structural features seen in micrographs that do not exist in the living cell. Unlike light microscopes, all images from electron microscopes are black and white, although artists often add false color to highlight specific objects in electron micrographs using a computer.

The stains used to improve the contrast of biological specimens for electron microscopy contain heavy metal atoms, which stop the passage of electrons. The resulting picture is like an X-ray photograph, with the more densely stained parts of the specimen appearing blacker.

Electron microscopes have revealed many subcellular structures that were impossible to resolve with the light microscope. But the light microscope offers advantages, especially in studying living cells.

Cell theory is the unifying foundation of cell biology

Cells are the basic structural and functional units of every organism. The **cell theory** was proposed to explain the observation that all organisms are composed of cells. In its modern form, the cell theory includes the following principles:

- A cell is the basic structural and functional unit of living organisms. When you define cell properties, you define the properties of life.

- The activity of an organism depends on both the individual and the combined activities of its cells.

- According to the principle of complementarity of structure and function, the biochemical activities of cells are dictated by their shapes or forms, and by the relative number of the subcellular structures they contain.

- Cells can only arise from other cells.

Although life likely evolved spontaneously in the environment of early Earth, biologists have concluded that no additional cells are originating spontaneously at present. Rather, life on Earth represents a continuous line of descent from those early cells.

Some types of organisms are **unicellular**. Unicellular organisms include almost all bacteria and archaeans; some protists, such as amoebas; and some fungi, such as yeasts. Each of these cells is a functionally independent organism capable of carrying out all activities necessary for its life. In more complex **multicellular** organisms, including plants and animals, the activities of life are divided among varying numbers of specialized cells. However, individual cells of multicellular organisms are potentially capable of surviving by themselves if placed in a chemical medium that can sustain them. The average adult human body consists of nearly 40 trillion cells, according to the best available estimate.

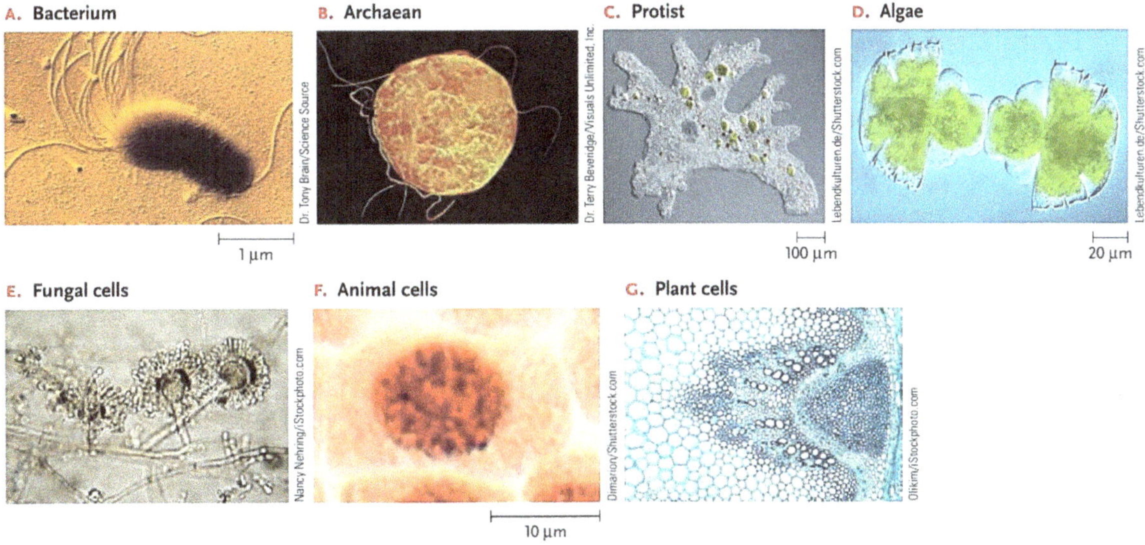

▲ **Figure 3.7** Examples of the varied kinds of cells:

(A) and (B) are prokaryotes, the others are eukaryotes.

life as we know does not exist in units more simple than individual cells. Viruses are tiny 'particles' which are much smaller than bacteria and are on the boundary between what we think of as living and non-living. Unlike prokaryotes and eukaryotes, viruses do not have a cell structure. They consist only of a nucleic acid molecule surrounded by a protein coat so cannot carry out most of the activities of life. Their only capacity is to infect living cells and direct them to make more virus particles of the same kind, so all viruses are parasitic. The protein coat (or **capsid**) is made up of separate protein molecules, each of which is called a **capsomere**. Some viruses only have a membrane-like outer layer, called the **envelope**, that is made of phospholipids. Proteins may project from the envelope.

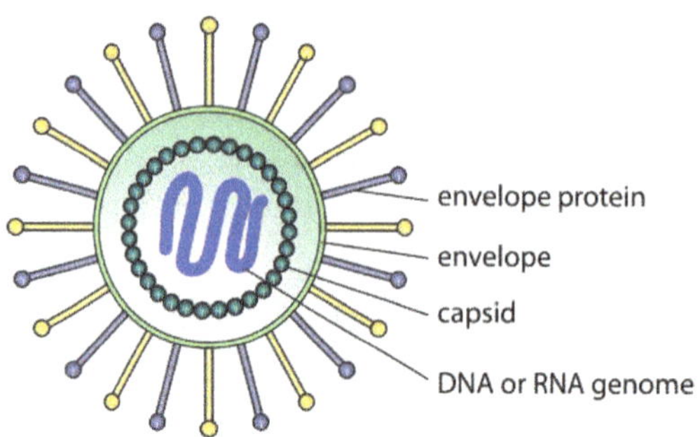

▲ **Figure 3.8** The structure of a virus with an envelope

Cell size is limited

Most cells are relatively small for reasons related to the diffusion of substances into and out of them. The rate of diffusion is affected by a number of variables, including (1) surface area available for diffusion, (2) temperature, (3) concentration gradient of diffusing substance, and (4) the distance over which diffusion must occur.

Doubling the diameter of a cell increases its volume by eight times but increases its surface area by only four times. The significance of this relationship is that the volume of a cell determines the amount of chemical activity that can take place within it, whereas the surface area determines the amount of substances that can be exchanged between the inside of the cell and the outside environment.

So as the size of a cell increases, the length of time for diffusion from the outside membrane to the interior of the cell increases as well. Larger cells need to synthesize more macromolecules, have correspondingly higher energy requirements, and produce a greater quantity of waste. Molecules used for energy and biosynthesis must be transported through the membrane. Any metabolic waste produced must be removed, also passing through the membrane. The rate at which this transport occurs depends on both the distance to the membrane and the area of membrane available. For this reason, an organism made up of many relatively small cells has an advantage over one composed of fewer, larger cells.

The advantage of small cell size is readily apparent in terms of the **surface area-to-volume ratio**. The cell surface provides the only opportunity for interaction with the environment, because all substances enter and exit a cell via this surface. The membrane surrounding the cell plays a key role in controlling cell function. Because small cells have more surface area per unit of volume than large ones, control over cell contents is more effective when cells are relatively small.

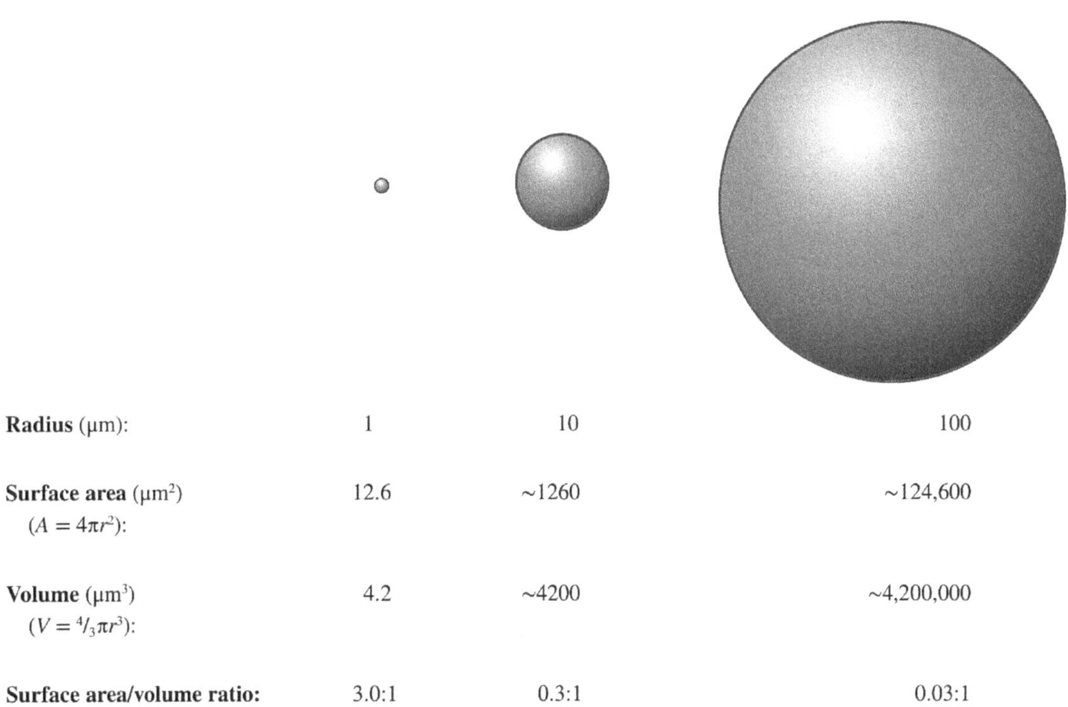

Radius (µm):	1	10	100
Surface area (µm²) ($A = 4\pi r^2$):	12.6	~1260	~124,600
Volume (µm³) ($V = \frac{4}{3}\pi r^3$):	4.2	~4200	~4,200,000
Surface area/volume ratio:	3.0:1	0.3:1	0.03:1

▲ **Figure 3.9** Relationship between cell size and the surface area/volume (SA/V) ratio. As cells get larger, the SA/V ratio gets smaller.

Some cells have adaptations that allow them to circumvent the surface area limitation just described. For instance, eggs of some species, such as birds and frogs, are much larger than typical cells, meaning that they have a low surface area-to volume ratio. In this case the eggs contain a large store of nutrients so no nutrients need to be brought into the cells. In addition, once fertilized, the eggs divide rapidly to produce a multi celled embryo with each cell of typical cell size. Another adaptation to circumvent the surface area limitation is for cells to be long and thin, or skinny and flat, both of which increase surface area. Examples are nerve cells and muscle cells. Yet another adaptation is seen in human intestinal cells, which have closely packed, fingerlike extensions that increase their surface area.

At the boundary of every cell, the plasma membrane functions as a selective barrier that allows passage of enough oxygen, nutrients, and wastes to service the entire cell.

For each square micrometer of membrane, only a limited amount of a particular substance can cross per second, so the ratio of surface area to volume is critical. As a cell (or any other object) increases in size, its surface area grows proportionately less than its volume. (Area is proportional to a linear dimension squared, whereas volume is proportional to the linear dimension cubed.) Thus, a smaller cell has a greater **ratio of surface area to volume**. The need for a surface area large enough to accommodate the volume helps explain the microscopic size of most cells and the narrow, elongated shapes of some cells, such as nerve cells. Larger organisms do not generally have larger cells than smaller organisms—they simply have more cells.

A sufficiently high ratio of surface area to volume is especially important in cells that exchange a lot of material with their surroundings, such as intestinal cells. Such cells may have many long, thin projections from their surface called **microvilli**, which increase surface area without an appreciable increase in volume.

► **Figure 3.10** Geometric relationships between surface area and volume. In this diagram, cells are represented as boxes. Using arbitrary units of length, we can calculate the cell's surface area (in square units, or units^2), volume (in cubic units, or units^3), and ratio of surface area to volume. A high surface-to-volume ratio facilitates the exchange of materials between a cell and its environment.

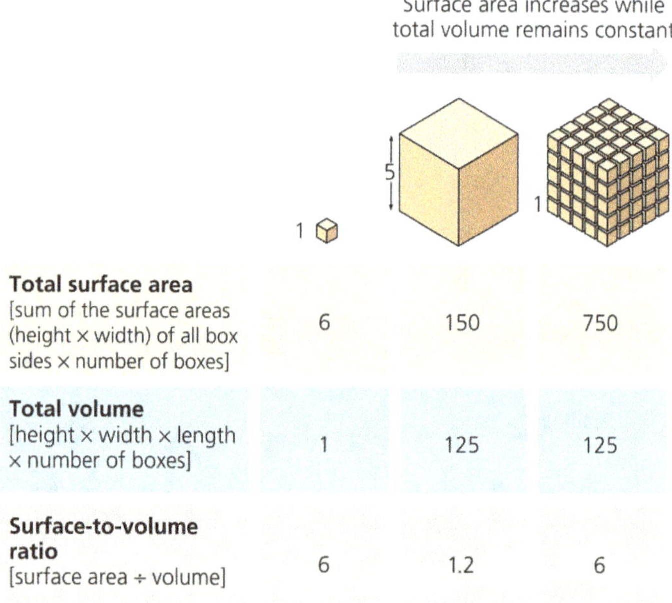

Total surface area [sum of the surface areas (height × width) of all box sides × number of boxes]	6	150	750
Total volume [height × width × length × number of boxes]	1	125	125
Surface-to-volume ratio [surface area ÷ volume]	6	1.2	6

The sizes and shapes of cells are adapted to the particular functions they perform. Nerve cells may be long (up to a meter or so), but they are also extremely thin, so the ratio of surface area to volume (SA/V) remains high. The flattened shape of a red blood cell maximizes its ability to carry oxygen, and the many microscopic extensions of an amoeba's membrane provide a large surface area for absorbing oxygen and capturing food.

Comparing Prokaryotic and Eukaryotic Cells

Biologists organized life into just two categories with three domains: **prokaryotic** and **eukaryotic**. Organisms of the domains **Bacteria** or eubacteria and **Archaea** or archaebacteria, consist of prokaryotic cells the simplest and most ancient forms of life that lack a nucleus . Organisms of the domain Eukarya— protists, fungi, animals, and plants—all consist of eukaryotic cells that contain a nucleus and other membranous organelles.

Prokaryotes are thought to have been the first living organisms on Earth. The earliest known fossil prokaryotes are about 3.5 billion years old (the Earth was formed about 4.5 billion years ago). Most biologists believe that eukaryotes evolved from prokaryotes about 2 billion years ago. Four major features all cells have in common: (1) a nucleoid or nucleus where genetic material is located, (2) cytoplasm, (3) ribosomes to synthesize proteins, and (4) a plasma membrane.

The **plasma membrane** encloses a cell and separates its contents from its surroundings. All cells contain **chromosomes**, which carry genes in the form of DNA. And all cells have **ribosomes**, tiny complexes that make proteins according to instructions from the genes. A major difference between prokaryotic and eukaryotic cells is the location of their DNA. In a eukaryotic cell, most of the DNA is in an organelle called the **nucleus**, which is bounded by a double membrane. In a prokaryotic cell, the DNA is concentrated in a region that is not membrane-enclosed, called the **nucleoid**.

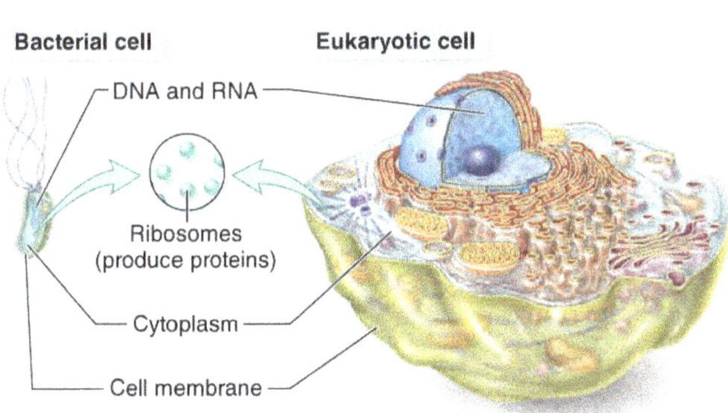

The interior of either type of cell is called the **cytoplasm**; in eukaryotic cells, this term refers only to the region between the nucleus and the plasma membrane. The part of the cytoplasm that contains organic molecules and ions in solution is called the **cytosol**. Within the cytoplasm of a eukaryotic cell, suspended in cytosol, are a variety of **organelles** of specialized form and function. These membrane-bounded structures are absent in almost all prokaryotic cells, another distinction between prokaryotic and eukaryotic cells. The term cytoplasm includes both the cytosol and all the organelles other than the nucleus.

Eukaryotic cells also differ from prokaryotic cells in having a supporting framework, or cytoskeleton, important in maintaining shape and transporting materials within the cell. Eukaryotic cells are generally much larger than prokaryotic cells. Typical bacteria are 1–5 μm in diameter. Eukaryotic cells are typically 10–100 μm in diameter. In fact, the average prokaryotic cell is only about 1/10 the diameter of the average eukaryotic cell.

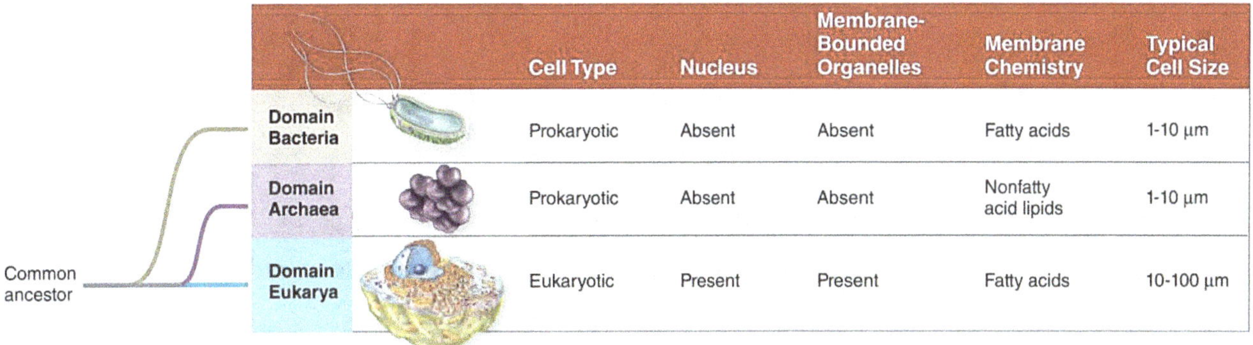

		Cell Type	Nucleus	Membrane-Bounded Organelles	Membrane Chemistry	Typical Cell Size
Domain Bacteria		Prokaryotic	Absent	Absent	Fatty acids	1-10 μm
Domain Archaea		Prokaryotic	Absent	Absent	Nonfatty acid lipids	1-10 μm
Domain Eukarya		Eukaryotic	Present	Present	Fatty acids	10-100 μm

Prokaryotic cells have relatively simple organization

Prokaryotes are the simplest organisms. Prokaryotic cells are small. They consist of cytoplasm surrounded by a plasma membrane and are encased within a rigid **cell wall**. They have no distinct interior compartments. Prokaryotes are very important in the ecology of living organisms. Some harvest light by photosynthesis, others break down dead organisms and recycle their components. Still others cause disease or have uses in many important industrial processes.

Although prokaryotic cells do contain ribosomes, which carry out protein synthesis, most lack the membrane-bounded organelles characteristic of eukaryotic cells. They also have molecules related to both actin and tubulin, which form two of the cytoskeletal elements. The strength and shape of the cell is determined by the cell wall and not these cytoskeletal elements.

The plasma membrane of a prokaryotic cell carries out some of the functions organelles perform in eukaryotic cells. For example, some photosynthetic bacteria, such as the cyanobacterium prochloron, have an extensively folded plasma membrane, with the folds extending into the cell's interior. These membrane folds contain the bacterial pigments connected with photosynthesis.

Because a prokaryotic cell contains no membrane-bounded organelles, the DNA, enzymes, and other cytoplasmic constituents have access to all parts of the cell. Reactions are not compartmentalized as they are in eukaryotic cells, and the whole prokaryote operates as a single unit.

The three shapes most common among prokaryotes are spherical, rodlike, and spiral. **Escherichia coli** (E. coli), a normal inhabitant of the mammalian intestine that has been studied extensively as a model organism in genetics, molecular biology, and genomics research, is rodlike in shape. Prokaryotes have two main domains: **archaea** and **bacteria**.

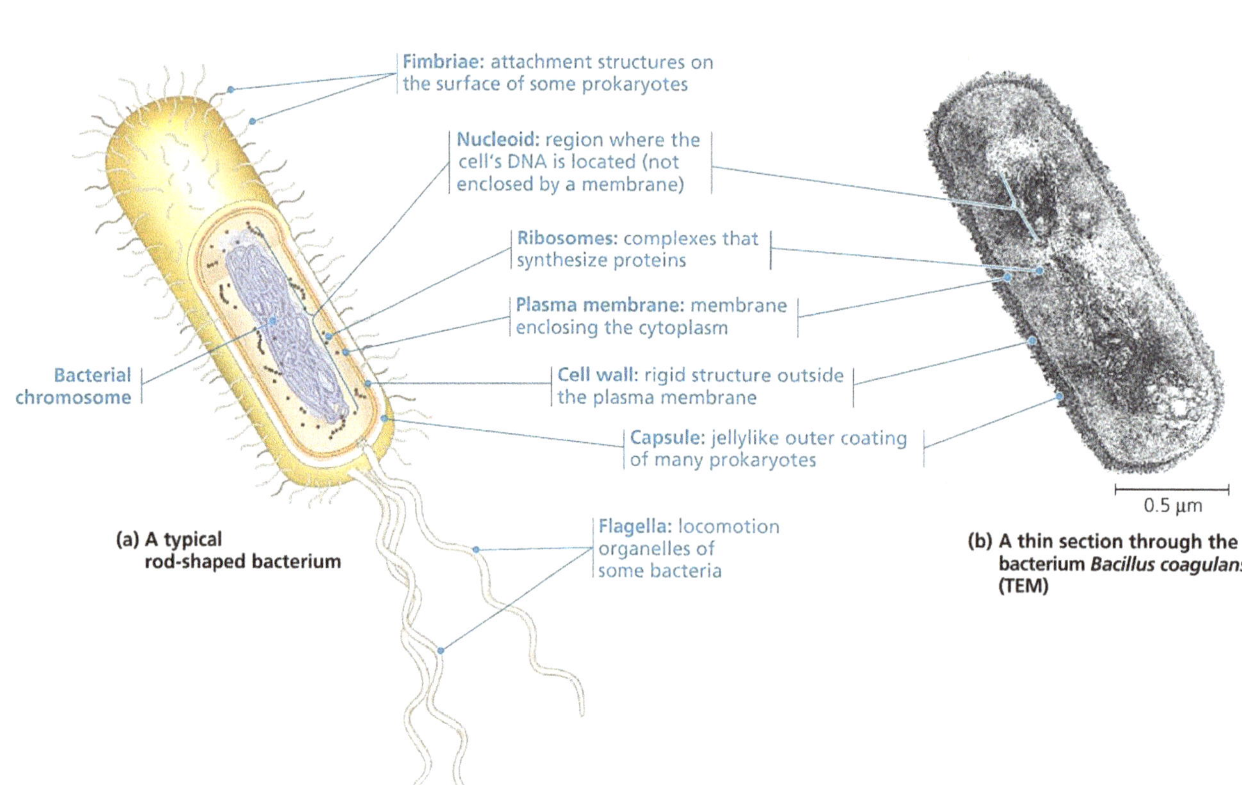

▲ **Figure 3.11** A prokaryotic cell. Lacking a true nucleus and the other membrane-enclosed organelles of the eukaryotic cell, the prokaryotic cell is much simpler in structure. Prokaryotes include bacteria and archaea; the general cell structure of the two domains is essentially the same.

Most bacterial cells are encased by a strong **cell wall**. This cell wall is composed of peptidoglycan, **murein**, which consists of a carbohydrate matrix (polymers of sugars) that is cross-linked by short polypeptide units. Cell walls protect the cell, maintain its shape, and prevent excessive uptake or loss of water. The exception is the class Mollicutes, which includes the common genus *Mycoplasma*, which lack a cell wall. The susceptibility of bacteria to antibiotics often depends on the structure of their cell walls.

Some bacteria are surrounded by an extra layer outside the cell wall. This may take the form of a **capsule** or a slime layer. A capsule is a definite structure, made mostly of polysaccharides. A slime layer is more diffuse and is easily washed off. Both help to protect the bacterium from drying out and may have other protective functions. For example, a capsule helps protect some bacteria from antibiotics. Some capsules prevent white blood cells known as phagocytes from engulfing disease-causing bacteria. Many disease-causing bacteria have such a capsule, which enables them to adhere to teeth, skin, food—or to practically any surface that can support their growth.

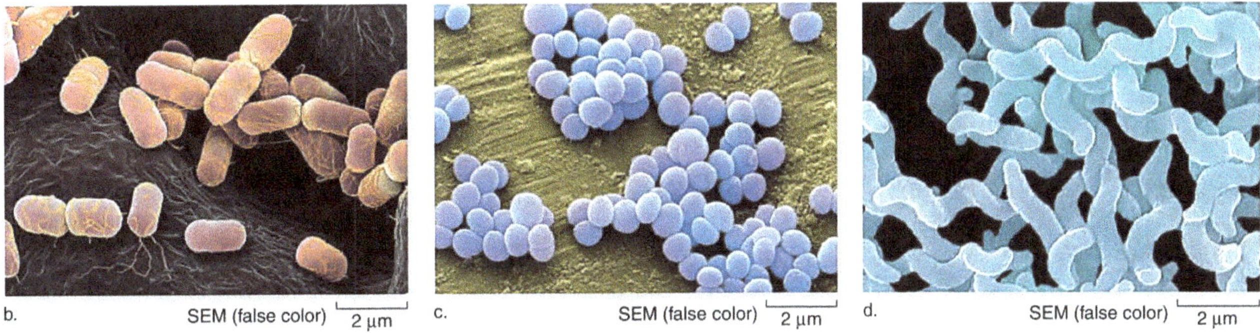

▲ **Figure 3.12 Anatomy of a Bacterium.**

(b) Rod-shaped cells of E. coli inhabit human intestines.

(c) Spherical Staphylococcus aureus cells cause "staph" infections that range from mild to deadly.

(d) These corkscrew shaped bacteria live in the digestive tract of many animals.

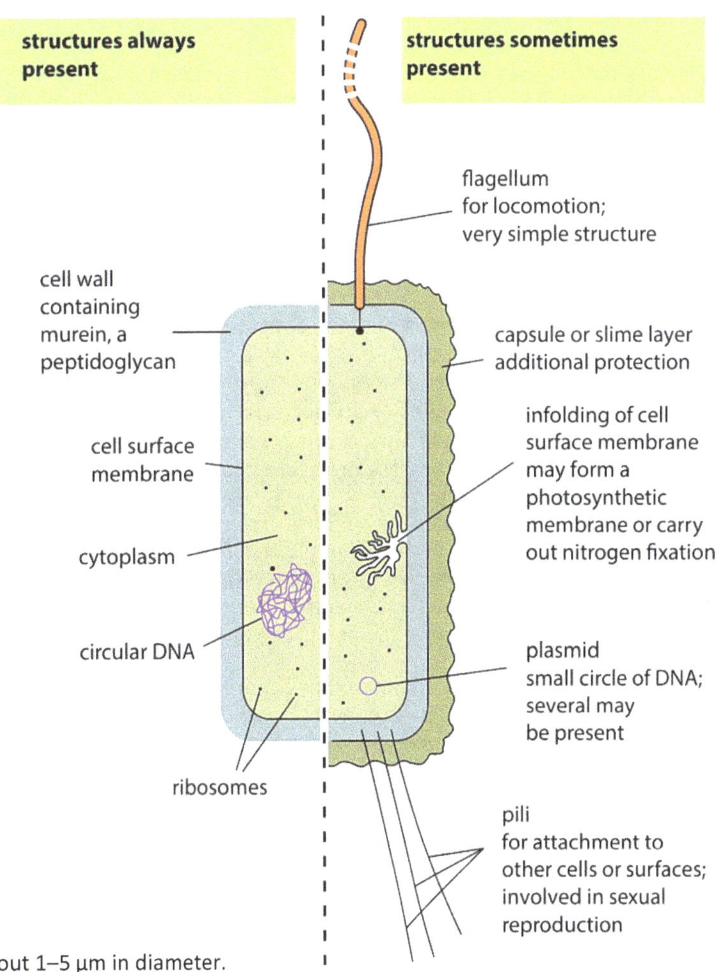

▶ **Figure 3.13** Diagram of a bacterium. Cells are generally about 1–5 μm in diameter.

Flagella (singular, flagellum) are long threadlike structures protruding from the surface of a cell (anchored in the cell wall and underlying cell membrane) that are used in locomotion (moving the cell forward or backward). Their structure is different from that of flagella found in eukaryotic cells. The bacterial flagellum is a simple hollow cylinder made of identical protein molecules. It is a rigid structure, so it does not bend, unlike the flagella in eukaryotes. It is wave-shaped and works by rotating at its base like a propeller to push the bacterium through its liquid environment. As a result, the bacterium moves forward with a corkscrew-shaped motion.

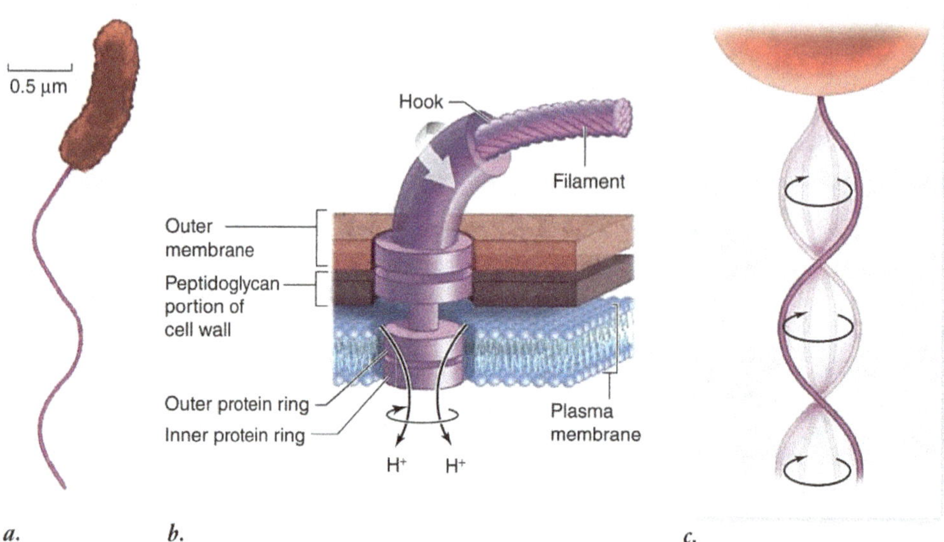

The rotary motor uses the energy stored in a gradient that transfers protons across the plasma membrane to power the movement of the flagellum. Interestingly, the same principle, in which a proton gradient powers the rotation of a molecule, is used in eukaryotic mitochondria and chloroplasts by an enzyme that synthesizes ATP. Some bacteria and archaeans have hairlike shafts of protein called **pili** (singular, pilus) extending from their cell walls. There are two types of pili. **Common pili** (also known as **fimbriae**), as their name suggests, are common among bacteria. They are relatively short, there are many per cell, and there are many subvarieties. Common pili are important for attaching a bacterial cell to surfaces or other cells, **biofilm** formation (A communal living arrangement in which single-celled organisms live in a shared mass of slime is called a biofilm), cell motility, and the transport of protein and DNA across membranes.

Sex pili are specialized structures on a bacterium that attach that cell to another bacterium lacking sex pili during **conjugation**. In conjugation, DNA, including plasmids is transferred in one direction from the cell with the sex pili to the cell without. Sex pili are long structures, there are only two or three of them per cell, and only particular bacterial cells can produce them. The DNA molecule in bacteria is circular. It is found in a region called the **nucleoid**, which also contains proteins and small amounts of RNA. It is not surrounded by a double membrane, unlike the nucleus of eukaryotes. There may be more than one copy of the DNA molecule in a given cell. Bacterial ribosomes are **70S ribosomes**, slightly smaller than the 80S ribosomes of eukaryotes.

In some bacteria, the cell surface membrane folds into the cell forming an extra surface on which certain biochemical reactions can take place. In blue–green bacteria, for example, the infolded membrane contains photosynthetic pigments which allow photosynthesis to take place. In some bacteria, nitrogen fixation takes place on the infolded membrane. Nitrogen fixation is the ability to convert nitrogen in the air to nitrogen containing compounds, such as ammonia, inside the cell.

All life depends on nitrogen fixation. Eukaryotes cannot carry out nitrogen fixation. A **plasmid** is a small circle of DNA separate from the main DNA of the cell. It contains only a few genes. Many plasmids may be present in a given cell. The genes have various useful functions. Commonly, plasmids contain genes that give resistance to particular antibiotics, such as penicillin. Plasmids can copy themselves independently of the chromosomal DNA and can spread rapidly from one bacterium to another. Plasmid DNA is not associated with protein and is referred to as 'naked' DNA.

Cell membranes, cell walls, and flagella are all chemically unique in archaea. The cell walls of archaea are composed of various chemical compounds, including polysaccharides and proteins, and possibly even inorganic components. A common feature distinguishing archaea from bacteria is the nature of their membrane lipids. The chemical structure of archaeal lipids is distinctly different from that of lipids in bacteria and can include saturated hydrocarbons that are covalently attached to glycerol at both ends, such that their membrane is a mono layer. These features seem to confer greater thermal stability to archaeal membranes, although they cannot adapt to changing environmental temperatures. Archaeal flagella function similarly to bacterial flagella, but the two types differ significantly in their structures and mechanisms of action. Both types of prokaryotic flagella are also fundamentally different from the much larger and more complex flagella of eukaryotic cells.

The cellular machinery that replicates DNA and synthesized proteins in archaea is more closely related to eukaryotic systems than to bacterial systems. Even though they share a similar overall cellular architecture with prokaryotes, archaea appear to be more closely related on a molecular basis to eukaryotes. Evidence widely supports the hypothesis that eukaryotic cells evolved from the archaea. The first members of Archaea to be described were methanogens, microbes that use carbon dioxide and hydrogen from the environment to produce methane. Archaea subsequently became famous as "extremophiles" because scientists discovered many of them in habitats that are extremely hot, acidic, or salty. This characterization is somewhat misleading, however, because bacteria also occupy the same environments. Moreover, researchers have now discovered archaea in a variety of moderate habitats, including soil, swamps, rice paddies, oceans, and even the human mouth.

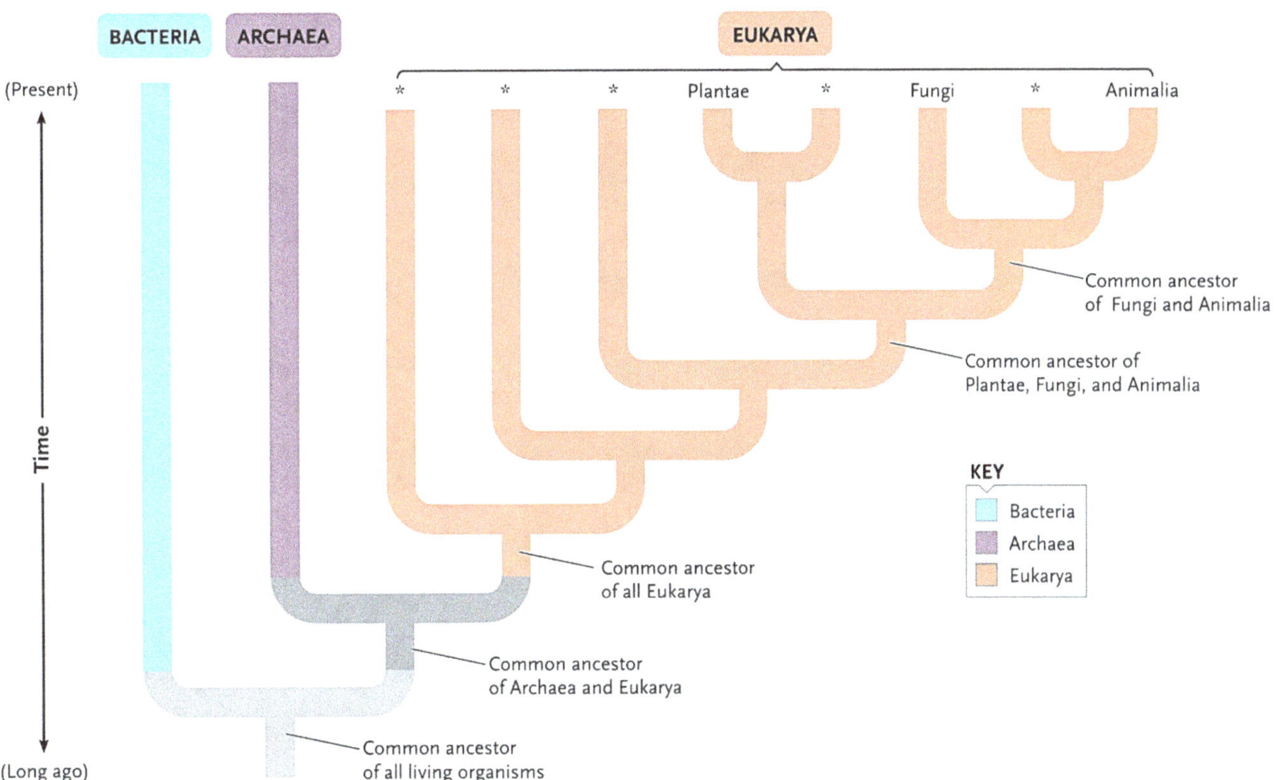

▲ **Figure 3.14 PHYLOGENETIC VIEW OF THE TREE OF LIFE.** Phylogenetic trees contain more information than simple hierarchical classifications do because the trees illustrate which ancestors gave rise to which descendants, as well as when those evolutionary events occurred. Each fork between trunks, branches, and twigs on the phylogenetic tree represents an evolutionary event in which one ancestral species gave rise to two descendant species. Detailed phylogenetic trees illustrate how, over time, descendant species gave rise to their own descendants, producing the great diversity of life. This phylogenetic tree of Life illustrates the relationships between the three domains. Bacteria, Archaea and Eukarya that include three well-defined kingdoms (Plantae, Fungi, and Animalia), as well as five groups of organisms that were once collectively described as protists (marked with *).

Eukaryotic cells have internal membranes that compartmentalize their functions

The eukaryotes make up the Domain **Eukarya**. In addition to the plasma membrane at its outer surface, a eukaryotic cell has extensive, elaborately arranged internal membranes that divide the cell into compartments—the organelles mentioned earlier. The cell's compartments provide different local environments that support specific metabolic functions, so incompatible processes can occur simultaneously in a single cell.

Animal Cell (cutaway view of generalized cell)

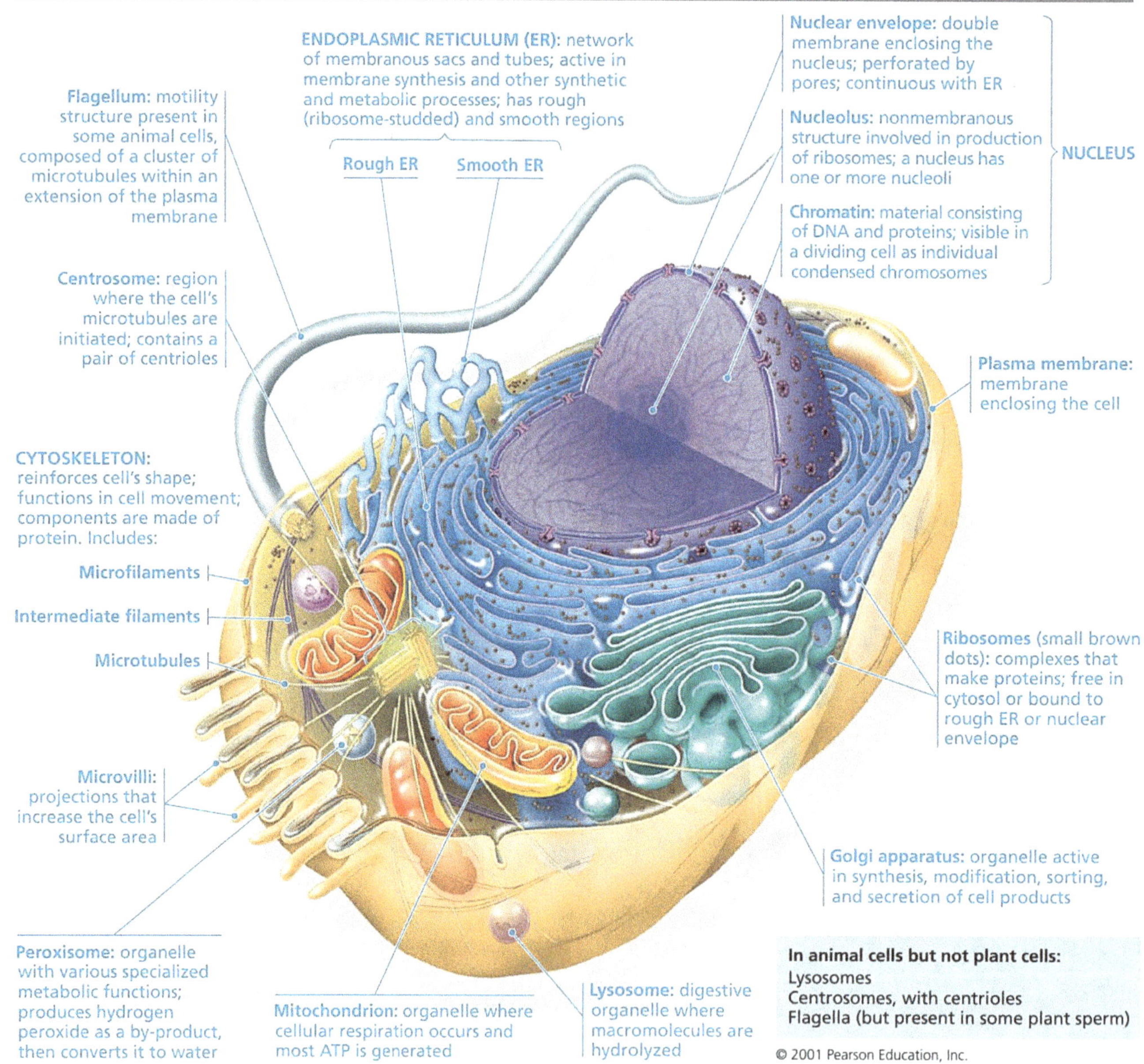

Plant Cell (cutaway view of generalized cell)

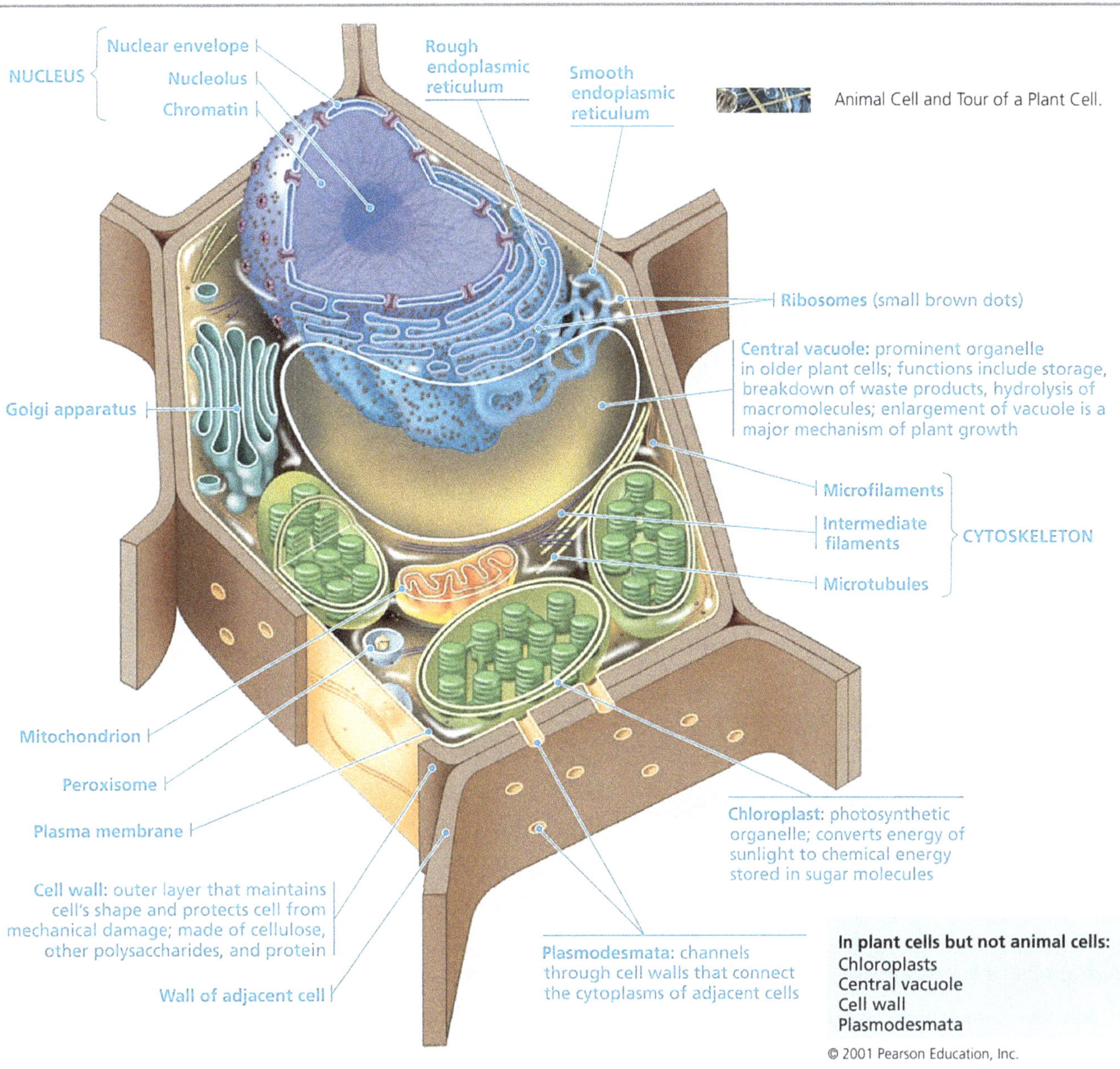

- **Plasma membrane**

The plasma membrane (or the cell surface membrane) is a selective barrier encloses a cell and separates its contents from its surroundings. The membrane is an essential feature of all cells because it controls exchange between the cell and its environment. It can act as a barrier, but it can also control movement of materials across the membrane in both directions. The membrane is therefore described as **partially permeable** or **selectively permeable**. If it were freely permeable, life could not exist, because the chemicals of the cell would simply mix with the surrounding chemicals by diffusion and the inside of the cell would be the same as the outside.

The plasma membrane is a phospholipid bilayer and thick, with proteins embedded in it so we describe it as a **fluid mosaic**. Viewed in cross section with the electron microscope, such membranes appear as two dark lines separated by a lighter area. This distinctive appearance arises from the tail-to-tail packing of the phospholipid molecules that make up the membrane. The cell membrane is often called a fluid mosaic because many of the proteins and phospholipids are free to move laterally within the bilayer. Sterols maintain the membrane's fluidity as the temperature fluctuates.

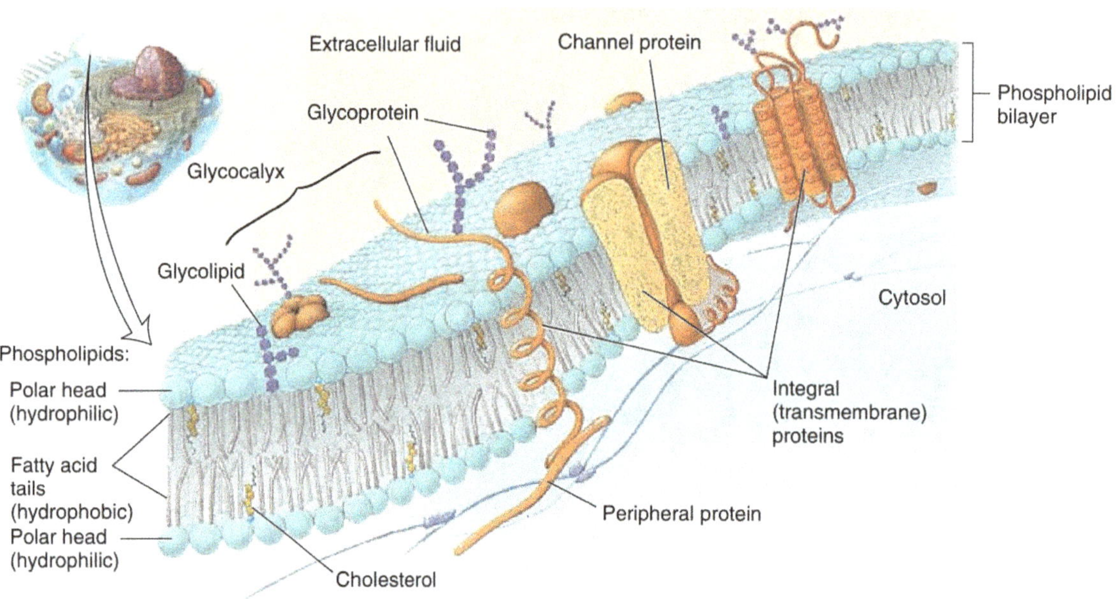

▲ **Figure 3.15** The plasma membrane, the outer boundary of a cell's cytoplasm.

Functions of Plasma Membrane

1. Physical barrier: Establishes a flexible boundary, protects cellular contents, and supports cell structure. Phospholipid bilayer separates substances inside and outside the cell.

2. Selective permeability: Regulates entry and exit of ions, nutrients, and waste molecules through the membrane.

3. Electrochemical gradients: Establishes and maintains an electrical charge difference across the plasma membrane.

4. Communication: Contains receptors that recognize and respond to molecular signals.

5. Connection: Link adjacent cells together by membrane junctions. Anchor cells to the extracellular matrix.

The proteins of the plasma membrane are generally responsible for a cell's ability to interact with the environment. Cells have multiple types of membrane proteins:

- **Transport proteins**: Transport proteins embedded in the phospholipid bilayer create passageways through which ions, glucose, and other polar substances pass into or out of the cell.

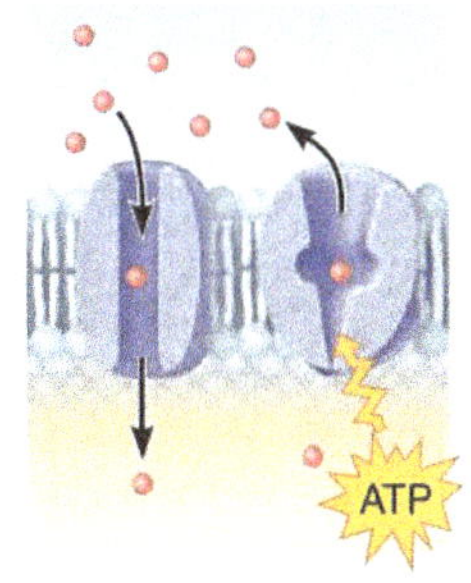

(a) Transport

- A protein (left) that spans the membrane may provide a hydrophilic channel across the membrane that is selective for a particular solute.
- Some transport proteins (right) hydrolyze ATP as an energy source to actively pump substances across the membrane.

- **Enzymes**: These proteins facilitate chemical reactions that otherwise would proceed too slowly to sustain life.

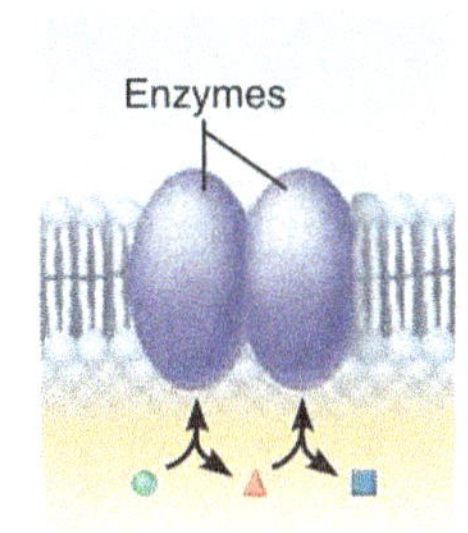

(d) Enzymatic activity

- A membrane protein may be an enzyme with its active site exposed to substances in the adjacent solution.
- A team of several enzymes in a membrane may catalyze sequential steps of a metabolic pathway as indicated (left to right) here.

- **Recognition proteins**: Carbohydrates attached to cell surface proteins serve as "name tags" that help the body recognize its own cells. The immune system attacks cells with unfamiliar surface molecules, which is why transplant recipients often reject donated organs. Surface proteins also distinctively mark specialized cells within an individual, so a bone cell's surface is different from that of a nerve cell or a muscle cell.

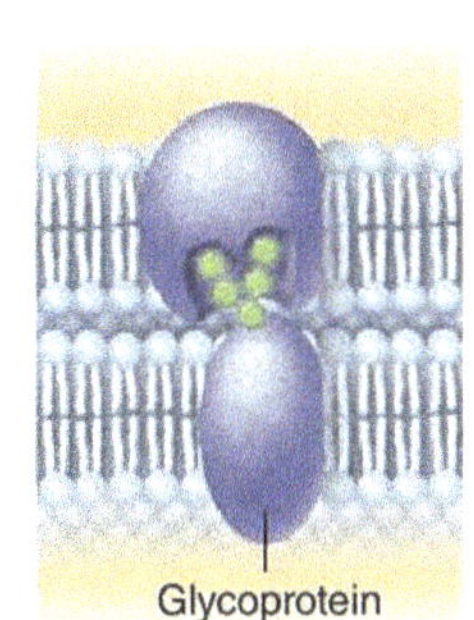

(f) Cell-cell recognition

- Some glycoproteins (proteins bonded to short chains of sugars which help to make up the glycocalyx) serve as identification tags that are specifically recognized by other cells.

- **Adhesion proteins**: These membrane proteins enable cells to stick to one another.

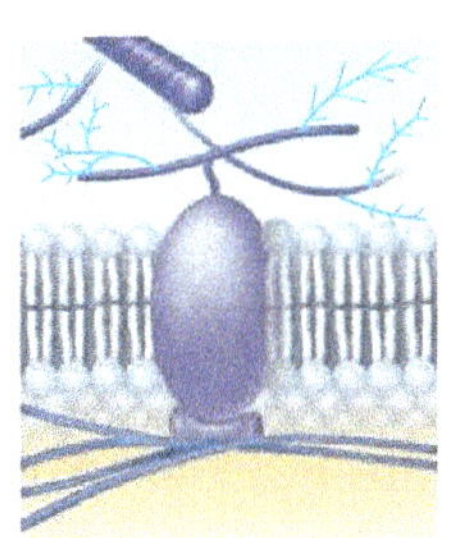

(c) Attachment to the cytoskeleton and extracellular matrix

- Elements of the cytoskeleton (cell's internal supports) and the extracellular matrix (fibers and other substances outside the cell) may anchor to membrane proteins, which helps maintain cell shape and fix the location of certain membrane proteins.
- Others play a role in cell movement or bind adjacent cells together.

- **Receptor proteins**: Receptor proteins bind to molecules outside the cell and trigger an internal response, a process called signal transduction. For example, when a hormone binds to a receptor, the resulting chain reaction produces the hormone's effects on the cell.

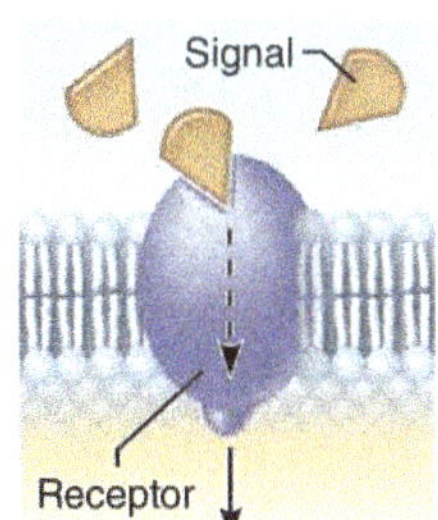

(b) Receptors for signal transduction

- A membrane protein exposed to the outside of the cell may have a binding site that fits the shape of a specific chemical messenger, such as a hormone.
- When bound, the chemical messenger may cause a change in shape in the protein that initiates a chain of chemical reactions in the cell.

- **The Cytoplasm**

Cytoplasm, the cellular material between the plasma membrane and the nucleus, is the site of most cellular activities. All the living material inside the cell is called **protoplasm**.

Therefore, cytoplasm + nucleus = protoplasm.

The electron microscope reveals that it consists of three major elements: the **cytosol**, **cytoskeleton**, **organelles**, and **inclusions**.

The cytosol is the aqueous (water), semitransparent and viscous or semifluid medium in which the other cytoplasmic elements are suspended. It is a complex mixture with properties of both a colloid and a true solution. Dissolved in the cytosol, which is largely water, are proteins, salts, sugars, and a variety of other solutes. The term **intracellular fluid (ICF)** refers to all the fluid inside a cell, in other words, cytosol plus the fluid inside all the organelles, including the nucleus. The chemical compositions of the fluids in cell organelles may differ from that of the cytosol. The cytosol is by far the largest intracellular fluid compartment.

Fluid outside body cells is called **extracellular fluid (ECF).** The ECF in the microscopic spaces between the cells of tissues is **interstitial fluid**. The ECF in blood vessels is called **blood plasma**, and that in lymphatic vessels is called **lymph**. The ECF within and around the brain and spinal cord is called **cerebrospinal fluid (CSF).**

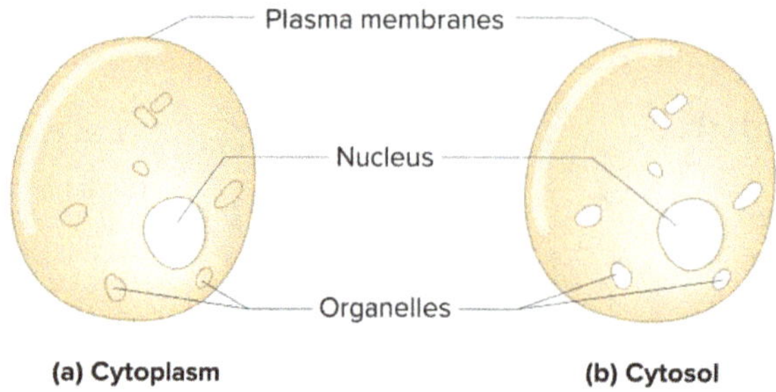

▲ **Figure 3.16 Comparison of cytoplasm and cytosol.**

(a) Cytoplasm (shaded area) is the region of the cell outside the nucleus.

(b) Cytosol (shaded area) is the fluid portion of the cytoplasm outside the cell organelles.

The cytoskeleton is a protein-based framework of filamentous structures that, among other things, helps maintain proper cell shape and plays key roles in cell division and chromosome segregation from cell generation to cell generation. The cytoskeleton was once thought to be specific to eukaryotes, but research has shown that all major eukaryotic cytoskeletal proteins have functional equivalents in prokaryotes.

Inclusions are chemical substances that may or may not be present, depending on cell type. Examples include stored nutrients, such as the glycogen granules in liver and muscle cells; lipid droplets in fat cells; and pigment (melanin) granules in certain skin and hair cells, keratin for waterproofing and carotenes, which are precursors to vitamin A. The organelles are the metabolic machinery of the cell. Originally the term organelle referred to only membranous structures but it usually is defined as a functionally and structurally distinct part of a cell. Eukaryotic cells have many types of organelles that allow for the **compartmentalization** of the cell. This keeps the various cellular activities separated from one another.

- **Nucleus, the information center**

The largest and most easily seen organelle within a eukaryotic cell is the **nucleus.** The primary functions of the nucleus are the protection, organization, replication, and expression of the genetic material. Another important function is the assembly of ribosomal subunits.

Nuclei are roughly spherical in shape, and in animal cells, they are typically located in the central region of the cell. The nuclear envelope encloses a jellylike fluid called **nucleoplasm** in which other nuclear elements are suspended. Like the cytosol, the nucleoplasm contains dissolved salts, nutrients, and other essential solutes. Inside the nucleus also are the **chromosomes** and a filamentous network of proteins called the **nuclear matrix**. The nuclear matrix consists of two parts: the nuclear lamina, which is composed of intermediate filaments that line the inner nuclear membrane, and an internal nuclear matrix, which is connected to the lamina and fills the interior of the nucleus. The nuclear matrix organizes the chromosomes within the nucleus.

In some cells, a network of fine cytoplasmic filaments seems to cradle the nucleus in this position. Most eukaryotic cells possess a single nucleus, although the cells of fungi and some other groups may have from several to many nuclei. Mammalian erythrocytes (red blood cells) lose their nuclei when they mature but some, including skeletal muscle cells, bone destruction cells, and some liver cells, are multinucleate that is, they have many nuclei. The presence of more than one nucleus usually signifies that a larger-than-usual cytoplasmic mass must be regulated. The nucleus contains most of the genes in the eukaryotic cell. (Some genes are located in mitochondria and chloroplasts).

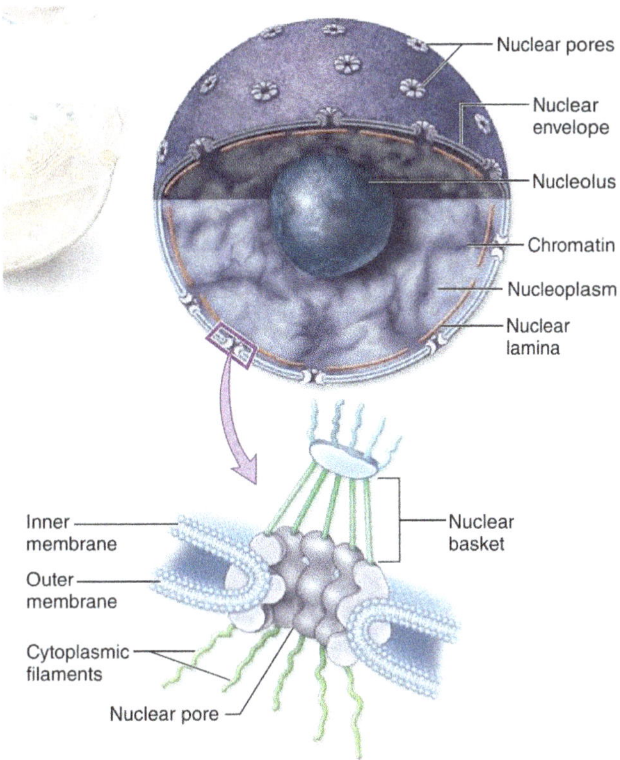

The **nuclear envelope** encloses the nucleus, separating its contents from the cytoplasm. The nuclear envelope is a double membrane. The two membranes, each a lipid bilayer with associated proteins, are separated by a space. The envelope is perforated by pore structures. At the lip of each pore, the inner and outer membranes of the nuclear envelope are continuous. An intricate protein structure called a **nuclear pore complex (NPC),** each contain approximately 34 nucleoporin proteins (probably the largest protein complex in the cell), lines each pore and plays an important role in the cell by regulating the entry and exit of proteins and RNAs, as well as large complexes of macromolecules.

Each protein from the cytoplasm that must be actively transported through the pore contains a nuclear localization signal (**NLS**) as part of its amino acid sequence. Special proteins called **importins** bind with the NLS sequence, forming a "cargo complex" that can then be captured by the nuclear pore machinery and transported through the pore into the nucleus. These sequences are not only found on soluble proteins synthesized by ribosomes, but also on proteins that are embedded in cytoplasmic membranes and must then be moved to the inner nuclear membrane.

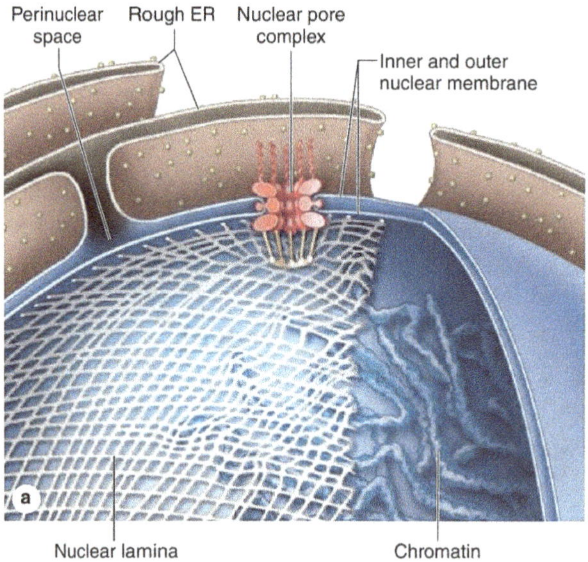

Except at the pores, the nuclear side of the envelope is lined by the **nuclear lamina**, a netlike array of protein filaments (in animal cells, called intermediate filaments) called nuclear lamins that maintains the shape of the nucleus by mechanically supporting the nuclear envelope and is also involved in the deconstruction and reconstruction of the nuclear envelope that accompanies cell division. Within the nucleus, the DNA is organized into discrete units called **chromosomes**, structures that carry the genetic information. Each chromosome contains one long DNA molecule associated with many proteins, including small basic proteins called histones. Some of the proteins help coil the DNA molecule of each chromosome, reducing its length and allowing it to fit into the nucleus.

Most of the space inside the nucleus is filled with deeply staining granular mass called **chromatin**. The terms chromatin and chromosome are similar but have distinct meanings. Chromatin refers to any collection of eukaryotic DNA molecules with their associated proteins. Chromosome refers to one complete DNA molecule with its associated proteins.

Chromatin is composed of approximately:

- 30% DNA, our genetic material

- 60% globular histone proteins, which package and regulate the DNA

- 10% RNA chains, newly formed or forming

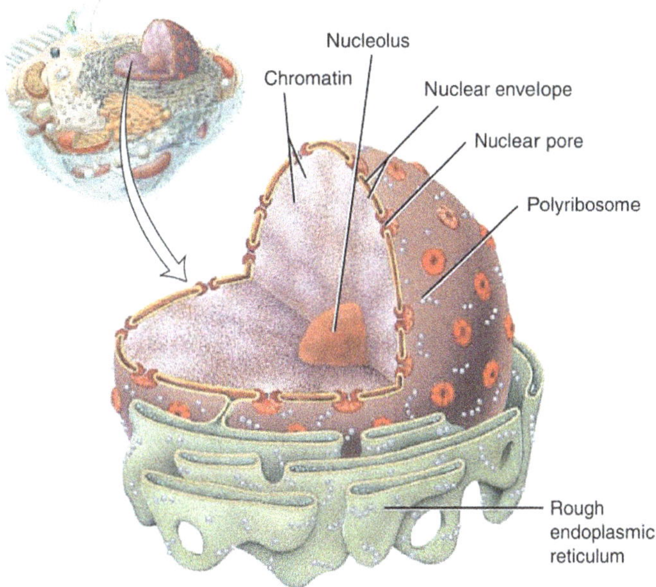

▲ **Figure 3.17** The nucleus is composed of a double membrane called the nuclear envelope, enclosing a fluid-filled interior containing chromatin.

Eukaryotic nuclei contain much more DNA than do prokaryotic nucleoids. For example, the entire complement of 46 chromosomes in the nucleus of a human cell has a total DNA length of about 2 meters (m), compared with about 1,500 mm in prokaryotic cells with the most DNA. Some eukaryotic cells contain even more DNA; for example, a single frog nucleus, although of microscopic diameter, is packed with about 10 m of DNA.

The genes for most of the proteins that the organism can make are found within the chromatin, as are the genes for specialized RNA molecules such as rRNA molecules. Expression of these genes is carefully controlled as required for the function of each cell. The other proteins in the cell are specified by genes in the DNA of mitochondria and chloroplasts.

When a cell is not dividing, stained chromatin appears as a diffuse mass in micrographs, and the chromosomes cannot be distinguished from one another, even though discrete chromosomes are present. As a cell prepares to divide, however, the chromosomes form loops and coil, condensing and becoming thick enough to be distinguished under a microscope as separate structures. Each eukaryotic species has a characteristic number of chromosomes. For example, a typical human cell has 46 chromosomes in its nucleus. A prominent structure within the nondividing nucleus is the **nucleolus** (plural, nucleoli), which appears through the electron microscope as a mass of densely stained granules. Nucleoli are associated with **nucleolar organizer regions**, which contain the DNA that issues genetic instructions for synthesizing ribosomal RNA (rRNA).

As rRNA molecules are synthesized, they are combined with proteins to form the two kinds of ribosomal subunits (The proteins are manufactured on ribosomes in the cytoplasm and "imported" into the nucleus). Most of these subunits leave the nucleus through the nuclear pores and enter the cytoplasm where a large and a small subunit can assemble into a ribosome. Sometimes there are two or more nucleoli; the number depends on the species and the stage in the cell's reproductive cycle, one to fjve being common in mammals. The nucleoli may also play a role in controlling cell division and the life span of a cell. The nucleus directs protein synthesis by synthesizing messenger RNA (mRNA) that carries information from the DNA. The mRNA is then transported to the cytoplasm via nuclear pores. Once an mRNA molecule reaches the cytoplasm, ribosomes translate the mRNA's genetic message into the primary structure of a specific polypeptide.

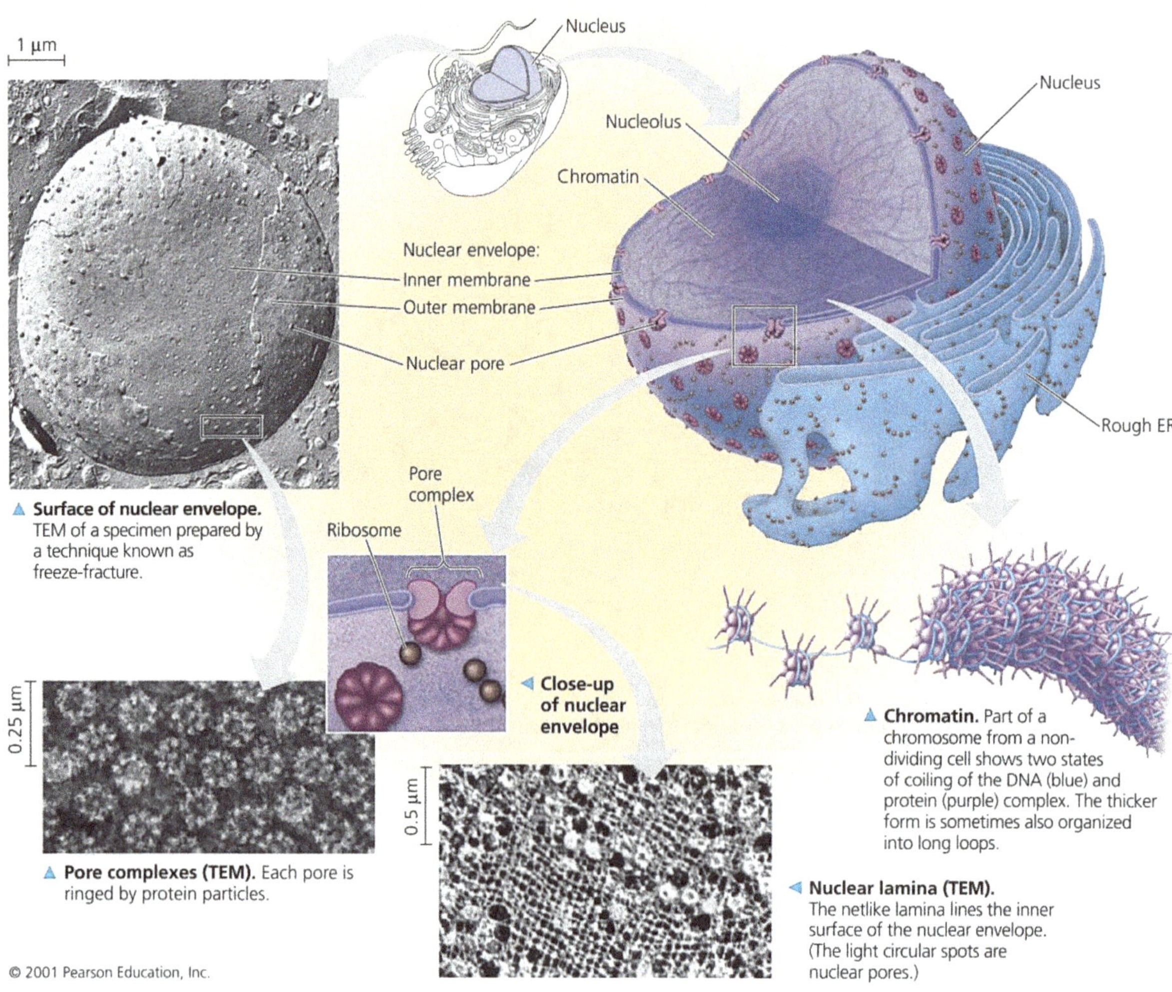

▲ **Figure 3.18** The nucleus and its envelope. Within the nucleus are the chromosomes, which appear as a mass of chromatin (DNA and associated proteins), and one or more nucleoli (singular, nucleolus), which function in ribosome synthesis. The nuclear envelope, which consists of two membranes separated by a narrow space, is perforated with pores and lined by the nuclear lamina.

- **Ribosomes: The cell's protein synthesis machinery**

Ribosomes, which are complexes made of ribosomal RNAs and proteins, are the cellular components that carry out protein synthesis (Note that ribosomes are not membrane bounded and thus are not considered organelles.) Ribosomes are molecular machines made up of both proteins and RNA molecules. A typical cell may contain as many as 10 million ribosomes. The sizes of structures this small are often quoted in S units (Svedberg units). S units are a measure of how rapidly substances sediment in a high speed centrifuge (an ultracentrifuge). The faster they sediment, the higher the S number. Eukaryotic ribosomes are 80S ribosomes. The ribosomes of prokaryotes are 70S ribosomes, so are slightly smaller. Mitochondria and chloroplasts contain 70S ribosomes, revealing their prokaryotic origins.

The two subunits of each eukaryotic ribosome actually consist of more than 80 different proteins and four types of rRNA molecules. The core of the small ribosomal subunit is a highly folded ribosomal RNA (rRNA) chain associated with more than 30 unique proteins; the core of the large subunit has three other rRNA molecules and nearly 50 other basic proteins. The rRNA molecules in the ribosomal subunits not only provide structural support but also position transfer RNAs (tRNA) molecules bearing amino acids in the correct "reading frame" and catalyze the formation of the peptide bonds. The more peripheral proteins of the ribosome seem to function primarily to stabilize the catalytic RNA core.

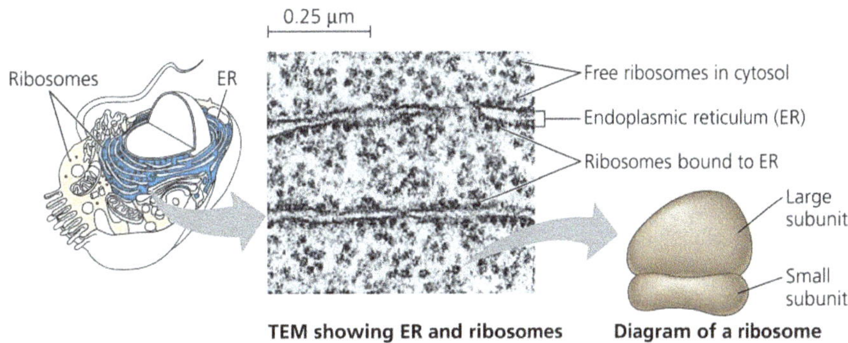

▲ **Figure 3.19 Ribosomes**. This electron micrograph of part of a pancreas cell shows many ribosomes, both free (in the cytosol) and bound (to the endoplasmic reticulum). The simplified diagram of a ribosome shows its two subunits.

The mRNAs that encode the ribosomal proteins are first copied from their respective genes in the nucleus. They must then be transported through the nuclear pores into the cytoplasm, where they are translated by ribosomes to synthesize their corresponding protein. Then, these newly formed ribosomal proteins must be imported back into the nucleus. Many of these proteins include an NLS (nuclear localization signal) sequence that is recognized by the nuclear import machinery. It is then packaged as cargo bound to carrier proteins that allow it to be translocated through the pore. In the nucleus, these ribosomal proteins are assembled with rRNAs into ribosomal subunits. Once assembled, the newly formed ribosome subunit is then transported from the nucleolar region to the inside of the nuclear envelope, where it exits the nucleus through a nuclear pore.

Cells with high rates of protein synthesis have particularly large numbers of ribosomes as well as prominent nucleoli, which makes sense, given the role of nucleoli in ribosome assembly. For example, a human pancreas cell, which makes many digestive enzymes, has a few million ribosomes.

Ribosomes build proteins in two cytoplasmic regions: At any given time, **free ribosomes** are suspended in the cytosol, while **bound ribosomes** are attached to the outside of the endoplasmic reticulum or nuclear envelope. Ribosomes can switch back and forth between these two functions, attaching to and detaching from the membranes of the endoplasmic reticulum, according to the type of protein they are making at a given time. Bound and free ribosomes are structurally identical, and ribosomes can play either role at different times. Free ribosomes synthesize cytosolic and cytoskeletal proteins and proteins for import into the nucleus, mitochondria, and peroxisomes.; examples are enzymes that catalyze the first steps of sugar breakdown.

Proteins that are to be incorporated into membranes, stored in lysosomes, or eventually secreted from the cell are made on ribosomes attached to the membranes of ER. The proteins produced by these ribosomes are segregated during translation into the interior of the ER's membrane cisternae. Cells that specialize in protein secretion—for instance, the cells of the pancreas that secrete digestive enzymes—frequently have a high proportion of bound ribosomes. Ribosomes can be thought of as "universal organelles" because they are found in all cell types from all three domains of life.

▶ **Figure 3.20 Ribosomes.**

(a) Free ribosomes produce proteins used in the cell's cytoplasm.

(b) Proteins produced by ribosomes attached to the rough ER's membrane are typically used in specialized organelles or in the cell membrane; they may also be secreted outside the cell.

During protein synthesis many ribosomes typically bind the same strand of mRNA to form larger complexes, called **Polyribosomes** or **polysomes.** Proper folding of new proteins is guided by protein **chaperones.** Denatured proteins or those that cannot be refolded properly are conjugated to the protein ubiquitin that targets them for breakdown by **proteasomes,** protein complexes that direct the destruction of defective proteins.

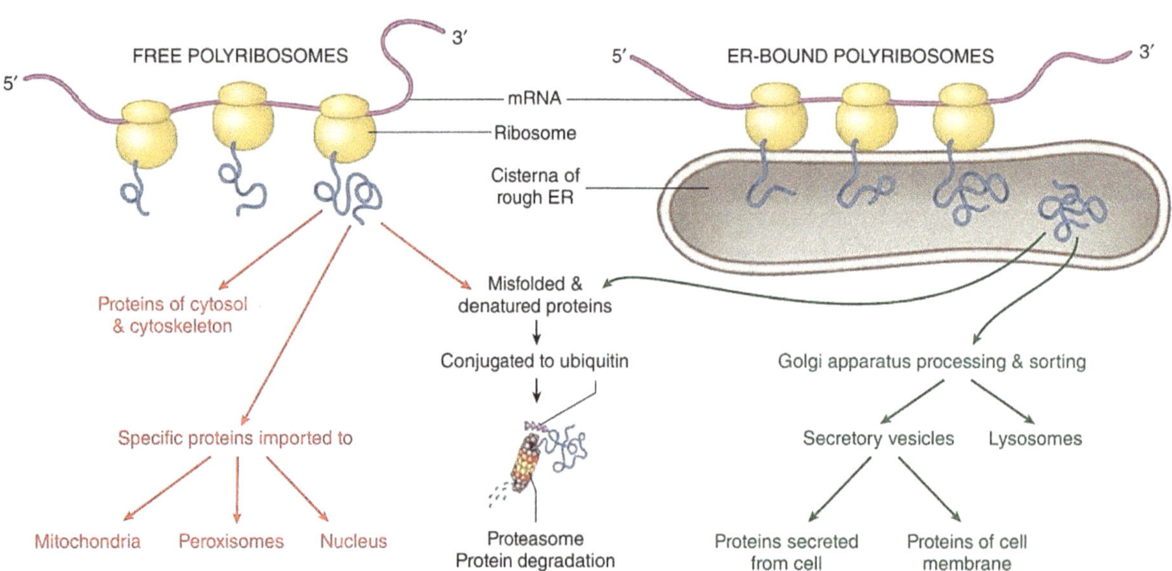

Many of the different membrane-bounded organelles of the eukaryotic cell are part of the **endomembrane system** that is a network of organelles (including the plasma membrane) that exchange materials through **transport vesicles** which are small membrane-bound compartments that transfer substances between parts of the system. Researchers refer to the substances as cargo. Endomembrane system includes the nuclear envelope, the endoplasmic reticulum, the Golgi apparatus, lysosomes, various kinds of vesicles and vacuoles, and the plasma membrane—that is, all of the membranous elements that are either structurally connected or arise via forming or fusing transport vesicles

In living cells, there is a constant flow of transport vesicle "traffic" between components of the endomembrane system. These vesicles, which are formed from "buds" on the surface membrane of the "donor" organelle, contain "cargo" proteins derived from the organelle's internal compartment or lumen. Each vesicle has proteins embedded in its membrane that are specific routing signals for its destination organelle.

The vesicle is then transported on cytoskeletal "tracks" by molecular motors (which we will discuss later in this chapter) to the surface of its "target" organelle membrane. On contact, the transport vesicle fuses with the target membrane, releasing its contents into the compartment of the acceptor organelle. This fusion of the two membranes results from interactions between targeting proteins in the vesicle and receptor proteins on the acceptor membrane. As the two membranes fuse, the surface area of the acceptor membrane is expanded by the integration of the vesicle membrane.

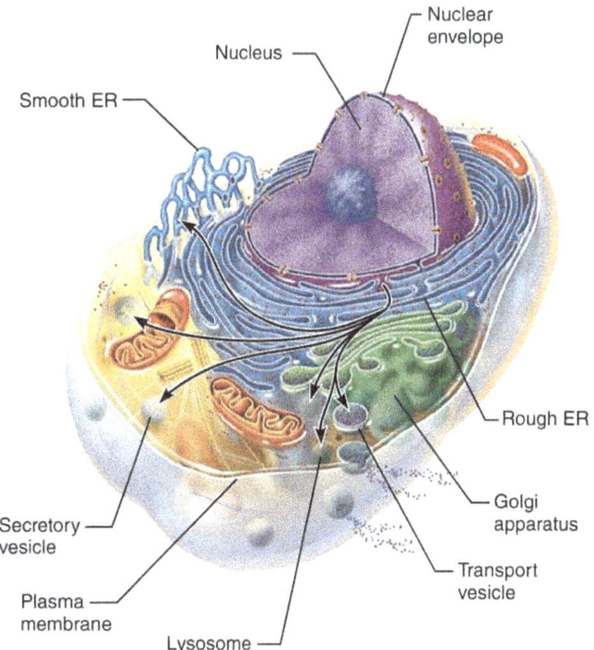

▶ **Figure 3.21** The endomembrane systeme

Membranes serve as important work surfaces. For example, many chemical reactions in cells are carried out by enzymes that are bound to membranes. Because the enzymes that carry out successive steps of a series of reactions are organized close together on a membrane surface, certain series of chemical reactions occur more rapidly.

Membranes allow cells to store energy. There is both an electric charge difference and a concentration difference of ions on the two sides of certain cell membranes. These differences constitute an electrochemical gradient. Such gradients store energy and so have potential energy. As particles of a substance move across the membrane from the side of higher concentration to the side of lower concentration, the cell can convert some of this potential energy to the chemical energy of ATP molecules. This process of energy conversion is a basic mechanism that cells use to capture and convert the energy necessary to sustain life.

- **The Endoplasmic Reticulum: Biosynthetic Factory**

The largest of the internal membranes is called the endoplasmic reticulum (ER). The endoplasmic reticulum (ER) is such an extensive network of membranes that it accounts for more than half the total membrane in many eukaryotic cells. The ER consists of a network of membranous tubules and sacs called **cisternae**. The ER membrane separates the internal compartment of the ER, called the **ER lumen** (cavity) or cisternal space, from the cytosol. And because the ER membrane is continuous with the nuclear envelope, the space between the two membranes of the envelope is continuous with the lumen of the ER.

There are two distinct, though connected, regions of the ER that differ in structure and function: **smooth or agranular ER** and **rough or granular ER**. Smooth ER is so named because its outer surface lacks ribosomes. Rough ER is studded with ribosomes on the outer surface of the membrane and thus appears rough through the electron microscope. As already mentioned, ribosomes are also attached to the cytoplasmic side of the nuclear envelope's outer membrane, which is continuous with rough ER. The nuclear envelope is directly connected to the rough and smooth ER.

The ratio of SER to RER is not fixed but depends on a cell's function. In multicellular animals such as ourselves, great variation exists in this ratio. Cells that carry out extensive lipid synthesis, such as those in the testes, intestine, and brain, have abundant SER. Cells that synthesize proteins that are secreted, such as antibodies, have much more extensive RER.

► **Figure 3.22 Endoplasmic reticulum (ER).**

A membranous system of interconnected tubules and flattened sacs called cisternae, the ER is also continuous with the nuclear envelope. The membrane of the ER encloses a continuous compartment called the ER lumen (or cisternal space).

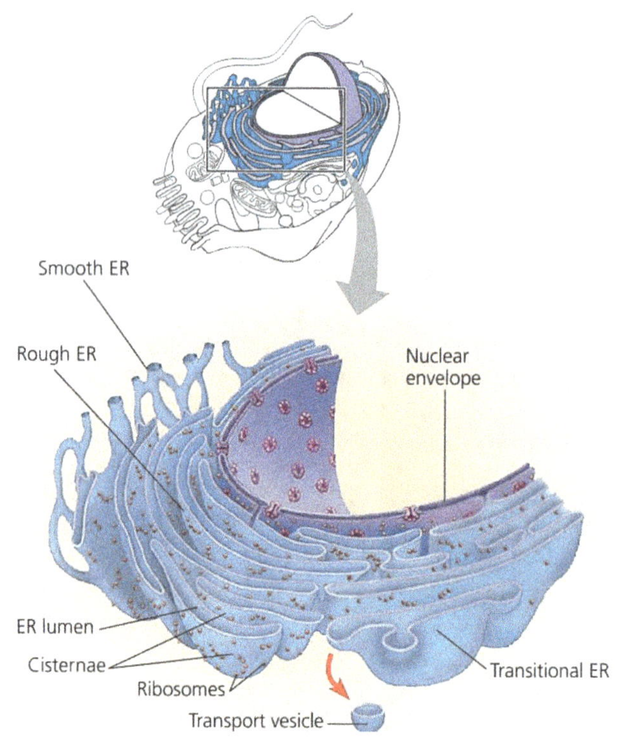

Functions of Endoplasmic Reticulum

1. Synthesis: Provides a place for chemical reactions

a. Smooth ER is the site of lipid synthesis and carbohydrate metabolism

b. Rough ER synthesizes proteins for secretion, incorporation into the plasma, membrane, and as enzymes within lysosomes

2. Transport: Moves molecules through cisternal space from one part of the cell to another, sequestered away from the cytoplasm

3. Storage: Stores newly synthesized molecules

4. Detoxification: Smooth ER detoxifies both drugs and alcohol

Functions of Smooth ER (SER)

The smooth ER is continuous with the rough ER and consists of tubules arranged in a looping network but it does not have attached ribosomes. The smooth ER functions in diverse metabolic processes, which vary with cell type.

- Synthesis of lipids, cholesterol and phospholipids, and synthesize the lipid components of lipoproteins (in liver cells). Synthesize steroid-based hormones such as sex hormones (The cells that synthesize and secrete these hormones—in the testes, ovaries and adrenal glands, for example—are rich in smooth ER, a structural feature that fits the function of these cells).

- Absorb, synthesize, and transport fats (in intestinal cells)

- Detoxify drugs, poisons, and cancer-causing chemicals (in liver and kidneys). Detoxification usually involves adding hydroxyl groups to drug molecules, making them more soluble and easier to flush from the body.

- Metabolism of carbohydrates (break down stored glycogen to form free glucose in liver cells especially)

- Removes the phosphate group from glucose-6-phosphate

- Additionally, skeletal and cardiac muscle cells have an elaborate smooth ER (called the **sarcoplasmic reticulum**) that plays an important role in storing and releasing calcium ions during muscle contraction. The smooth ER membrane pumps calcium ions from the cytosol into the ER lumen. When a muscle cell is stimulated by a nerve impulse, calcium ions rush back across the ER membrane into the cytosol and trigger contraction of the muscle cell. In other cells, Ca^{2+} release from SER stores is involved in diverse signaling pathways.

Except for the examples given above, most body cells contain relatively little, if any, smooth ER.

B. Smooth ER

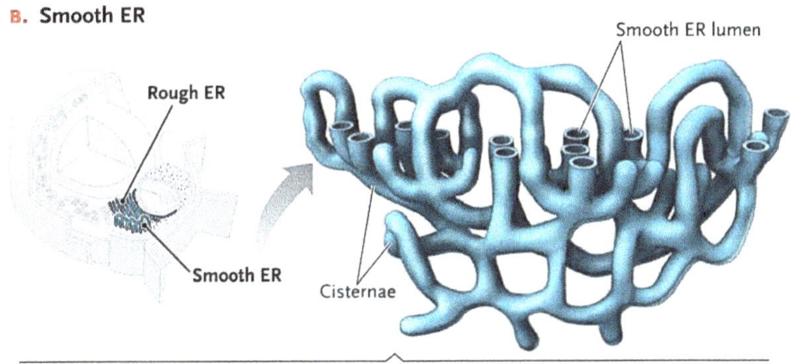

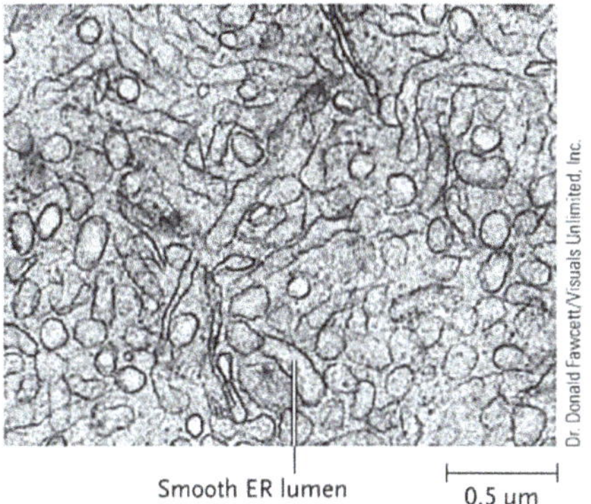

Among the functions of the smooth ER are the synthesis of lipids for cell membranes, and enzymatic conversion of certain toxic molecules to safer molecules.

Functions of Rough ER (RER)

The proteins synthesized on the surface of the RER are destined to be exported from the cell, sent to lysosomes or vacuoles or embedded in the plasma membrane. The nuclear envelope is one functional domain of the rough ER. The outer and inner nuclear membranes are actually continuous with the rough ER. Ribosomes are typically present on the outer (cytosolic) surface of the nuclear envelope, but are not found on the inner (nucleoplasmic) surface.

Protein synthesis begins on polyribosomes in the cytosol. The 5'ends of mRNAs for proteins destined to be segregated in the ER encode an N-terminal **signal sequence** of 15-40 amino acids. The newly translated signal sequence is bound by a protein complex called the **signal recognition particle** (SRP). The SRP–ribosome–nascent peptide complex binds to SRP receptors on the ER membrane. Another receptor in the ER membrane binds a structural protein of the large ribosomal subunit, more firmly attaching the ribosome to the ER. SRP then releases the hydrophobic signal peptide, allowing translation to continue with the nascent polypeptide chain transferred to a **translocator complex** (also called a **translocon**) through the ER membrane.

Inside the lumen of the RER, the signal sequence is removed by an enzyme, **signal peptidase**. With the ribosome docked at the ER surface, translation continues with the growing polypeptide pushing itself while chaperones and other proteins serve to "pull" the nascent polypeptide through the translocator complex. Upon release from the ribosome, post translational modifications and proper folding of the polypeptide continue. Most secretory proteins are **glycoproteins**, proteins that have carbohydrates covalently bonded to them. The carbohydrates are attached to the proteins in the ER by enzymes built into the ER membrane.

Other enzymes, called molecular chaperones, in the ER lumen catalyze the efficient folding of proteins into proper conformations. Proteins that are not processed correctly—for example, proteins that are misfolded—are transported back to the cytosol. There they are degraded by proteasomes, protein complexes that direct the destruction of defective proteins.

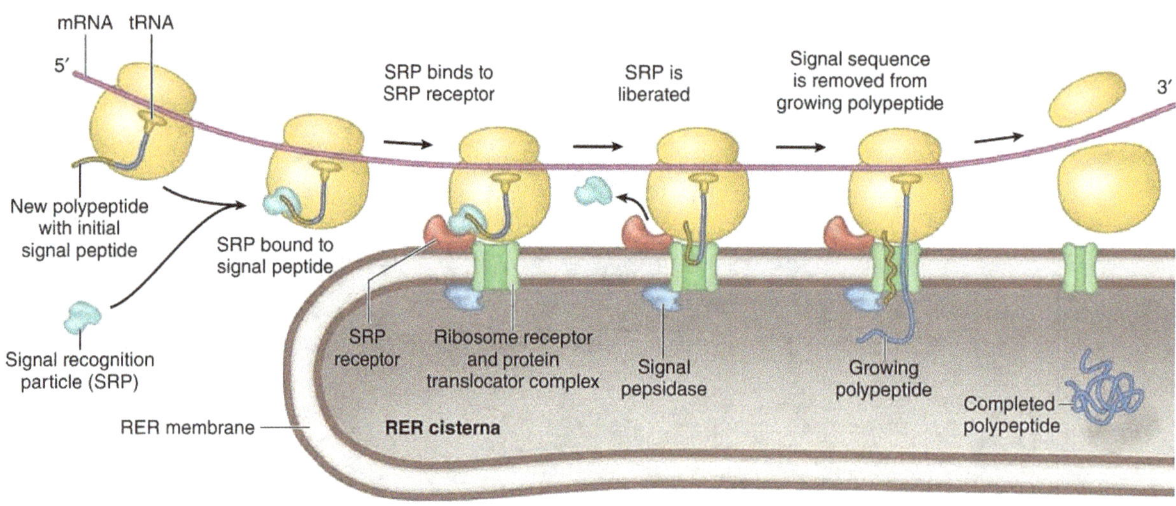

▲ **Figure 3.23** Movement of polypeptides into the RER.

After secretory proteins are formed, the ER membrane keeps them separate from proteins that are produced by free ribosomes and that will remain in the cytosol. Secretory proteins depart from the ER wrapped in the membranes of vesicles that bud like bubbles from a specialized region called transitional ER. Vesicles in transit from one part of the cell to another are called **transport vesicles**.

In addition to making secretory proteins, rough ER is a membrane factory for the cell; it grows in place by adding membrane proteins and phospholipids to its own membrane. As polypeptides destined to be membrane proteins grow from the ribosomes, they are inserted into the ER membrane itself and anchored there by their hydrophobic portions. Like the smooth ER, the rough ER also makes membrane phospholipids; enzymes built into the ER membrane assemble phospholipids from precursors in the cytosol. The ER membrane expands, and portions of it are transferred in the form of transport vesicles to other components of the endomembrane system.

Most cell membranes are first assembled in the endoplasmic reticulum. The core of biological membranes consists of a phospholipid bilayer. These phospholipids are synthesized on the cytosolic surface of the smooth ER and then integrated into the membrane bilayer, causing the surfaces of the membrane to expand. Proteins that are to be embedded into newly formed membrane are synthesized by ribosomes attached to the cytosolic surface of the rough ER membrane. By contrast, membranes in other organelles within the endomembrane system grow when transport vesicles from the ER fuse with them.

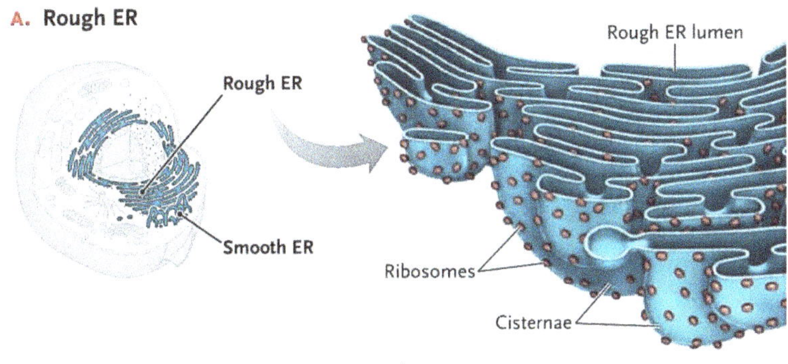

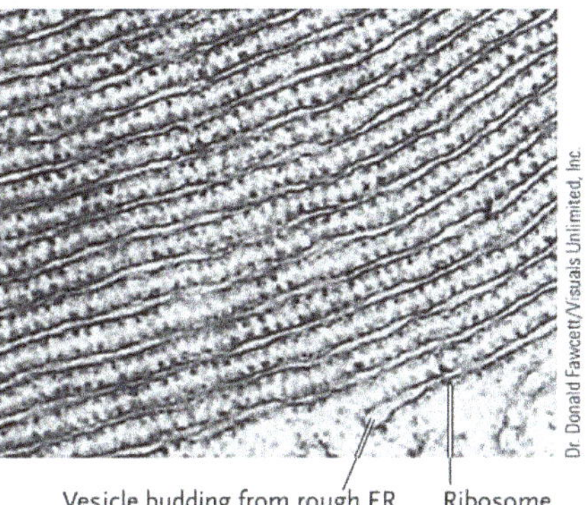

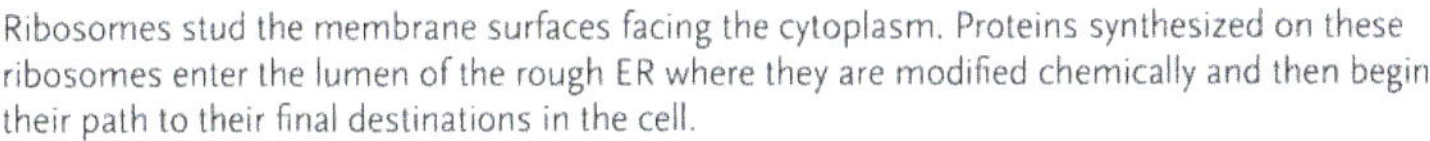
Ribosomes stud the membrane surfaces facing the cytoplasm. Proteins synthesized on these ribosomes enter the lumen of the rough ER where they are modified chemically and then begin their path to their final destinations in the cell.

- **The Golgi Apparatus: Shipping and Receiving Center**

The use of special stains containing silver resulted in the Golgi apparatus being discovered in 1898 by Camillo Golgi. After leaving the ER, many transport vesicles travel to the Golgi apparatus or Golgi body or Golgi complex. We can think of the Golgi as a warehouse for receiving, sorting, shipping, and even some manufacturing. Here, products of the ER, such as proteins, are modified and stored and then sent to other destinations.

Not surprisingly, the Golgi apparatus is especially extensive in cells specialized for secretion. Three steps in this process :

1. Transport vesicles that bud off from the rough ER move to and fuse with the membranes at the convex cis face, the "receiving" side, of the Golgi apparatus.

2. Enzymes in the Golgi apparatus process, or modify, certain proteins and lipids. As mentioned earlier, carbohydrates can be attached to proteins and lipids in the endoplasmic reticulum. **Glycosylation** continues in the Golgi. For this to occur, a protein or lipid is transported via vesicles from the ER to the cis Golgi. Most glycosylation occurs in the medial Golgi. A second type of processing event is **proteolysis**, whereby enzymes called proteases cut proteins into smaller polypeptides. For example, the hormone insulin is first made as a large precursor protein termed proinsulin. In the Golgi apparatus, proinsulin is packaged with proteases into vesicles. The proteases cut out a portion of the proinsulin to create a smaller insulin molecule that is a functional hormone. In some cases, phosphate groups are added to molecules.

3. Various proteins are "tagged" for delivery to a specific address, sorted, and packaged in at least three types of vesicles that bud from the concave trans face (the "shipping" side) of the Golgi stack: **Secretory vesicles**, or granules, containing proteins destined for export migrate to the plasma membrane and discharge their contents from the cell by **exocytosis**. The pathway from the ER to the Golgi to the plasma membrane is termed the **secretory pathway**. Vesicles containing lipids and transmembrane proteins are destined for the plasma membrane or for other membranous organelles. Vesicles containing digestive enzymes are packaged into membranous lysosomes that remain in the cell.

The Golgi apparatus consists of a group of associated, flattened membranous sacs. The individual stacks of membrane are called cisternae or saccules, and they vary in number within the Golgi body from 1 or a few in protists, to 20 or more in animal cells and to several hundred in plant cells. In vertebrates individual Golgi are linked to form a Golgi ribbon. They are especially abundant in glandular cells, which manufacture and secrete substances.

Vesicles concentrated in the vicinity of the Golgi apparatus are engaged in the transfer of material between parts of the Golgi and other structures. Each Golgi stack has three areas, referred to as the **cis face**, the **trans face**, and a **medial region** in between. Typically, the cis face (the entry or concave face) is located nearest the nucleus and receives materials from transport vesicles bringing molecules from the ER. The trans face (the exit or convex face) is closest to the plasma membrane. It packages molecules in vesicles and transports them out of the Golgi.

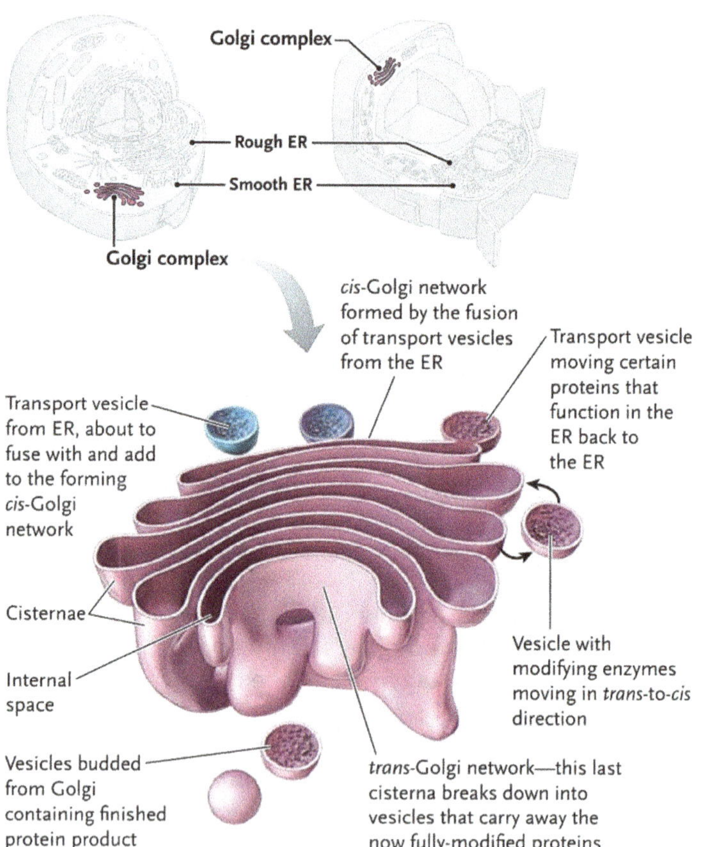

Transport vesicles that bud off from the ER and move to the Golgi. Transport vesicles fuse with one another to form the cis-Golgi network, the cisterna that is closest to the ER. Certain proteins that function in the ER are moved to the ER in transport vesicles that bud off from the cis-Golgi network. The remaining proteins are transported across the stack of cisternae to the trans-Golgi network, the last cisterna, which faces the plasma membrane. During that transport, the proteins are modified to their final functional forms by enzymes within the cisternae. Protein modification includes the addition of carbohydrates and removing segments of the polypeptide chain.

The favored model for protein transport in the Golgi complex is **cisternal maturation**. In this model, the Golgi cisternae form anew from fusion of transport vesicles from the ER, and then they mature progressively until they dissipate in the form of vesicles formed from the trans-Golgi network. In other words, a new cisterna that forms gradually becomes the next cisterna in the stack as yet another cisterna is formed and the last (trans) cisterna breaks down into vesicles. The contents of the vesicles are kept separate from the cytosol by the vesicle membrane.

But as **vesicular transport model**, materials are transported between the Golgi cisternae via membrane vesicles that bud from one compartment in the Golgi (for example, the cis Golgi) and fuse with another compartment (for example, the medial Golgi). Vesicles also bud from cisternae that are internally located within the stack and move to and fuse with cisternae closer to the cis side. Their cargo consists of modifying enzymes of the Golgi complex. In this way, the modifying enzymes are moved constantly to their locations within the maturing Golgi complex where they are needed to modify the ER-derived proteins that are passing from the cis to the trans side.

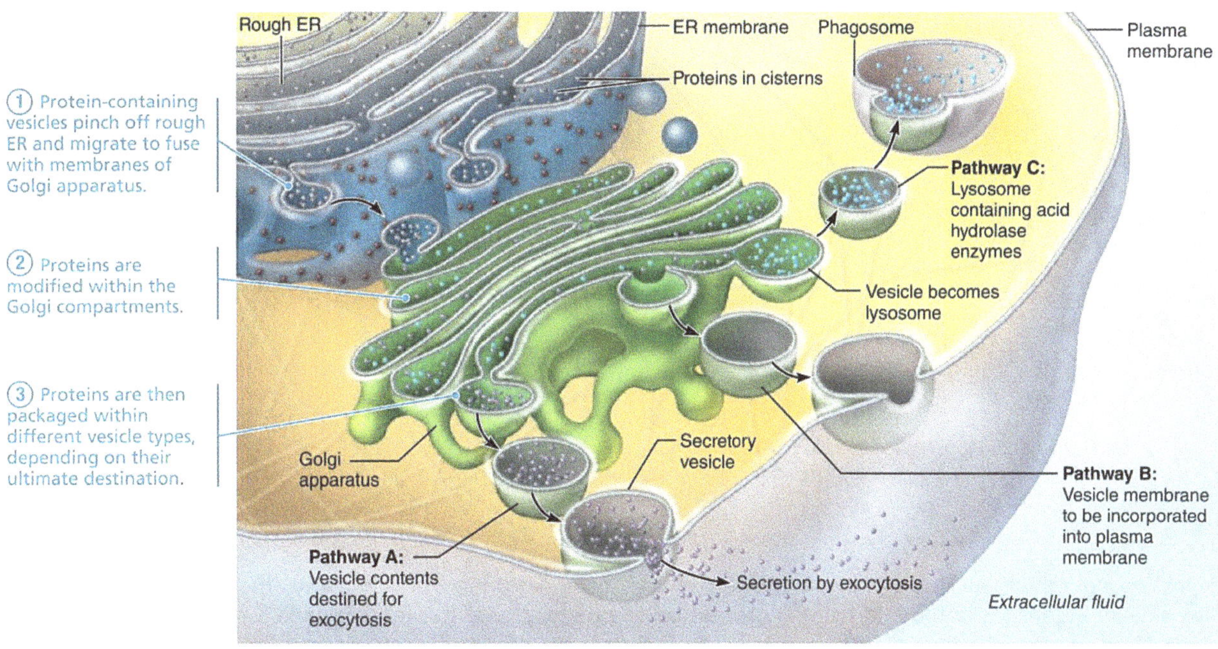

▲ **Figure 3.25** The sequence of events from protein synthesis on the rough ER to the final distribution of those proteins.

A process called **endocytosis**, brings molecules into the cell from the exterior. In this process, the plasma membrane forms a pocket, which bulges inward and pinches off into the cytoplasm as an endocytic vesicle. Once in the cytoplasm, **endocytic vesicles**, which contain segments of the plasma membrane as well as proteins and other molecules, are carried to the Golgi complex or to other destinations such as lysosomes in animal cells. The substances carried to the Golgi complex are sorted and placed into vesicles for routing to other locations, which may include lysosomes. Those routed to lysosomes are digested into molecular subunits that may be recycled as building blocks for the biological molecules of the cell.

Lysosomes are found in animals and protists. A lysosome is a spherical membranous sac of **hydrolytic enzymes** that many eukaryotic cells use to digest (hydrolyze) macromolecules. Lysosomal enzymes work best in the acidic environment found in lysosomes. If a lysosome breaks open or leaks its contents, the released enzymes are not very active because the cytosol has a near-neutral pH (~7.2). The lysosomal membrane is adapted to serve lysosomal functions in two important ways. First, it contains H+ (proton) "pumps," which are ATPases that gather hydrogen ions from the surrounding cytosol to maintain the organelle's acidic pH. Second, it retains the dangerous lysosomal enzymes (acid hydrolases) while permitting the final products of digestion to escape so that they can be used by the cell or excreted.

In this way, lysosomes provide sites where digestion can proceed safely within a cell. However, excessive leakage from a large number of lysosomes can destroy a cell by self-digestion. They contain high levels of degrading enzymes, which catalyze the rapid breakdown of proteins, nucleic acids, lipids, and carbohydrates. The lysosomal membrane is ordinarily quite stable, but it becomes fragile when the cell is injured or deprived of oxygen and when excessive amounts of vitamin A are present. Hydrolytic enzymes and lysosomal membrane are made by rough ER and then transferred to the Golgi apparatus for further processing. At least some lysosomes probably arise by budding from the trans face of the Golgi apparatus.

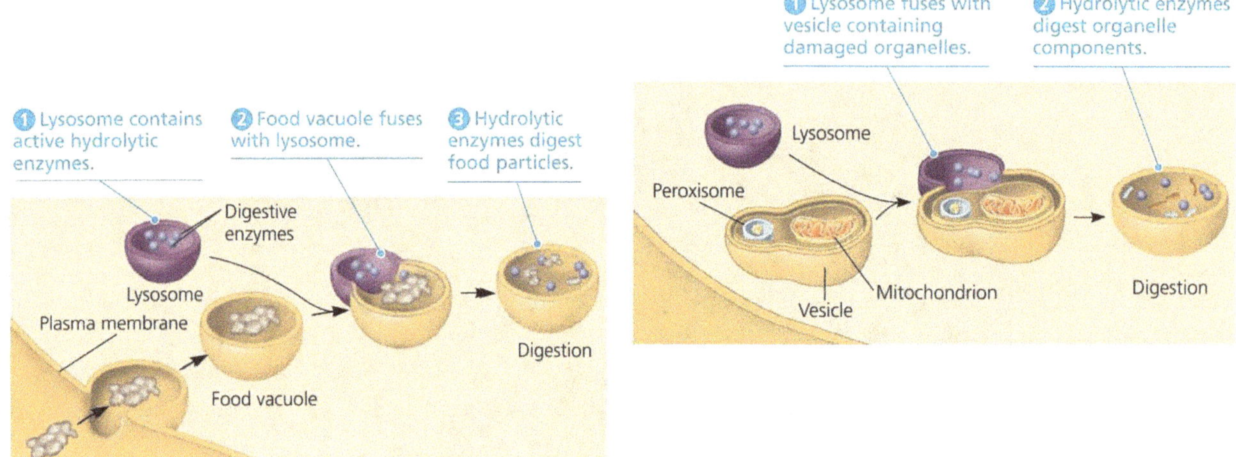

▲ **Figure 3.26 Lysosomes: Phagocytosis.** In phagocytosis, lysosomes digest (hydrolyze) materials taken into the cell. **Autophagy.** In autophagy, lysosomes recycle intracellular materials.

Lysosomes carry out intracellular digestion in a variety of circumstances. Amoebas and many other unicellular protists eat by engulfing smaller organisms or food particles, a process called **phagocytosis**. The food vacuole formed in this way then fuses with a lysosome, whose enzymes digest the food. The fusion event activates proton pumps in the lysosomal membrane, resulting in a lower internal pH. Some human cells also carry out phagocytosis. Among them are macrophages, a type of white blood cell that helps defend the body by engulfing and destroying bacteria and other invaders.

Primary lysosomes are formed by budding from the Golgi complex. Their hydrolytic enzymes are synthesized in the rough ER. As these enzymes pass through the lumen of the ER, sugars attach to each molecule, identifying it as bound for a lysosome. This signal permits the Golgi complex to sort the enzyme to the lysosomes rather than to export it from the cell.

Lysosomes act by fusing their membranes with vesicles that contain material to be digested. For example, bacteria (or cellular debris) that are engulfed by scavenger cells are captured from the exterior by **endocytotic vesicles** that form from the plasma membrane. One or more primary lysosomes fuse with a vesicle containing the ingested material, forming a larger vesicle called a **secondary lysosome**. Powerful enzymes in the secondary lysosome come in contact with the ingested molecules and degrade them into their components.

Lysosomes function as a cell's "demolition crew" by:

- Digesting particles taken in by endocytosis, particularly ingested bacteria, viruses, and toxins

- Degrading stressed or dead cells and worn-out or nonfunctional organelles, a process more specifically called autophagy ("self-eating")

- Performing metabolic functions, such as glycogen breakdown and release

- Breaking down bone to release calcium ions into the blood

Digestion products, including simple sugars, amino acids, and other monomers, pass into the cytosol and become nutrients for the cell. Lysosomes also use their hydrolytic enzymes to recycle the cell's own organic material, the process by which wornout organelles are digested is called **autophagy.**

In this way, a human liver cell, for example, recycles about half its contents every week. Mitochondria are replaced in some tissues every 10 days. During autophagy, a damaged organelle or small amount of cytosol becomes surrounded by a double membrane (of unknown origin), and a lysosome fuses with the outer membrane of this vesicle.

The lysosomal membrane contains carrier proteins that allow the final products of digestion, such as monosaccharides, fatty acids, and amino acids, to be transported into the cytosol, but indigestible molecules remain in a membrane-enclosed **residual body**, which may accumulate in long-lived cells as lipofuscin.

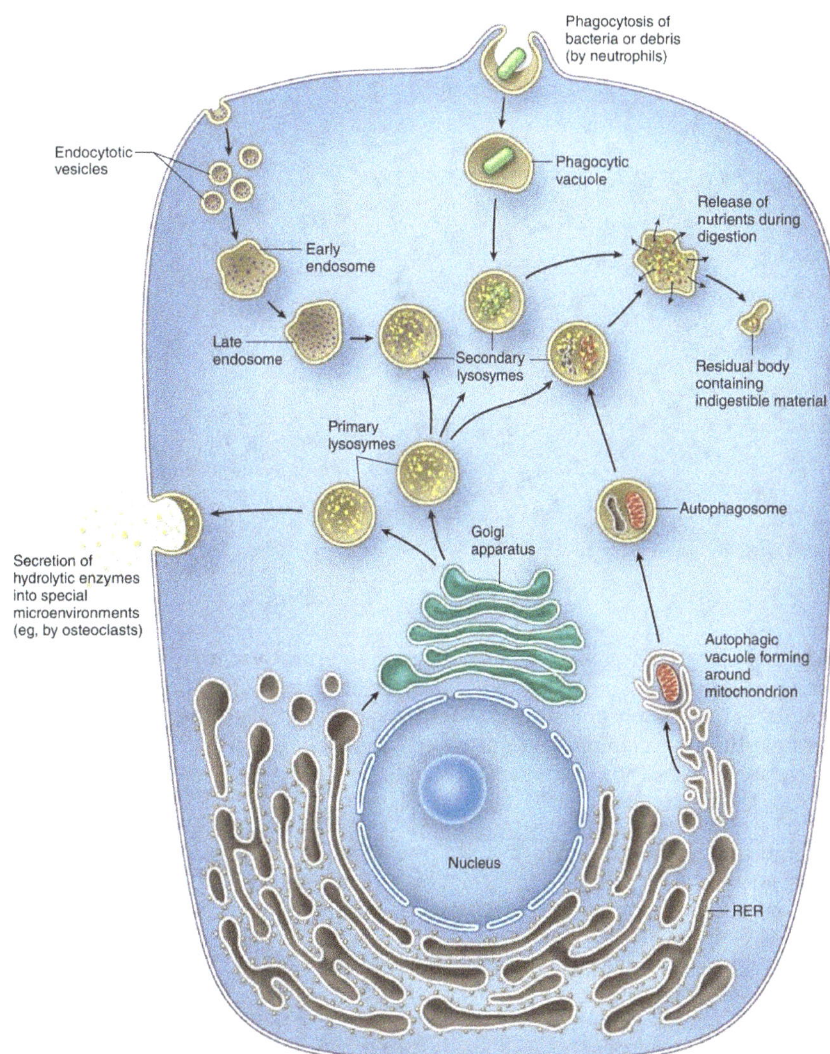

The contents of lysosomes are sometimes released into the cytoplasm. This results in the whole cell being digested (a process called **autolysis**). This may be part of normal development, as when a tadpole tail is reabsorbed during metamorphosis or when a uterus is restored to its normal size after pregnancy. It also occurs after the death of an individual as membranes lose their partial permeability. With the help of lysosomes, the cell continually renews itself. A human liver cell, for example, recycles half of its macromolecules each week. The cells of people with inherited lysosomal storage diseases lack a functioning hydrolytic enzyme normally present in lysosomes.

In Tay-Sachs disease, for example, a lipid-digesting enzyme is missing or inactive, and the brain becomes impaired by an accumulation of lipids in the cells. Although lysosomes degrade proteins delivered to them in vesicles, proteins in the cytosol also require disposal at certain times in the life of a cell. Continuous destruction of unneeded, damaged, or faulty proteins is the function of tiny barrel-shaped structures called **proteasomes.** A typical body cell contains many thousands of proteasomes, in both the cytosol and the nucleus.

- **Vacuoles: Diverse Maintenance Compartments**

Most plant cells lack lysosomes, but they do have an organelle that serves a similar function. Vacuoles are large vesicles derived from the endoplasmic reticulum and Golgi apparatus. Thus, vacuoles are an integral part of a cell's endomembrane system. Like all cellular membranes, the vacuolar membrane is selective in transporting solutes; as a result, the solution inside a vacuole differs in composition from the cytosol. Vacuoles perform a variety of functions in different kinds of cells.

Macrophages, a type of cell found in animals' immune systems, engulf bacterial cells into **phagocytic vacuoles**, which then fuse with lysosomes, where the bacteria are destroyed. Some protists engulf their food into large **food vacuoles** in the process of phagocytosis. Food vacuoles contain hydrolytic enzymes that break down macromolecules within food. Many unicellular protists living in fresh water have **contractile vacuoles** that pump excess water out of the cell, thereby maintaining a suitable concentration of ions and molecules inside the cell.

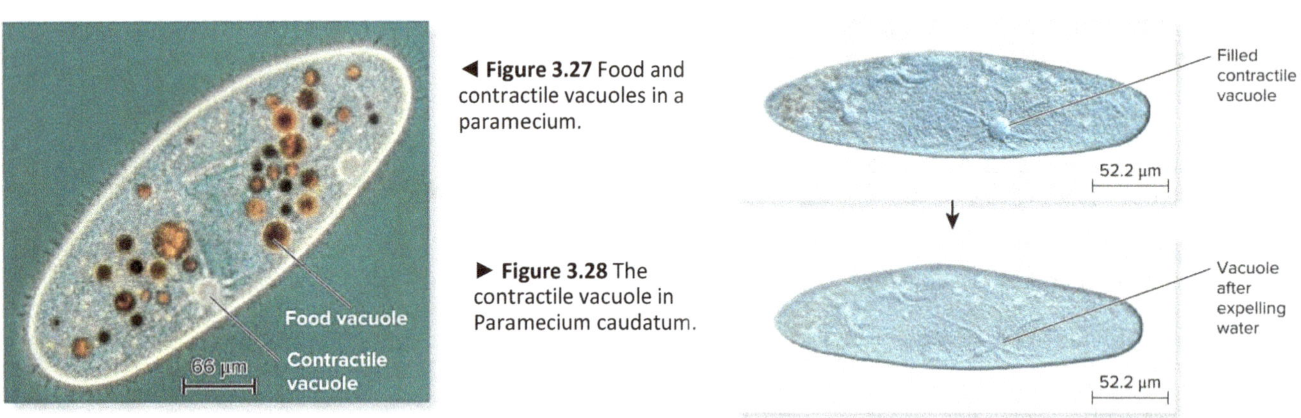

◄ **Figure 3.27** Food and contractile vacuoles in a paramecium.

► **Figure 3.28** The contractile vacuole in Paramecium caudatum.

Mature plant cells generally contain a large **central vacuole**, which develops by the coalescence of smaller vacuoles. The solution inside the central vacuole, called **cell sap**, is the plant cell's main repository of inorganic ions, including potassium and chloride. The membrane surrounding this vacuole is called the **tonoplast** because it contains channels for water that are used to help the cell maintain its tonicity, or osmotic balance. The central vacuole plays a major role in the growth of plant cells, which enlarge as the vacuole absorbs water, enabling the cell to become larger with a minimal investment in new cytoplasm.

The cytosol often occupies only a thin layer between the central vacuole and the plasma membrane, so the ratio of plasma membrane surface to cytosolic volume is sufficient, even for a large plant cell. Because the vacuole contains a high concentration of solutes (dissolved materials), it takes in water and pushes outward on the cell wall. This hydrostatic pressure, called **turgor pressure**, provides much of the mechanical strength of plant cells.

Turgor pressure is important in maintaining the structure of plant cells and the plant itself, and it helps to drive the expansion of the cell wall, which is necessary for growth. Turgid tissues help to support the stems of plants that lack wood. Food reserves, such as sucrose in sugar beet, or mineral salts, may be stored in the vacuole. Protein-storing vacuoles are common in seeds. Waste products, such as crystals of calcium oxalate, may be stored in vacuoles. So the central vacuole serves two important purposes. First, it stores a large amount of water, enzymes, and inorganic ions such as calcium, as well as other materials, including proteins, pigments and secondary metabolites. Second, it performs a space-filling function.

Plant vacuoles may contain hydrolases and act as lysosomes. Secondary metabolites although not essential for growth and development, contribute to survival in various ways. These are often stored in vacuoles. Examples are: Anthocyanins (pigments that are responsible for most of the red, purple, pink and blue colours of flowers and fruits. They attract pollinators and seed dispersers), Certain alkaloids and tannins (deter herbivores from eating the plant) and Latex (a milky fluid, can accumulate in vacuoles, for example in rubber trees).

- **Chloroplasts: Capture of Light Energy**

Chloroplasts, found in plants and algae, are the sites of **photosynthesis**. This process in chloroplasts converts solar energy to chemical energy by absorbing sunlight and using it to drive the synthesis of organic compounds such as sugars from carbon dioxide and water. A leaf cell may have 20 to 100 chloroplast. Chloroplasts tend to be somewhat larger than mitochondria. They are found in the green parts of the plant, mainly in the leaves. Chloroplasts contain the green pigment **chlorophyll**, along with enzymes and other molecules that function in the photosynthetic production of sugar. These lens-shaped organelles, about 3–6 μm in length, are found in leaves and other green organs of plants and in algae. Chloroplasts also contain a variety of light-absorbing yellow and orange pigments known as **carotenoids.**

The contents of a chloroplast are partitioned from the cytosol by an **envelope** consisting of two membranes separated by a very narrow **intermembrane space**. Inside the chloroplast is another membranous system in the form of flattened, interconnected sacs called **thylakoids**. In some regions, thylakoids are stacked like poker chips; each stack is called a **granum** (plural, grana). The fluid outside the thylakoids is the **stroma**, which contains the **chloroplast DNA** and **ribosomes** as well as many **enzymes**. The sugars made may be stored in the form of starch grains in the stroma. Lipid droplets are also seen in the stroma. They appear as black spheres in electron micrographs. They are reserves of lipid for making membranes or are formed from the breakdown of internal membranes as the chloroplast ages. The membranes of the chloroplast divide the chloroplast space into three compartments: the intermembrane space, the stroma, and the thylakoid space.

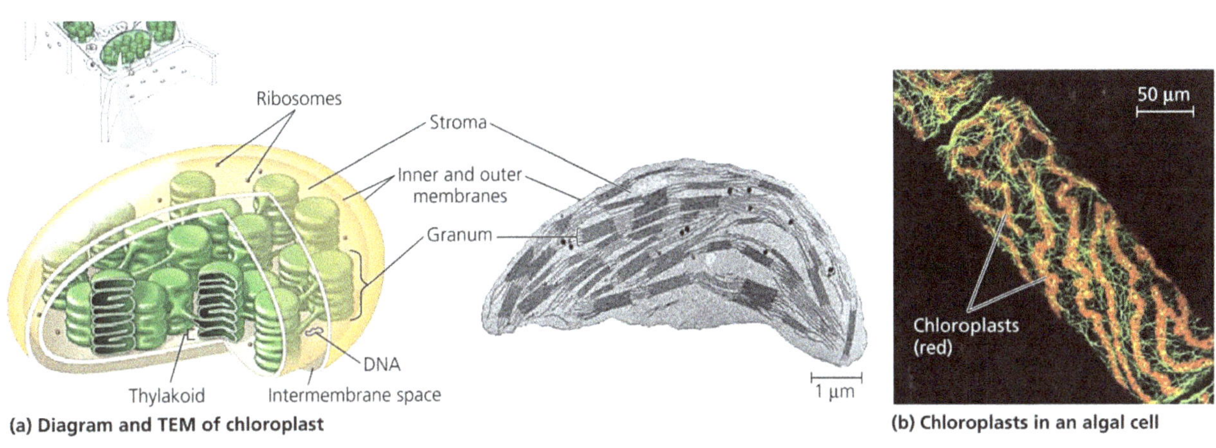

▲ **Figure 3.29 The chloroplast, site of photosynthesis.** A typical chloroplast has three compartments: the intermembrane space, the stroma, and the thylakoid space. (b) *Spirogyra crassa*, which is named for its spiral chloroplasts.

This compartmental organization enables the chloroplast to convert light energy to chemical energy during photosynthesis. Their shape is changeable, and they grow and occasionally pinch in two, reproducing themselves. They are mobile and, with mitochondria and other organelles, move around the cell along tracks of the cytoskeleton. Chloroplasts belong to a group of organelles, known as **plastids**, that produce and store food materials in cells of plants and algae. All plastids develop from **proplastids**, precursor organelles found in less specialized plant cells, particularly in growing, undeveloped tissues.

Depending on the specific functions a cell will eventually have, its proplastids can develop into a variety of specialized mature plastids. They are extremely versatile organelles; in fact, under certain conditions even mature plastids can convert from one form to another. Chloroplasts are produced when proplastids are stimulated by exposure to light. **Chromoplasts** contain pigments that give certain flowers and fruits their characteristic colors, and these colors attract animals that serve as pollinators or as seed dispersers. In autumn, chromoplasts also give many leaves their yellow, orange, and red colors. **Leukoplasts** are unpigmented plastids; they include **amyloplasts** which store starch in the cells of many seeds, roots, and tubers (such as white potatoes).

- **Mitochondrion: Chemical Energy Conversion**

Mitochondria (singular, mitochondrion) are the sites of **cellular respiration**, the metabolic process that uses oxygen to drive the generation of **ATP** (adenosine triphosphate, the energy-carrying molecule found in all living cells that is known as the universal energy carrier) by extracting energy from sugars, fats, and other fuels. In addition to providing most of the energy required to power physiological events such as muscle contraction, mitochondria also function in the synthesis of certain lipids, such as the hormones estrogen and testosterone. Mitochondria are found in nearly all eukaryotic cells, including those of plants, animals, fungi, and most protists. Some cells have a single large mitochondrion, but more often a cell has hundreds or even thousands of mitochondria.

The number of mitochondria in a cell is very variable and correlates with the cell's level of metabolic activity so it is not surprising that cells with a high demand for energy, such as liver and muscle cells, contain large numbers of mitochondria. A liver cell may contain as many as 2000 mitochondria. If you exercise regularly, your muscles will make more mitochondria. Each of the two membranes enclosing the mitochondrion is a phospholipid bilayer with a unique collection of embedded proteins.

The **outer membrane** is smooth, contains many transmembrane proteins called **porins** that form channels through which small molecules such as pyruvate and other metabolites readily pass from the cytoplasm to the intermembrane space. The **inner membrane** is convoluted, with infoldings called **cristae**. Cristae greatly increase the surface area of the inner mitochondrial membrane, providing a surface for the chemical reactions that transform the chemical energy in food molecules into the energy of ATP and contains the enzymes and other proteins needed for these reactions.

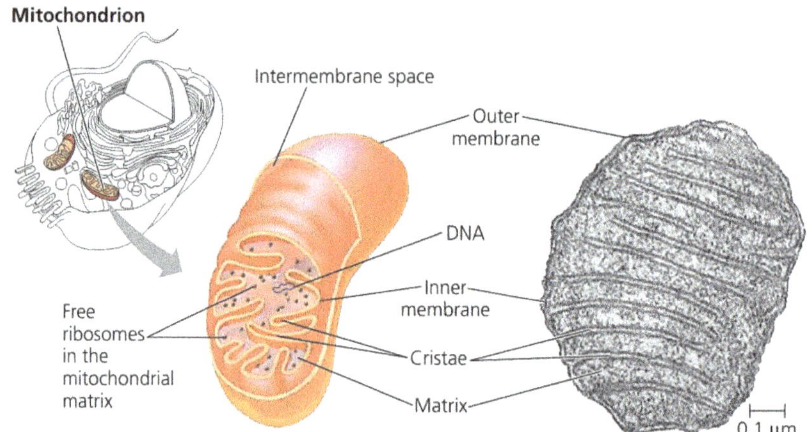

▶ **Figure 3.30 Mitochondria.** The inner membrane of a mitochondrion is shaped into folds called cristae that greatly increase the surface area for oxidative metabolism.

The inner membrane divides the mitochondrion into two internal compartments. The first is the **intermembrane space**, the narrow region between the inner and outer membranes. The second compartment, the mitochondrial **matrix**, is enclosed by the inner membrane. The matrix contains many different **enzymes** as well as the **mitochondrial DNA** and **ribosomes**. Enzymes in the matrix catalyze some of the steps of cellular respiration. Mitochondria play an important role in programmed cell death, or apoptosis.

Unlike **necrosis**, which is uncontrolled cell death that causes inflammation and damages other cells, **apoptosis** is a normal part of development and maintenance. For example, during the metamorphosis of a tadpole to a frog, the cells of the tadpole tail must die. The hand of a human embryo is webbed until apoptosis destroys the tissue between the fingers. Cell death also occurs in the adult. For example, cells that are no longer functional because they have aged or become damaged are destroyed by apoptotic mechanisms and replaced by new cells.

when the protein cytochrome c is released from the inner membrane's electron-transport chain, this protein activates enzymes known as **caspases**, which cut up vital compounds in the cell. Inappropriate inhibition of apoptosis may contribute to a variety of diseases, including cancer. On the other hand, too much apoptosis may deplete needed cells and lead to death of brain cells associated with Alzheimer's disease, Parkinson's disease, and stroke. Mutations that promote apoptosis may be an important mechanism in mammalian aging. Pharmaceutical companies are developing drugs that block apoptosis. However, cell dynamics are extremely complex, and blocking apoptosis could lead to a worse fate, including cancer.

Mutations in mitochondrial DNA have been associated with certain genetic diseases, including a form of young adult blindness, and certain types of progressive muscle degeneration. Mitochondria also affect health and aging by leaking electrons. These electrons form **free radicals**, which are toxic, highly reactive compounds with unpaired electrons. The electrons bond with other compounds in the cell, interfering with normal function. Mitochondria may fuse, or split apart, or even destroy themselves as the energetic demands of cells change.

New mitochondria originate by growth and division of preexisting mitochondria (**Binary fission**). In most mammals, mitochondria are inherited from the female parent only. (This is because the mitochondria in a sperm cell degenerate after fertilization.) Mitochondrial DNA is therefore useful for tracking inheritance through female lines in a family. For the same reason, genetic mutations that cause defective mitochondria also pass only from mother to offspring. Mitochondrial illnesses are most serious when they affect the muscles or brain, because these energy-hungry organs depend on the functioning of many thousands of mitochondria in every cell.

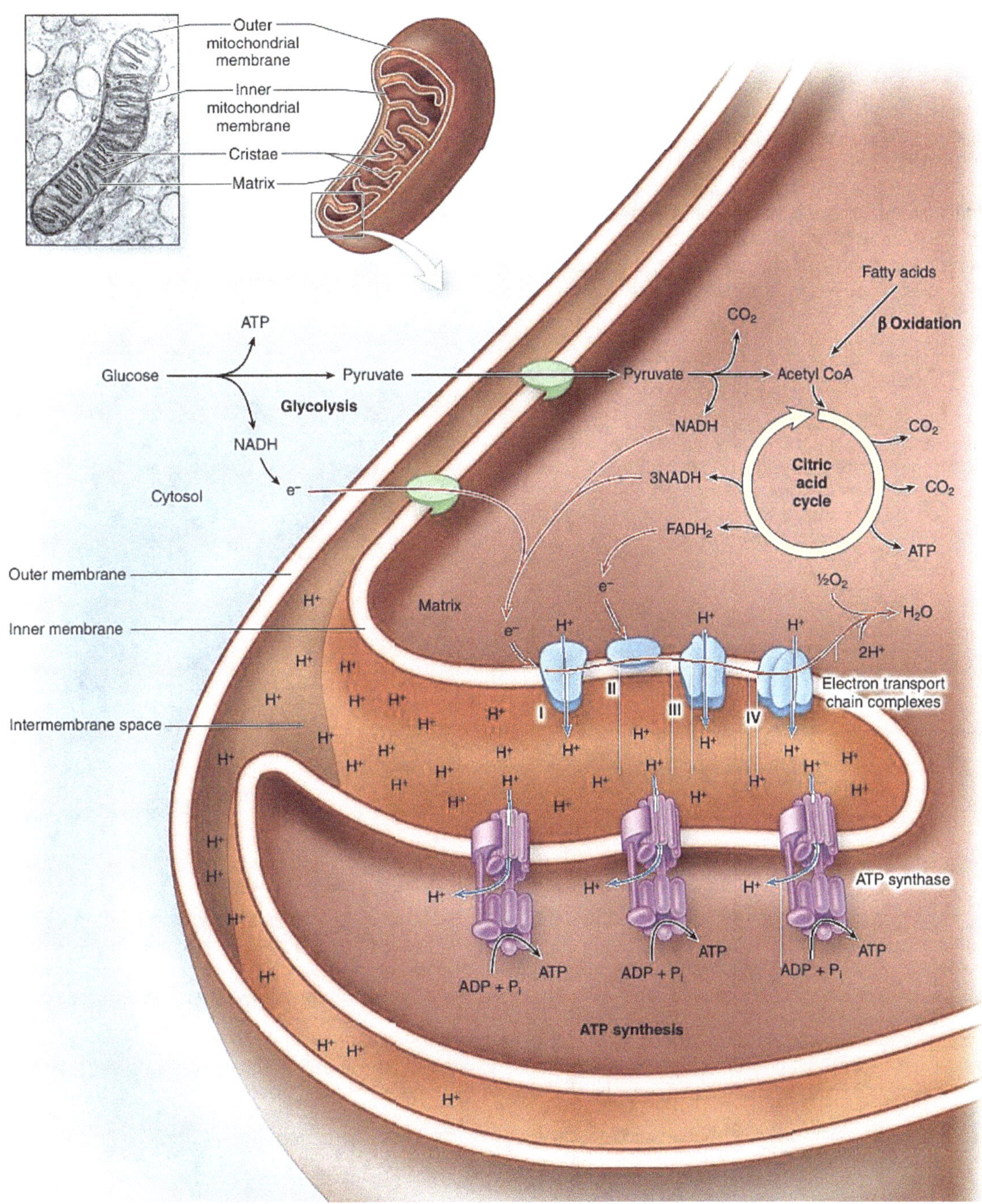

▲ **Figure 3.31** Mitochondrial structure and ATP formation

The Evolutionary Origins of Mitochondria and Chloroplasts

Mitochondria and chloroplasts display similarities with bacteria that led to the **endosymbiont** theory. This theory states that an early ancestor of eukaryotic cells (a host cell) engulfed an oxygen using nonphotosynthetic prokaryotic cell. Eventually, the engulfed cell formed a relationship with the host cell in which it was enclosed, becoming an endosymbiont (a cell living within another cell). Indeed, over the course of evolution, the host cell and its endosymbiont merged into a single organism, a eukaryotic cell with the endosymbiont having become a mitochondrion. At least one of these cells may have then taken up a photosynthetic prokaryote, becoming the ancestor of eukaryotic cells that contain chloroplasts.

A symbiotic relationship is beneficial to one or both species. According to the endosymbiosis theory, such a relationship provided eukaryotic cells with useful cellular characteristics. Chloroplasts, which were derived from cyanobacteria, have the ability to carry out photosynthesis. This benefits plant cells by giving them the ability to use the energy from sunlight. By comparison, mitochondria are thought to have been derived from a different type of bacteria known as proteobacteria. In this case, the endosymbiotic relationship enabled eukaryotic cells to synthesize greater amounts of ATP. How the relationship would have been beneficial to a cyanobacterium or proteobacterium is less clear, though the cytosol of a eukaryotic cell may have provided a stable environment with an adequate supply of nutrients.

During the evolution of eukaryotic species, many genes that were originally found in the genomes of the primordial proteobacteria and cyanobacteria have been transferred from the organelles to the nucleus. In modern cells, hundreds of different proteins that make up these organelles are coded by genes that have been transferred to the nucleus. These proteins are made in the cytosol and then taken up into mitochondria or chloroplasts.

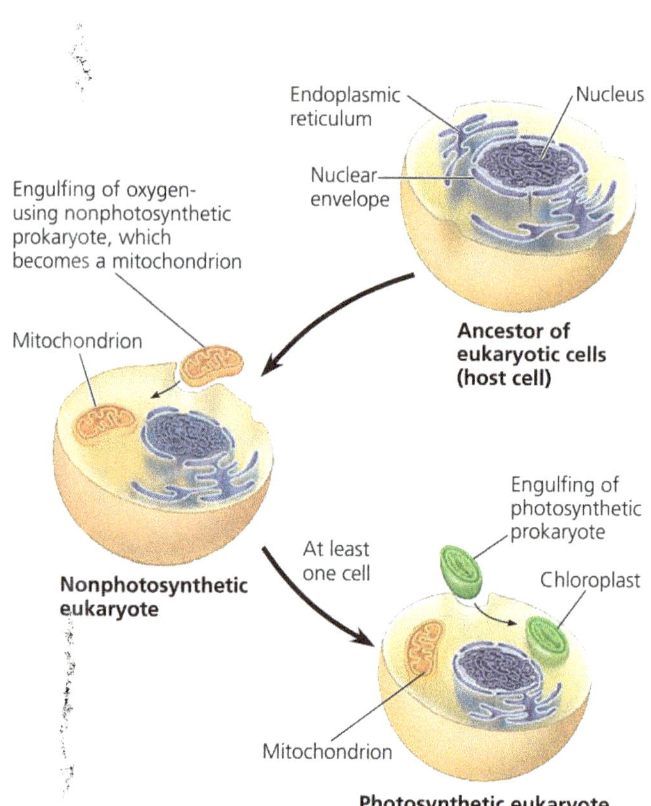

▶ **Figure 3.32** The endosymbiont theory of the origin of mitochondria and chloroplasts in eukaryotic cells. According to this theory, the proposed ancestors of mitochondria were oxygen using nonphotosynthetic prokaryotes, while the proposed ancestors of chloroplasts were photosynthetic prokaryotes.

This theory is consistent with many structural features of mitochondria and chloroplasts. **First**, rather than being bounded by a single membrane like organelles of the endomembrane system, mitochondria and typical chloroplasts have two membranes surrounding them. (Chloroplasts also have an internal system of membranous sacs.) There is evidence that the ancestral engulfed prokaryotes had two outer membranes, which became the double membranes of mitochondria and chloroplasts.

Second, like prokaryotes, mitochondria and chloroplasts contain ribosomes, as well as circular DNA molecules—like bacterial chromosomes—associated with their inner membranes. The DNA in these organelles programs the synthesis of some organelle proteins on ribosomes that have been synthesized and assembled there as well.

Third, also consistent with their probable evolutionary origins as cells, mitochondria and chloroplasts are autonomous (somewhat independent) organelles that grow and reproduce within the cell.

- **Peroxisomes: Oxidation**

Eukaryotic cells contain a variety of enzyme-bearing, membrane enclosed vesicles called **microbodies**. Research has shown that the ER is involved in microbody production. Proteins and phospholipids are continuously imported into microbodies. The phospholipids are used for new membrane synthesis, leading to growth of the microbody. Division of a microbody then produces new microbodies.

Microbodies have various functions that are often specific to an organism or cell type. Commonly, they contain enzymes that conduct preparatory or intermediate reactions linking major biochemical pathways. For example, the series of reactions that allows cells to use fats as an energy source begins in microbodies and continues in mitochondria. Beginning or intermediate steps in the breakdown of some amino acids and alcohols also take place in microbodies, including about half of the ethyl alcohol that a human may consume.

Many types of microbodies produce as a by-product the toxic substance hydrogen peroxide (H_2O_2), which is broken down into water and oxygen by the enzyme catalase. Microbodies with this reaction are often termed **peroxisomes**.

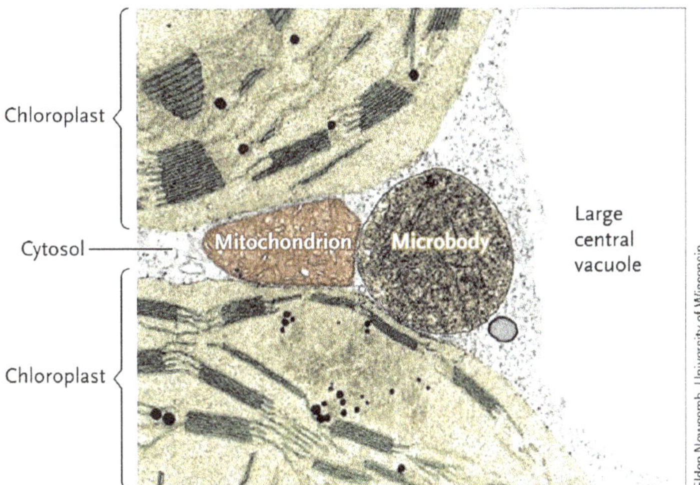

◄ **Figure 3.33 Microbody.** EM of a microbody in the cytoplasm of a tobacco leaf.

► **Figure 3.34 A peroxisome.** Peroxisomes are roughly spherical and often have a granular or crystalline core that is thought to be a dense collection of enzyme molecules. This peroxisome is in a leaf cell (TEM). Notice its proximity to two chloroplasts and a mitochondrion. These organelles cooperate with peroxisomes in certain metabolic functions.

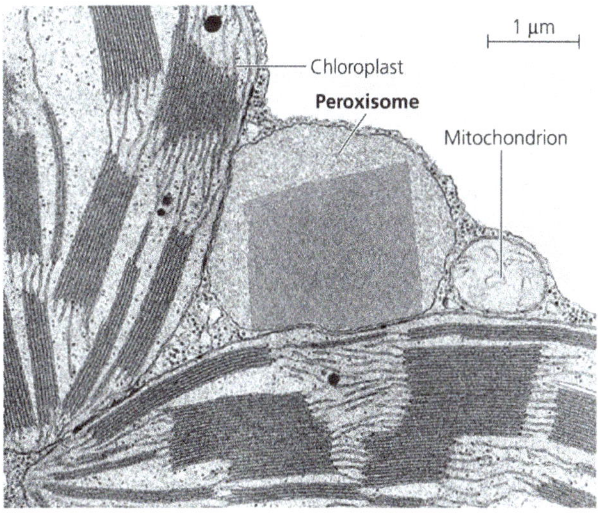

The peroxisome is a specialized metabolic compartment bounded by a single membrane. Their most important function is to neutralize free radicals, highly reactive chemicals with unpaired electrons that can scramble the structure of biological molecules. Peroxisomes contain **oxidases** that remove hydrogen atoms from various substrates and transfer them to oxygen (O_2), producing hydrogen peroxide (H_2O_2 from which the organelle derives its name) to detoxify harmful substances, including alcohol and formaldehyde. Amino acids and fatty acids are oxidized in peroxisomes as part of normal metabolism.

$$RH_2 + O_2 \longrightarrow R + H_2O_2$$
$$\text{(toxin)}$$

Peroxisomes in the liver detoxify alcohol and other harmful compounds by transferring hydrogen from the poisonous compounds to oxygen. Hydrogen peroxide has the potential to be highly toxic. In the presence of metals such as iron (Fe^{2+}), which are found naturally in living cells, H_2O_2 can be broken down to form a hydroxide ion (OH^-) and a molecule called a hydroxide free radical ($\bullet OH$):

$$Fe^{2+} + H_2O_2 \rightarrow Fe^{3+} + OH^- + \cdot OH \text{ (hydroxide free radical)}$$

The $\bullet OH$ is highly reactive and can damage proteins, lipids, and DNA. Therefore, it is beneficial for cells to break down H_2O_2 in an alternative manner that does not form $\bullet OH$. Peroxisomes contain an enzyme called **catalase** that breaks down hydrogen peroxide to make water and oxygen gas (hence the name peroxisome):

$$2\ H_2O_2 \xrightarrow{\text{Catalase}} 2\ H_2O + O_2$$

Other diverse enzymes in peroxisomes complement certain functions of the SER and mitochondria in the metabolism of lipids and other molecules. Thus, the **β-oxidation** of long-chain fatty acids (18 carbons and longer are broken down into two-carbon fragments) is preferentially accoplished by peroxisomal enzymes that differ from their mitochondrial counterparts. Peroxisomes synthesize cholesmterol and certain phospholipids that are components of the insulating covering of nerve cells. Peroxisomes in human liver and kidney cells detoxify certain toxic compounds, including ethanol. Some peroxisomes are formed when existing peroxisomes simply pinch in half.

But recent evidence suggests that most new peroxisomes form by budding off of the endoplasmic reticulum via a special ER machinery that differs from that used for vesicles destined for modification in the Golgi apparatus. Microbodies in plants convert oils or fats to sugars that can be used directly for energy-releasing reactions in mitochondria or for reactions that require sugars as chemical building blocks. These microbody reactions are particularly important in plant embryos that develop from oily seeds, such as those of the peanut or soybean. Depending on the particular reaction pathways they carry out, plant microbodies are called peroxisomes, **glyoxysomes** or **glycosomes**. Animal cells lack glyoxysomes and cannot convert fatty acids into sugars.

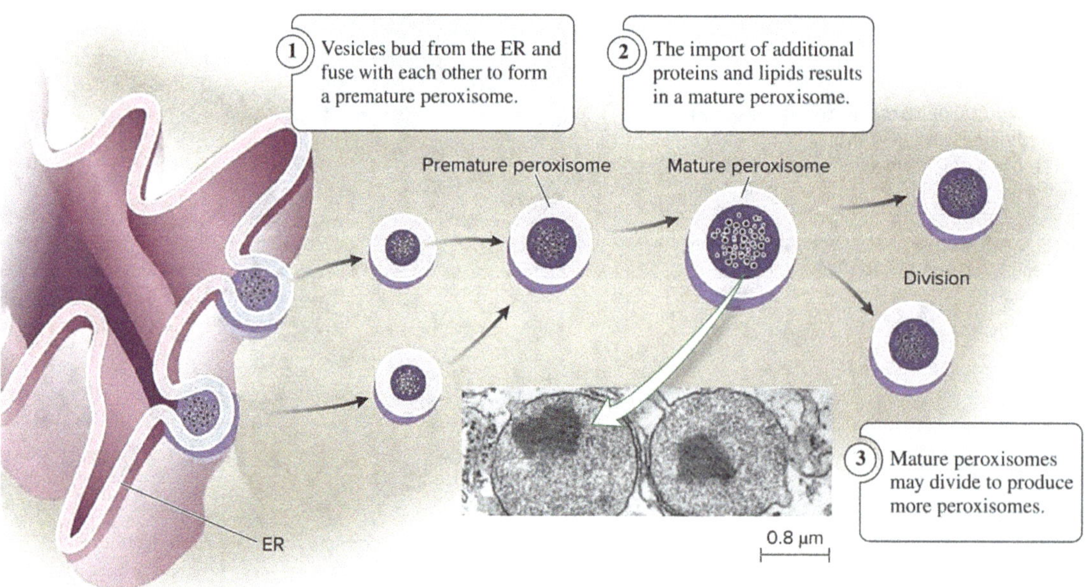

▲ **Figure 3.35** Formation of peroxisomes.

The cytoskeleton is a network of fibers that organizes structures and activities in the cell

The cytoskeleton or "cell skeleton,", is a dense network of protein fibers, gives cells mechanical strength, shape, and their ability to move. The cytoskeleton also functions in cell division and in the transport of materials within the cell. Although cytoskeletal structures are also present in plant cells, the fibers and tubes of the system are less prominent; much of cellular support in plants is provided by the cell wall and a large central vacuole. Bacterial cells also have fibers that form a type of cytoskeleton, constructed of proteins similar to eukaryotic ones.

The most obvious function of the cytoskeleton is to give mechanical support to the cell and maintain its shape. This is especially important for animal cells, which lack walls. The cytoskeleton is stabilized by a balance between opposing forces exerted by its elements. And just as the skeleton of an animal helps fix the positions of other body parts, the cytoskeleton provides anchorage for many organelles and even cytosolic enzyme molecules. The cytoskeleton is more dynamic than an animal skeleton, however. It can be quickly dismantled in one part of the cell and reassembled in a new location, changing the shape of the cell.

In both animals and fungi, the ER is associated with both microtubules and actin filaments. In animal cells, if we visualize cellular structures with fluorescent labels, the tubules of the ER align closely with microtubules. This may be involved in the growth and distribution of the ER. In yeast, microfilaments are involved in inheritance of the ER during cell division. The cytoskeleton also manipulates the plasma membrane, bending it inward to form food vacuoles or other phagocytic vesicles.

Some types of cell motility (movement) also involve the cytoskeleton. The term cell motility includes both changes in cell location and movements of cell parts. Cell motility generally requires interaction of the cytoskeleton with **motor proteins**. There are many such examples: Cytoskeletal elements and motor proteins work together with plasma membrane molecules to allow whole cells to move along fibers outside the cell.

Inside the cell, vesicles and other organelles often use motor protein "feet" to "walk" to their destinations along a track provided by the cytoskeleton. For example, this is how vesicles containing neurotransmitter molecules migrate to the tips of axons, the long extensions of nerve cells that release these molecules as chemical signals to adjacent nerve cells.

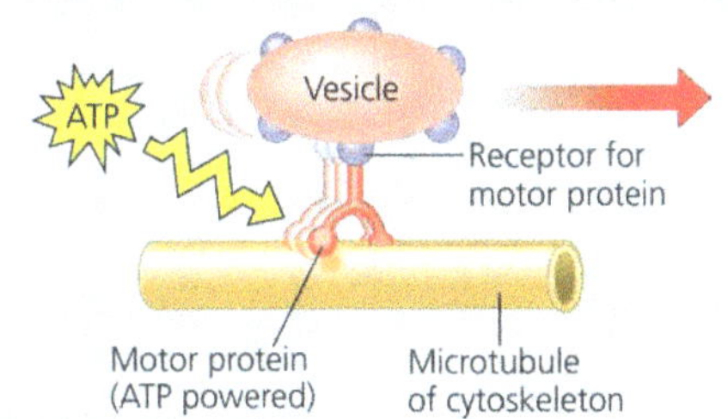

▲ **Figure 3.36 Motor proteins and the cytoskeleton.**

Motor proteins that attach to receptors on vesicles can "walk" the vesicles along the cytoskeletal fibers called microtubules or, in some cases, along microfilaments. ATP powers the movement.

Components of the Cytoskeleton

Three main types of fibers make up the cytoskeleton: **Microtubules** are the thickest of the three types; **microfilaments** (also called actin filaments) are the thinnest; and **intermediate filaments** are fibers with diameters in a middle range.

Property	Microtubules (Tubulin Polymers)	Microfilaments (Actin Filaments)	Intermediate Filaments
Structure	Hollow tubes; wall consists of 13 columns of tubulin molecules	Two intertwined strands of actin, each a polymer of actin subunits	Fibrous proteins supercoiled into thicker cables
Diameter	25 nm with 15-nm lumen	7 nm	8–12 nm
Protein subunits	Tubulin, a dimer consisting of α-tubulin and β-tubulin	Actin	One of several different proteins (such as keratins), depending on cell type
Main functions	Maintenance of cell shape Cell motility (as in cilia or flagella) Chromosome movements in cell division (see Figure 9.7) Organelle movements	Maintenance of cell shape Changes in cell shape Muscle contraction (see Figure 39.4) Cytoplasmic streaming in plants Cell motility (as in amoeboid movement) Division of animal cells (see Figure 9.10)	Maintenance of cell shape Anchorage of nucleus and certain other organelles Formation of nuclear lamina

▲ **Table 3.1** The Structure and Function of the Cytoskeleton

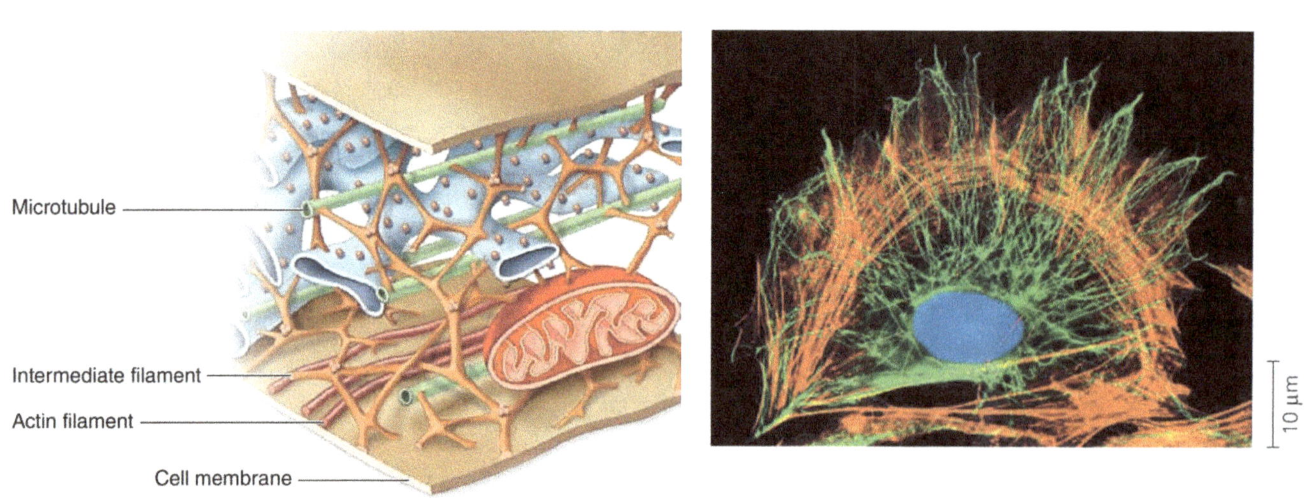

▲ **Figure 3.37 The cytoskeleton.** As shown in this fluorescence micrograph, the cytoskeleton extends throughout the cell. The cytoskeletal elements have been tagged with different fluorescent molecules: green for microtubules and red for microfilaments (which look orangish here). A third component of the cytoskeleton, intermediate filaments, is not evident. (The blue area is DNA in the nucleus.)

- **Microtubules**

Microtubules shape and support the cell and also serve as tracks along which organelles equipped with motor proteins can move. In addition microtubules guide vesicles from the ER to the Golgi apparatus and from the Golgi to the plasma membrane. Microtubules are also involved in the separation of chromosomes during cell division. All eukaryotic cells have microtubules, hollow rods constructed from globular proteins called tubulins. Each tubulin protein is a dimer, a molecule made up of two components. A tubulin dimer consists of two slightly different polypeptides, α-tubulin and β-tubulin. Globular proteins consisting of dimers of α- and β-tubulin subunits polymerize to form the 13 protofilaments. The protofilaments are arrayed side by side around a central core, giving the microtubule its characteristic tube shape.

Microtubules grow in length by adding tubulin dimers; they can also be disassembled and their tubulins used to build microtubules elsewhere in the cell. Because of the orientation of tubulin dimers, the two ends of a microtubule are slightly different. One end can accumulate or release tubulin dimers at a much higher rate than the other, thus growing and shrinking significantly during cellular activities. (This is called the "plus end," not because it can only add tubulin proteins but because it's the end where both "on" and "off" rates are much higher). The end with the β-tubulin subunits is designated as plus (+) (distal to the centrosome) and the end with α-tubulin subunits is minus (−) (proximal to the centrosome).

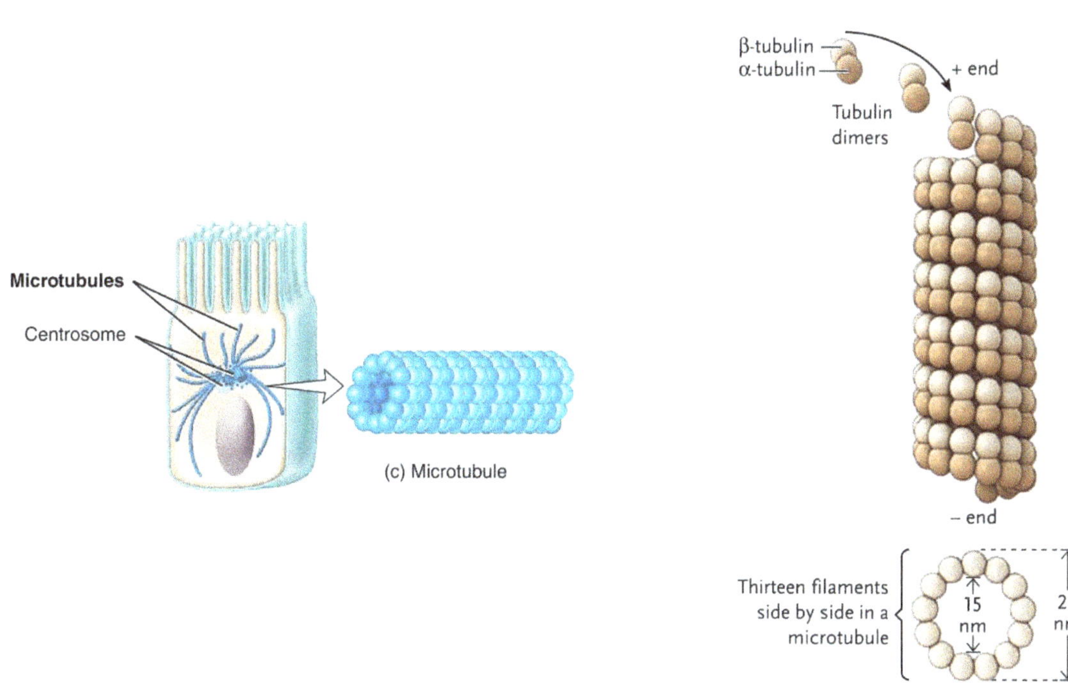

▲ Figure 3.38 Microtubules

Other proteins are important in microtubule function. Microtubule-associated proteins (MAPs) are classified into two groups: **structural MAPs** and **motor MAPs**. Structural MAPs may help regulate microtubule assembly, and they cross-link microtubules to other cytoskeletal polymers. Motor MAPs use ATP energy to produce movement. A motor protein consists of three domains:

Head: the site where ATP binds and is hydrolyzed to adenosine diphosphate (ADP) and inorganic phosphate (P_i).

Hinge: the site that bends in response to ATP binding and hydrolysis. This bending is what causes movement to occur.

Tail: an elongated region that is attached to other proteins or to other kinds of cellular molecules. Together, the hinge and tail make up a structure called the lever arm.

Four components are required to move material along microtubules: (1) a vesicle or organelle that is to be transported, (2) a motor protein that provides the energy-driven motion, (3) a connector molecule that connects the vesicle to the motor molecule, and (4) microtubules on which the vesicle will ride like a train on a rail.

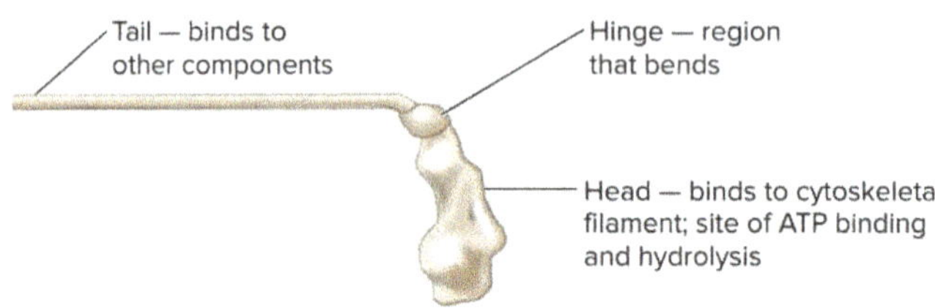

▲ **Figure 3.39** Three-domain structure of myosin, a motor protein

Cells utilize the actions of motor proteins to promote three kinds of movements:

- **Movement of cargo:** The filament is fixed in place and the motor protein moves a cargo from one location to another.

- **Movement of a filament:** The motor protein is fixed in place. The action of the motor protein moves the filament.

- **Bending of a filament:** The motor protein and filament are both fixed in place due to linking proteins, so the action of the motor protein causes the filament to bend.

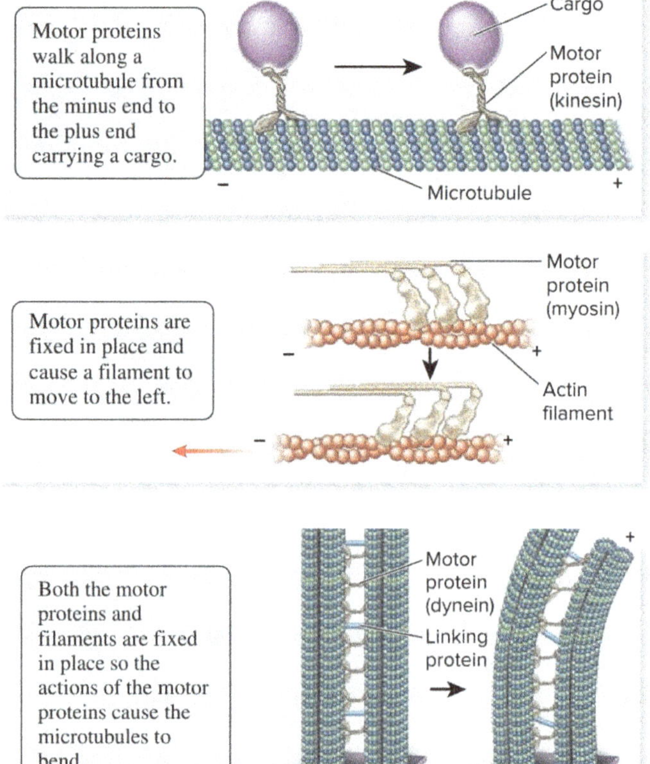

The direction a vesicle is moved depends on the type of motor protein involved and the fact that microtubules are organized with their plus ends toward the periphery of the cell. For example, the ER is transported through the cytoplasm by kinesin, which moves toward the + ends of microtubules, which are distal from the centrosome. As a result, the ER becomes distributed throughout the cytoplasm. Similarly, secretory vesicles budded off of the trans side of the Golgi complex and destined to fuse with the plasma membrane are moved along microtubules by kinesin. In one case, a protein called **kinectin** binds vesicles to the motor protein **kinesin**. Kinesin uses ATP to power its movement toward the cell periphery, dragging the vesicle with it as it travels along the microtubule toward the plus end.

▶ **Figure 3.40 A model of a kinesin motor.** A kinesin molecule attaches to a specific receptor on the vesicle. Energy from ATP powers the kinesin molecule so that it changes its conformation and "walks" along the microtubule, carrying the vesicle along.

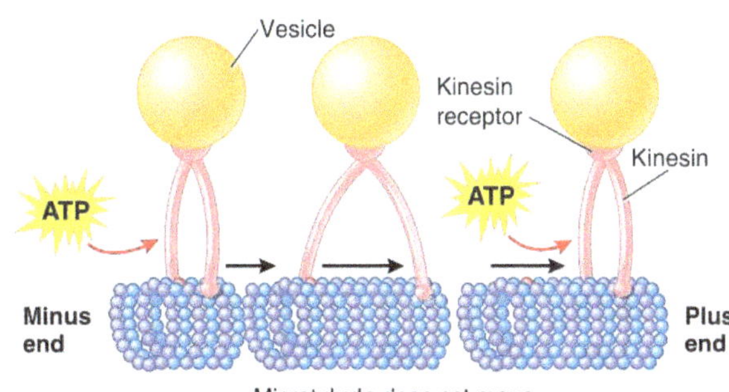

Another set of vesicle proteins, called the **dynactin** complex, binds vesicles to the motor protein **dynein** which directs movement in the opposite direction along microtubules toward the minus end, inward toward the cell's center. For example, dynein transports the Golgi complex toward the - ends of microtubules so that the organelle becomes located near the centrosome.

This dynein movement is referred to as **retrograde transport**. Dynein is also involved in the movement of eukaryotic flagella. The destination of a particular transport vesicle and its content is thus determined by the nature of the linking protein embedded within the vesicle's membrane. At times, for example, during peroxisome transport, kinesins and dyneins may work together.

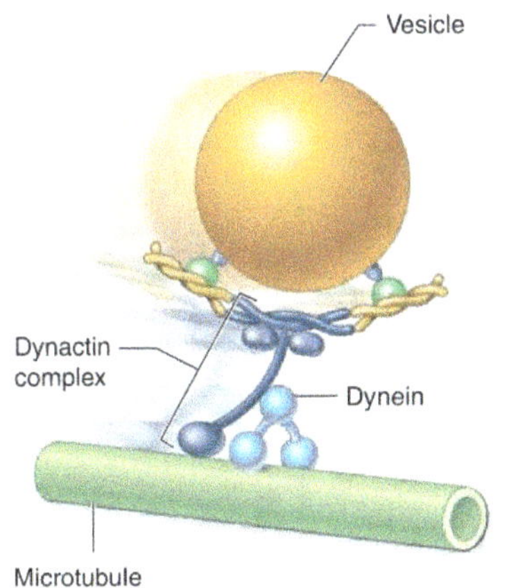

◀ **Figure 3.41 A Molecular motors.** Vesicles can be transported along microtubules using motor proteins that use ATP to generate force. The vesicles are attached to motor proteins by connector molecules, such as the dynactin complex shown here. The motor protein dynein moves the connected vesicle along microtubules.

The sites where microtubules form within a cell vary among different types of organisms. A microtubule-organizing center (**MTOC**) is a site in a eukaryotic cell from which microtubules grow. Nondividing animal cells contain a single MTOC near their nucleus called the **centrosome** or **cell center** (the main microtubule organising centre in animal cells). Within the centrosome is a pair of **centrioles** usually located at right angles to each other, each composed of nine sets of triplet microtubules arranged in a ring. Surrounding the centrioles is the pericentriolar material, containing hundreds of tubulins.

In animal cells, microtubule growth starts at the centrosome in such a way that the minus end is anchored there. In contrast, most plant cells and many protists lack centrosomes and centrioles. The centrosomes of plants and fungi lack centrioles, but still contain microtubule-organizing centers. which suggests that centrioles are not essential to most microtubule assembly processes and that alternative assembly mechanisms are present. Microtubules are created at many sites that are scattered throughout a plant cell. In plants, sites along the nuclear membrane appear to function as MTOCs.

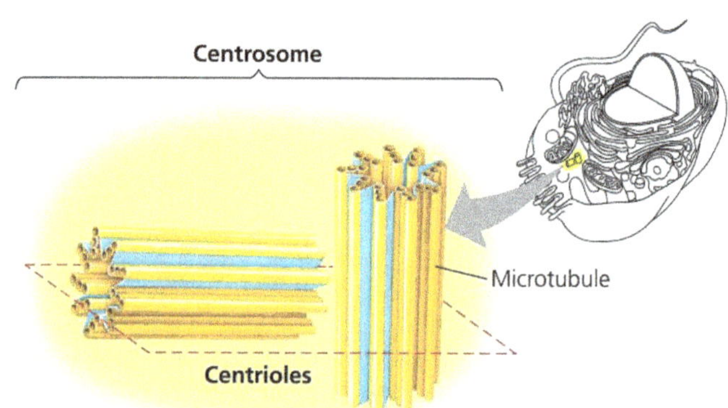

▶ **Figure 3.42** Centrosome containing a pair of centrioles.

Some eukaryotic cells have **flagella** (singular, flagellum) and **cilia** (singular, cilium), cellular extensions that contain microtubules. Eukaryotic cilia and flagella are structurally alike (but different from bacterial flagella) and have the same internal structure called the **axoneme**. The axoneme contains microtubules, the motor protein dynein, and linking proteins (The whole cylindrical structure inside the cell surface membrane is called the axoneme.)

In the cilia and flagella of most eukaryotic organisms, the microtubules form an arrangement called a **9 + 2** array. Each of the two central microtubules is a single microtubule. A ring of nine microtubule doublets (MTDs) located around the outside. Each MTD contains an A and a B microtubule. The wall of the A microtubule is a complete ring of 13 protofilaments and the B microtubule attached is an incomplete ring with only 10 protofilaments. Each A microtubule has inner and outer arms. These are made of the protein dynein. They connect with the B microtubules of neighbouring MTDs during beating. So there are two rows of several hundred dynein arms along the outside of each A microtubule. Radial spokes project from the outer doublet microtubules toward the central pair.

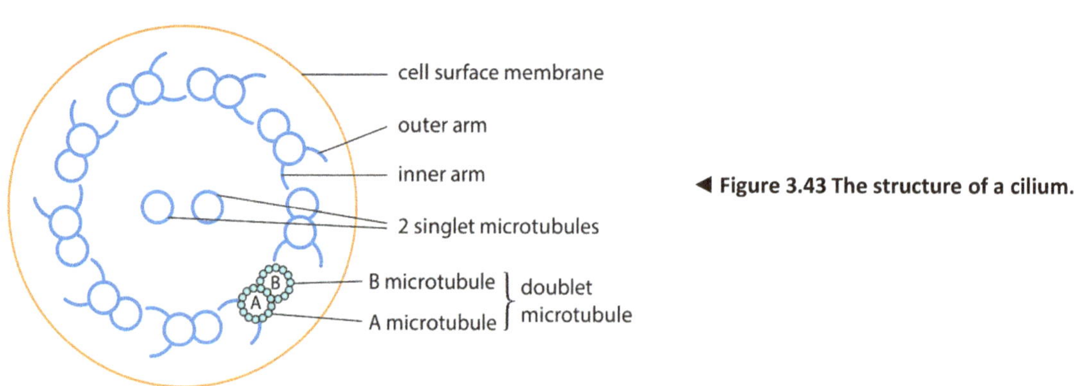

◀ **Figure 3.43 The structure of a cilium.**

The microtubules in flagella and cilia emanate from **basal bodies**, which are anchored to the cytoplasmic side of the plasma membrane. At the basal body, the microtubules form a triplet structure. Much like the centrosome of animal cells, the basal bodies provide a site for microtubules to grow.

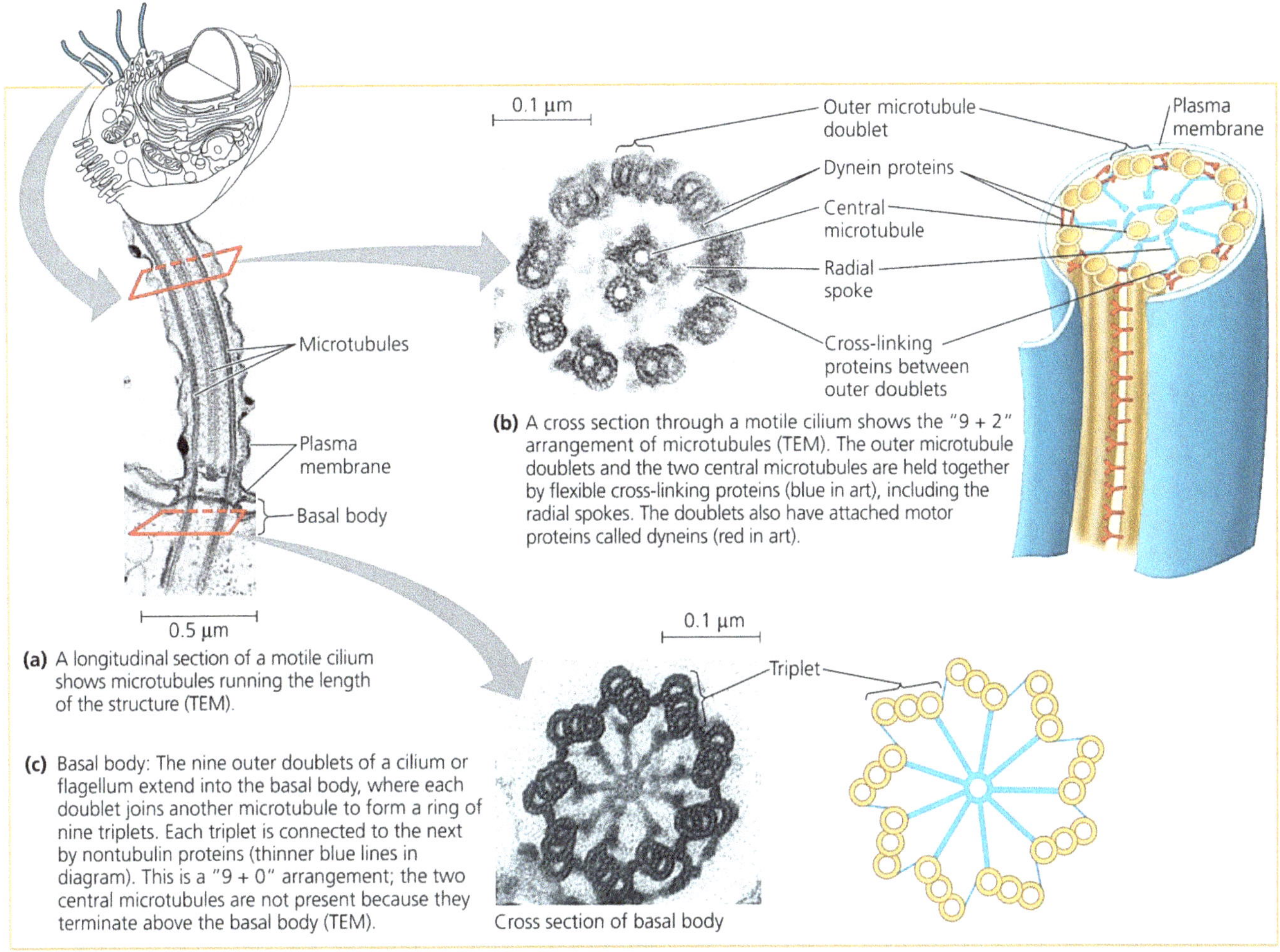

(a) A longitudinal section of a motile cilium shows microtubules running the length of the structure (TEM).

(b) A cross section through a motile cilium shows the "9 + 2" arrangement of microtubules (TEM). The outer microtubule doublets and the two central microtubules are held together by flexible cross-linking proteins (blue in art), including the radial spokes. The doublets also have attached motor proteins called dyneins (red in art).

(c) Basal body: The nine outer doublets of a cilium or flagellum extend into the basal body, where each doublet joins another microtubule to form a ring of nine triplets. Each triplet is connected to the next by nontubulin proteins (thinner blue lines in diagram). This is a "9 + 0" arrangement; the two central microtubules are not present because they terminate above the basal body (TEM).

▲ **Figure 3.44** Structure of a flagellum or motile cilium.

A specialized arrangement of the microtubules is responsible for the beating of these structures. Many unicellular protists are propelled through water by cilia or flagella that act as locomotor appendages, and the sperm of animals, algae, and some plants have flagella. When cilia or flagella extend from cells that are attached tightly together in a sheet that is part of a tissue layer, they can move fluid over the surface of the tissue. For example, the ciliated lining of the trachea (windpipe) sweeps mucus containing trapped debris out of the lungs.

Cilia movement is paralyzed by nicotine in cigarette smoke. For this reason, smokers cough often to remove foreign particles from their airways. In a woman's reproductive tract, the cilia lining the oviducts help move an egg toward the uterus. Cilia on cells lining the ventricles (cavities) of the brain circulate fluid through the brain. Motile cilia usually occur in large numbers on the cell surface.

Flagella are usually limited to just one or a few per cell, and they are longer than cilia. Flagella and cilia differ in their beating patterns. A flagellum has an undulating motion like the tail of a fish. In contrast, cilia have alternating power and recovery strokes, much like the oars of a racing crew boat. They exert a force that is parallel to the cell surface while a flagellum moves like a whip, exerting a force perpendicular to the cell surface. Notice that cilia propel other substances across a cell's surface, whereas a flagellum propels the cell itself.

Almost every vertebrate cell has a **primary or nonmotile cilium**, a single cilium on the cell surface that serves as a cellular **antenna**. The primary cilium has receptors on its surface that bind with specific molecules outside the cell or on surfaces of other cells. Cilium-based signaling appears to be crucial to brain function and to embryonic development. A good example is the nonmotile cilia found in the olfactory (smell) sensory neurons in the nose; these cilia contain in their membranes odor-detecting proteins that initiate the sense of smell. (Nonmotile primary cilia have a "9 +0" pattern, lacking the central pair of microtubules). Although the purpose of the eukaryotic flagellum is the same as that of prokaryotic flagella, the genes that encode the components of the flagellar apparatus of cells of Domains Bacteria, Archaea, and Eukarya are different in each case. Thus, the three types of flagella are analogous, not homologous, structures, and they must have evolved independently.

During the formation of a flagellum or cilium, a centriole moves to a position just under the plasma membrane. Then two of the three microtubules of each triplet grow outward from one end of the centriole to form the ring of nine double microtubules. The two central microtubules of the 9+2 complex also grow from the end of the centriole, but without direct connection to any centriole microtubules. The centriole remains at the innermost end of a flagellum or cilium when its development is complete as the **basal body** of the structure. In fact, the microtubule assembly of a cilium or flagellum is anchored in the cell by a basal body, which is structurally very similar to a centriole, with microtubule triplets in a **"9 +0"** pattern. In many animals (including humans), the basal body of the fertilizing sperm's flagellum enters the egg and becomes a centriole.

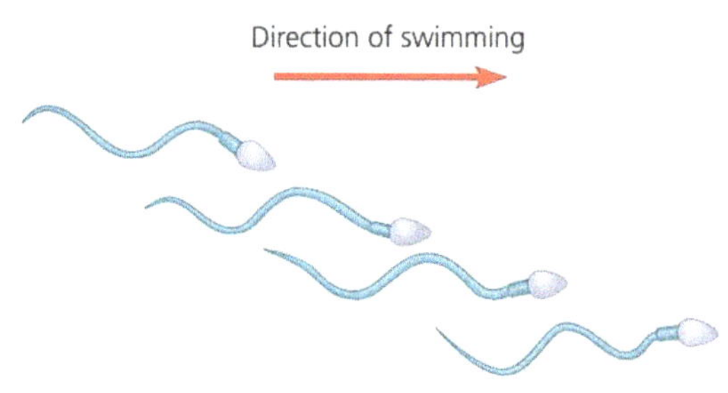

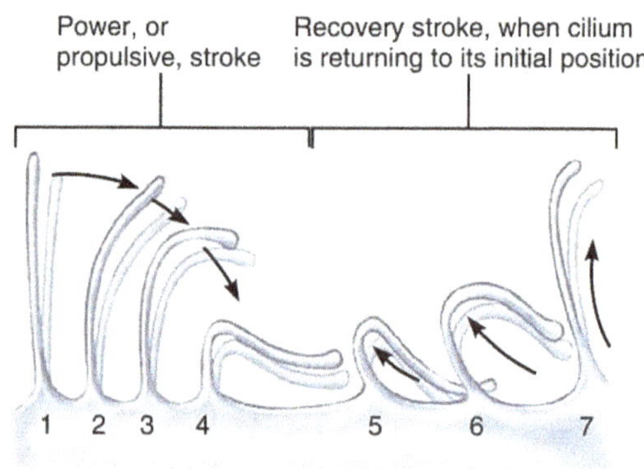

(a) **Phases of ciliary motion.**

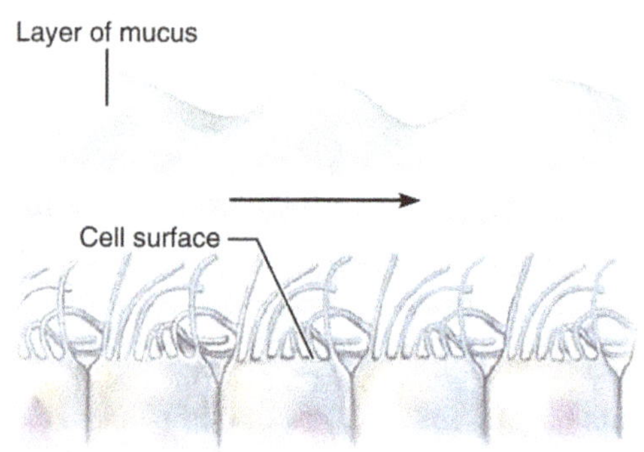

(b) **Traveling wave created by the activity of many cilia acting together propels mucus across cell surfaces.**

▲ **Figure 3.45 A comparison of the beating of flagella and motile cilia.** A flagellum usually undulates, its snakelike motion driving a cell in the same direction as the axis of the flagellum. Cilia have a back-and-forth motion. The rapid power stroke moves the cell in a direction perpendicular to the axis of the cilium. Then, during the slower recovery stroke, the cilium bends and sweeps sideways, closer to the cell surface.

Bending involves large motor proteins called dyneins that are attached along each outer microtubule doublet. A typical dynein protein has two "feet" that "walk" along the microtubule of the adjacent doublet, using ATP for energy. One foot maintains contact, while the other releases and reattaches one step farther along the microtubule. The outer doublets and two central microtubules are held together by flexible cross-linking proteins and the walking movement is coordinated so that it happens on one side of the circle at a time. Dynein molecules shift in a way that slides adjacent microtubules against each other cause the microtubules—and the organelle as a whole—to bend.

- ## Microfilaments (Actin Filaments)

Microfilaments are thin solid rods. They are also called actin filaments because they are built from molecules of **actin**, a globular protein. A microfilament is a twisted double chain of actin subunits. Besides occurring as linear filaments, microfilaments can form structural networks when certain proteins bind along the side of such a filament and allow a new filament to extend as a branch.

Like microtubules actin filaments (**F-actin**) are highly dynamic. **G-actin** (or "globular actin") subunits are added rapidly at the (+) or barbed end, with ATP hydrolysis at each addition; at the same time G-actin subunits dissociate at the (−) or pointed end. This leads to migration of subunits through the polymer, which occurs rapidly in a process called **treadmilling**.

Like microtubules, microfilaments seem to be present in all eukaryotic cells. Each cell has its own unique arrangement of microfilaments, so no two cells are alike.

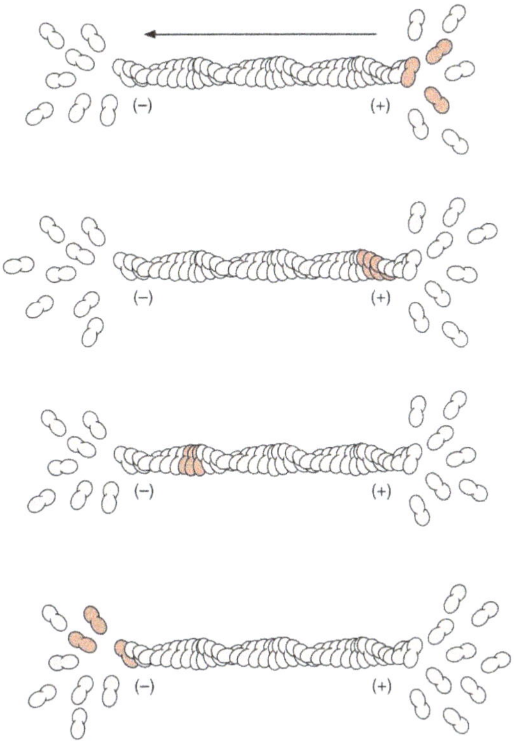

▶ **Figure 3.46 Actin filament treadmilling.**

In contrast to the compression-resisting role of microtubules, the structural role of microfilaments in the cytoskeleton is to bear tension (pulling forces). Actin is very abundant in all cells and is concentrated in networks of actin filaments and as free G-actin subunits near the cell membrane, a cytoplasmic region often called the **cell cortex**. A three-dimensional network formed by microfilaments just to the inside of the plasma membrane (cortical microfilaments or terminal web) helps support the cell's shape.

The sides of actin filaments are often anchored to other proteins near the plasma membrane, which explains why they are typically found there. The plus ends grow toward the plasma membrane and play a key role in cell shape and movement.

In some kinds of animal cells, such as nutrient-absorbing intestinal cells, bundles of microfilaments that extend into the terminal web of the cell make up the core of **microvilli**, delicate projections that increase the cell's surface area. Actin is sometimes a contractile protein, but in microvilli it appears to function as a mechanical "stiffener."

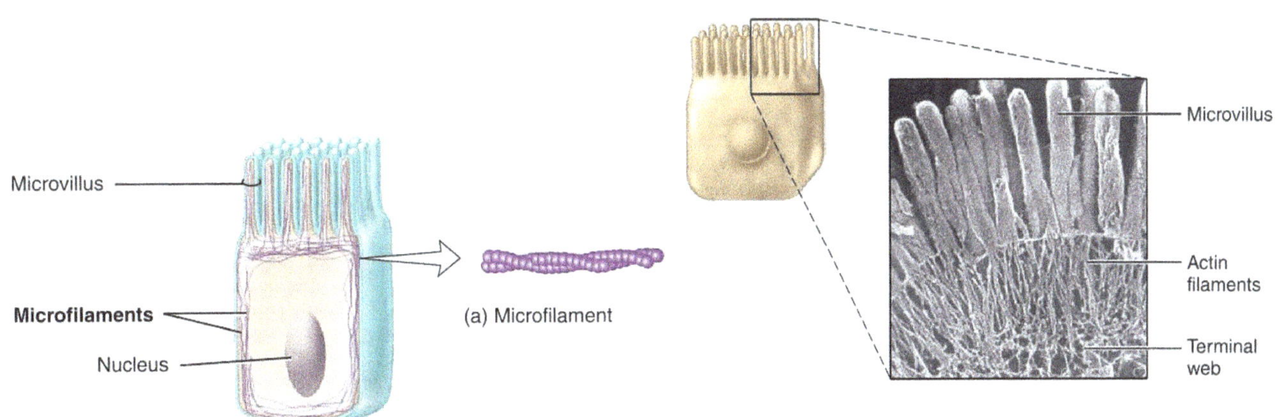

▲ **Figure 3.47 A structural role of microfilaments.** The surface area of this intestinal cell is increased by its many microvilli (singular, microvillus), cellular extensions reinforced by bundles of microfilaments (TEM).

Microfilaments are well known for their role in cell motility. Thousands of actin filaments and thicker filaments made of a protein called **myosin** interact to cause **contraction** of muscle cells. When ATP is hydrolyzed to ADP, myosin binds to actin and causes the microfilament to slide. **Contractile rings** of microfilaments with myosin constricts to produce two cells by cytokinesis during mitosis. Membrane-associated molecules of myosin whose movements along microfilaments produce the cell surface changes during **endocytosis**.

In the unicellular protist Amoeba and some of our white blood cells, localized contractions brought about by actin and myosin are involved in the amoeboid (crawling) movement of the cells. The cell crawls along a surface by extending cellular extensions called **pseudopodia** and moving toward them. In plant cells, actin-protein interactions contribute to **cytoplasmic streaming**, a circular flow of cytoplasm within cells. This movement, which is especially common in large plant cells, speeds the movement of organelles and the distribution of materials within the cell.

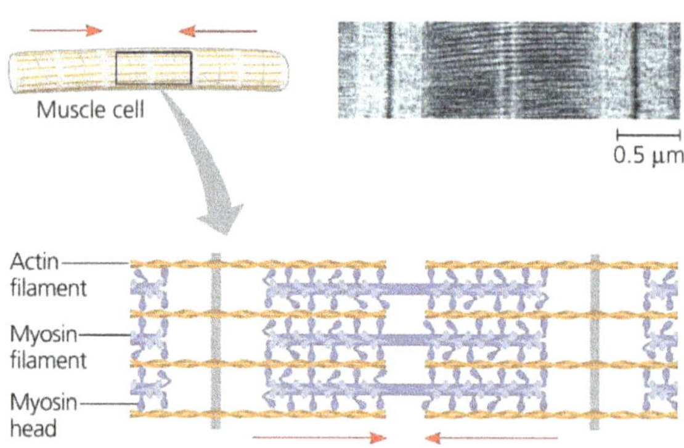

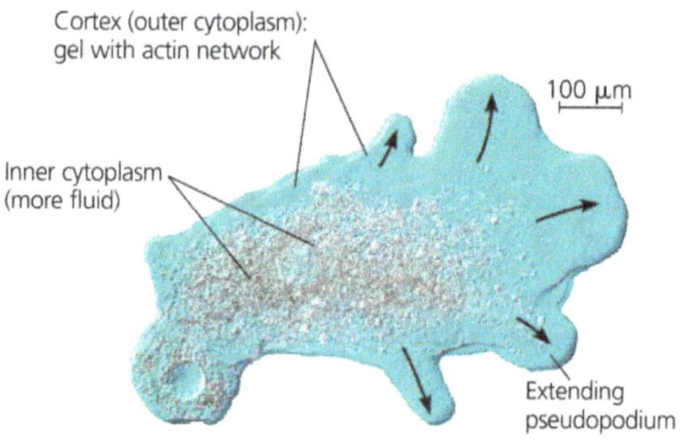

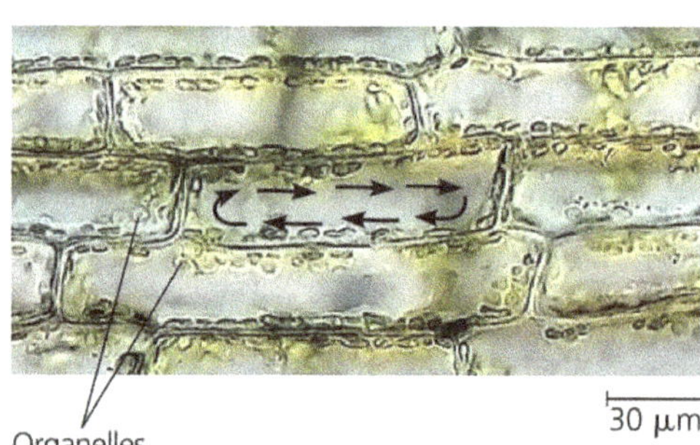

▲ **Figure 3.48 Microfilaments and motility.**

- Muscle cell contraction.

- Amoeboid movement.

- Cytoplasmic streaming in plant cells.

- **Intermediate Filaments**

Intermediate filaments are tough, insoluble protein fibers that resemble woven ropes and are the most durable element of the cytoskeleton in animal cells. Made of twisted units of tetramer (4) fibrils. These intermediate filaments are characteristically 8 to 10 nm in diameter—between the size of actin filaments and microtubules. Once formed, intermediate filaments are stable and usually do not break down. They are found in parts of cells subject to tension (such as stretching), help hold organelles such as the nucleus in place, and help attach cells to one another. While microtubules and microfilaments are found in all eukaryotic cells, intermediate filaments are only found in the cells of some animals, including vertebrates.

Specialized for bearing tension (like microfilaments), intermediate filaments are a diverse class of cytoskeletal elements. Microtubules and microfilaments, in contrast, are consistent in diameter and composition in all eukaryotic cells. **Keratin**, a class of intermediate filament, is found in epithelial cells (cells that line organs and body cavities) and associated structures such as hair and finger nails. **Vimentin** is the most common class III intermediate filament protein and is found in most cells derived from embryonic mesenchyme. **Lamins** are a family of seven isoforms present in the cell nucleus, where they form a structural framework called the nuclear lamina just inside the nuclear envelope. The intermediate filaments of nerve cells are called **neurofilaments**.

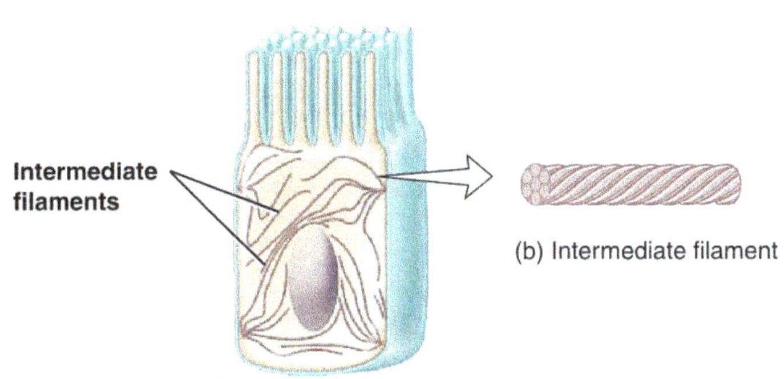
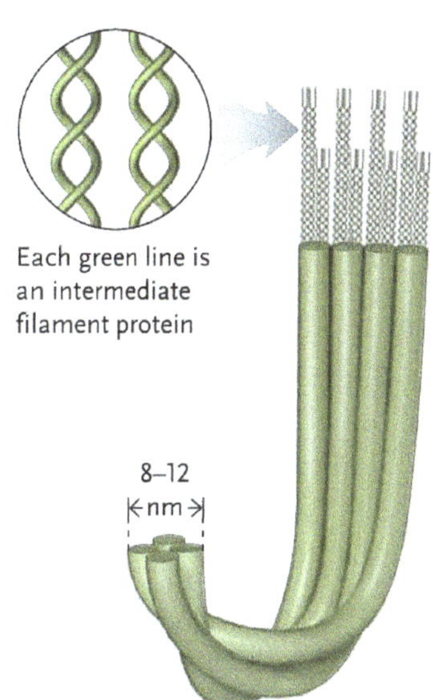

► **Figure 3.49 Intermediate filament**

Intermediate filaments are more permanent fixtures of cells than are microfilaments and microtubules, which are often disassembled and reassembled in various parts of a cell. Even after cells die, intermediate filament networks often persist; for example, the outer layer of our skin consists of dead skin cells full of keratin filaments. Intermediate filaments are especially sturdy and play an important role in reinforcing the shape of a cell and fixing the position of certain organelles.

Cell Surface Specializations (cell coverings and cell junctions)

A basic feature of multicellular animals is the formation of diverse kinds of tissue, such as skin, blood, or muscle, where cells are organized in specific ways. Cells must also be able to communicate with each other and have markers of individual identity. All of these functions—**connections between cells, markers of cellular identity, and cell communication**—involve membrane proteins and proteins secreted by cells. As an organism develops, the cells acquire their identities by carefully controlling the expression of those genes, turning on the specific set of genes that encode the functions of each cell type.

Cell Coverings:

- **Glycocalyx:** Many cells are surrounded by a **glycocalyx**, or cell coat, formed by branching sugar (oligosaccharide chains) groups covalently attached to proteins and lipids that are part of the plasma membrane. The glycocalyx protects the cell and may help keep other cells at a distance. Certain molecules of the glycocalyx enable cells to recognize one another, to make contact, and in some cases to form adhesive or communicating associations. Other molecules of the cell coat contribute to the mechanical strength of multicellular tissues.

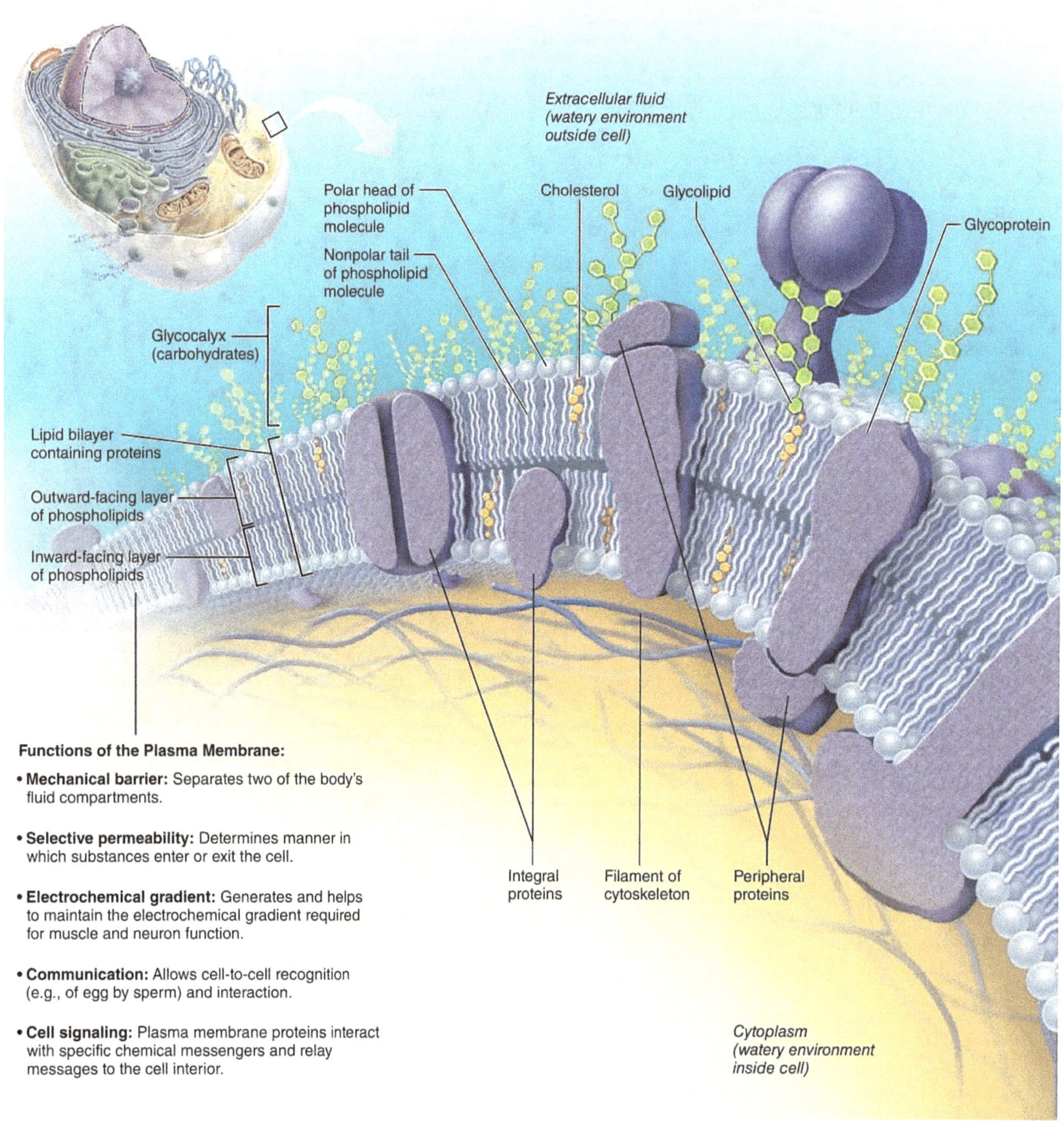

Most tissue-specific cell-surface markers are **glycolipids**—that is, lipids with carbohydrate heads. The glycolipids on the surface of red blood cells are also responsible for the A, B, and O blood types. One example of the function of cell-surface markers is the recognition of "self " and "nonself " cells by the immune system. This function is vital for multicellular organisms, which need to defend themselves against invading or malignant cells. The immune system of vertebrates uses a particular set of markers to distinguish self from nonself cells, encoded by genes of the major histocompatibility complex (**MHC**). The best-understood glycocalyx molecules fall into two large families:

1. Cell adhesion molecules

Thousands of cell adhesion molecules (**CAMs**) are found on almost every cell in the body. CAMs play key roles in embryonic development and wound repair (situations where cell mobility is important) and in immunity. These sticky glycoproteins (cadherins and integrins) act as:

- The molecular "Velcro" that cells use to anchor themselves to molecules in the extracellular space and to each other.

- The "arms" that migrating cells use to haul themselves past one another.

- SOS signals sticking out from the blood vessel lining that rally protective white blood cells to a nearby infected or injured area.

- Mechanical sensors that respond to changes in local tension or fluid movement at the cell surface by stimulating synthesis or degradation of tight junctions.

- Transmitters of intracellular signals that direct cell migration, proliferation, and specialization.

2. Plasma membrane receptors

A huge and diverse group of integral proteins and glycoproteins that serve as binding sites are collectively known as membrane receptors. Some function in contact signaling, and others in chemical signaling.

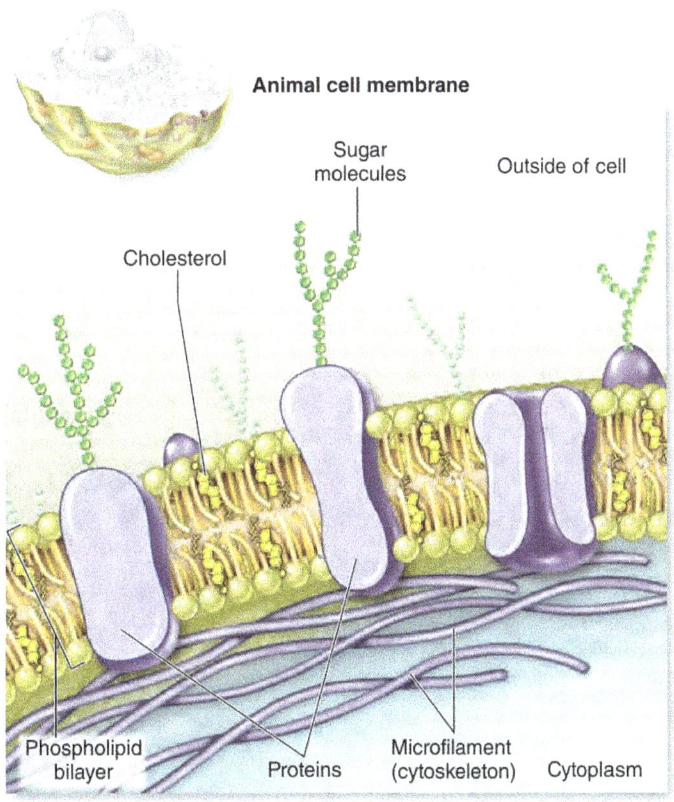

• **Cell Walls of Plants Are Strong, Flexible, and porous**

Prokaryotes, plants, fungi, and many types of protists have cell walls, which protect and support the cells. The cell wall is an extracellular structure because it lies outside the plasma membrane. The cell walls of eukaryotes are chemically and structurally different from prokaryotic cell walls. In plants and protists, the cell walls are composed of fibers of the polysaccharide **cellulose**, whereas in fungi, the cell walls are composed of **chitin**.

The cell wall is an extracellular structure of plant cells. This is one of the features that distinguishes plant cells from animal cells. The wall provides structural support, protects the plant cell, maintains its shape, prevents it from bursting due to increased hydrostatic pressure and prevents excessive uptake of water. At the level of the whole plant, the strong walls of specialized cells hold the plant up against the force of gravity.

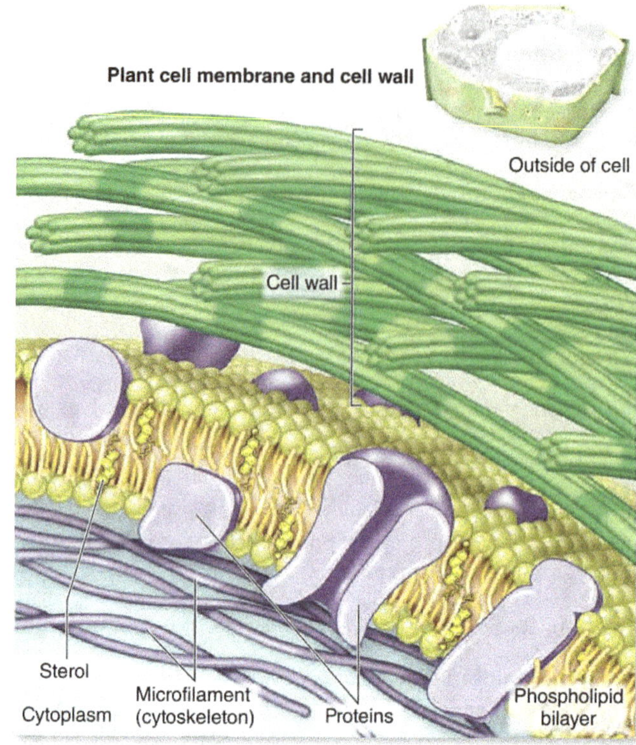

Plant cell walls are much thicker than the plasma membrane. Microfibrils made of the polysaccharide cellulose are synthesized by an enzyme called cellulose synthase and secreted to the extracellular space, where they become embedded in a matrix of other polysaccharides and proteins. A young plant cell first secretes a relatively thin and flexible wall called the **primary cell wall**. It is usually very flexible and allows new cells to increase in size. The following are the main components of the primary cell wall:

Cellulose: The main macromolecule of the plant cell wall is cellulose, a polysaccharide made of repeating molecules of glucose attached end to end. These glucose polymers associate with each other via hydrogen bonding to form microfibrils that provide great tensile strength. Cellulose fibres are inelastic.

Hemicellulose: Hemicellulose is another linear polysaccharide, with a structure similar to that of cellulose, but it contains sugars other than glucose in its structure and usually forms thinner microfibrils.

Glycans: Polysaccharides with branching structures are also important in cell-wall structure. The crosslinking glycans bind to cellulose and provide organization to the cellulose microfibrils.

Pectins: Highly negatively charged polysaccharides, called pectins, attract water and have a gel-like character. They provide the cell wall with the ability to resist compression.

▶ **Figure 3.50 Structure of cellulose**

Between primary walls of adjacent cells is the **middle lamella**, a thin layer rich in sticky polysaccharides called **pectins**. The middle lamella glues adjacent cells together. When the cell matures and stops growing, it strengthens its wall. Some plant cells do this simply by secreting hardening substances into the primary wall. Other cells add a **secondary cell wall** between the plasma membrane and the primary wall. The secondary wall, often deposited in several laminated layers, has a strong and durable matrix that affords the cell protection and support. In a given layer the cellulose fibres are parallel, but the fibres of different layers run in different directions forming a cross-ply structure which is stronger as a result.

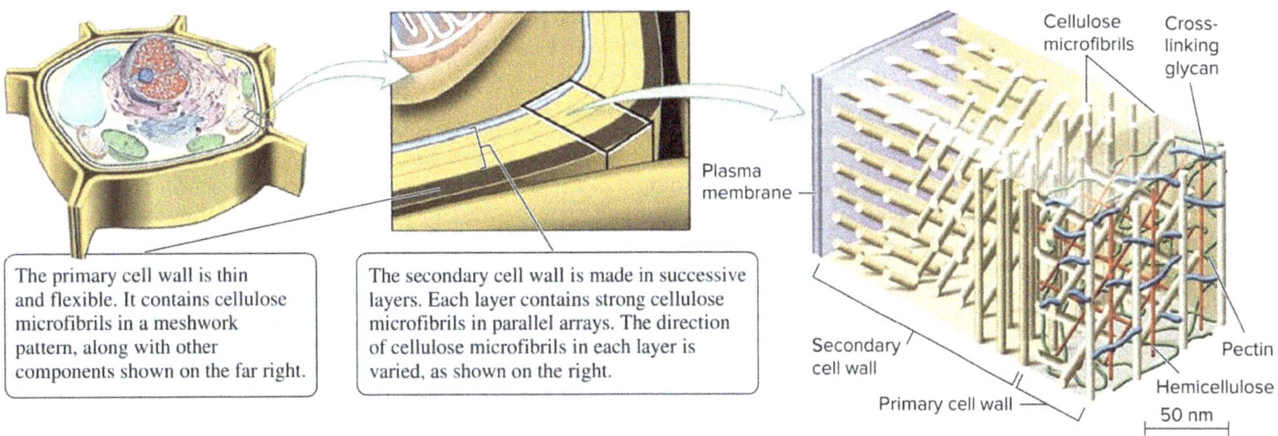

Lignin is a hard and highly resistant substance assembled from complex alcohols, surrounds the cellulose fibers in secondary wall. Lignin-impregnated cell walls are actually stronger than reinforced concrete by weight; hence, trees can grow to substantial size. Wood (secondary xylem) strength is needed for support in shrubs and trees. The system of interconnected cell walls in a plant is called the **apoplast**. It is a major transport route for water, inorganic ions and other materials.

Living connections through neighbouring cell walls, the **plasmodesmata**, help form another transport pathway through the plant known as the **symplast**. The cell walls of the root endodermis are impregnated with **suberin**, a waterproof substance that forms a barrier to the movement of water, thus helping in the control of water and mineral ion uptake by the plant. Epidermal cells often have a waterproof layer of waxy cutin, the **cuticle**, on their outer walls. This helps reduce water loss by evaporation.

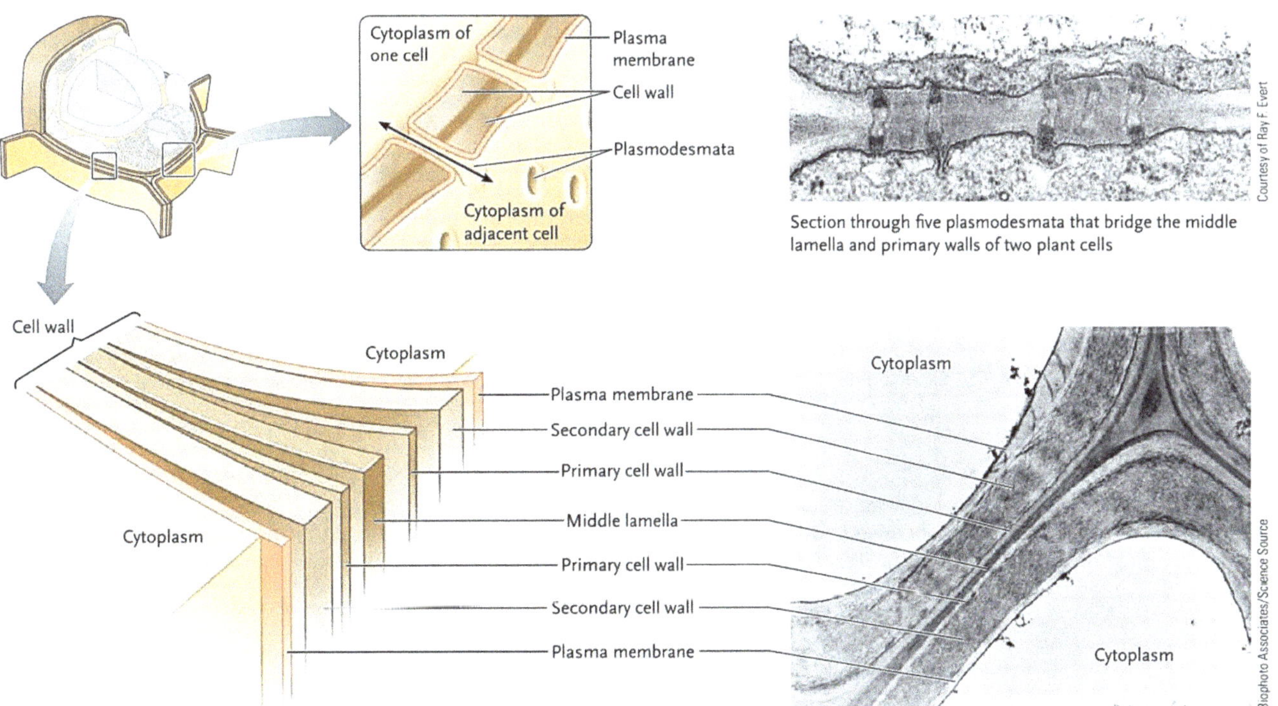

Section through five plasmodesmata that bridge the middle lamella and primary walls of two plant cells

- **The Extracellular Matrix (ECM) of Animal Cells**

Animal cells that are not in direct contact with one another have an elaborate **extracellular matrix (ECM),** a complex mixture of molecules that often includes polysaccharides and fibrous proteins. The composition and function of ECM vary by the type of cell that secretes it. A **cuticle** is a type of ECM secreted by cells at a body surface. In plants, a cuticle of waxes and proteins helps stems and leaves fend off insects and retain water. Crabs, spiders, and other arthropods have a cuticle that consists mainly of chitin, a tough polysaccharide.

A cell wall is also a type of ECM. Bacteria and archaea secrete a wall around their plasma membrane, as do fungi, plants, and some protists. Cell wall structure differs among these groups, but in all cases it protects, supports, and imparts shape to the cell. Animal cells have no walls, but some secrete an extracellular matrix called basement membrane. **Basement membrane** is a sheet of fibrous material that structurally supports and organizes tissues, and it has roles in cell signaling. It is not considered to be a cell membrane because it does not consist of a lipid bilayer.

The primary function of the ECM is protection and support. The ECM forms the mass of skin, bones, and tendons; it also forms many highly specialized extracellular structures such as the cornea of the eye and filtering networks in the kidney. The ECM also affects cell division, adhesion, motility, and embryonic development, and it takes part in reactions to wounds and disease. The proteins found in the ECM are grouped into adhesive proteins, such as **fibronectin** and **laminin**, and structural proteins, such as **collagen** and **elastin.**

The main ingredients of the ECM are **glycoproteins** and other carbohydrate-containing molecules secreted by the cells (glycoproteins are proteins with covalently bonded carbohydrates). The most abundant glycoprotein in the ECM of most animal cells is collagen, which forms strong fibers outside the cells. In fact, collagen accounts for about 40% of the total protein in the human body. A key function of collagen is to impart tensile strength, which is a measure of how much stretching force a material can bear without tearing apart. Elasticity is needed in regions of the body, such as the lungs and blood vessels, that regularly expand and return to their original shape. In these places, the ECM contains elastic fibers composed primarily of the protein elastin.

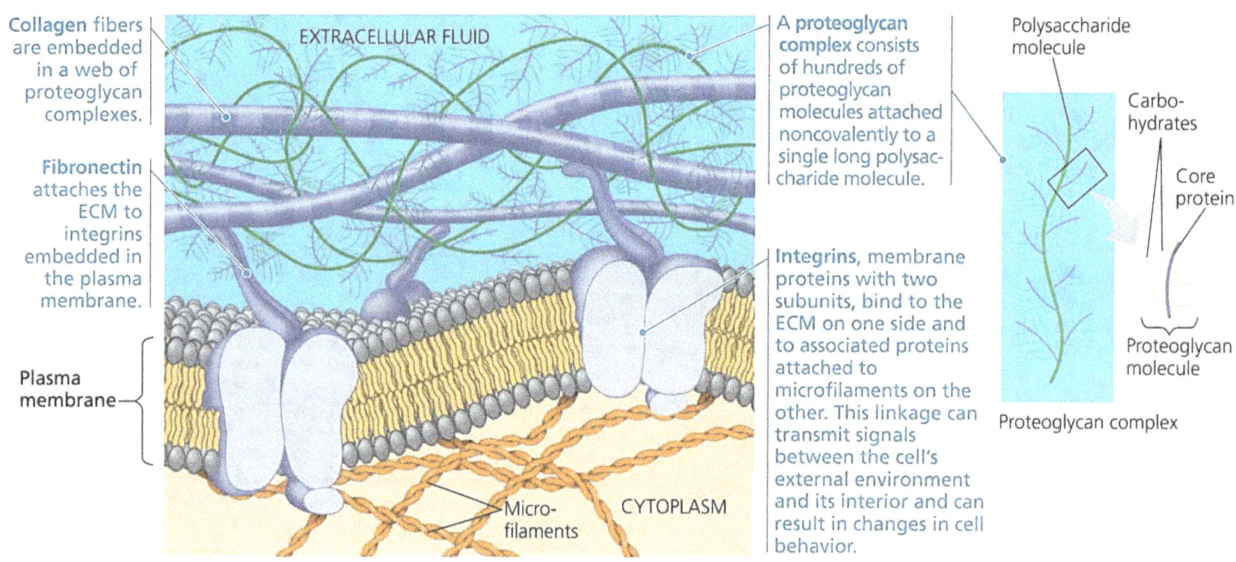

▲ **Figure 3.51 Extracellular matrix (ECM) of an animal cell.** proteoglycans, collagen, and fibronectin.

Fibronectin and laminin (Adhesive Proteins) have multiple binding sites that bind to other components in the ECM, such as protein fibers and polysaccharides. The same proteins also have binding sites for receptors on the surfaces of cells. Therefore, adhesive proteins are so named because they make ECM components adhere to one another and to the cell surface. They provide organization to the ECM and facilitate the attachment of cells to the ECM.

Fibronectin and other ECM proteins bind to cell surface receptor proteins called **integrins** that are built into the plasma membrane. Integrins span the membrane and bind on their cytoplasmic side to associated proteins attached to microfilaments of the cytoskeleton. The name integrin is based on the word integrate: Integrins are in a position to transmit signals between the ECM and the cytoskeleton and thus to integrate changes occurring outside and inside the cell.

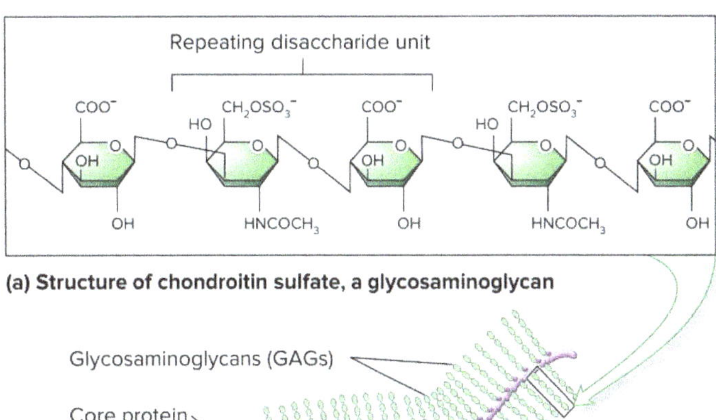

(a) Structure of chondroitin sulfate, a glycosaminoglycan

(b) General structure of a proteoglycan

▶ **Figure 3.52** Structures of glycosaminoglycans and proteoglycans.

By communicating with a cell through integrins, the ECM can regulate a cell's behavior. For example, some cells in a developing embryo migrate along specific pathways by matching the orientation of their microfilaments to the "grain" of fibers in the extracellular matrix. Researchers have also learned that the extracellular matrix around a cell can influence the activity of genes in the nucleus. Information about the ECM probably reaches the nucleus by a combination of mechanical and chemical signaling pathways.

Mechanical signaling involves fibronectin, integrins, and microfilaments of the cytoskeleton. Changes in the cytoskeleton may in turn trigger chemical signaling pathways inside the cell, leading to changes in the set of proteins being made by the cell and therefore changes in the cell's function. In this way, the extracellular matrix of a particular tissue may help coordinate the behavior of all the cells of that tissue.

The collagen and elastin are embedded in a network woven of secreted **proteoglycans**. Polysaccharides are the second major component of the ECM of animals. As discussed before, polysaccharides are polymers of many simple sugars. Among vertebrates, the most abundant types of polysaccharides in the ECM are **glycosaminoglycans** (GAGs). These macromolecules are long, unbranched polysaccharides containing a repeating disaccharide unit. GAGs are highly negatively charged molecules that tend to attract positively charged ions and water.

The majority of GAGs in the ECM are covalently attached to core proteins, forming proteoglycans. Large proteoglycan complexes can form when hundreds of proteoglycan molecules become noncovalently attached to a single long polysaccharide molecule. Two examples of GAGs are **chondroitin sulfate**, which is a major component of cartilage, and **hyaluronic acid**, which is found in the skin, eyes, and joint fluid. The primary function of GAGs and proteoglycans is to resist compression. Once secreted from cells, these macromolecules form a gel-like component in the ECM. How is this gel-like property important? Due to its high water content, the ECM is difficult to compress and thereby protects cells. GAGs and proteoglycans are found abundantly in regions of the body that are subjected to harsh mechanical forces, such as the joints of the human body.

The consistency of the matrix, which may range from soft and jellylike to hard and elastic, depends on a network of proteoglycans that surrounds the collagen fibers. Matrix consistency depends on the number of interlinks in this network, which determines how much water can be trapped in it. For example, cartilage, which contains a high proportion of interlinked glycoproteins, is relatively soft. Tendons, which are almost pure collagen, are tough and elastic. In bone, the glycoprotein network that surrounds collagen fibers is impregnated with mineral crystals, producing a dense and hard—but still elastic—structure that is about as strong as fiberglass or reinforced concrete.

Cell Junctions:

In multicelled species, cells interact with one another and their surroundings by way of cell junctions, which are structures that connect a cell to other cells and to its environment. Cells send and receive substances and signals through some junctions. Other kinds help cells recognize and stick to each other and to ECM. However, in tissues, there is usually a space between the plasma membranes of adjacent cells. This space, filled with extracellular (**interstitial**) fluid, provides a pathway for substances to pass between cells on their way to and from the blood.

- **Adhesive junctions** (Adherens junctions, Desmosomes, Hemidesmosomes and focal adhesions)

- **Septate or Tight junctions** (Septate junctions and Tight junctions)

- **Communicating junctions** (Gap junctions in animals and Plasmodesmata in plants)

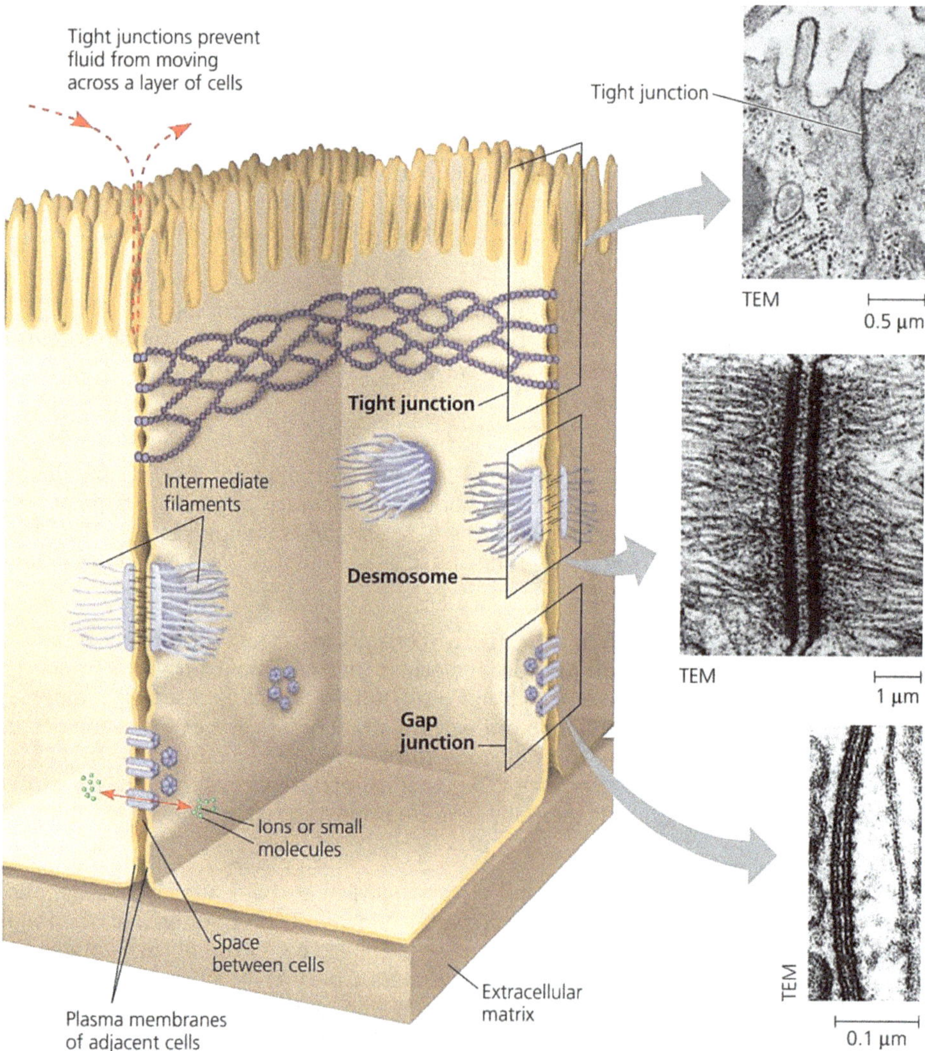

▲ **Figure 3.53 Cell Junctions in Animal Tissues.**

Septate or Tight junctions

Septate junctions are found in both invertebrates and vertebrates and form a barrier that can seal off a sheet of cells. The proteins found at these junctions have been given different names in different systems; in Drosophila, the proteins include Discs Large and **Neurexin**.

Tight junctions (Although called "impermeable" junctions, some tight junctions are leaky and may allow certain ions to pass.) are unique to vertebrates and contain proteins called **Claudins** because of their ability to occlude or block substances from passing between cells. At tight junctions, the plasma membranes of neighboring cells are very tightly pressed against each other, bound together by specific proteins. In a tight junction, a series of integral protein molecules in the plasma membranes of adjacent cells fuse together. Proteins anchored in membranes connect to actin in the cytoskeleton and join cells into sheets, such as those lining the inside of the human digestive tract and the tubules of the kidneys.

These connections allow the body to control where biochemicals move, since fluids cannot leak between the joined cells. For example, tight junctions prevent stomach acid from seeping into the tissues surrounding the stomach. Likewise, tight junctions create the "blood–brain barrier," densely packed cells that prevent many harmful substances from entering the brain. However, this barrier readily admits lipid-soluble drugs such as heroin and cocaine across its cell membranes, accounting for the rapid action of these drugs. Tight junctions between skin cells make us watertight.

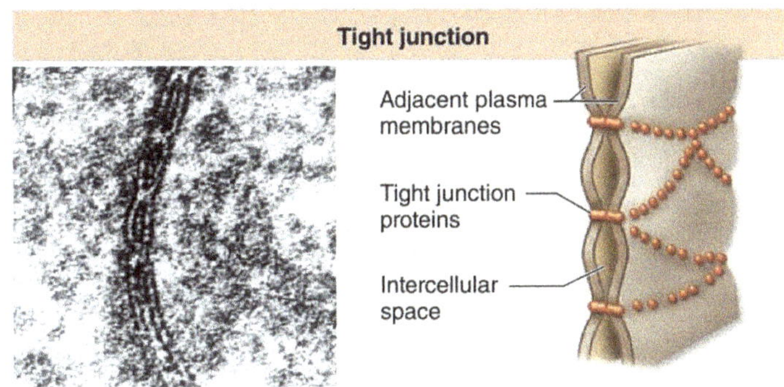

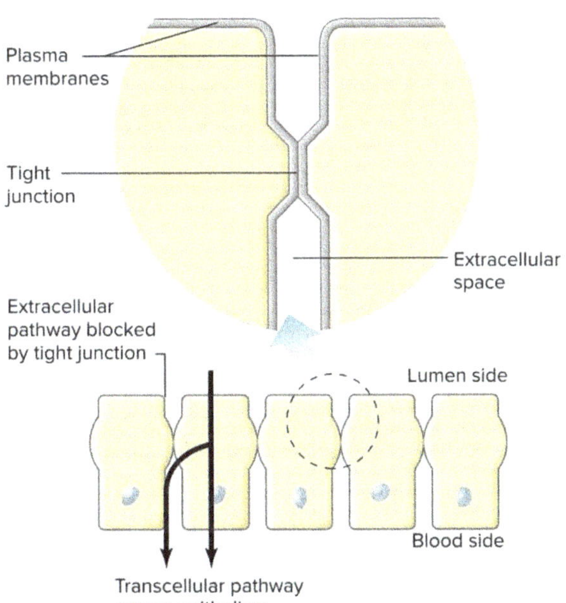

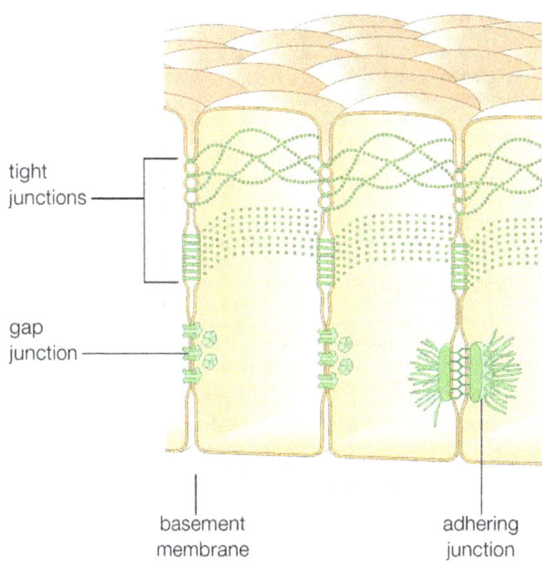

Communicating junctions

1. Gap junctions (or nexus):

Provide cytoplasmic channels from one cell to an adjacent cell and in this way are similar in their function to the plasmodesmata in plants. Gap junctions are found in both invertebrates and vertebrates. In invertebrates they are formed by proteins known as **pannexins**. In vertebrates pannexin-base gap junctions exist, but there is an additional type based on similar proteins called **connexons**. In each case, a structure is formed by complexes of six identical transmembrane proteins.

The proteins are arranged in a circle to create a channel through the plasma membrane that protrudes several nanometers from the cell surface. A gap junction forms when the connexons/pannexins of two cells align perfectly, creating an open channel that spans the plasma membranes of both cells. These proteins create pores through which ions, sugars, amino acids, and other small molecules may pass. Gap junctions are necessary for communication between cells in many types of tissues, such as heart muscle, and in animal embryos.

Gap junctions link heart muscle cells to one another, allowing groups of cells to contract together. Similarly, the muscle cells that line the digestive tract coordinate their contractions to propel food along its journey, courtesy of countless gap junctions. Gap junction channels are dynamic structures that can open or close in response to a variety of factors, including Ca^{2+} and H^+ ions. This gating serves at least one important function. When a cell is damaged, its plasma membrane often becomes leaky. Ions in high concentrations outside the cell, such as Ca^{2+}, flow into the damaged cell and close its gap junction channels. This isolates the cell and prevents the damage from spreading.

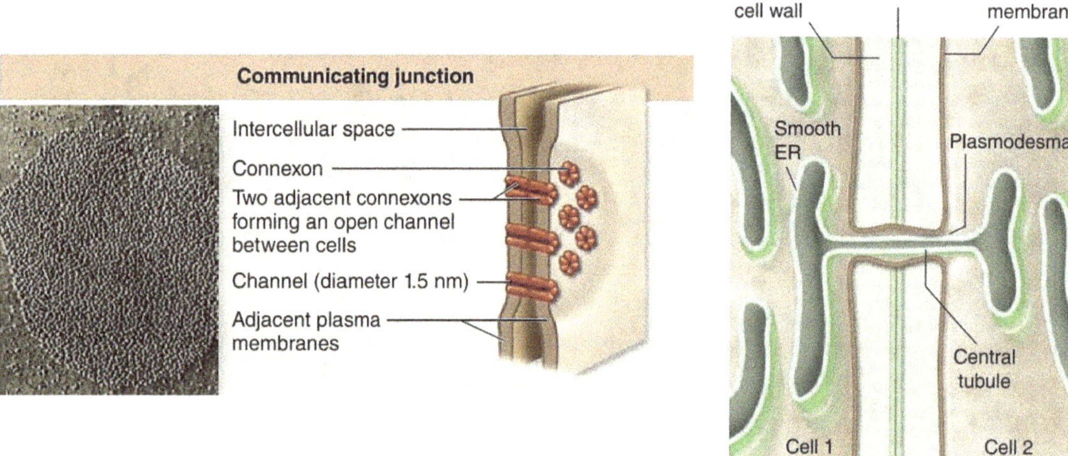

2. Plasmodesmata in Plant Cells

It might seem that the nonliving cell walls of plants would isolate plant cells from one another. But in fact, many plant cell walls are perforated with plasmodesmata (singular, plasmodesma), channels that connect cells. The plasma membranes of adjacent cells line the channel of each plasmodesma and thus are continuous. Because the channels are filled with cytosol, the cells share the same internal chemical environment. By joining adjacent cells, plasmodesmata unify most of the plant into one living continuum. Most plasmodesmata also contain a narrow tubelike structure derived from the smooth endoplasmic reticulum of the connected cells.

Water and small solutes can pass freely from cell to cell, certain proteins and RNA molecules can do this as well. Plasmodesmata, however, may also play a role in the spread of disease within a plant; viruses use them as conduits to pass from cell to cell. The macromolecules transported to neighboring cells appear to reach the plasmodesmata by moving along fibers of the cytoskeleton.

Adhesive junctions

They mechanically attach the cytoskeleton of a cell to the cytoskeletons of other cells or to the extracellular matrix. Protein complexes at each anchoring junction span the cell membrane and link to each cell's cytoskeleton. These junctions are found in tissues subject to mechanical stress, such as muscle and skin epithelium.

- **Adherens junctions** are based on the protein **cadherin**, which is a Ca^{2+} dependent adhesion molecule with very wide phylogenetic distribution. Cadherin is a single-pass transmembrane protein with an extracellular domain that can interact with the extracellular domain of a cadherin in an adjacent cell to join the cells together. On the cytoplasmic side, the cadherins interact indirectly through other proteins with actin to form flexible connections between cells.

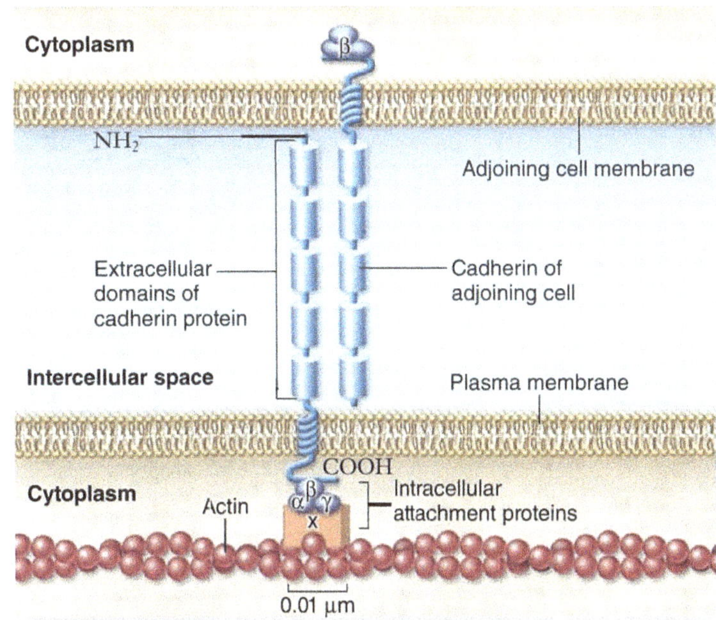

▶ **Figure 3.54 A cadherin-mediated junction.** The cadherin molecule is anchored to actin in the cytoskeleton and passes through the membrane to interact with the cadherin of an adjoining cell.

- **Desmosomes** (one type of anchoring junction) are a cadherin-based junction unique to vertebrates. On the cytoplasmic face of each plasma membrane is a buttonlike thickening called a plaque. Adjacent cells are held together by thin linker protein filaments (cadherins) that extend from the plaques and fit together like the teeth of a zipper in the intercellular space.They contain the cadherins **desmocollin** and **desmoglein**, which interact with intermediate filaments of cytoskeletons instead of actin. Desmosomes join adjacent cells. These connections support tissues against mechanical stress. Desmosomes attach muscle cells to each other in a muscle. Some "muscle tears" involve the rupture of desmosomes.

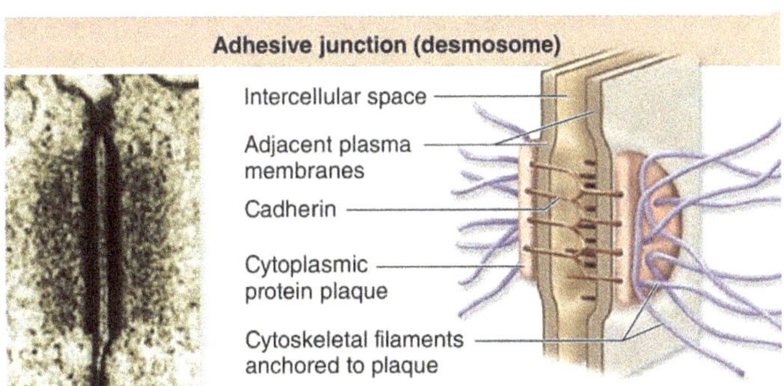

- **Hemidesmosomes** and **focal adhesions** connect cells to the basal lamina or other ECM. In this case the proteins that interact with the ECM are called **integrins**. The integrins are members of a large superfamily of cell-surface receptors that bind to a protein component of the extracellular matrix. At least 20 different integrins exist, each with a differently shaped binding domain. These junctions also connect to the cytoskeleton of cells: actin filaments at focal adhesions and intermediate filaments at hemidesmosomes.

CELL PART*	STRUCTURE	FUNCTIONS
Plasma Membrane (Figure 3.3)		
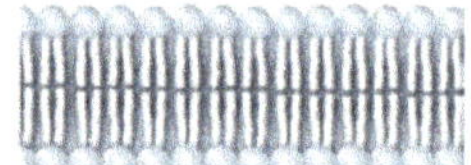	Membrane made of a double layer of lipids (phospholipids, cholesterol, and so on) within which proteins are embedded. Proteins may extend entirely through the lipid bilayer or protrude on only one face. Most externally facing proteins and some lipids have attached sugar groups.	Serves as an external cell barrier, and acts in transport of substances into or out of the cell. Maintains a resting potential that is essential for functioning of excitable cells. Externally facing proteins act as receptors (for hormones, neurotransmitters, and so on), transport proteins, and in cell-to-cell recognition.
Cytoplasm		
	Cellular region between the nuclear and plasma membranes. Consists of fluid **cytosol** containing dissolved solutes, **organelles** (the metabolic machinery of the cytoplasm), and **inclusions** (stored nutrients, secretory products, pigment granules).	
Organelles		
• Mitochondria (Figure 3.15) 	Rodlike, double-membrane structures; inner membrane folded into projections called cristae.	Site of ATP synthesis; powerhouse of the cell.
• Ribosomes (Figures 3.32, 3.33, Focus Figure 3.4)   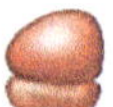	Dense particles consisting of two subunits, each composed of ribosomal RNA and protein. Free or attached to rough endoplasmic reticulum.	The sites of protein synthesis.
• Rough endoplasmic reticulum (Figures 3.16, 3.33) 	Membranous system enclosing a cavity, the cistern, and coiling through the cytoplasm. Externally studded with ribosomes.	Sugar groups are attached to proteins within the cisterns. Proteins are bound in vesicles for transport to the Golgi apparatus and other sites. External face synthesizes phospholipids.
• Smooth endoplasmic reticulum (Figure 3.16) 	Membranous system of sacs and tubules; free of ribosomes.	Site of lipid and steroid (cholesterol) synthesis, lipid metabolism, and drug detoxification.
• Golgi apparatus (Figures 3.17, 3.18) 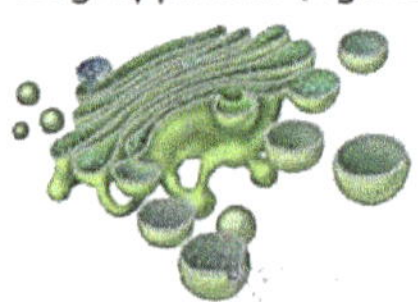	A stack of flattened membranes and associated vesicles close to the nucleus.	Packages, modifies, and segregates proteins for secretion from the cell, inclusion in lysosomes, and incorporation into the plasma membrane.
• Peroxisomes (Figure 3.2) 	Membranous sacs of catalase and oxidase enzymes.	The enzymes detoxify a number of toxic substances. The most important enzyme, catalase, breaks down hydrogen peroxide.

CELL PART*	STRUCTURE	FUNCTIONS
Cytoplasm		
• Lysosomes (Figure 3.19)	Membranous sacs containing acid hydrolases.	Sites of intracellular digestion.
• Microtubules (Figures 3.21–3.23)	Cylindrical structures made of tubulin proteins.	Support the cell and give it shape. Involved in intracellular and cellular movements. Form centrioles and cilia and flagella, if present.
• Intermediate filaments (Figure 3.21)	Protein fibers; composition varies.	The stable cytoskeletal elements; resist mechanical forces acting on the cell.
• Microfilaments (Figure 3.21)	Fine filaments composed of the protein actin.	Involved in muscle contraction and other types of intracellular movement, help form the cell's cytoskeleton.
• Centrioles (Figure 3.22)	Paired cylindrical bodies, each composed of nine triplets of microtubules.	As part of the centrosome, organize a microtubule network during mitosis (cell division) to form the spindle and asters. Form the bases of cilia and flagella.
Inclusions	Varied; includes stored nutrients such as lipid droplets and glycogen granules, protein crystals, pigment granules.	Storage for nutrients, wastes, and cell products.
Cellular Extensions		
• Cilia (Figures 3.23, 3.24)	Short cell-surface projections; each cilium composed of nine pairs of microtubules surrounding a central pair.	Coordinated movement creates a unidirectional current that propels substances across cell surfaces.
• Flagellum	Like a cilium, but longer; only example in humans is the sperm tail.	Propels the cell.
• Microvilli (Figure 3.25)	Tubular extensions of the plasma membrane; contain a bundle of actin filaments.	Increase surface area for absorption.
Nucleus (Figures 3.2, 3.26)		
	Largest organelle. Surrounded by the nuclear envelope; contains fluid nucleoplasm, nucleoli, and chromatin.	Control center of the cell; responsible for transmitting genetic information and providing the instructions for protein synthesis.

CELL PART*	STRUCTURE	FUNCTIONS

Nucleus _(continued)_ (Figures 3.2, 3.26)

CELL PART*	STRUCTURE	FUNCTIONS
• Nuclear envelope (Figure 3.26) 	Double-membrane structure pierced by pores. Outer membrane continuous with the endoplasmic reticulum.	Separates the nucleoplasm from the cytoplasm and regulates passage of substances to and from the nucleus.
• Nucleolus (Figure 3.26) 	Dense spherical (non-membrane-bounded) bodies, composed of ribosomal RNA and proteins.	Site of ribosome subunit manufacture.
• Chromatin (Figure 3.27) 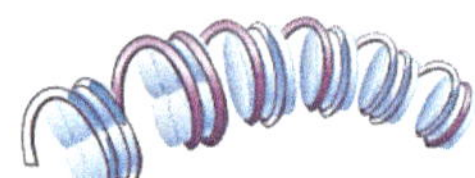	Granular, threadlike material composed of DNA and histone proteins.	DNA constitutes the genes.

Prokaryotes	Eukaryotes
Prokaryotes are thought to have evolved about 3.5 billion years ago.	Eukaryotes are thought to have evolved about 1.5 billion years ago.
Their typical diameter is 1–5 μm.	Cells are up to 40 μm diameter and up to 1000 times the volume of prokaryotic cells.
DNA is circular and free in the cytoplasm; it is not surrounded by a double membrane.	DNA is not circular and is contained in a nucleus. The nucleus is surrounded by a double membrane (the nuclear envelope).
70S ribosomes are present (smaller than those of eukaryotes).	80S ribosomes are present (larger than those of prokaryotes).
Very few types of cell organelle are present. No separate membrane-bound organelles are present.	Many types of cell organelle are present. • Some organelles are surrounded by a single membrane (e.g. lysosomes, Golgi apparatus, vacuoles, ER). • Some are surrounded by an envelope of two membranes (e.g. nucleus, mitochondrion, chloroplast). • Some have no membrane (e.g. ribosomes, centrioles, microtubules).
The cell wall contains peptidoglycan (a polysaccharide combined with amino acids).	A cell wall is sometimes present (e.g. in plants and fungi); it contains cellulose or lignin in plants, and chitin (a nitrogen-containing polysaccharide similar to cellulose) in fungi.
Flagella are simple and lack microtubules; they project outside the cell surface membrane so they are extracellular (outside the cell).	Flagella (and cilia) are complex with a '9 + 2' arrangement of microtubules; they are surrounded by the cell surface membrane so they are intracellular (inside the cell).
Cell division occurs by binary fission (the cell splits into two); it does not involve a spindle (see Chapter 6).	Cell division takes place by mitosis or meiosis and involves a spindle (see Chapter 6).
Some carry out nitrogen fixation.	None carries out nitrogen fixation.

Type of Connection	Structure	Function	Example
Surface markers	Variable, integral proteins or glycolipids in plasma membrane	Identify the cell	MHC complexes, blood groups, antibodies
Septate junctions Tight junctions	Tightly bound, leakproof, fibrous claudin protein seal that surrounds cell	Hold cells together such that materials pass through but not between the cells	Junctions between epithelial cells in the gut
Adhesive junction (desmosome)	Variant cadherins, desmocollins, bind to intermediate filaments of cytoskeleton	Creates strong flexible connections between cells. Found in vertebrates	Epithelium
Adhesive junction (adherens junction)	Classical cadherins, bind to microfilaments of cytoskeleton	Connects cells together. Oldest form of cell junction, found in all multicellular organisms	Tissues with high mechanical stress, such as the skin
Adhesive junction (hemidesmosome, focal adhesion)	Integrin proteins bind cell to extracellular matrix	Provides attachment to a substrate	Involved in cell movement and important during development
Communicating junction (gap junction)	Six transmembrane connexon/pannexin proteins creating a pore that connects cells	Allows passage of small molecules from cell to cell in a tissue	Excitable tissue such as heart muscle
Communicating junction (plasmodesmata)	Cytoplasmic connections between gaps in adjoining plant cell walls	Communicating junction between plant cells	Plant tissues

	Bacteria	Animal cells	Plant cells
Type of cell	Prokaryotic	Eukaryotic	Eukaryotic
Structures on cell surface			
Cell wall*	Present	Absent	Present
Flagella/cilia	Flagella sometimes present	Cilia or flagella present on certain cell types	Rarely present[†]
Plasma membrane	Present	Present	Present
Interior structures			
Cytoplasm	Usually a single compartment inside the plasma membrane	Composed of membrane-bound organelles that are surrounded by the cytosol	Composed of membrane-bound organelles that are surrounded by the cytosol
Ribosomes	Present	Present	Present
Chromosomes	Typically one circular chromosome per nucleoid; a nucleoid is not a separate compartment	Multiple linear chromosomes in the nucleus, which is surrounded by a double membrane; mitochondria also have chromosomes	Multiple linear chromosomes in the nucleus, which is surrounded by a double membrane; mitochondria and chloroplasts also have chromosomes
Endomembrane system	Absent	Present	Present
Mitochondria	Absent	Present	Present
Chloroplasts	Absent	Absent	Present

*Note that the biochemical composition of bacterial cell walls is very different from that of plant cell walls.
†Some plant species produce sperm cells with flagella, but flowering plants produce sperm within pollen grains that lack flagella.

- All organisms are composed of one or more cells. Cells arise only by division of preexisting cells.

- Cell size is constrained by the diffusion distance. As cell size increases, diffusion becomes inefficient.

- Magnification gives better resolution than is possible with the naked eye. Staining with chemicals enhances contrast of structures.

- All cells have centrally located DNA, a semifluid cytoplasm, and an enclosing plasma membrane.

- Prokaryotic cells contain DNA and ribosomes, but they lack a nucleus, an internal membrane system, and membrane-bounded organelles. A rigid cell wall surrounds the plasma membrane.

- Peptidoglycan is composed of carbohydrate cross-linked with short peptides.

- Archaeal cell walls do not contain peptidoglycan, and they have unique plasma membranes.

- Prokaryotic flagella rotate because of proton transfer across the plasma membrane.

- The nucleus in eukaryotic cells is surrounded by an envelope of two phospholipid bilayers; the outer layer is contiguous with the ER. Pores allow exchange of small molecules. The nucleolus is a region of the nucleoplasm where rRNA is transcribed and ribosomes are assembled.

- In most prokaryotes, DNA is organized into a single circular chromosome. In eukaryotes, numerous chromosomes are present.

- Ribosomes translate mRNA to produce polypeptides. They are found in all cell types.

- The rough ER (RER), studded with ribosomes, synthesizes and modifies proteins and manufactures membranes.

- The smooth endoplasmic reticulum (SER) lacks ribosomes; it is involved in carbohydrate and lipid synthesis and detoxification.

- The Golgi apparatus receives vesicles from the ER, modifies and packages macromolecules, and transports them.

- Lysosomes break down macromolecules and recycle the components of old organelles.

- Microbodies are a diverse category of organelles.

- Plants use vacuoles for storage and water balance.

- Mitochondria and chloroplasts have a double-membrane structure, contain their own DNA, and can divide independently.

- The inner membrane of mitochondria is extensively folded into layers called cristae. Proteins on the surface and in the inner membrane carry out metabolism to produce ATP.

- Chloroplasts capture light energy via thylakoid membranes arranged in stacks called grana, and use it to synthesize glucose.

- The endosymbiont theory proposes that mitochondria and chloroplasts were once prokaryotes engulfed by another cell.

- The cytoskeleton consists of crisscrossed protein fibers that support the shape of the cell and anchor organelles.

- Actin filaments, or microfilaments, are long, thin polymers involved in cellular movement. Microtubules are hollow structures that move materials within a cell. Intermediate filaments serve a wide variety of functions.

- Centrosomes help assemble the nuclear division apparatus of animal cells.

- Molecular motors move vesicles along microtubules, like a train on a railroad track. Kinesin and dynein are two motor proteins.

- Cell crawling occurs as actin polymerization forces the cell membrane forward, while myosin pulls the cell body forward.

- Eukaryotic flagella have a 9 +2 structure and arise from a basal body. Cilia are shorter and more numerous than flagella.

- Plants have cell walls composed of cellulose fibers. The middle lamella, between cell walls, holds adjacent cells together.

- Glycoproteins are the main component of the extracellular matrix (ECM) of animal cells.

- Glycolipids and MHC proteins on cell surfaces help distinguish self from nonself.

- Cell junctions include tight junctions, adhesive junctions, and communicating junctions. In animals, gap junctions allow the passage of small molecules between cells. In plants, plasmodesmata penetrate the cell wall and connect cells.

CHAPTER 4

Ramón Andrade, 3Dciencia/Science Source

Membrane Transport and Cell Signaling

Chapter Contents:

- Cellular membranes consist of three component groups
- Cellular membranes are fluid mosaics of lipids and proteins
- Membrane structure results in selective permeability
- Passive transport
- Active transport
- Bulk transport
- Cell Communication
- Signal reception
- Signal transduction
- Different receptors can produce the same second messengers
- Receptor subtypes can lead to different effects in different cells
- Cellular response
- Apoptosis requires integration of multiple cell-signaling pathways
- Key concepts

Cellular membranes consist of three component groups

A eukaryotic cell contains many membranes. Although they are not all identical, they share the same fundamental architecture. Cell membranes are assembled from three components:

- **Lipid.** Every cell membrane is composed of phospholipids in a bilayer. The other components of the membrane are embedded within the bilayer, which provides a flexible matrix and, at the same time, imposes a barrier to permeability. Animal cell membranes also contain cholesterol, a steroid with a polar hydroxyl group (–OH). Plant cells have other sterols, but little or no cholesterol.

- **Protein.** A major component of every membrane is a collection of proteins that float in the lipid bilayer. These proteins have a variety of functions, including transport and communication across the membrane. Many integral membrane proteins are not fixed in position. They can move about, just as the phospholipid molecules do. Some membranes are crowded with proteins, but in others, the proteins are more sparsely distributed. Membranes are structurally supported by intracellular proteins that reinforce the membrane's shape. For example, a red blood cell has a characteristic biconcave shape because a scaffold made of a protein called spectrin links proteins in the plasma membrane with actin filaments in the cell's cytoskeleton. Membranes use networks of other proteins to control the lateral movements of some key membrane proteins, anchoring them to specific sites.

- **Carbohydrate:** membrane sections assemble in the endoplasmic reticulum, transfer to the Golgi apparatus, and then are transported to the plasma membrane. The ER adds chains of sugar molecules to membrane proteins and lipids, converting them into glycoproteins and glycolipids. Different cell types exhibit different varieties of these glycoproteins and glycolipids, which act as cell identity markers.

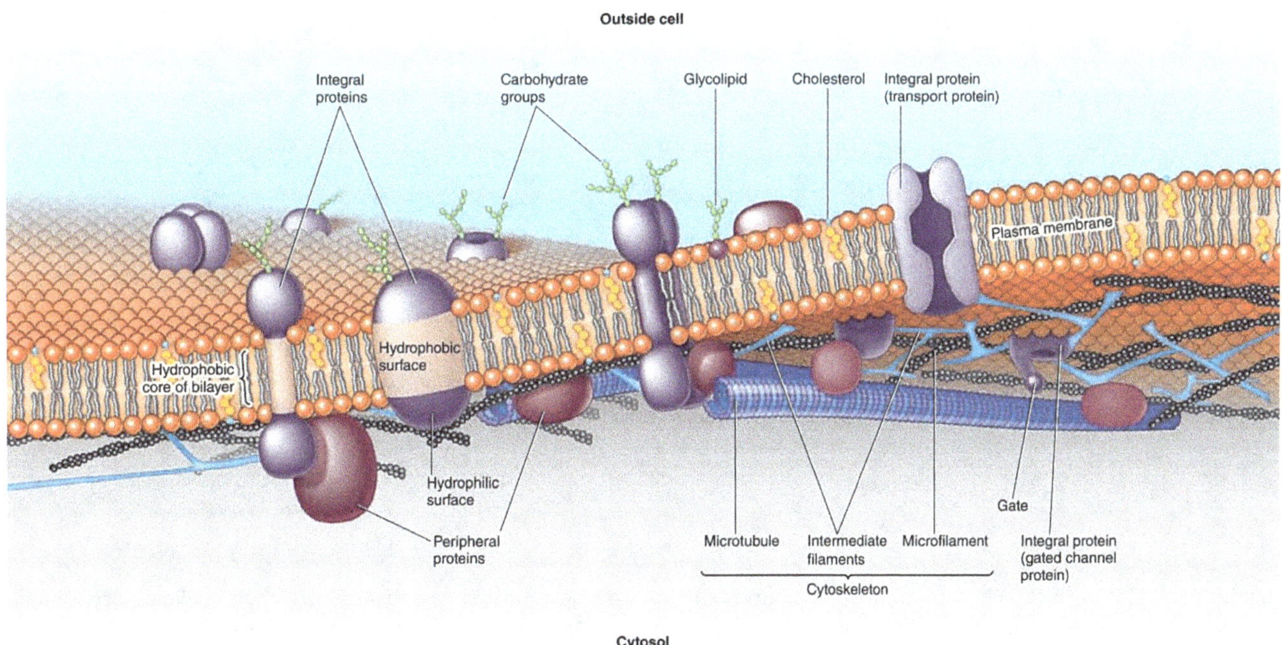

▲ **Figure 4.1** Cell membrane

Cellular membranes are fluid mosaics of lipids and proteins

Lipids and proteins are the staple ingredients of membranes, although carbohydrates are also important. We can organize membrane lipids into three classes: **Phospholipids** (**Glycerol phospholipids** and **Sphingolipids**), **glycolipids** and **Sterols** (such as cholesterol). The most abundant lipids in most membranes are phospholipids. These phospholipids include primarily the glycerol phospholipids and the sphingolipids such as sphingomyelin. Their ability to form membranes is inherent in their molecular structure.

A phospholipid is an **amphipathic** molecule, meaning it has both a hydrophilic ("water-loving") region and a hydrophobic ("water-fearing") region. A phospholipid bilayer can exist as a stable boundary between two aqueous compartments because the molecular arrangement shelters the hydrophobic tails of the phospholipids from water while exposing the hydrophilic heads to water. Because the tails of phospholipids are non-polar (hydrophobic), it is difficult for polar molecules or ions to pass through membranes. Membranes therefore act as a barrier to most water-soluble substances. This means that water-soluble molecules such as sugars, amino acids and proteins cannot leak out of the cell, and unwanted water-soluble molecules cannot enter the cell. Some phospholipids can be modified to act as signalling molecules.

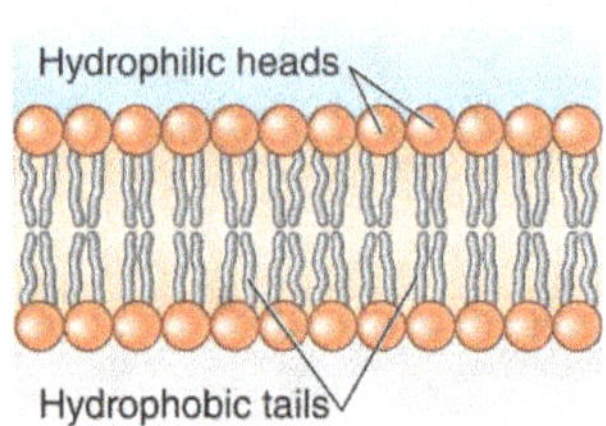

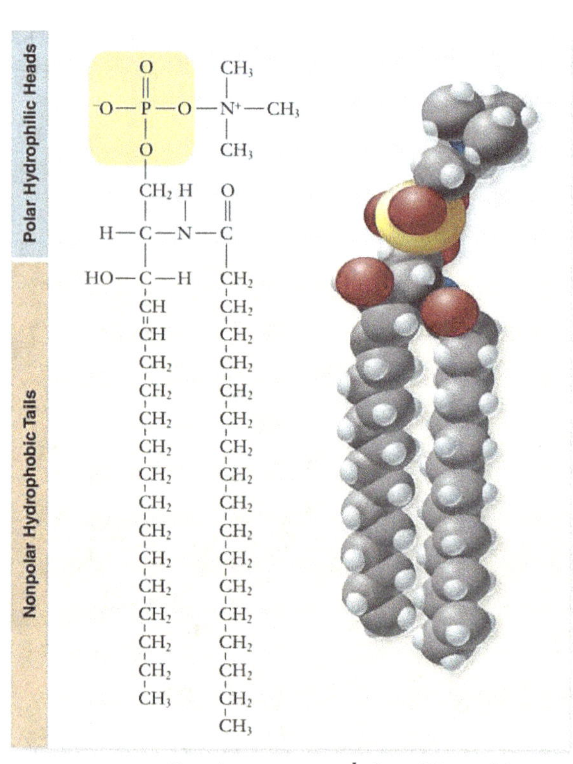

► **Figure 4.2** Phospholipids are composed of glycerol (pink) linked to two fatty acids and a phosphate group. The phosphate group (yellow) can have additional molecules attached, such as the positively charged choline (green) shown. Phosphatidylcholine is a common component of membranes. It is shown in (a) with its chemical formula, (b) as a space-filling model

Glycolipids are lipids with attached sugar groups. Found only on the outer plasma membrane surface, glycolipids account for about 5% of total membrane lipids. Their sugar groups, like the phosphate-containing groups of phospholipids, make that end of the glycolipid molecule polar, whereas the fatty acid tails are nonpolar. The carbohydrate chains help the glycolipids to act as receptor molecules.

The function of receptor molecules is to bind with particular substances at the cell surface. Different cells have different receptors, depending on their function. One group of receptors are called 'signalling receptors', because they are part of a signalling system that coordinates the activities of cells. Signalling receptors recognise messenger molecules like hormones and neurotransmitters. Some glycolipids act as cell markers allowing cells to recognise each other.

Cholesterol is a relatively small molecule. Like phospholipids, cholesterol molecules have hydrophilic heads and hydrophobic tails. They fit between the phospholipid molecules with their heads at the membrane surface. Cell surface membranes in animal cells contain almost as much cholesterol as phospholipid. Cholesterol is important for the mechanical stability of membranes.

Like membrane lipids, most membrane proteins are amphipathic. Such proteins can reside in the phospholipid bilayer with their hydrophilic regions protruding. This molecular orientation maximizes contact of hydrophilic regions of a protein with water in the cytosol and extracellular fluid, while providing their hydrophobic parts with a nonaqueous environment.

Membranes have distinct inside and outside faces. The two lipid layers may differ in lipid composition, and each protein has directional orientation in the membrane. The asymmetrical arrangement of proteins, lipids, and their associated carbohydrates in the plasma membrane is determined as the membrane is being built. The plasma membrane is not homogeneous and contains **microdomains** with distinct lipid and protein composition. This was first observed in epithelial cells in which the lipid composition of the apical and basal membranes was shown to be distinctly different. Theoretical work also showed that lipids can exist in either a disordered or an ordered phase within a bilayer. This led to the idea of lipid microdomains called **lipid rafts** that are heavily enriched in cholesterol and sphingolipids.

A lipid raft is a group of lipids that float together as a unit within a larger sea of lipids. Lipid rafts have a lipid composition that differs from the surrounding membrane. For example, they usually have a high amount of cholesterol. In addition, lipid rafts may contain unique sets of lipid-anchored proteins and transmembrane proteins. Lipid rafts may play an important role in endocytosis and cell signaling.

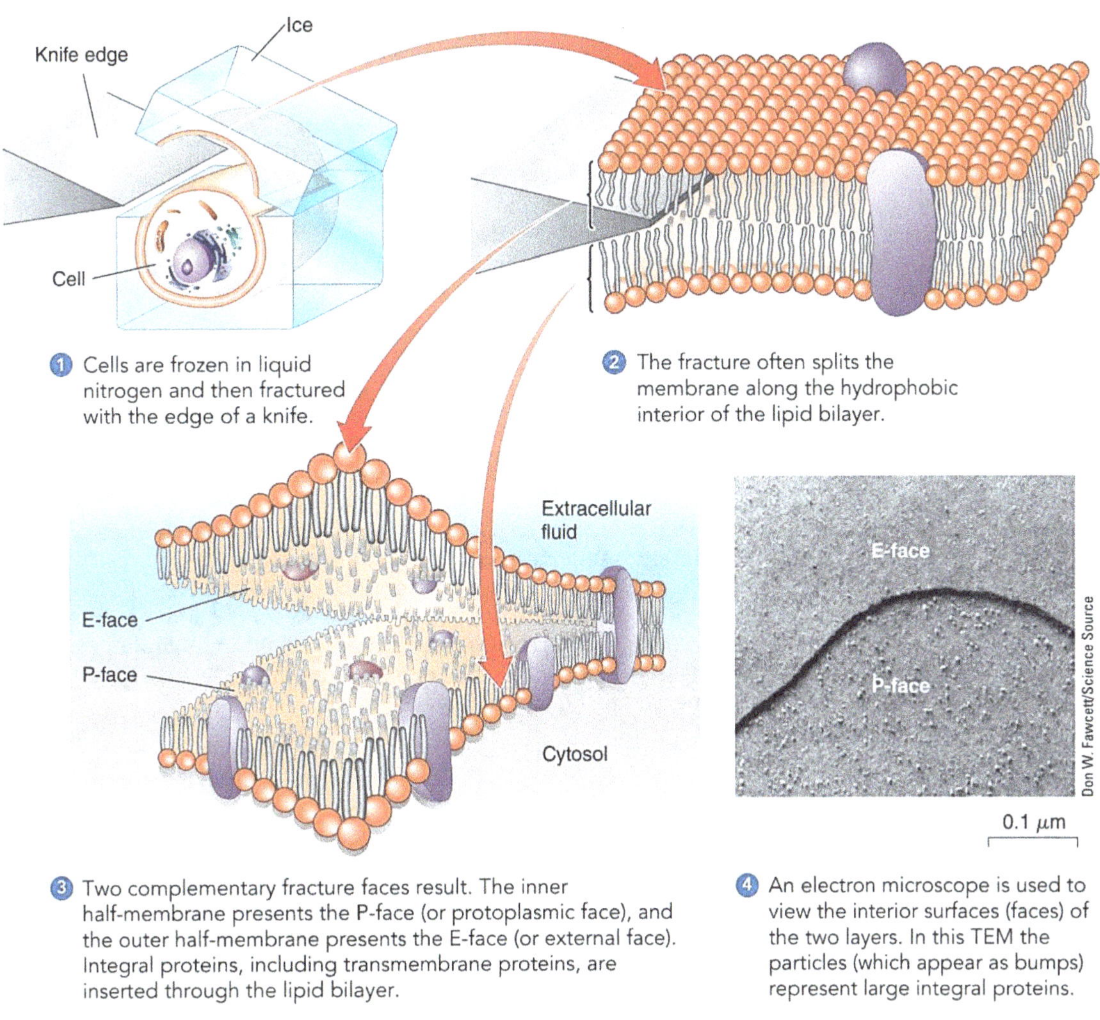

▲ **Figure 4.3** Viewing a plasma membrane with Freeze–fracture method.

The Fluidity of Membranes

In 1972, S. Jonathan Singer and Garth J. Nicolson revised the model in a simple but profound way: They proposed that the globular proteins are inserted into the lipid bilayer, with their nonpolar segments in contact with the nonpolar interior of the bilayer and their polar portions protruding out from the membrane surface. In this model, called the **fluid mosaic model**, a mosaic of proteins floats in or on the fluid lipid bilayer like boats on a pond.

Membranes are not static sheets of molecules locked rigidly in place. A membrane is held together mainly by hydrophobic interactions, which are much weaker than covalent bonds. Most of the lipids and some proteins can shift about sideways. Very rarely, also, a lipid may flip-flop across the membrane, switching from one phospholipid layer to the other. In contrast to rotational and lateral movements, the "flip-flop" of lipids from one leaflet to the opposite leaflet does not occur spontaneously. Flip-flop is energetically unfavorable because the polar head of a phospholipid has to travel through the hydrophobic interior of the membrane.

The transport of lipids between leaflets is due to the action of the enzyme flippase, which requires energy input in the form of ATP. The sideways movement of phospholipids within the membrane is rapid. Proteins are much larger than lipids and move more slowly, when they do move. Many membrane proteins seem to be held immobile by their attachment to the cytoskeleton or to the extracellular matrix.

► **Figure 4.4** Membrane fluidity.

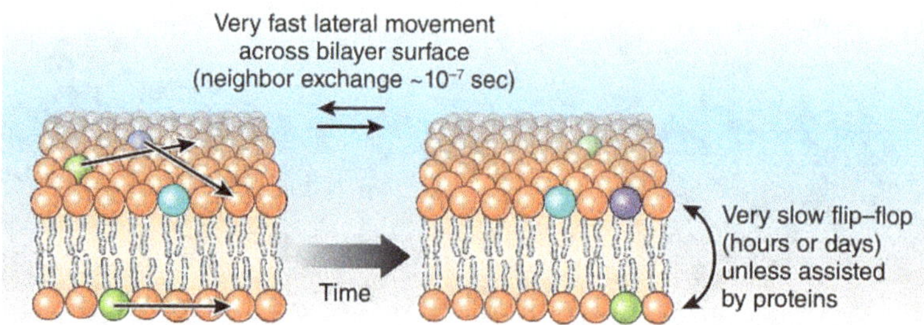

As the temperature decreases, the membrane remains fluid to a lower temperature if it is rich in phospholipids with unsaturated hydrocarbon tails. Because of kinks in the tails where double bonds are located, unsaturated hydrocarbon tails cannot pack together as closely as saturated hydrocarbon tails, making the membrane more fluid.

The steroid cholesterol, which is wedged between phospholipid molecules in the plasma membranes of animal cells, has different effects on membrane fluidity at different temperatures. At relatively high temperatures, at 37°C, the body temperature of humans, for example, cholesterol makes the membrane less fluid by restraining phospholipid movement. However, because cholesterol also hinders the close packing of phospholipids, it lowers the temperature required for the membrane to solidify. Thus, cholesterol can be thought of as a "**fluidity buffer**" for the membrane, resisting changes in membrane fluidity that can be caused by changes in temperature. Compared to animals, plants have very low levels of cholesterol; rather, related steroid lipids buffer membrane fluidity in plant cells.

► **Figure 4.5** The position taken by cholesterol in bilayers. The hydrophilic OOH group at one end of the molecule extends into the polar regions of the bilayer; the ring structure extends into the nonpolar membrane interior.
© Cengage Learning 2017

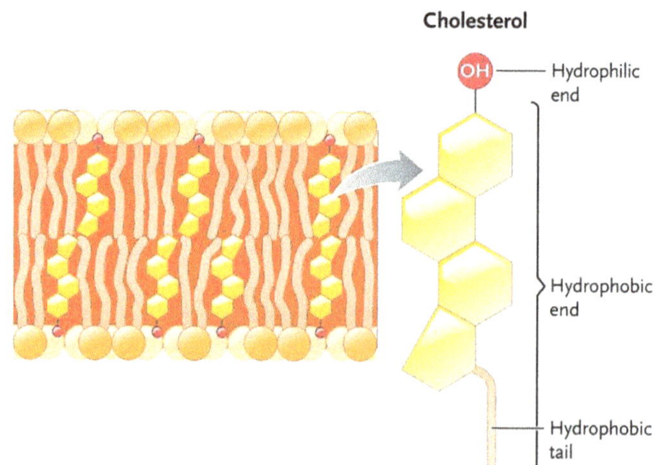

The biochemical properties of phospholipids affect the fluidity of the phospholipid bilayer. These include the following:

Length of the nonpolar tails: The tails of phospholipids typically range from 14 to 24 carbon atoms, with 18 to 20 carbons being the most common. Longer tails are able to form stronger van der Waals dispersion forces with each other and, therefore, make a membrane less fluid. In contrast, shorter tails are less likely to interact with each other, which makes the membrane more fluid.

Presence of double bonds: Double bonds make a membrane more fluid. When a double bond is present, the lipid is said to be unsaturated with respect to the number of hydrogens that are bound to the carbon atoms. A double bond creates a kink in the lipid tail, making it more difficult for neighboring tails to interact and making the bilayer more fluid.

Presence of cholesterol: Cholesterol, which is found in animal cells, tends to stabilize membranes because it is a short, rigid molecule. Its effects depend on temperature. At higher temperatures, such as those observed in mammals that maintain a constant body temperature, cholesterol makes the membrane less fluid. At lower temperatures, such as icy water, cholesterol has the opposite effect. It makes the membrane more fluid and prevents it from freezing. Plant cell membranes contain **phytosterols** that resemble cholesterol in their chemical structure.

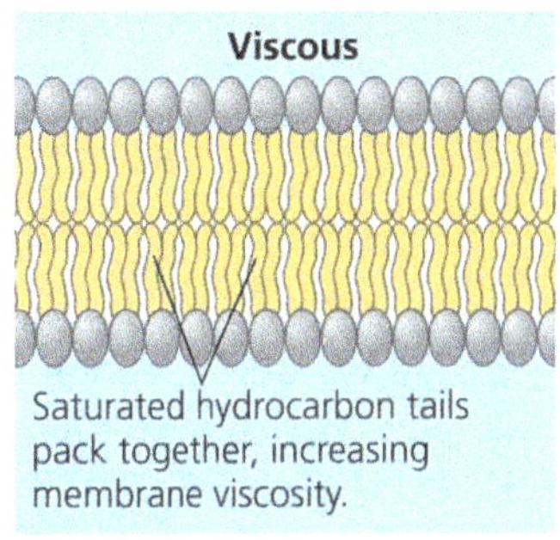

(a) Unsaturated versus saturated hydrocarbon tails.

▶ **Figure 4.6** Factors that affect membrane fluidity.

(b) Cholesterol within the animal cell membrane. Cholesterol reduces membrane fluidity at moderate temperatures by reducing phospholipid movement, but at low temperatures it hinders solidification by disrupting the regular packing of phospholipids.

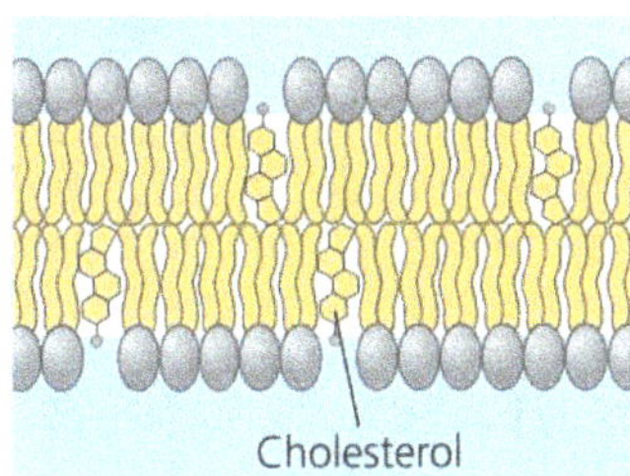

Variations in the cell membrane lipid compositions of many species appear to be evolutionary adaptations that maintain the appropriate membrane fluidity under specific environmental conditions. For instance, fishes that live in extreme cold have membranes with a high proportion of unsaturated hydrocarbon tails, enabling their membranes to remain fluid in spite of the low temperature. At the other extreme, some bacteria and archaea thrive at temperatures greater than 90°C in thermal hot springs and geysers. Their membranes include unusual lipids that may prevent excessive fluidity at such high temperatures.

The ability to change the lipid composition of cell membranes in response to changing temperatures has evolved in organisms that live where temperatures vary. In many plants that tolerate extreme cold, such as winter wheat, the percentage of unsaturated phospholipids increases in autumn, an adjustment that keeps the membranes from solidifying during winter. Some bacteria and archaea contain enzymes called fatty acid desaturases that can introduce double bonds into fatty acids in membranes. So they exhibit different proportions of unsaturated phospholipids in their cell membranes, depending on the temperature at which they are growing. Overall, natural selection has apparently favored organisms whose mix of membrane lipids ensures an appropriate level of membrane fluidity for their environment.

Membrane Proteins and Their Functions

Different types of cells contain different sets of membrane proteins, and the various membranes within a cell each have a unique collection of proteins. there are two major populations of membrane proteins: **integral** proteins and **peripheral** proteins.

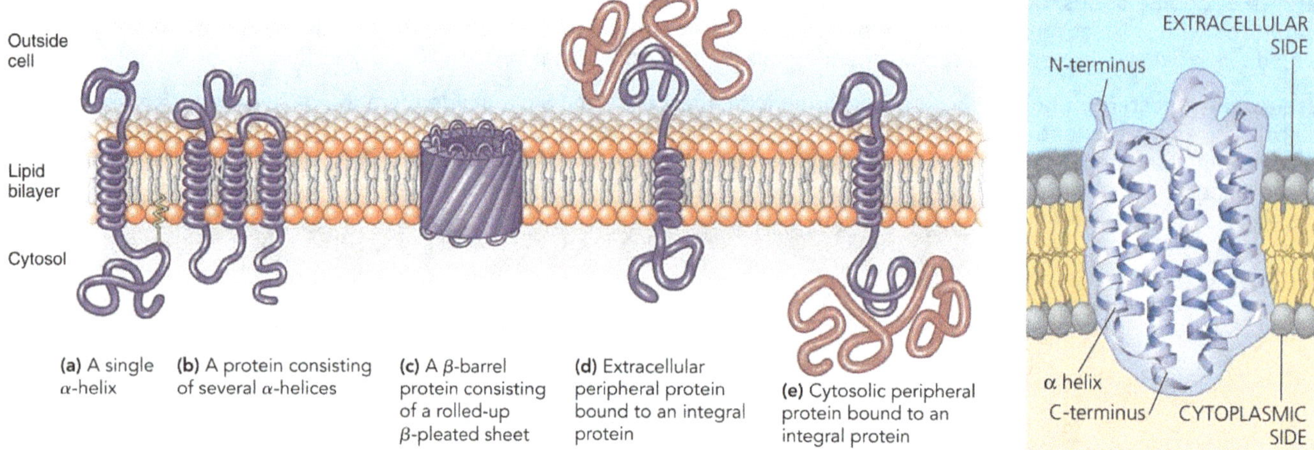

▲ **Figure 4.7** Membrane proteins. Three transmembrane proteins are shown in (a), (b), and (c). Water molecules and ions pass through specific types of protein pores in the membrane formed between groups of a-helices or through b-barrel structures formed as in (c). Peripheral proteins are bound to integral proteins by noncovalent interactions (d, e).

▶ **Figure 4.8** The structure of a transmembrane protein. Bacteriorhodopsin. The protein includes seven transmembrane helices. The nonhelical hydrophilic segments are in contact with the aqueous solutions on the extracellular and cytoplasmic sides of the membrane.

Integral proteins penetrate the hydrophobic interior of the lipid bilayer. The majority are **transmembrane proteins**, which span the membrane; other integral proteins extend only partway into the hydrophobic interior (**lipid anchored proteins**). The hydrophobic regions of an integral protein consist of one or more stretches of nonpolar amino acids, typically 20–30 amino acids in length, usually coiled into α helices. The hydrophilic parts of the molecule are exposed to the aqueous solutions on either side of the membrane. Some proteins also have one or more hydrophilic channels that allow passage through the membrane of hydrophilic substances (even of water itself). Peripheral proteins are not embedded in the lipid bilayer at all; they are loosely bound to the surface of the membrane, often to exposed parts of integral proteins.

On the cytoplasmic side of the plasma membrane, some membrane proteins are held in place by attachment to the cytoskeleton and on the extracellular side, certain membrane proteins may attach to materials outside the cell.

Proteins on a cell's surface are important in the medical field. For example, a protein called CD4 on the surface of immune cells helps the human immunodeficiency virus (HIV) infect these cells, leading to acquired immune deficiency syndrome (AIDS).

▶ **Figure 4.9** Attachment of transmembrane proteins to the cytoskeleton and ECM of an animal cell.

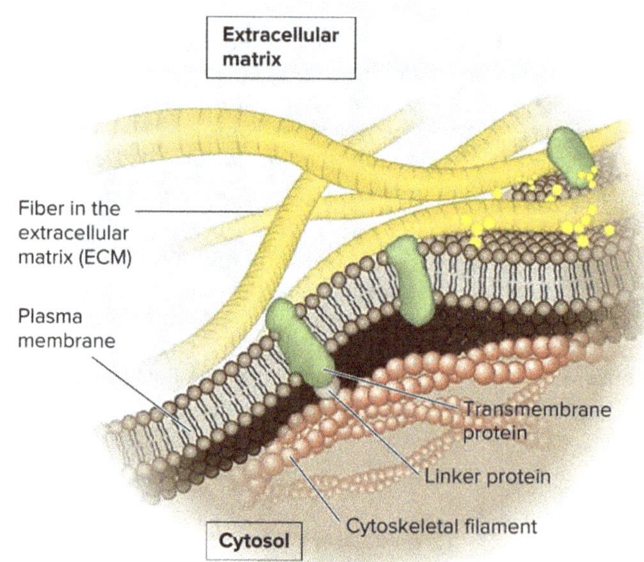

Some functions of membrane proteins:

- **Transport.** Left: A protein that spans the membrane may provide a hydrophilic channel across the membrane that is selective for a particular solute. Right: Other transport proteins shuttle a substance from one side to the other by changing shape. Some of these proteins hydrolyze ATP as an energy source to actively pump substances across the membrane.

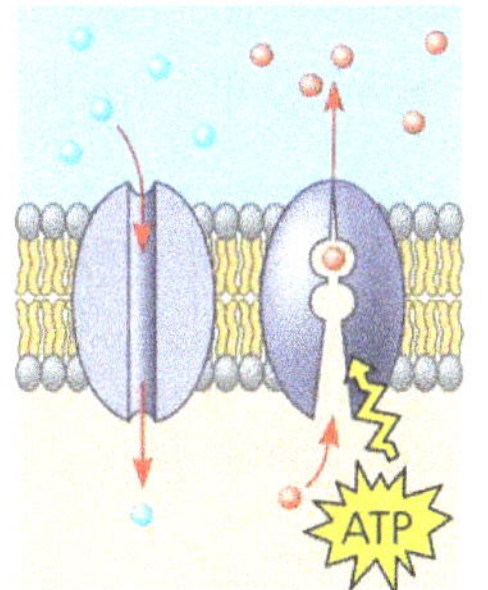

- **Enzymatic activity.** A protein built into the membrane may be an enzyme with its active site (where the reactant binds) exposed to substances in the adjacent solution. In some cases, several enzymes in a membrane are organized as a team that carries out sequential steps of a metabolic pathway.

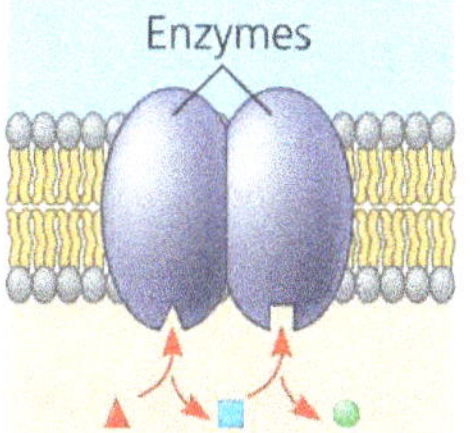

- **Signal transduction.** A membrane protein (receptor) may have a binding site with a specific shape that fits the shape of a chemical messenger, such as a hormone. The external messenger (signaling molecule) may cause the protein to change shape, allowing it to relay the message to the inside of the cell, usually by binding to a cytoplasmic protein.

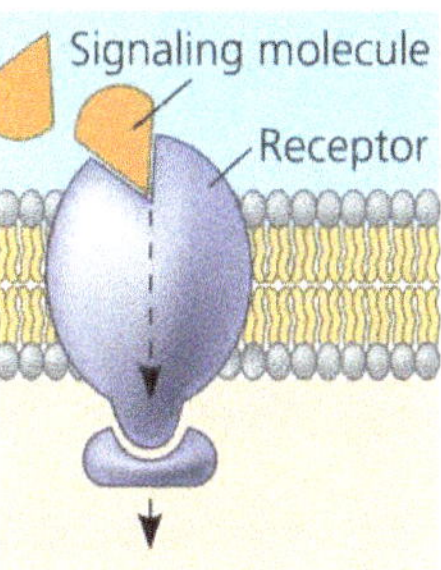

- **Cell-cell recognition.** Some glycoproteins serve as identification tags that are specifically recognized by membrane proteins of other cells.

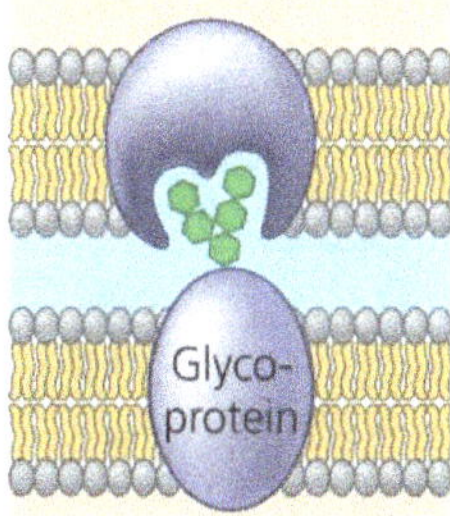

- **Intercellular joining.** Membrane proteins of adjacent cells may hook together in various kinds of junctions, such as gap junctions or tight junctions.

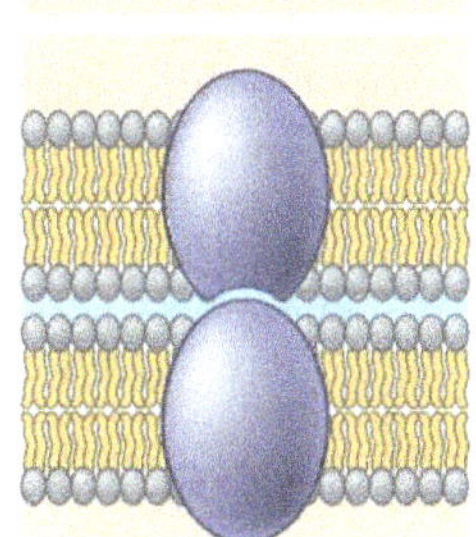

- **Attachment to the cytoskeleton and extracellular matrix (ECM).** Microfilaments or other elements of the cytoskeleton may be noncovalently bound to membrane proteins, a function that helps maintain cell shape and stabilizes the location of certain membrane proteins. Proteins that can bind to ECM molecules can coordinate extracellular and intracellular changes.

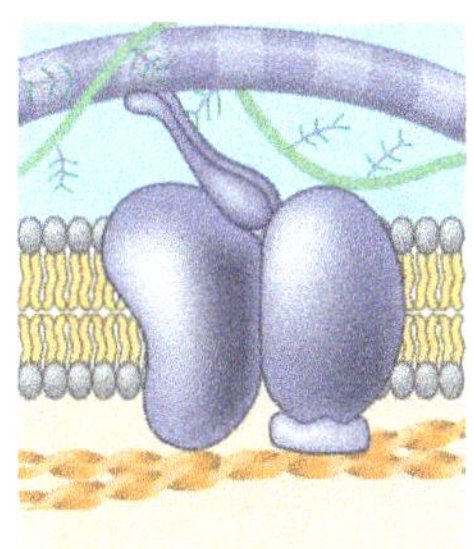

The Role of Membrane Carbohydrates in Cell-Cell Recognition

Cell-cell recognition, a cell's ability to distinguish one type of neighboring cell from another, is crucial to the functioning of an organism. It is important, for example, in the sorting of cells into tissues and organs in an animal embryo. It is also the basis for the rejection of foreign cells by the immune system, an important line of defense in vertebrate animals. Cells recognize other cells by binding to molecules, often containing carbohydrates, on the extracellular surface of the plasma membrane.

Membrane carbohydrates are usually short, branched chains of fewer than 15 sugar units. Some are covalently bonded to lipids, forming molecules called **glycolipids**. However, most are covalently bonded to proteins, which are thereby **glycoproteins**.

The carbohydrates on the extracellular side of the plasma membrane vary from species to species, among individuals of the same species, and even from one cell type to another in a single individual. The diversity of the molecules and their location on the cell's surface enable membrane carbohydrates to function as markers that distinguish one cell from another. For example, the four human blood types designated A, B, AB, and O reflect variation in the carbohydrate part of glycoproteins on the surface of red blood cells.

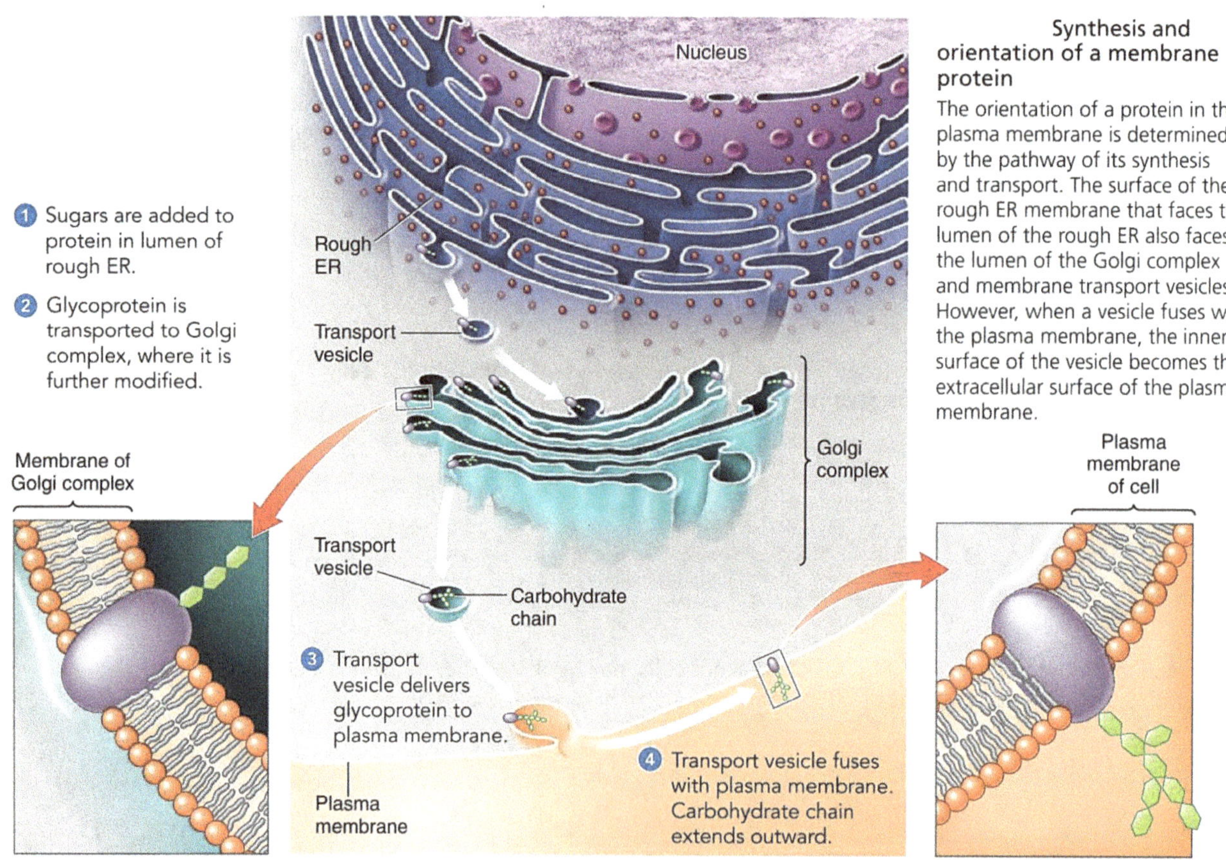

▲ **Figure 4.10** Synthesis and orientation of a membrane protein.

Membrane structure results in selective permeability

A membrane exhibits selective permeability; that is, it allows some substances to cross more easily than others. The ability to regulate transport across cellular boundaries is essential to the cell's existence. Sugars, amino acids, and other nutrients enter the cell, and metabolic waste products leave it. The cell takes in O_2 for use in cellular respiration and expels CO_2. Also, the cell regulates its concentrations of inorganic ions, such as Na^+, K^+, Ca^{2+} and Cl^-, by shuttling them one way or the other across the plasma membrane.

Nonpolar molecules, such as hydrocarbons, CO2, and O2, are hydrophobic, as are lipids. They can all therefore dissolve in the lipid bilayer of the membrane and cross it easily, without the aid of membrane proteins. However, the hydrophobic interior of the membrane impedes direct passage through the membrane of ions and polar molecules, which are hydrophilic.

Polar molecules such as glucose and other sugars pass only slowly through a lipid bilayer, and even water, a very small polar molecule, does not cross rapidly relative to nonpolar molecules. A charged atom or molecule and its surrounding shell of water are even less likely to penetrate the hydrophobic interior of the membrane. Furthermore, the lipid bilayer is only one aspect of the gatekeeper system responsible for a cell's selective permeability. Proteins built into the membrane play key roles in regulating transport. Movement across the membrane can be either **passive** or **active**:

Passive movement includes **filtration**, **diffusion**, **facilitated diffusion** and **osmosis**. None of these activities requires the cell to expend energy. Diffusion, facilitated diffusion and osmosis move molecules "down" their concentration gradients. Filtration is the movement of solutes in response to fluid pressure. Your kidneys separate waste products from the blood via filtration.

When energy is consumed to move a molecule or an ion against the concentration gradient, we call the process **active transport**. Active transport accounts for the almost complete uptake of digested nutrients from the small intestine, the collecting of iodine in thyroid gland cells, and the return to the blood of the vast majority of sodium ions filtered from the blood by the kidneys.

Sometimes cells need to transport materials across their cell surface membranes on a much large scale. The materials include large molecules such as proteins or polysaccharides, parts of cells or even whole cells. As a result, mechanisms have evolved for the **bulk transport** of large quantities of materials into and out of cells. Bulk transport of materials into cells is called **endocytosis**. Bulk transport out of cells is called **exocytosis**. These processes require energy.

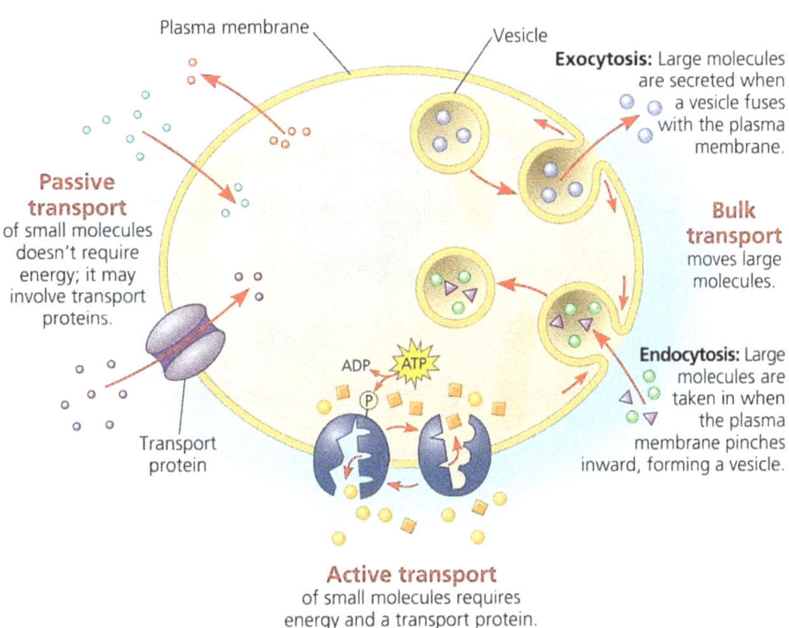

Passive transport

• Simple diffusion

Molecules have a type of energy called thermal energy, due to their constant motion. One result of this motion is **diffusion**, the movement of particles of any substance so that they spread out into the available space. Each molecule moves randomly, yet diffusion of a population of molecules may be directional. In the absence of any other forces, a substance will diffuse from where it is more concentrated to where it is less concentrated. Diffusion is a spontaneous process, needing no input of energy. Each substance diffuses **down** or **with** its own concentration gradient ("**downhill**"), unaffected by the concentration gradients of other substances. Solutes moving from a low-concentration area to a high-concentration area are said to move **up** or **against** the concentration gradient ("**uphill**").

When a substance is more concentrated on one side of a membrane than on the other, there is a tendency for it to diffuse across, down its **concentration gradient** (assuming that the membrane is permeable to that substance). A concentration gradient is a difference in concentration between two different areas, for example, the ICF (intracellular fluid) and ECF (extracellular fluid). Even after their concentration is the same in all regions, molecules or ions still move constantly from one space to another, but there is no net change in concentration on either side. This condition is an example of a **dynamic equilibrium** (dynamic with respect to the continuous movement, and equilibrium with respect to the exact balance between opposing forces).

One important example is the uptake of oxygen by a cell performing cellular respiration. Dissolved oxygen diffuses into the cell across the plasma membrane. As long as cellular respiration consumes the O_2 as it enters, diffusion into the cell will continue because the concentration gradient favors movement in that direction. Water can diffuse very rapidly across the membranes of cells with aquaporins, compared with diffusion in the absence of aquaporins. The diffusion of a substance across a biological membrane is called **passive transport** because it requires no energy. The concentration gradient itself represents potential energy and drives diffusion.

Lipid-soluble (nonpolar and uncharged) substances that move across membranes by simple diffusion through the lipid bilayer include oxygen, carbon dioxide, and nitrogen gases; fatty acids; steroids; and fat-soluble vitamins (A, D, E, and K). Polar molecules such as water and urea also move through the lipid bilayer. The rate of simple diffusion is directly related to the concentration of the solute; the more concentrated the solute, the more rapid the diffusion.

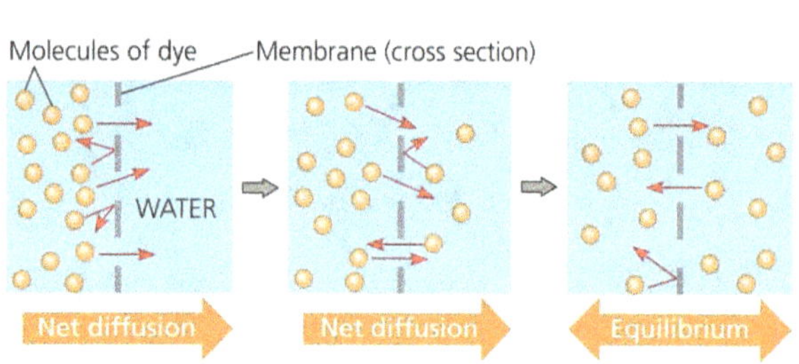

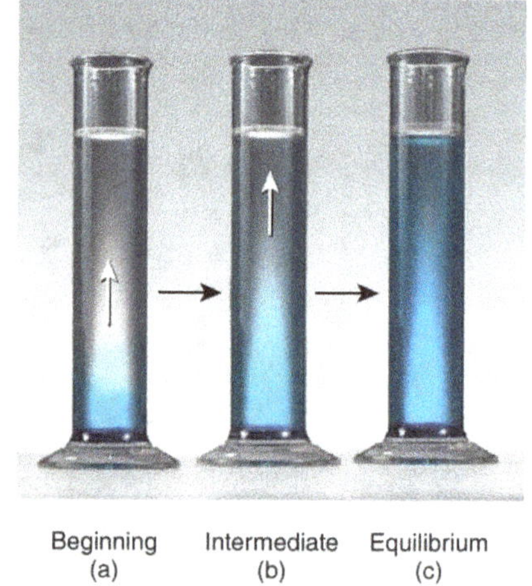

▲ **Figure 4.11** Diffusion of a solute. The membrane has pores large enough for molecules of dye to pass through. The dye diffuses from where it is more concentrated to where it is less concentrated (called diffusing down a concentration gradient). This leads to a dynamic equilibrium: The solute molecules continue to cross the membrane, but at equal rates in both directions.

Four factors greatly affect the ability of solutes to cross a phospholipid bilayer by simple diffusion:

- **Size.** Small substances diffuse faster than larger ones.
- **Polarity.** Nonpolar substances diffuse faster than polar ones.
- **Charge.** Noncharged substances diffuse faster than charged ones.
- **Concentration.** The rate of movement of a solute across a membrane will be higher when its concentration is higher.

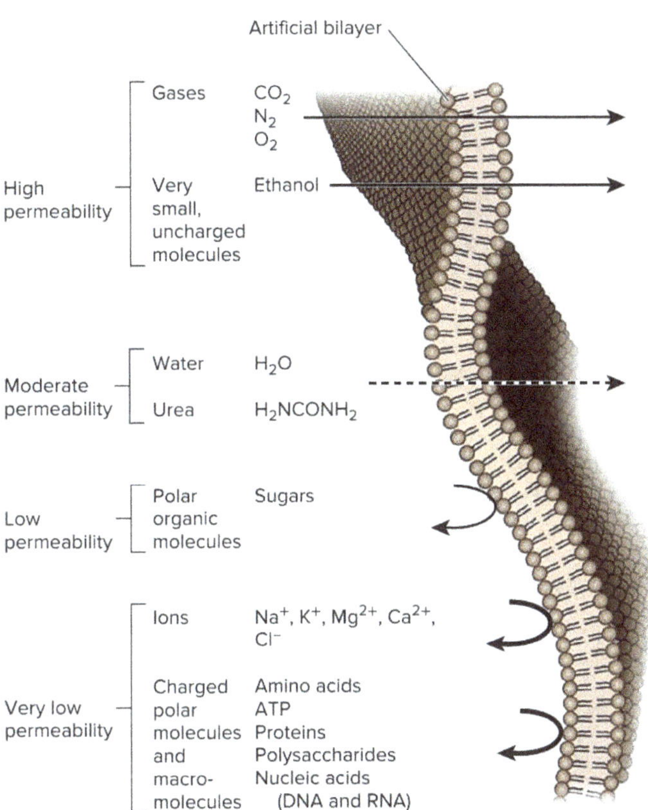

▲ **Figure 4.12** Relative permeability of an artificial phospholipid bilayer to a variety of solutes. Solutes that easily penetrate are shown with a straight arrow that passes through the bilayer. The dashed arrow indicates solutes that have moderate permeability. The remaining solutes, shown at the bottom, are relatively impermeable.

- **Osmosis**

Osmosis is a special type of diffusion involving only water molecules.

solute + solvent = solution (In a sugar solution, for example, the solute is sugar and the solvent is water.)

The diffusion of free water across a selectively permeable membrane, whether artificial or cellular, is called osmosis. Water diffuses across the membrane from the region of higher free water concentration (lower solute concentration) to that of lower free water concentration (higher solute concentration) until the solute concentrations on both sides of the membrane are more nearly equal. The movement of water across cell membranes and the balance of water between the cell and its environment are crucial to organisms. The total concentration of all solute particles in a solution is referred to as the solution's **osmolarity**. **Osmotic pressure**, defined as the force needed to stop osmotic flow.

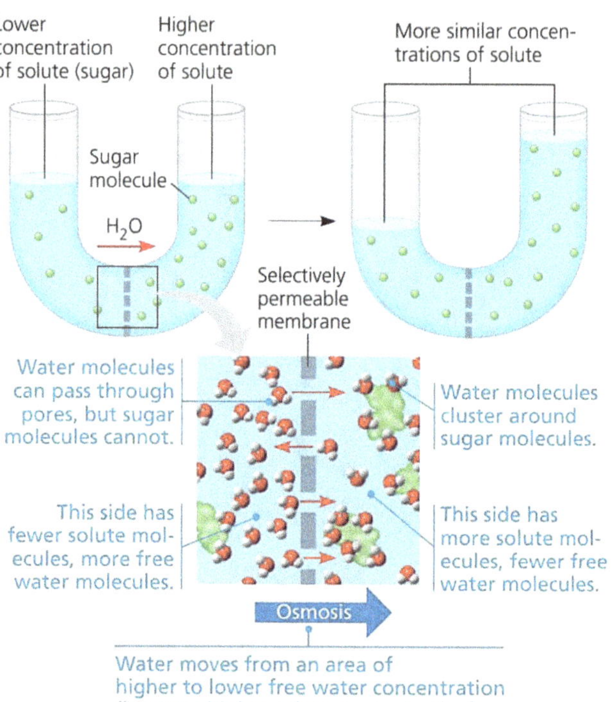

▶ **Figure 4.13** Osmosis. Two sugar solutions of different concentrations are separated by a membrane that the solvent (water) can pass through but the solute (sugar) cannot. Water molecules move randomly and may cross in either direction, but overall, water diffuses from the solution with less concentrated solute to that with more concentrated solute. This passive transport of water, called osmosis, reduces the difference in sugar concentrations.

Maintaining osmotic balance:

Extrusion. Some single-celled eukaryotes, such as the protist Paramecium, use organelles called contractile vacuoles to remove water. Each vacuole collects water from various parts of the cytoplasm and transports it to the central part of the vacuole, near the cell surface. The vacuole possesses a small pore that opens to the outside of the cell. By contracting rhythmically, the vacuole pumps out (extrudes) through this pore the water that is continuously drawn into the cell by osmotic forces.

Isosmotic Regulation. Some organisms that live in the ocean adjust their internal concentration of solutes to match that of the surrounding seawater. Because they are isosmotic with respect to their environment, no net flow of water occurs into or out of these cells. The blood in your body, contains a high concentration of the protein albumin, which elevates the solute concentration of the blood to match that of your cells' cytoplasm.

Turgor. Most plant cells are hypertonic to their immediate environment, containing a high concentration of solutes in their central vacuoles. The resulting internal hydrostatic pressure, known as turgor pressure, presses the plasma membrane firmly against the interior of the cell wall, making the cell rigid. Most green plants depend on turgor pressure to maintain their shape, and thus they wilt when they lack sufficient water.

The effect a solution has on cell volume when the solution surrounds the cell is the **tonicity** of the solution:

- If a cell without a cell wall, such as an animal cell, is immersed in an environment that is **isotonic** to the cell (iso means "same"), there will be no net movement of water across the plasma membrane. Water diffuses across the membrane, but at the same rate in both directions. In an isotonic environment, the volume of an animal cell is stable. In fact, because the osmotic pressure of cytosol and interstitial fluid is the same, cell volume remains constant. Cells neither shrink due to water loss by osmosis nor swell due to water gain by osmosis.

- If we place the cell in a solution that is **hypertonic** to the cell (hyper means "more"), the cell will lose water, shrivel, and probably die. This is why an increase in the salinity (saltiness) of a lake can kill the animals there; if the lake water becomes hypertonic to the animals' cells, they might shrivel in a process called crenation and die.

- If we place the cell in a solution that is **hypotonic** to the cell (hypo means "less"), water will enter the cell faster than it leaves, and the cell will swell and lyse (burst) like an overfilled water balloon. As water diffuses into the cell, the point is finally reached where the hydrostatic pressure (the back pressure exerted by water against the membrane) in the cell is equal to its osmotic pressure (the tendency of water to move into the cell by osmosis). At this point, there is no further (net) water entry.

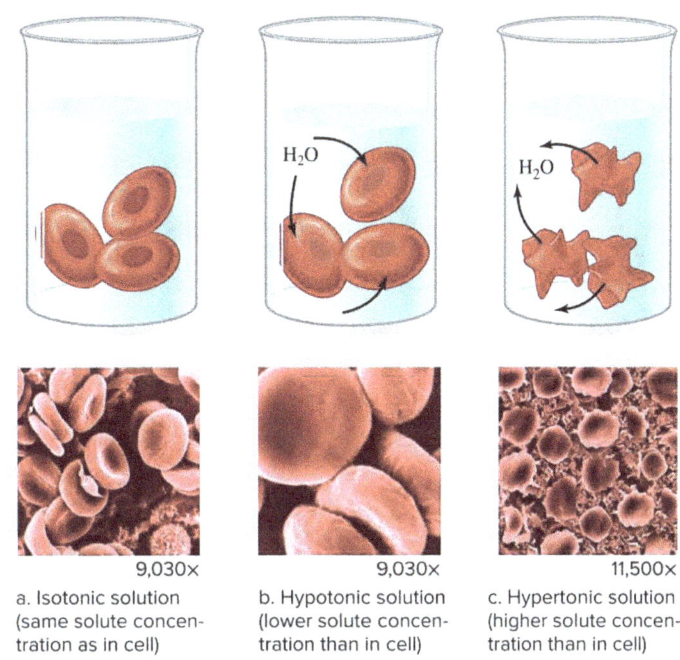

a. Isotonic solution (same solute concentration as in cell)

b. Hypotonic solution (lower solute concentration than in cell)

c. Hypertonic solution (higher solute concentration than in cell)

▲ **Figure 4.14** The water balance of living animal cell (without cell wall). An animal cell fares best in an isotonic environment unless it has special adaptations that offset the osmotic uptake or loss of water.

Another set of terms, **isoosmotic**, **hypoosmotic**, and **hyperosmotic**, denotes the osmolarity of a solution relative to that of normal extracellular fluid without regard to whether the solute is penetrating or nonpenetrating. The two sets of terms are therefore not synonymous.

The term **water potential** is very useful when considering osmosis. The Greek letter psi, ψ, can be used to mean water potential. water potential is a measure of the tendency of water to move from one place to another; water moves from a solution with higher water potential to one with lower water potential; water potential is decreased by the addition of solute, and increased by the application of pressure.

A cell without rigid cell walls can't tolerate either excessive uptake or excessive loss of water. This problem of water balance is automatically solved if such a cell lives in isotonic surroundings. In hypertonic or hypotonic environments, however, organisms that lack rigid cell walls must have other adaptations for **osmoregulation**, the control of solute concentrations and water balance.

The cells of plants, prokaryotes, fungi, and some protists are surrounded by cell walls. When such a cell is immersed in a hypotonic solution—bathed in rainwater, for example—the plant cell swells as water enters by osmosis. However, the relatively inelastic cell wall will expand only so much before it exerts a back pressure on the cell, called turgor pressure, that opposes further water uptake. At this point, the cell is **turgid** (very firm), which is the healthy state for most plant cells. Plants that are not woody, such as most houseplants, depend for mechanical support on cells kept turgid by a surrounding hypotonic solution.

If a plant's cells and surroundings are isotonic, there is no net tendency for water to enter and the cells become **flaccid** and in a hypertonic environment a plant cell protoplast (the living part of the cell inside the cell wall surrounded by the cell surface membrane) will lose water to its surroundings and shrink. As the plant cell shrivels its plasma membrane pulls away from the cell wall at multiple places. This phenomenon, called **plasmolysis**, causes the plant to wilt and can lead to plant death. The walled cells of bacteria and fungi also plasmolyze in hypertonic environments.

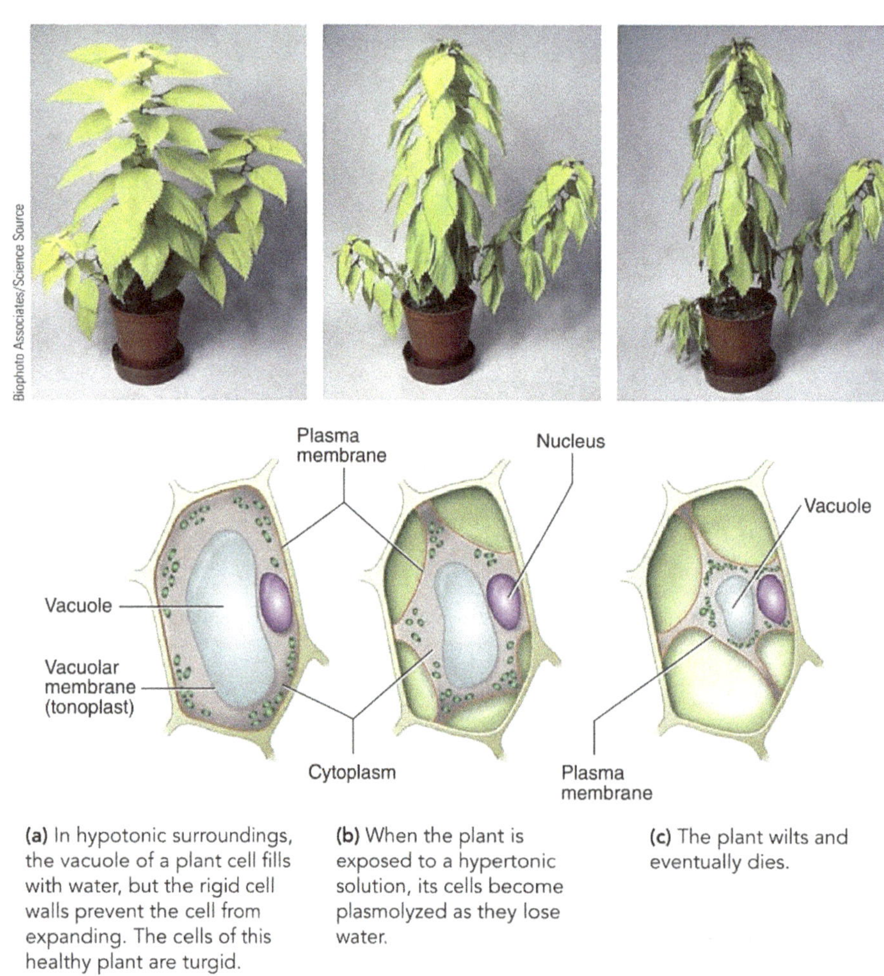

(a) In hypotonic surroundings, the vacuole of a plant cell fills with water, but the rigid cell walls prevent the cell from expanding. The cells of this healthy plant are turgid.

(b) When the plant is exposed to a hypertonic solution, its cells become plasmolyzed as they lose water.

(c) The plant wilts and eventually dies.

▲ **Figure 4.15** The water balance of living Plant cell. Plant cells are turgid (firm) and generally healthiest in a hypotonic environment, where the uptake of water is eventually balanced by the wall pushing back on the cell.

- **Facilitated Diffusion: Passive Transport Aided by Proteins**

Many polar molecules and ions blocked by the lipid bilayer of the membrane diffuse passively with the help of transport proteins that span the membrane. This phenomenon is called **facilitated diffusion**. Cell biologists are still trying to learn exactly how various **transport proteins** facilitate diffusion. Most transport proteins are very specific: They transport some substances but not others. Facilitated diffusion by carrier proteins can become saturated when there are too few transport proteins to handle all the solute molecules.

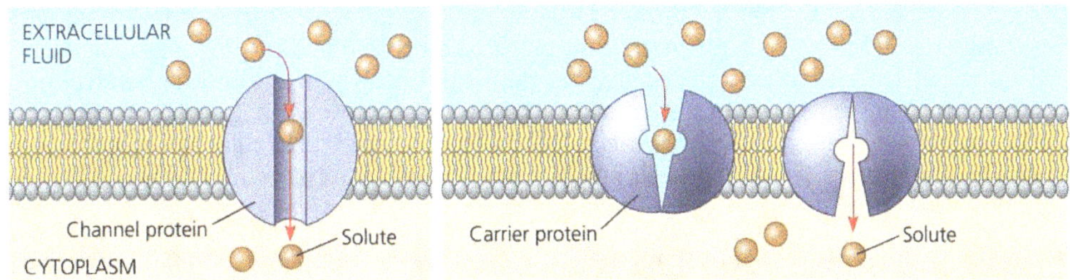

▲ **Figure 4.16** Two types of transport proteins that carry out facilitated diffusion. In both cases, the protein can transport the solute in either direction, but the net movement is down the concentration gradient of the solute.
(a) A channel protein (purple) has a channel through which water molecules or a specific solute can pass.
(b) A carrier protein alternates between two shapes, moving a solute across the membrane during the shape change.

Aquaporins, the water channel proteins, facilitate the massive levels of diffusion of water (osmosis) that occur in plant cells and in animal cells such as red blood cells. Certain kidney cells also have a high number of aquaporins, allowing them to reclaim water from urine before it is excreted.

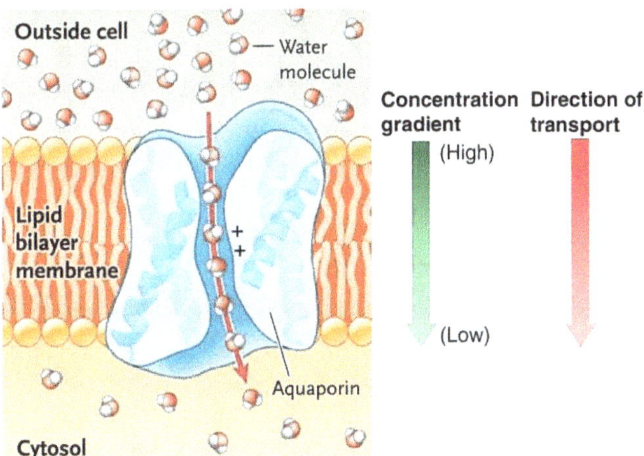

▲ **Figure 4.17** An aquaporin is a water channel. Water molecules move through the channel by being handed off to a succession of hydrogen-bonding sites on the channel in this protein. © Cengage Learning 2017

A transport protein is specific for the substance it translocates (moves), allowing only a certain substance (or a small group of related substances) to cross the membrane.

Two types of transport proteins are **channel proteins** and **carrier proteins (transporters)**. Channel proteins simply provide corridors that allow specific molecules or ions to cross the membrane and may be **gated** or **nongated**. The hydrophilic passageways provided by these proteins can allow water molecules or small ions to diffuse very quickly from one side of the membrane to the other. **Porins** are transmembrane channel proteins that allow various solutes or water to pass through membranes. These channel proteins are rolled-up, barrel-shaped b-pleated sheets that form pores. Channel proteins maybe gated or nongated. Channel proteins that transport ions are called ion channels. Many ion channels function as gated channels, which open or close in response to a stimulus.

Three factors can alter the channel protein conformations, producing changes in how long or how often a channel opens. First, the binding of specific molecules to channel proteins may directly or indirectly produce either an allosteric or covalent change in the shape of the channel protein. A molecule that binds to a protein like this is called a ligand. Such channels are therefore termed **ligand-gated ion channels**, and the ligands that influence them are often chemical messengers, such as those released from the ends of neurons onto target cells. Second, changes in the membrane potential can cause movement of certain charged regions on a channel protein, altering its shape—these are **voltage-gated ion channels**. Third, physically deforming (stretching) the membrane may affect the conformation of some channel proteins—these are **mechanically gated ion channels.**

A single type of ion may pass through several different types of channels. For example, a membrane may contain ligand-gated K^+ channels, voltage-gated K^+ channels, and mechanically gated K^+ channels.

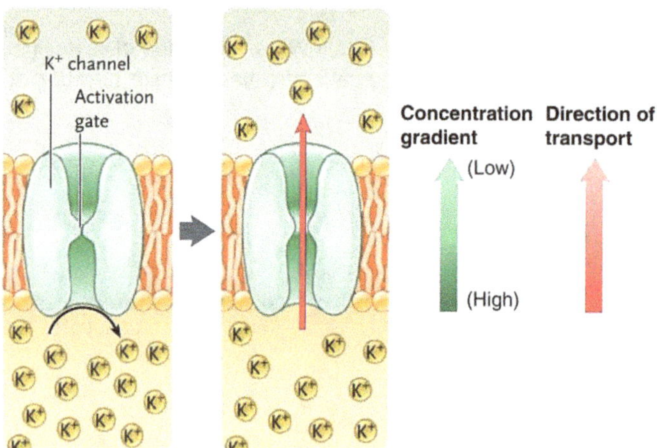

▲ **Figure 4.18** K^+ voltage-gated channel. © Cengage Learning 2017

Amino acids and glucose, are too polar to diffuse through the lipid bilayer and too large to diffuse through channels. The passage of these molecules and the nondiffusional movements of ions are mediated by integral membrane proteins known as transporters or Carrier proteins. The movement of substances through a membrane by any of these mechanisms is called **mediated transport**, which depends on conformational changes in these transporters.

Carrier proteins, such as the glucose transporter, seem to undergo a subtle change in shape that somehow translocates the solute-binding site across the membrane. Such a change in shape may be triggered by the binding and release of the transported molecule. Like ion channels, carrier proteins involved in facilitated diffusion result in the net movement of a substance down its concentration gradient. No energy input is thus required. This is passive transport.

Carrier proteins are called **uniporters** if they transport a single type of molecule and **cotransporters** if they transport two different molecules together. **Symporters** transport two molecules in the same direction, and **antiporters** transport two molecules in opposite directions.

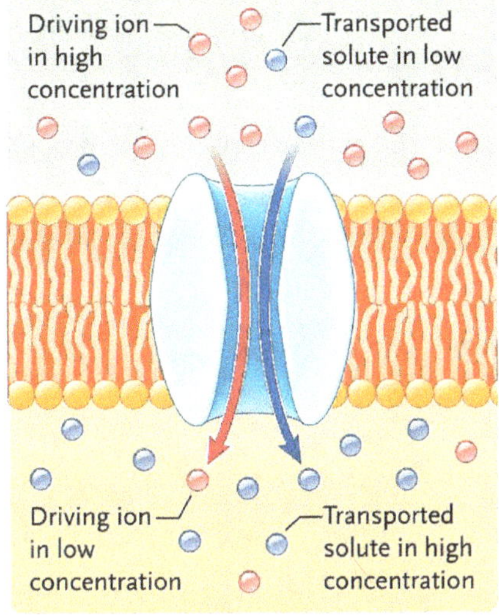

▲ **Figure 4.19 A symporter**
© Cengage Learning 2017

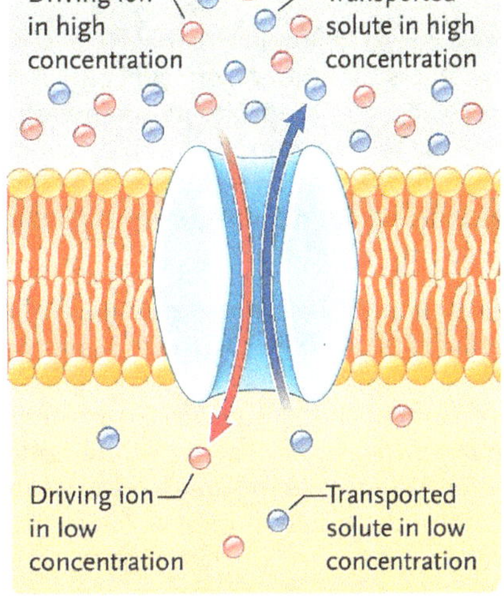

▲ **Figure 4.20 An antiporter**
© Cengage Learning 2017

Four factors determine the magnitude of solute flux through a mediated-transport system: (1) the solute concentration, (2) the affinity of the transporters for the solute, (3) the number of transporters in the membrane, and (4) the rate at which the conformational change in the transport protein occurs. The flux through a mediated-transport system can be altered by changing any of these four factors.

▶ **Figure 4.21** The flux of molecules diffusing into a cell across the lipid bilayer of a plasma membrane (green line) increases continuously in proportion to the extracellular concentration, whereas the flux of molecules through a mediated-transport system (purple line) reaches a maximal value.

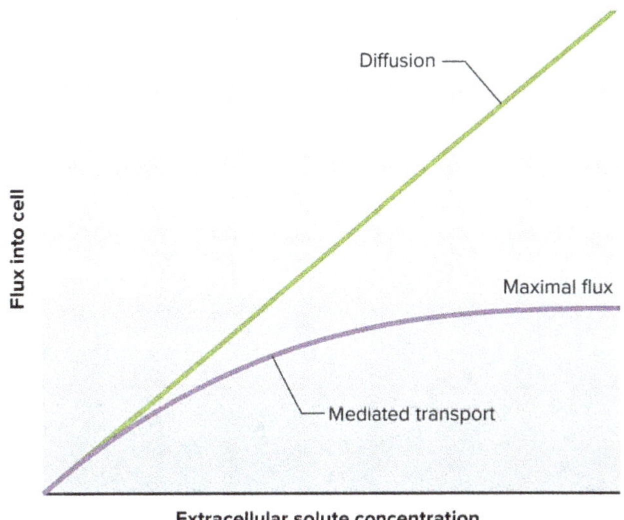

Active transport

Some transport proteins can use energy to move solutes "uphill" against their concentration gradients, across the plasma membrane from the side where they are less concentrated (whether inside or outside) to the side where they are more concentrated. To pump a solute across a membrane against its gradient the cell must expend energy. Therefore, this type of membrane traffic is called active transport. The use of energy from ATP in active transport may be direct or indirect.

The transport proteins that move solutes against their concentration gradients are all carrier proteins rather than channel proteins. Like carrier-mediated facilitated diffusion, active transport requires carrier proteins that combine specifically and reversibly with the transported substances.

Active transport enables a cell to maintain internal concentrations of small solutes that differ from concentrations in its environment. For example, compared with its surroundings, an animal cell has a much higher concentration of potassium ions (K^+) and a much lower concentration of sodium ions (Na^+). The plasma membrane helps maintain these steep gradients by pumping Na^+ out of the cell and K^+ into the cell.

As in other types of cellular work, ATP hydrolysis supplies the energy for most active transport. One way ATP can power active transport is when its terminal phosphate group is transferred directly to the transport protein. This can induce the protein to change its shape in a manner that translocates a solute bound to the protein across the membrane. One transport system that works this way is the **sodium-potassium pump**, which exchanges Na^+ for K^+ across the plasma membrane of animal cells and runs directly on ATP.

▶ **Figure 4.22** The sodium-potassium pump: a specific case of active transport. This transport system pumps ions against steep concentration gradients: Sodium ion concentration ($[Na^+]$) is high outside the cell and low inside, while potassium ion concentration ($[K^+]$) is low outside the cell and high inside. The pump oscillates between two shapes in a cycle that moves 3 Na^+ out of the cell for every 2 K^+ pumped into the cell. The two shapes have different affinities for Na^+ and K^+. ATP powers the shape change by transferring a phosphate group to the transport protein (phosphorylating the protein).

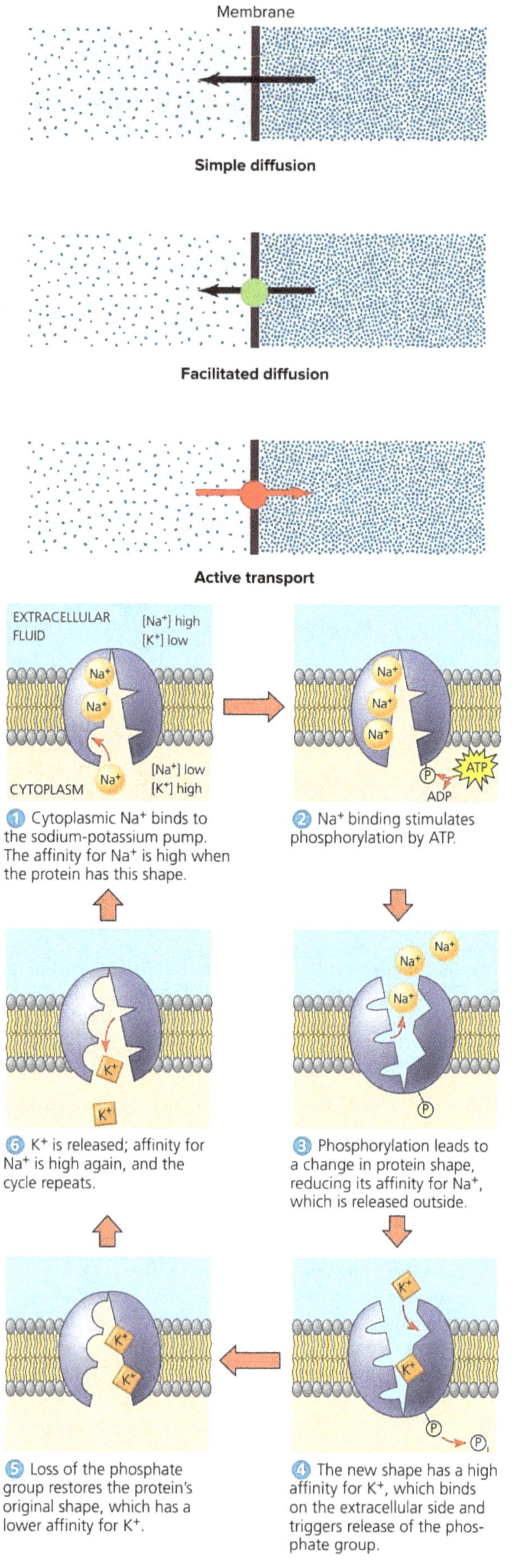

1 Cytoplasmic Na^+ binds to the sodium-potassium pump. The affinity for Na^+ is high when the protein has this shape.

2 Na^+ binding stimulates phosphorylation by ATP.

3 Phosphorylation leads to a change in protein shape, reducing its affinity for Na^+, which is released outside.

4 The new shape has a high affinity for K^+, which binds on the extracellular side and triggers release of the phosphate group.

5 Loss of the phosphate group restores the protein's original shape, which has a lower affinity for K^+.

6 K^+ is released; affinity for Na^+ is high again, and the cycle repeats.

How Ion Pumps Maintain Membrane Potential

All cells have voltages across their plasma membranes. Voltage is electrical potential energy, a separation of opposite charges. The cytoplasmic side of the membrane is negative in charge relative to the extracellular side because of an unequal distribution of anions and cations on the two sides. The voltage across a membrane, called a **membrane potential**, ranges from about -50 to -200 millivolts (mV). The minus sign indicates that the inside of the cell is negative relative to the outside. For this reason, all cells are said to be **polarized**.

The membrane potential acts like a battery, an energy source that affects the traffic of all charged substances across the membrane. Because the inside of the cell is negative compared with the outside, the membrane potential favors the passive transport of cations into the cell and anions out of the cell. Thus, two forces drive the diffusion of ions across a membrane: a **chemical force** (the ion's concentration gradient) and an **electrical force** (the effect of the membrane potential on the ion's movement). This combination of forces acting on an ion is called the **electrochemical gradient**. An ion diffuses not simply down its concentration gradient but, more exactly, down its electrochemical gradient.

For example, the concentration of Na^+ inside a resting nerve cell is much lower than outside it. When the cell is stimulated, gated channels open that facilitate Na^+ diffusion. Sodium ions then "fall" down their electrochemical gradient, driven by the concentration gradient of Na^+ and by the attraction of these cations to the negative side (inside) of the membrane. In this example, both electrical and chemical contributions to the electrochemical gradient act in the same direction across the membrane, but this is not always so. In cases where electrical forces due to the membrane potential oppose the simple diffusion of an ion down its concentration gradient, active transport may be necessary.

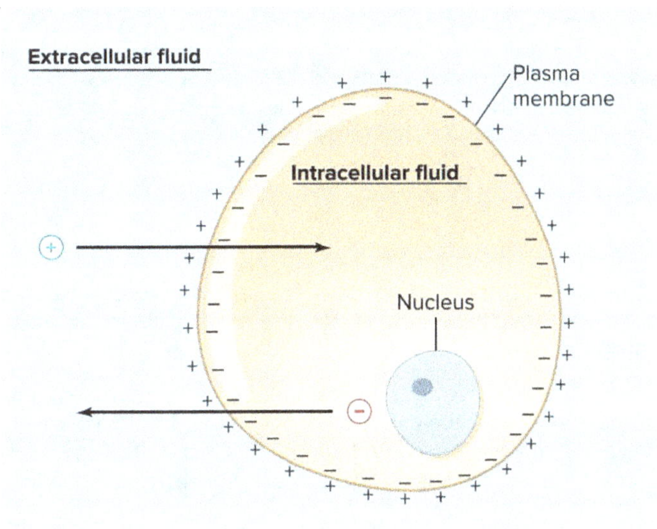

▲ **Figure 4.23** The separation of electrical charge across a plasma membrane (the membrane potential) provides the electrical force that tends to drive positive ions (+) into a cell and negative ions (−) out.

Some membrane proteins that actively transport ions contribute to the membrane potential. The pump does not translocate Na+ and K+ one for one, but pumps three sodium ions out of the cell for every two potassium ions it pumps into the cell. With each "crank" of the pump, there is a net transfer of one positive charge from the cytoplasm to the extracellular fluid, a process that stores energy as voltage. A transport protein that generates voltage across a membrane is called an **electrogenic pump**.

The **sodium-potassium pump** appears to be the major electrogenic pump of animal cells. The main electrogenic pump of plants, fungi, and bacteria is a **proton pump**, which actively transports protons (hydrogen ions, H$^+$) out of the cell. The pumping of H$^+$ transfers positive charge from the cytoplasm to the extracellular solution. By generating voltage across membranes, electrogenic pumps help store energy that can be tapped for cellular work. One important use of proton gradients in the cell is for ATP synthesis during cellular respiration.

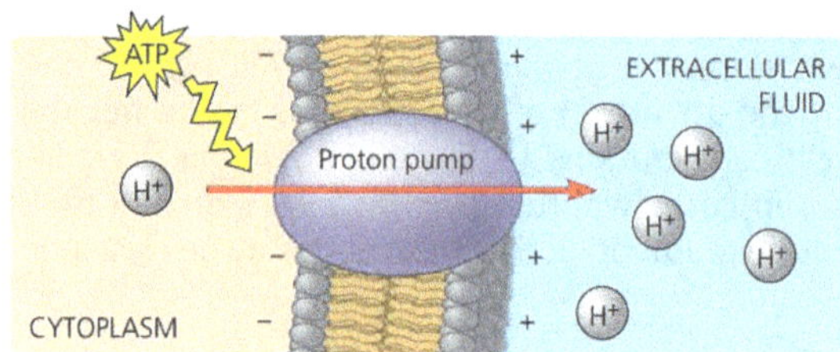

▲ **Figure 4.24** A proton pump. Proton pumps are electrogenic pumps that store energy by generating voltage (charge separation) across membranes. A proton pump translocates positive charge in the form of hydrogen ions (that is, protons). The voltage and H$^+$ concentration gradient represent a dual energy source that can drive other processes, such as the uptake of nutrients. Most proton pumps are powered by ATP.

There are two kinds of active transport:

1. In **primary active transport**, the same protein (pump) that transports a substance also hydrolyzes ATP to power the transport directly. In addition to the **Na$^+$/K$^+$ ATPase** transporter, the major primary active-transport proteins found in most cells are Ca^{2+} ATPase; H$^+$ ATPase; and H$^+$/K$^+$ ATPase.

Ca^{2+} ATPase is found in the plasma membrane and several organelle membranes, including the membranes of the endoplasmic reticulum. In the plasma membrane, the direction of active Ca^{2+} transport is from cytosol to extracellular fluid. In organelle membranes, it is from cytosol into the organelle lumen. Thus, active transport of Ca^{2+} out of the cytosol, via Ca^{2+} ATPase, is one reason that the cytosol of most cells has a very low Ca^{2+} concentration.

H$^+$ ATPase is in the plasma membrane and several organelle membranes, including the inner mitochondrial and lysosomal membranes. In the plasma membrane, H$^+$ ATPase moves H$^+$ produced by metabolism out of cells and in this way helps maintain cellular pH. All enzymes in the body require a narrow range of pH for optimal activity; consequently, this active-transport process is vital for cell metabolism and survival.

H$^+$/K$^+$ ATPase is in the plasma membranes of numerous cells, such as the acid-secreting cells in the stomach, where it pumps one H$^+$ out of the cell and moves one K$^+$ in for each molecule of ATP hydrolyzed. The hydrogen ions enter the stomach lumen where they have an important function in the digestion of proteins.

2. In **secondary active transport**, the transport is driven indirectly by ATP hydrolysis. That is, the transporter proteins do not break down ATP but use instead a favorable concentration gradient of ions, built up by primary active transport, as their energy source for active transport of a different ion or molecule.

A solute that exists in different concentrations across a membrane can do work as it moves across that membrane by diffusion down its concentration gradient. In a mechanism called cotransport, a transport protein (a cotransporter) can couple the "downhill" diffusion of the solute to the "uphill" transport of a second substance against its own concentration gradient. For instance, a plant cell uses the gradient of H$^+$ generated by its ATP-powered proton pumps to drive the active transport of amino acids, sugars, and several other nutrients into the cell. This protein can translocate sucrose into the cell against its concentration gradient. Plants use **H$^+$/sucrose symporter** to load sucrose produced by photosynthesis into cells in the veins of leaves.

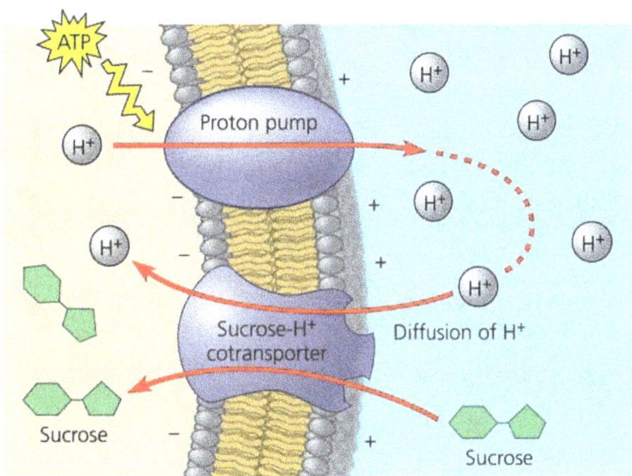

▲ **Figure 4.25** Cotransport: active transport driven by a concentration gradient. A carrier protein, such as this sucrose/H$^+$ cotransporter in a plant cell, is able to use the diffusion of H$^+$ down its electrochemical gradient into the cell to drive the uptake of sucrose. The H$^+$ gradient is maintained by an ATP-driven proton pump that concentrates H$^+$ outside the cell, thus storing potential energy that can be used for active transport, in this case of sucrose. Thus, ATP indirectly provides the energy necessary for cotransport.

In many tissues, a **Na$^+$/Ca^{2+} antiporter** is used to remove cytosolic calcium from cells, exchanging one Ca^{2+} for three Na$^+$. A cotransporter in animals transports Na$^+$ into intestinal cells together with glucose, which is moving down its concentration gradient into the cell. (The Na$^+$ is then pumped out of the cell into the blood on the other side by Na$^+$/K$^+$ pumps.) Our understanding of **Na$^+$/glucose symporter** has helped us find more effective treatments for diarrhea, a serious problem in developing countries.

Normally, sodium in waste is reabsorbed in the colon, maintaining constant levels in the body, but diarrhea expels waste so rapidly that reabsorption is not possible, and sodium levels fall precipitously. To treat this life-threatening condition, patients are given a solution to drink containing high concentrations of salt (NaCl) and glucose. The solutes are taken up by Na$^+$/glucose cotransporters on the surface of intestinal cells and passed through the cells into the blood. This simple treatment has lowered infant mortality worldwide.

Passive and active transport

- Passive transport: Substances diffuse spontaneously down their concentration gradients, crossing a membrane with no expenditure of energy by the cell. The rate of diffusion can be greatly increased by transport proteins in the membrane.

- Active transport: Some transport proteins act as pumps, moving substances across amembrane against their concentration (or electrochemical) gradients. Energy for this work is usually supplied by ATP.

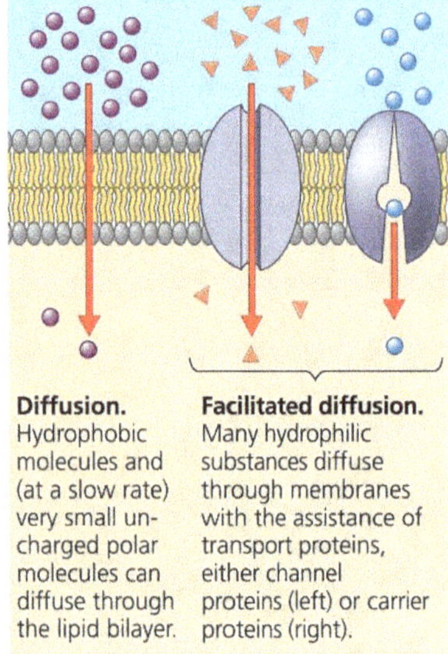

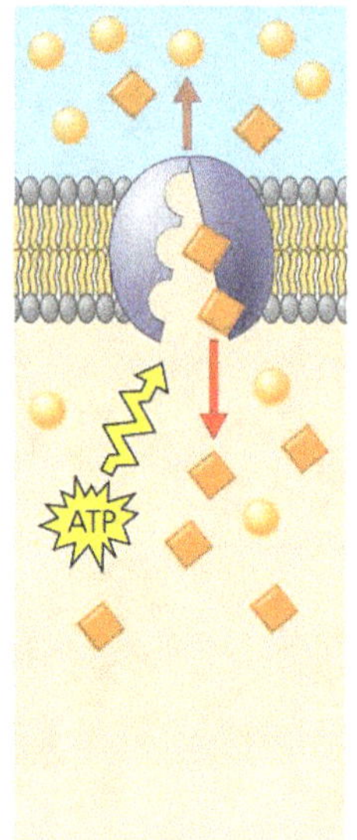

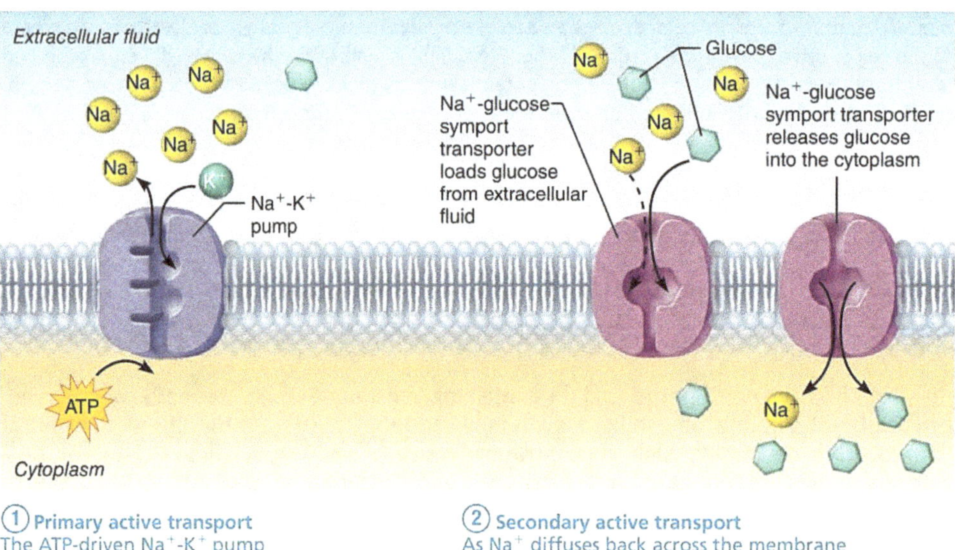

▲ **Figure 4.26** Secondary active transport is driven by the concentration gradient created by primary active transport.

Bulk transport

Large molecules, such as proteins and polysaccharides, generally don't cross the membrane by diffusion or transport proteins. Instead, they usually enter and leave the cell in bulk, packaged in vesicles.

• Exocytosis

The cell secretes certain molecules by the fusion of vesicles with the plasma membrane; this process is called exocytosis. The substance to be removed from the cell is first enclosed in a **protein-coated** membranous sac called a **secretory vesicle** that has budded from the Golgi apparatus moves along a microtubule of the cytoskeleton to the plasma membrane. When the vesicle membrane and plasma membrane come into contact, specific proteins in both membranes rearrange the lipid molecules of the two bilayers so that the two membranes fuse. The contents of the vesicle spill out of the cell, and the vesicle membrane becomes part of the plasma membrane.

Exocytosis, like other mechanisms in which vesicles are targeted to their destinations, involves a "docking" process in which transmembrane proteins on the vesicles, fancifully called v-SNAREs (v for vesicle), recognize certain plasma membrane proteins, called t-SNAREs (t for target), and bind with them. This binding causes the membranes to "corkscrew" together and fuse.

Exocytosis is typically stimulated by a cell-surface signal such as binding of a hormone to a membrane receptor or a change in membrane voltage. Exocytosis accounts for hormone secretion, neurotransmitter release, mucus secretion, and in some cases, ejection of wastes. For example, cells in the pancreas that make insulin secrete it into the extracellular fluid by exocytosis. In another example, nerve cells use exocytosis to release neurotransmitters that signal other neurons or muscle cells. When plant cells are making cell walls, exocytosis delivers some of the necessary proteins and carbohydrates from Golgi vesicles to the outside of the cell.

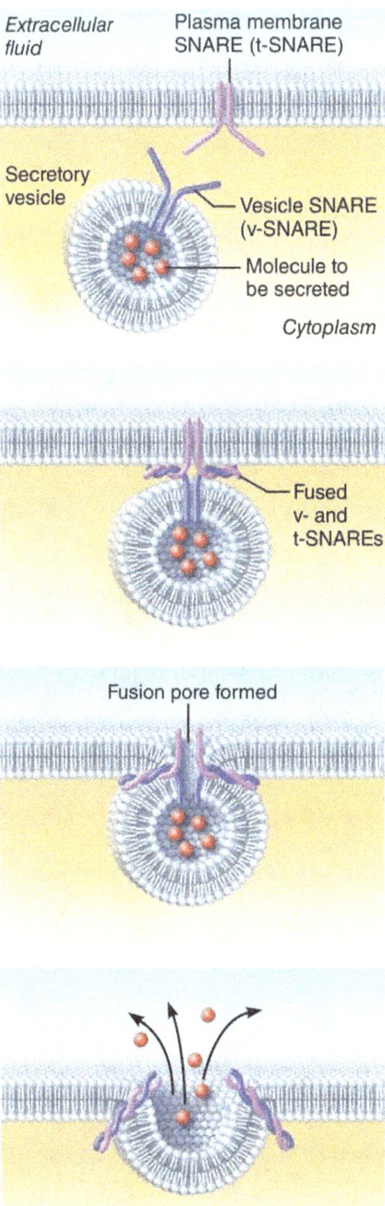

Exocytosis of macromolecules made by cells occurs via either of two pathways:

- **Constitutive secretion** is used for products that are released from cells continuously, as soon as synthesis is complete, such as collagen subunits for the ECM.

- **Regulated secretion** occurs in response to signals coming to the cells, such as the release of digestive enzymes from pancreatic cells in response to specific stimuli.

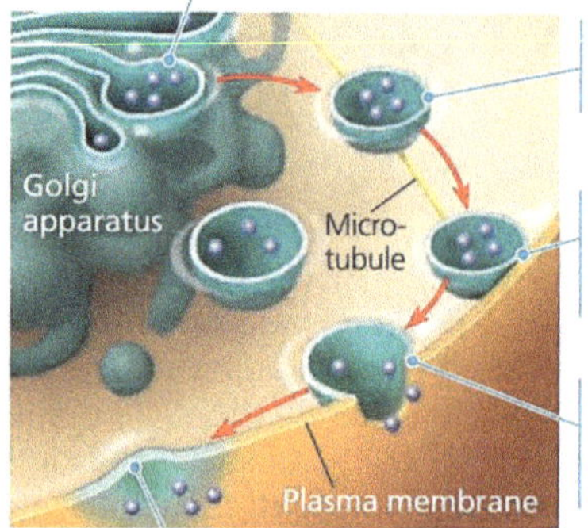

- **Endocytosis**

In endocytosis, the cell takes in molecules and particulate matter by forming new vesicles from the plasma membrane by budding off. Although the proteins involved in the processes are different, the events of endocytosis look like the reverse of exocytosis. First, a small area of the plasma membrane sinks inward to form a pocket. Then, as the pocket deepens, it pinches in, forming a vesicle containing material that had been outside the cell. Three types of endocytosis are:

Phagocytosis ("cellular eating"): In phagocytosis, a cell engulfs a particle by wrapping pseudopodia (singular, pseudopodium) around it and packaging it within a membranous sac called a food vacuole. Vesicle may or may not be protein coated but has receptors capable of binding to microorganisms or solid particles. The particle will be digested after the food vacuole fuses with a lysosome containing hydrolytic enzymes. For example, macrophages, which are cells of the immune system in mammals, kill bacteria via phagocytosis. Macrophages engulf bacterial cells into phagosomes. Once inside the cell, the phagosome fuses with a lysosome, and the digestive enzymes within the lysosome destroy the bacterium.

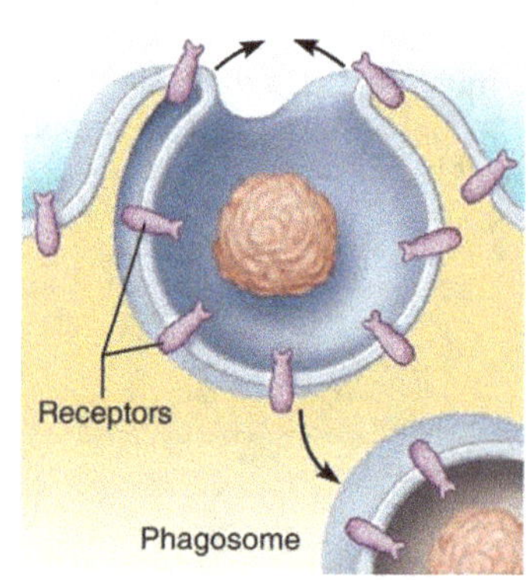

Pinocytosis ("cellular drinking"): In pinocytosis, a bit of infolding plasma membrane (which begins as a protein-coated pit) surrounds a very small volume of extracellular fluid containing dissolved molecules. In fact a cell continually "gulps" droplets of extracellular fluid into tiny vesicles. No receptors are used, so the process is nonspecific. In many cases, as above, the parts of the plasma membrane that form vesicles are lined on their cytoplasmic side by a fuzzy layer of coat protein; the "pits" and resulting vesicles are said to be "coated." Pinocytosis is particularly important in cells that are actively involved in nutrient absorption, such as cells that line the intestine in animals.

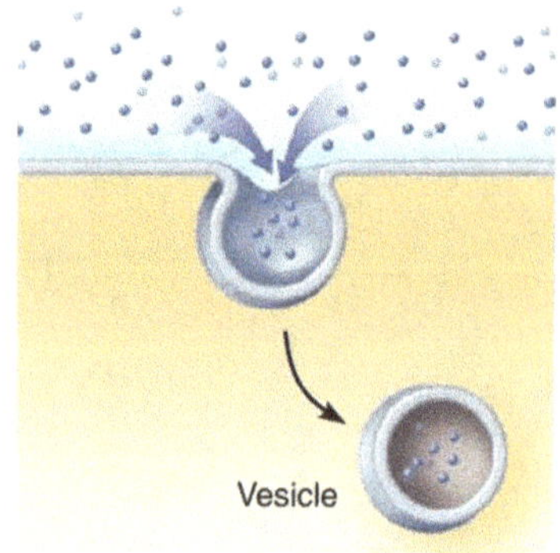

Receptor-mediated endocytosis: Receptor-mediated endocytosis is a specialized type of pinocytosis that enables the cell to specifically acquire bulk quantities of substances, even though those substances may not be very concentrated in the extracellular fluid. Embedded in the plasma membrane are proteins with receptor sites exposed to the extracellular fluid. Specific solutes bind to the sites. The receptor proteins then cluster in pits coated on the cytoplasmic side with the protein **clathrin**, and each coated pit forms a vesicle containing the bound molecules.

After the ingested material is liberated from the vesicle, the emptied receptors are recycled to the plasma membrane by the same vesicle. One type of molecule that is taken up by receptor mediated endocytosis is low-density lipoprotein (LDL). LDL molecules bring cholesterol into the cell where it can be incorporated into membranes. Mammalian cells take in many substances by receptor mediated endocytosis such as iron, and vitamin B$_{12}$.

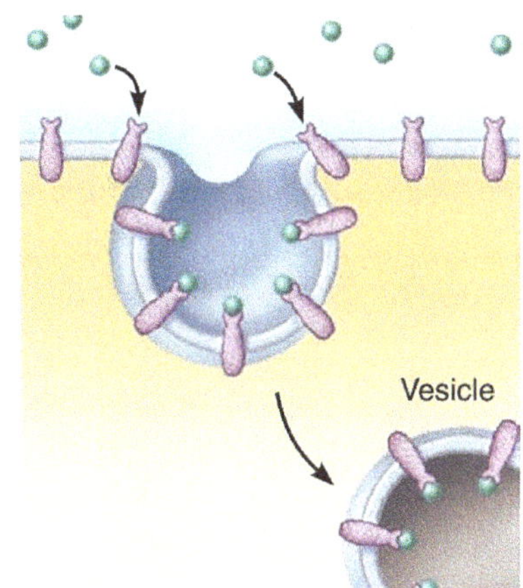

Another type of receptor-mediated endocytosis prominent in very thin cells produces invaginations called caveolae that involve a family of integral membrane proteins called caveolins associated with diverse peripheral proteins called **cavins**. Some viruses exploit the receptor-mediated exocytosis mechanism to enter cells. For instance, HIV, the virus that causes AIDS, binds to membrane receptors that function normally to internalize a needed molecule.

Human cells use receptor-mediated endocytosis to take in cholesterol for membrane synthesis and the synthesis of other steroids. Cholesterol travels in the blood in particles called low-density lipoproteins (LDLs), each a complex of lipids and a protein. LDLs bind to LDL receptors on plasma membranes and then enter the cells by endocytosis. In the inherited disease familial hypercholesterolemia, characterized by a very high level of cholesterol in the blood, LDLs cannot enter cells because the LDL receptor proteins are defective or missing. Cholesterol thus accumulates in the blood, contributing to early atherosclerosis, the buildup of lipids within blood vessel walls. This narrows the space in the vessels and impedes blood flow, potentially resulting in heart damage and stroke.

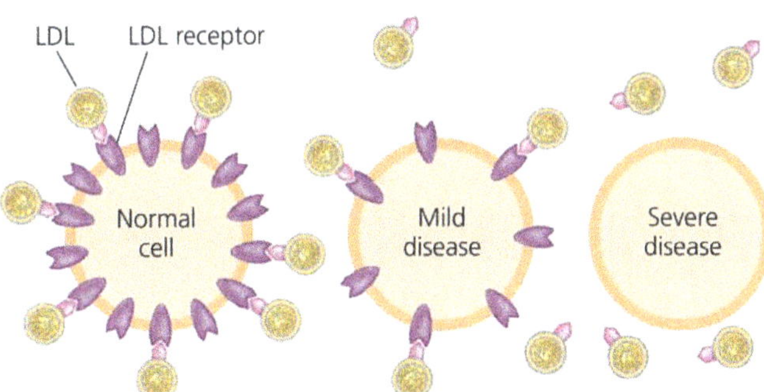

Exocytosis and endocytosis are coupled processes. Within seconds to minutes of exocytosis, endocytosis is initiated to retrieve vesicle membrane and its associated proteins. This coupling of the two processes maintains the surface area of the plasma membrane of a secretory cell at a controlled level. The balance between endocytosis and exocytosis keeps the surface area of a cell's plasma membrane relatively constant. This process of membrane movement and recycling is called **membrane trafficking.**

- **Transcytosis**

Moving substances into, across, and then out of the cell

- **Moving substances into, across, and then out of the cell**

Moving substances from one area (or membranous organelle) in the cell to another.

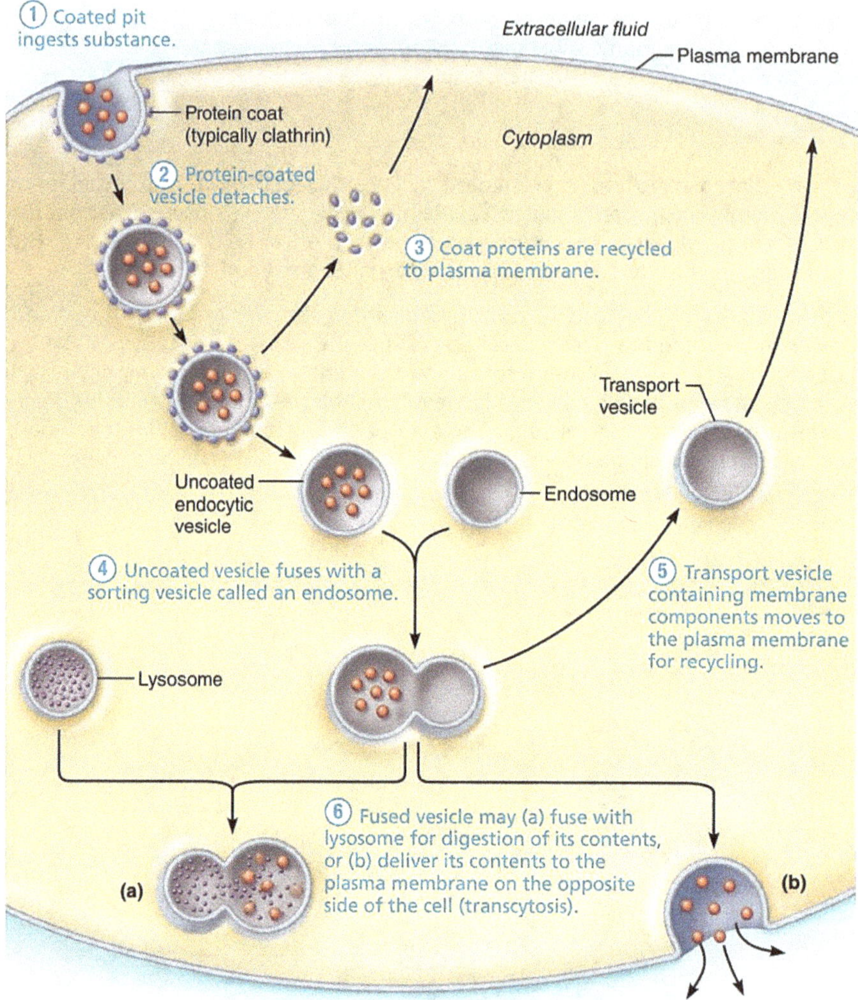

▲ **Figure 4.27** Events of endocytosis mediated by protein-coated pits. Note the three possible fates for a vesicle and its contents, shown in 5 and 6.

Cell Communication

Changes in the environment are a persistent feature of life, and living cells are continually faced with alterations in temperature and availability of nutrients, water, and light. All cells and organisms must be able to respond appropriately to their environments. This is made possible by means of signalling pathways which coordinate the activities of cells, even if they are large distances apart in the same body. Signalling pathways can be electrical (e.g. the nervous system) or chemical (e.g. the hormone system in animals). They involve a wide range of molecules such as neurotransmitters and hormones.

One of the best-coordinated, communication- rich events in a cell's life cycle is nuclear division, or mitosis. Another of the most significant communications in a cell's life cycle is the instruction it receives to die by a programmed death, called apoptosis.

Local and Long-Distance Signaling

Like bacteria or yeast cells, cells in a multicellular organism communicate by way of a signaling molecules targeted for cells that may or may not be immediately adjacent. Effective signaling requires a signaling molecule, called a **ligand**, and a molecule to which the signal binds, called a **receptor** protein.

The interaction of these two components initiates the process of **signal transduction**, which converts the information in the signal into a **cellular response**. Ligands include most neurotransmitters (nervous system signals), hormones (endocrine system signals), and paracrines (chemicals that act locally and are rapidly destroyed).

Cells can communicate through any of four basic mechanisms, depending primarily on the distance between the signaling and responding cells. These mechanisms are **(1) juxtacrine signaling (direct contact), (2) paracrine signaling, (3) endocrine signaling, and (4) synaptic signaling**. In addition to using these four basic mechanisms, some cells actually send signals to themselves, secreting signals that bind to specific receptors on their own plasma membranes. This process, called **autocrine signaling**, is thought to play an important role in reinforcing developmental changes, and it is an important component of signaling in the immune system.

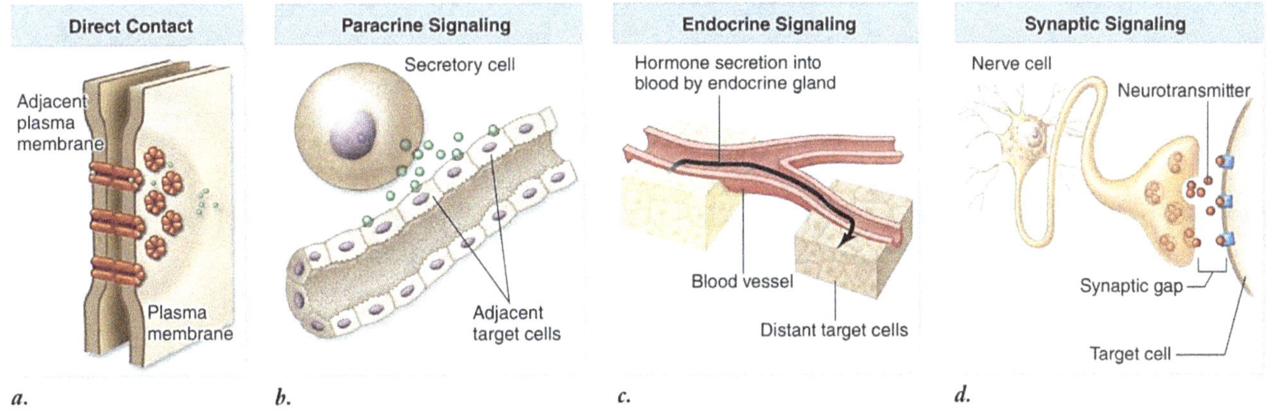

▲ **Figure 4.28** Four kinds of cell signaling.

Eukaryotic cells may communicate by direct contact, which is one type of **local signaling**. Many animal and plant cells have cell junctions that directly connect the cytoplasms of adjacent cells. In these cases, signaling substances dissolved in the cytosol can pass between neighboring cells. Moreover, some animal cells may communicate by direct contact between cell-surface molecules. This type of local signaling is especially important in embryonic development, the immune response, and in maintaining adult stem cell populations. Some local regulators are considered hormones. Local regulators include local chemical mediators such as growth factors, histamine, nitric oxide, and prostaglandins.

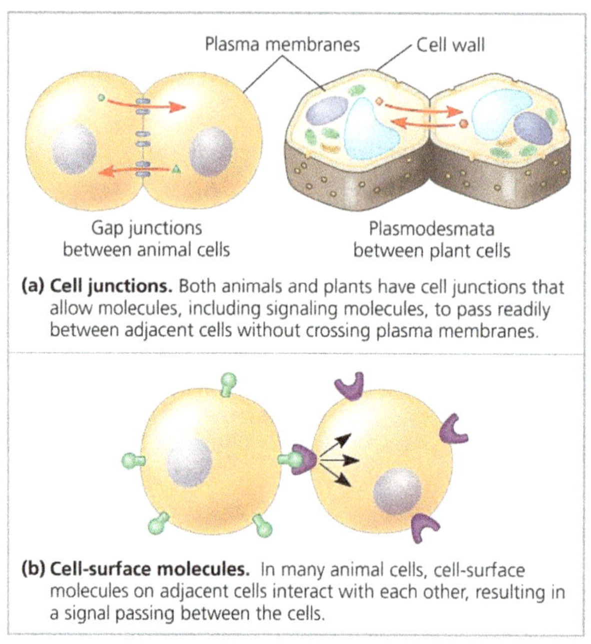

(a) **Cell junctions.** Both animals and plants have cell junctions that allow molecules, including signaling molecules, to pass readily between adjacent cells without crossing plasma membranes.

(b) **Cell-surface molecules.** In many animal cells, cell-surface molecules on adjacent cells interact with each other, resulting in a signal passing between the cells.

▲ **Figure 4.29** Communication requiring contact between cells.

In many other cases of local signaling, signaling molecules are secreted by the signaling cell. Some molecules travel only short distances; such local regulators influence cells that are nearby. This type of local signaling in animals is called **paracrine signaling**. One class of local regulators in animals, growth factors, consists of compounds that stimulate nearby target cells to grow and divide. Numerous cells can simultaneously receive and respond to the growth factors produced by a single cell in their vicinity.

A highly specialized type of local signaling called **synaptic signaling** occurs in the animal nervous system. An electrical signal along a nerve cell triggers the secretion of neurotransmitter molecules. These molecules act as chemical signals, diffusing across the synapse— the narrow space between the nerve cell and its target cell—triggering a response in the target cell.

Both animals and plants use molecules called hormones for **long-distance signaling**. In **hormonal signaling** in animals, also known as **endocrine signaling**, specialized cells release hormones, which travel through the circulatory system to other parts of the body, where they reach target cells that can recognize and respond to them.

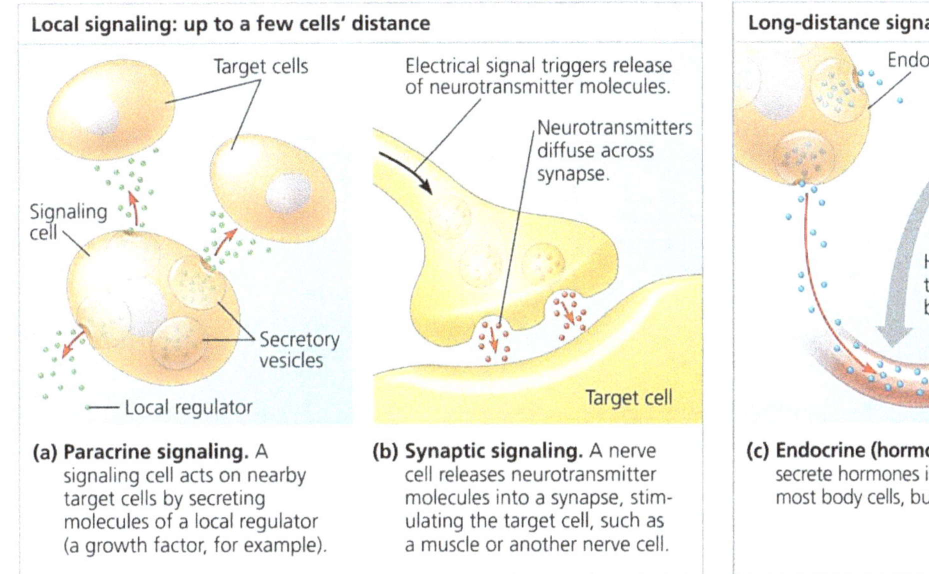

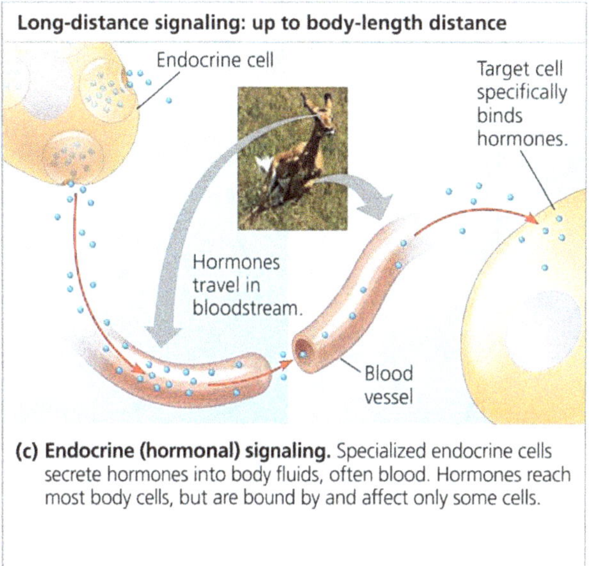

(a) Paracrine signaling. A signaling cell acts on nearby target cells by secreting molecules of a local regulator (a growth factor, for example).

(b) Synaptic signaling. A nerve cell releases neurotransmitter molecules into a synapse, stimulating the target cell, such as a muscle or another nerve cell.

(c) Endocrine (hormonal) signaling. Specialized endocrine cells secrete hormones into body fluids, often blood. Hormones reach most body cells, but are bound by and affect only some cells.

▲ **Figure 4.30** Local and long-distance cell signaling by secreted molecules in animals. In both local and long-distance signaling, only specific target cells that can recognize a given signaling molecule will respond to it.

What happens when a potential target cell is exposed to a secreted signaling molecule? The ability of a cell to respond is determined by whether it has a specific receptor molecule that can bind to the signaling molecule. The information conveyed by this binding, the signal, must then be changed into another form—transduced—inside the cell before the cell can respond.

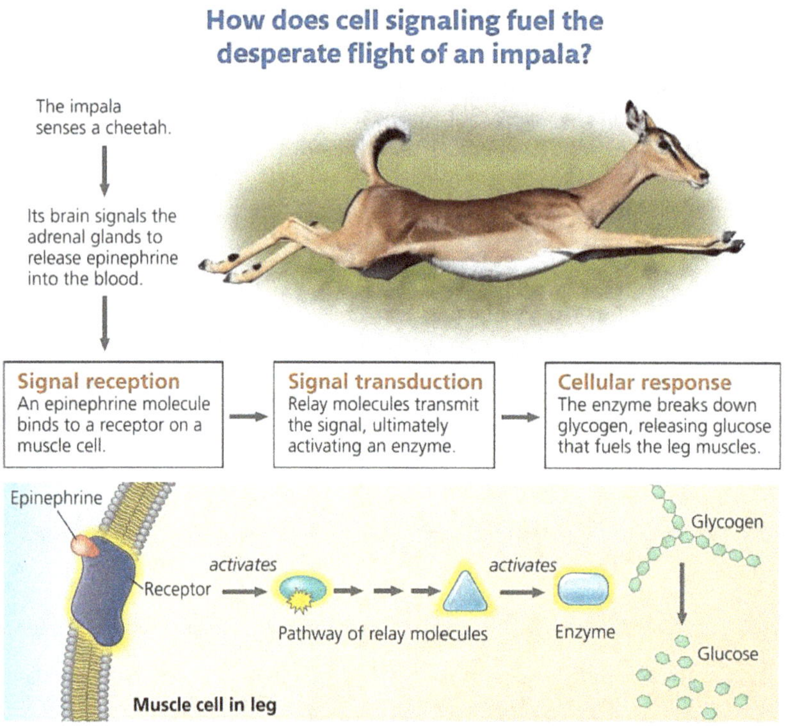

How does cell signaling fuel the desperate flight of an impala?

▲ **Figure 4.31** This impala is fleeing for its life, racing to escape the predatory cheetah nipping at its heels. The impala is breathing rapidly, its heart pounding and its legs pumping furiously. These physiological functions are all part of the impala's "fightor-flight" response, driven by hormones released from its adrenal glands at times of stress—in this case, upon sensing the cheetah

The Three Stages of Cell Signaling:

Signal reception. Reception is the target cell's detection of a signaling molecule coming from outside the cell. A chemical signal is "detected" when the signaling molecule binds to a receptor protein located at the cell's surface (or inside the cell, to be discussed later).

Signal transduction. The binding of the signaling molecule changes the receptor protein in some way, initiating the process of transduction. The transduction stage converts the signal to a form that can bring about a specific cellular response. Transduction sometimes occurs in a single step but more often requires a sequence of changes in a series of different molecules—a **signal transduction pathway**. The molecules in the pathway are often called relay molecules; three are shown as an example.

Cellular response. The transduced signal finally triggers a specific cellular response. The response may be almost any imaginable cellular activity—such as catalysis by an enzyme (for example, glycogen phosphorylase), rearrangement of the cytoskeleton, or activation of specific genes in the nucleus. The cell-signaling process helps ensure that crucial activities like these occur in the right cells, at the right time, and in proper coordination with the activities of other cells of the organism.

Signal reception: A signaling molecule binds to a receptor, causing it to change shape

The signaling molecule, which is called a ligand, binds noncovalently to the receptor molecule. The binding occurs when the ligand and receptor happen to collide in the correct orientation and with enough energy to form a ligand-receptor complex. Unlike enzymes, which convert their substrates into products, receptors do not usually alter the structure of their ligands. Instead, the ligands alter the structure of their receptors, causing a conformational change. For many receptors, this shape change directly activates the receptor, enabling it to interact with other molecules in or on the cell.

For other receptors, the immediate effect of ligand binding is to cause the aggregation of two or more receptor proteins, which leads to further molecular events inside the cell. Most signal receptors are plasma membrane proteins, but others are located inside the cell. Receptors can be differentiated based on their location and the chemical nature of their ligands. **Intracellular receptors** bind to hydrophobic ligands that can easily cross the plasma membrane to enter the cell. In contrast, cell-surface or **membrane receptors** bind to hydrophilic ligands outside the cell because these ligands cannot easily cross the membrane.

- **Receptors in the Plasma Membrane**

Cell-surface transmembrane receptors play crucial roles in the biological systems of animals. The largest family of human cell-surface receptors is that of the G protein-coupled receptors (GPCRs). Most water-soluble signaling molecules bind to specific sites on transmembrane receptor proteins that transmit information from the extracellular environment to the inside of the cell. Surface receptors bind various ligands such as peptide hormones, growth factors, and neurotransmitters. Three major types: **G protein-coupled receptors** (GPCRs), **enzyme-linked receptors**, and **ion channel receptors**.

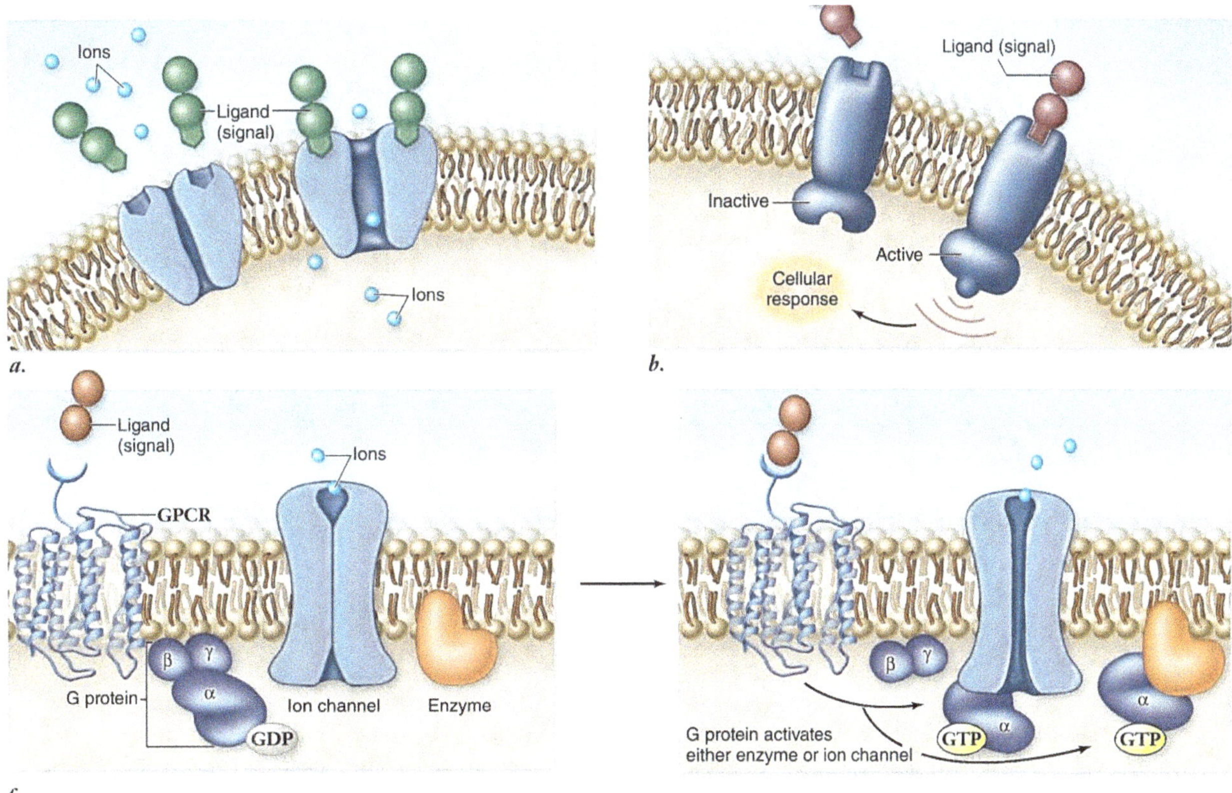

▲ **Figure 4.32 Cell-surface receptors. a.** Chemically gated ion channels form a pore in the plasma membrane that can be opened or closed by chemical signals. They are usually selective, allowing the passage of only one type of ion. **b.** Enzymatic receptors bind to ligands on the extracellular surface. A catalytic region on their cytoplasmic portion transmits the signal across the membrane by acting as an enzyme in the cytoplasm. **c.** G protein–coupled receptors (GPCRs) bind to ligands outside the cell and to G proteins inside the cell. The G protein then activates an enzyme or ion channel, transmitting signals from the cell's surface to its interior.

G Protein-Coupled Receptors

A G protein-coupled receptor (GPCR) is a cell-surface transmembrane receptor that works with the help of a G protein (GTP-binding protein), a protein that binds the energy-rich molecule GTP. Many different signaling molecules—including yeast mating factors, neurotransmitters, and epinephrine (adrenaline) and many other hormones—use GPCRs. The human genome contains approximately 1000 different genes that encode GPCRs. They are known to play many key roles, including those that affect our behavior, vision, and sense of smell.

GPCR-based signaling systems are extremely widespread and diverse in their functions, including roles in embryonic development and sensory reception. In humans, for example, vision, smell, and taste depend on GPCRs. Similarities in the structures of G proteins and GPCRs in diverse organisms suggest that G proteins and their associated receptors evolved very early among eukaryotes.

Malfunctions of the associated G proteins themselves are involved in many human diseases, including bacterial infections. Up to 60% of all medicines used today exert their effects by influencing G protein pathways.

G proteins are composed of three subunits, called α, β, and γ. As a result, they are often called heterotrimeric G proteins. When a ligand binds to a GPCR and activates its associated G protein, the G protein exchanges GDP for GTP and dissociates into two parts consisting of the G_α subunit bound to GTP, and the G_β and G_γ subunits together ($G_{\beta\gamma}$). The signal can then be transmitted by either the G_α or the $G_{\beta\gamma}$ components stimulating effector proteins. The hydrolysis of bound GTP to GDP by G_α causes reassociation of the heterotrimer, restoring the "off" state of the system.

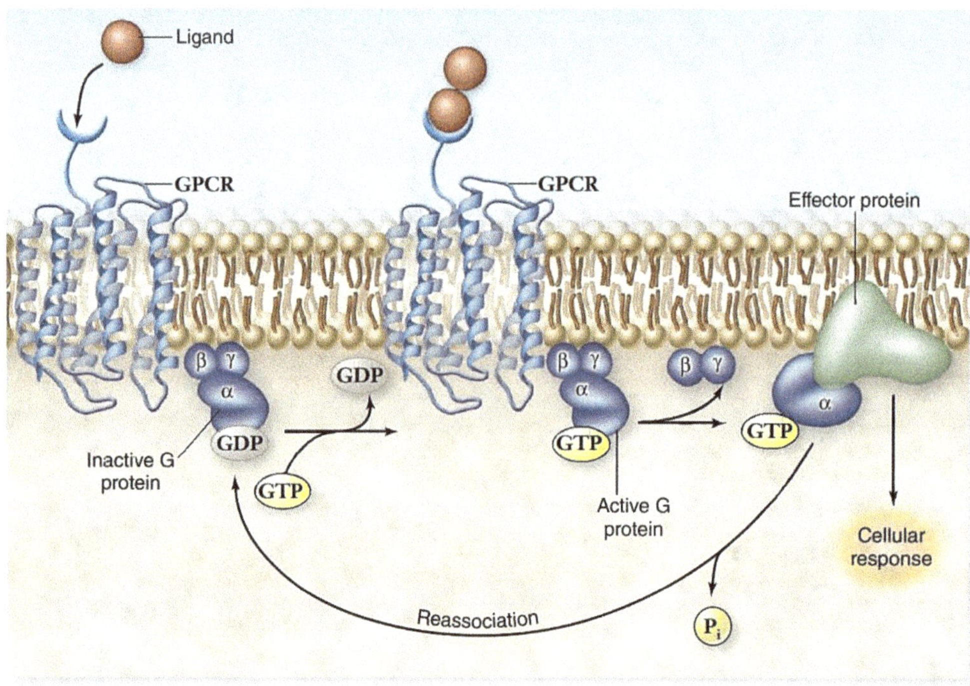

▲ **Figure 4.33** The action of G protein–coupled receptors. G protein–coupled receptors act through a heterotrimeric G protein that links the receptor to an effector protein. When ligand binds to the receptor, it activates an associated G protein, exchanging GDP for GTP. The active G protein complex dissociates into G_α and $G_{\beta\gamma}$. The G_α subunit (bound to GTP) is shown activating an effector protein. The effector protein may act directly on cellular proteins or produce a second messenger to cause a cellular response. G_α can hydrolyze GTP inactivating the system, then reassociate with $G_{\beta\gamma}$.

G protein-coupled receptors vary in the binding sites for their ligands and also for different types of G proteins inside the cell. Nevertheless, GPCR proteins are all remarkably similar in structure. In fact, they make up a large family of eukaryotic receptor proteins with a secondary structure in which the single polypeptide, has **seven transmembrane chelices**. Specific loops between the helices form binding sites for signaling molecules (outside the cell) and G proteins (on the cytoplasmic side).

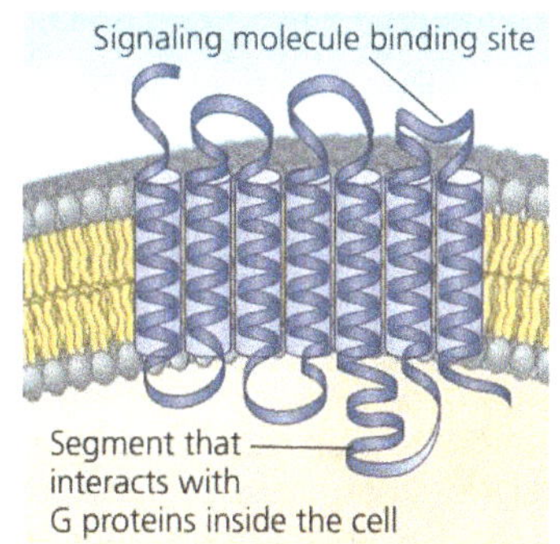

1. Attached but able to move along the cytoplasmic side of the membrane, a G protein functions as a molecular switch that is either on or off, depending on whether GDP or GTP is attached — hence the term G protein. (GTP, or guanosine triphosphate, is similar to ATP.) When GDP is bound to the G protein, the G protein is inactive. The receptor and G protein work together with another protein, usually an enzyme.

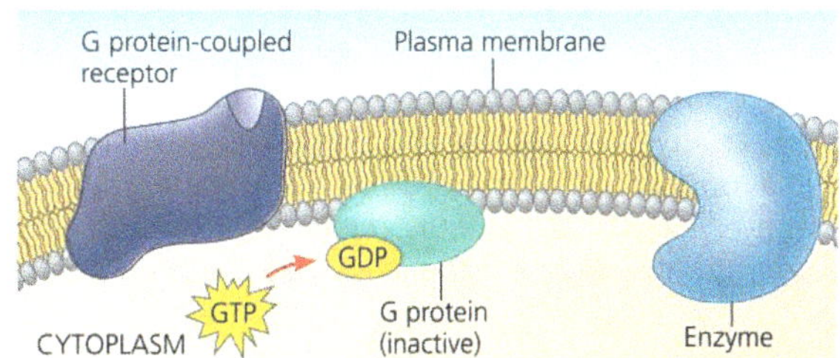

2. When the appropriate signaling molecule binds to the extracellular side of the receptor, the receptor is activated and changes shape. Its cytoplasmic side then binds an inactive G protein, causing a GTP to displace the GDP. This activates the G protein.

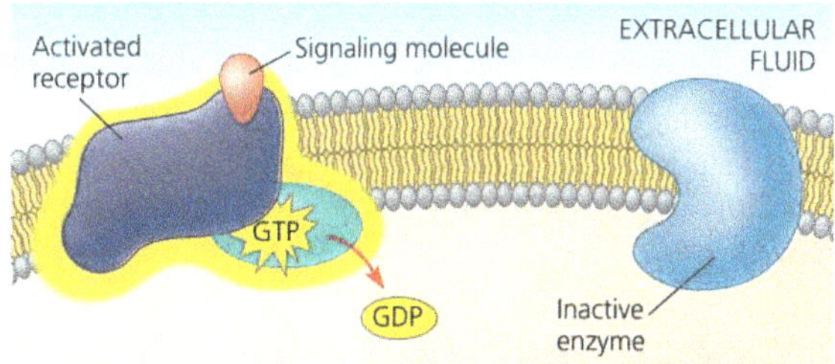

3. The activated G protein dissociates from the receptor, diffuses along the membrane, and then binds to an enzyme, altering the enzyme's shape and activity. Once activated, the enzyme can trigger the next step leading to a cellular response. Binding of signaling molecules is reversible: Like other ligands, they bind and dissociate many times. The ligand concentration outside the cell determines how often a ligand is bound and initiates signaling.

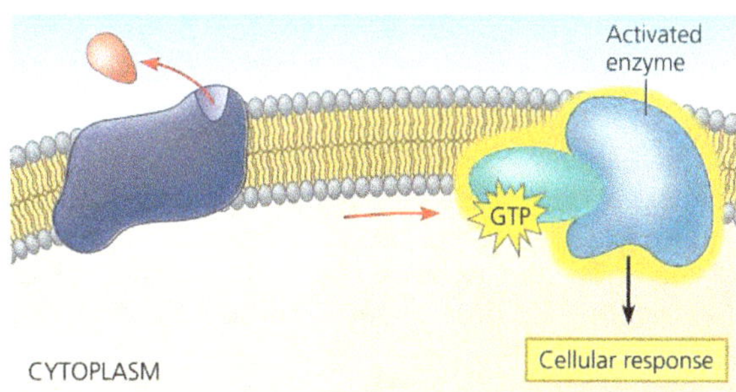

4. The changes in the enzyme and G protein are only temporary: The G protein also acts as a GTPase, hydrolyzing its bound GTP to GDP and ; as a result, it can no longer activate the enzyme. The G protein leaves the enzyme, which returns to its inactive state. The G protein is now available for reuse. Its GTPase function allows the pathway shut down rapidly when the signaling molecule is no longer present.

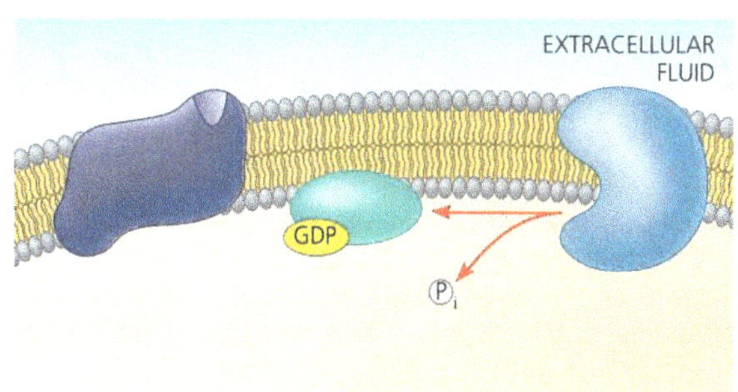

Enzyme-linked receptors function directly as enzymes or are linked to enzymes

Receptor **tyrosine kinases** (RTKs) belong to a major class of plasma membrane receptors characterized by having enzymatic activity. An RTK is a protein kinase—an enzyme that catalyzes the transfer of phosphate groups from ATP to another protein. The part of the receptor protein extending into the cytoplasm functions more specifically as a tyrosine kinase, an enzyme that catalyzes the transfer of a phosphate group from ATP to the amino acid tyrosine of a substrate protein. Thus, RTKs are membrane receptors that attach phosphates to tyrosines.

Upon binding a ligand such as a growth factor, one RTK may activate ten or more different transduction pathways and cellular responses. Often, more than one signal transduction pathway can be triggered at once, helping the cell regulate and coordinate many aspects of cell growth and cell reproduction. Tyrosine kinase receptors bind certain hormones that regulate many cellular processes and are important in development. They include insulin and growth factors, such as nerve growth factor. The ability of a single ligand-binding event to trigger so many pathways is a key difference between RTKs and GPCRs; GPCRs generally activate a single transduction pathway. Abnormal RTKs that function even in the absence of signaling molecules are associated with many kinds of cancer.

1. Many receptor tyrosine kinases have the structure depicted schematically here. Before the signaling molecule binds, the receptors exist as individual units referred to as monomers. Notice that each monomer has an extracellular ligand-binding site, an α helix spanning the membrane, and an intracellular tail containing multiple tyrosines.

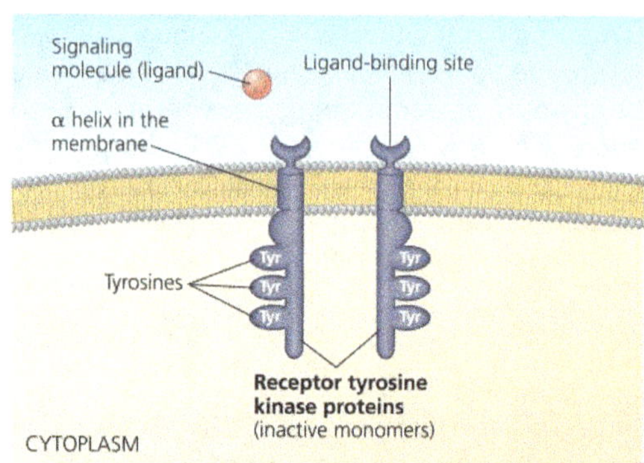

2. The binding of a signaling molecule (such as a growth factor) causes two receptor monomers to associate closely with each other, forming a complex known as a dimer, a process called **dimerization**.

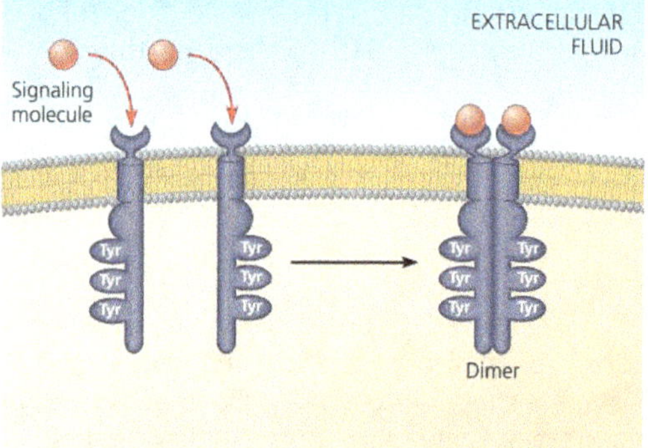

3. Dimerization activates the tyrosine kinase region of each monomer; each tyrosine kinase adds a phosphate from an ATP molecule to a tyrosine that is part of the tail of the other monomer, a process called **autophosphorylation**.

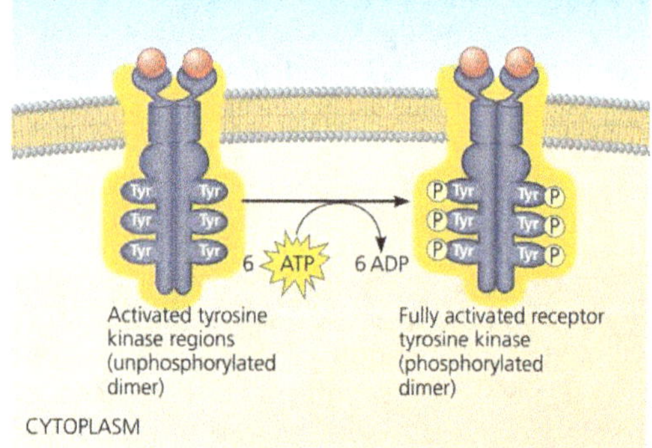

4. Now that the receptor is fully activated, it is recognized by specific relay proteins inside the cell. Each such protein binds to a specific phosphorylated tyrosine, undergoing a resulting structural change that activates the bound relay protein. Each activated protein triggers a transduction pathway, leading to a cellular response.

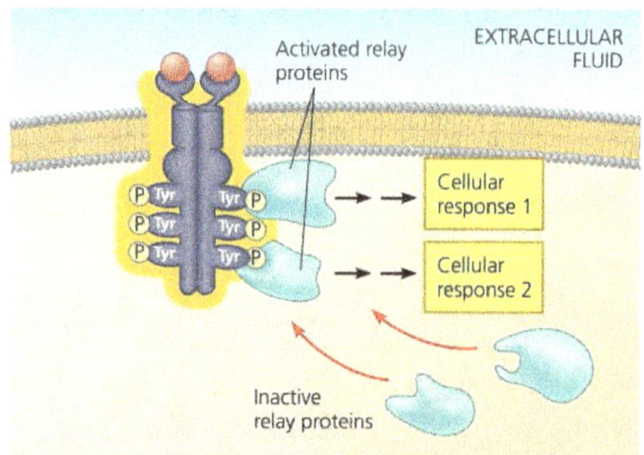

The insulin receptor is a receptor tyrosine kinase that initiates a variety of cellular responses related to glucose metabolism. One signal transduction pathway that this receptor mediates leads to the activation of the enzyme glycogen synthase. This enzyme converts glucose to glycogen thereby lowering blood glucose. Other proteins activated by the insulin receptor act to inhibit the synthesis of enzymes involved in making glucose, and to increase the number of glucose transporter proteins in the plasma membrane.

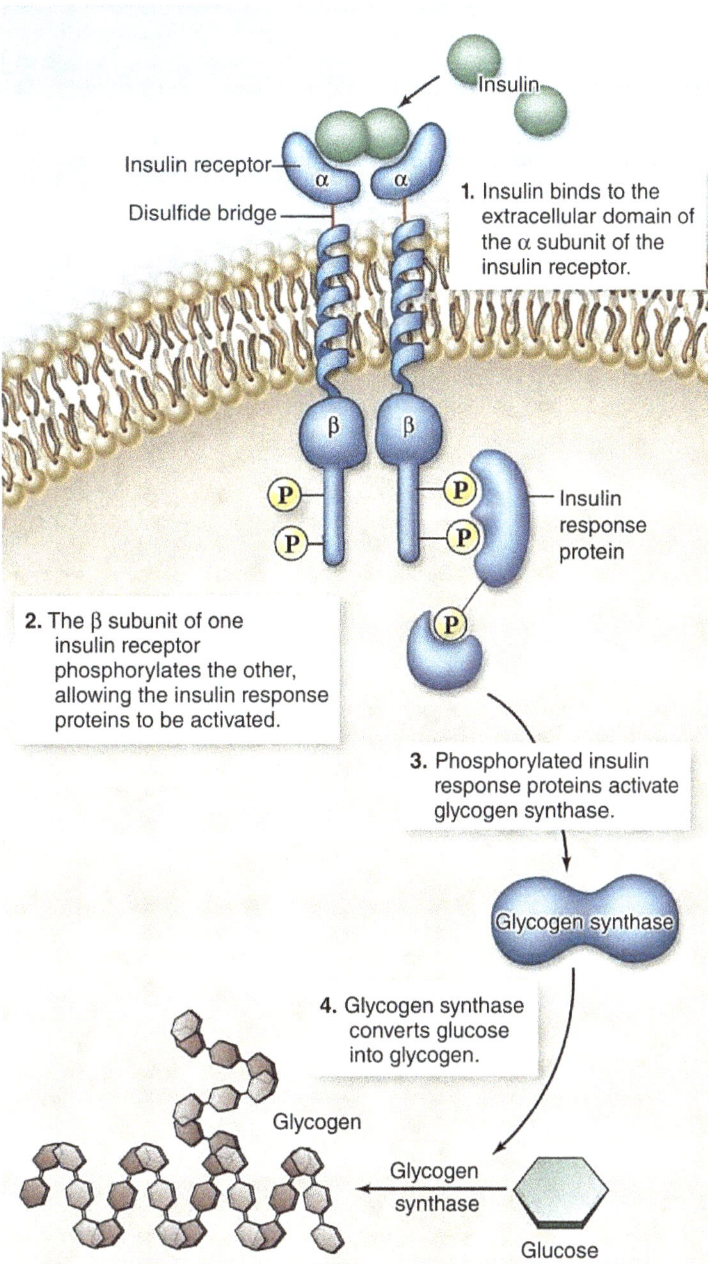

▲ **Figure 4.1** Factors that affect membrane fluidity.

Ion Channel Receptors

A ligand-gated ion channel is a type of membrane channel receptor containing a region that can act as a "gate," opening or closing the channel when the receptor changes shape. When a signaling molecule binds as a ligand to the channel receptor, the channel opens or closes, allowing or blocking the flow of specific ions, such as Na^+ or Ca^{2+}. Like the other receptors we have discussed, these proteins bind the ligand at a specific site on their extracellular sides. Ion channel receptors convert chemical signals into electrical signals.

In animals, **ligand-gated ion channels** (ionotropic receptors) are very important in the transmission of signals between neurons and muscle cells and between two neurons. For example, the neurotransmitter molecules released at a synapse between two nerve cells bind as ligands to some ion channels on the receiving cell, causing the channels to open. Ions flow in (or, in some cases, out), triggering an electrical signal that propagates down the length of the receiving cell.

Some of the gated ion channels are controlled by electrical signals instead of ligands; these **voltage-gated ion channels** are crucial to the functioning of the nervous system. Some ion channels are present on membranes of organelles, such as the ER.

1. A ligand-gated ion channel receptor in which the channel remains closed until a ligand binds to the receptor.

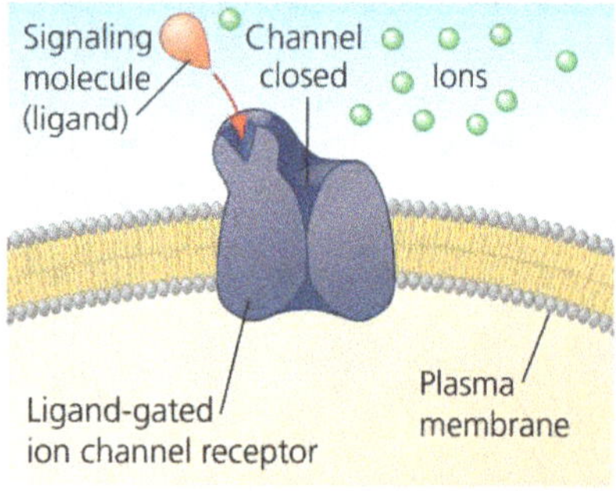

2. When the ligand binds to the receptor and the channel opens, specific ions can flow through the channel and rapidly change the local concentration of that ion inside the cell. This change may directly affect the activity of the cell in some way.

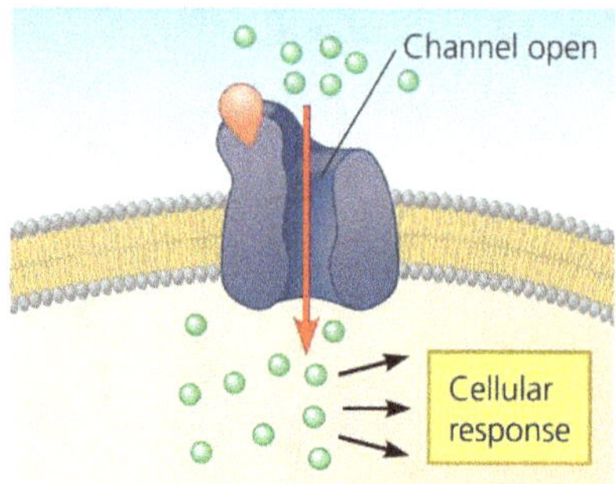

3. When the ligand dissociates from this receptor, the channel closes and ions no longer enter the cell.

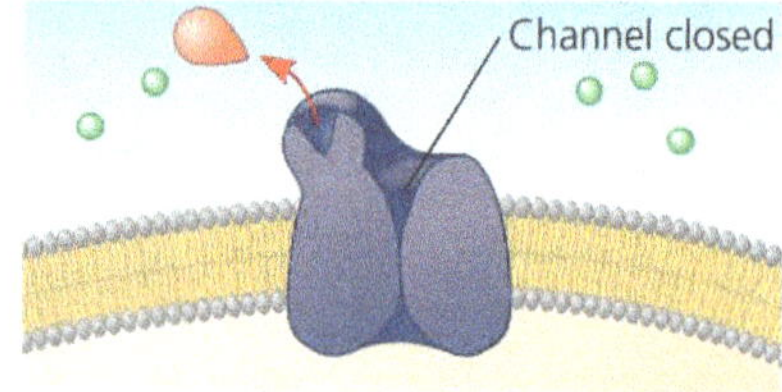

The acetylcholine receptor found in muscle cell membranes functions as an Na⁺ channel. When the receptor binds to its ligand, the neurotransmitter acetylcholine, the channel opens allowing Na⁺ to flow into the muscle cell. This is a critical step linking the signal from a motor neuron to muscle cell contraction.

The acetylcholine ligand-gated ion channel receptor is expressed in brain neurons that are important in learning and memory, as well as in skeletal muscle cells. Loss of acetylcholine receptors in the brain is seen in patients with epilepsy, Alzheimer disease, schizophrenia, and drug addiction. Because the receptor also binds nicotine, it might be involved in nicotine addiction in smokers.

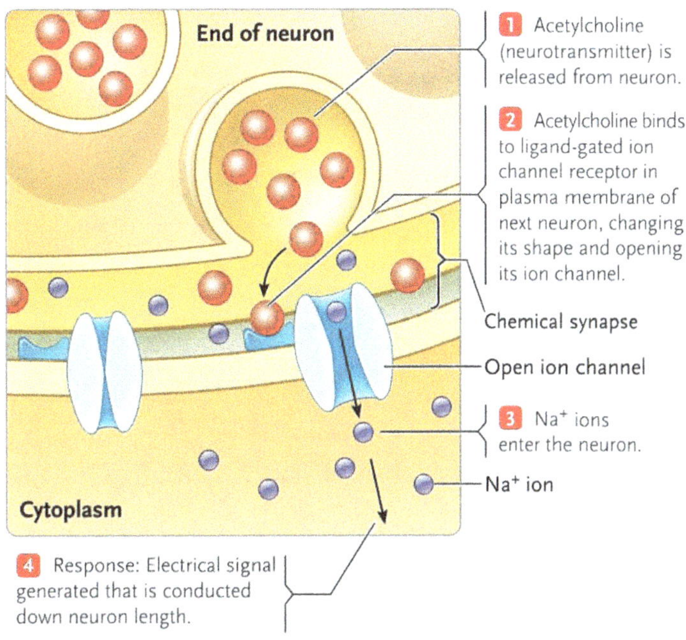

▲ **Figure 4.34** Operation of a ligand-gated ion channel, exemplified by synaptic signaling at a chemical synapse involving the neurotransmitter, acetylcholine. © Cengage Learning 2017

- **Intracellular Receptors**

Intracellular receptor proteins are found in either the cytoplasm or nucleus of target cells. To reach such a receptor, a signaling molecule passes through the target cell's plasma membrane. A number of important signaling molecules can do this because they are either hydrophobic enough or small enough to cross the hydrophobic interior of the membrane.

In animal cells most steroid hormones, such as cortisol in vertebrates, enter target cells and combine with receptor molecules in the cytosol. Vitamins A and D and nitric oxide (NO) also bind with intracellular receptors. Within a target cell, NO binds to and activates the enzyme guanylyl cyclase, which catalyzes the synthesis of cGMP (cyclic GMP) from GTP in a reaction analogous to the synthesis of cAMP from ATP. cGMP functions as a second messenger similar to cAMP.

After binding, the ligand–receptor complex moves into the nucleus. Thyroid hormones (which are not steroids) bind to receptors already bound to DNA inside the nucleus. As the ligand–receptor complex makes it all the way to the nucleus of the cell, these receptors are often called **nuclear receptors**.

The behavior of aldosterone is a representative example of how steroid hormones work. This hormone is secreted by cells of the adrenal gland (a gland that lies above the kidney), then travels through the blood and enters cells all over the body. However, a response occurs only in kidney cells because they alone contain receptors for aldosterone. In these cells, the hormone binds to and activates the receptor protein. With aldosterone attached, the active form of the receptor protein then enters the nucleus and turns on specific genes that control water and sodium flow in kidney cells, ultimately affecting blood volume.

How does the activated hormone-receptor complex turn on genes? The genes in a cell's DNA function by being transcribed and processed into messenger RNA (mRNA), which leaves the nucleus and is translated into a specific protein by ribosomes in the cytoplasm. Special proteins called transcription factors control which genes are turned on—that is, which genes are transcribed into mRNA—in a particular cell at a particular time. When the aldosterone receptor is activated, it acts as a transcription factor that turns on specific genes.

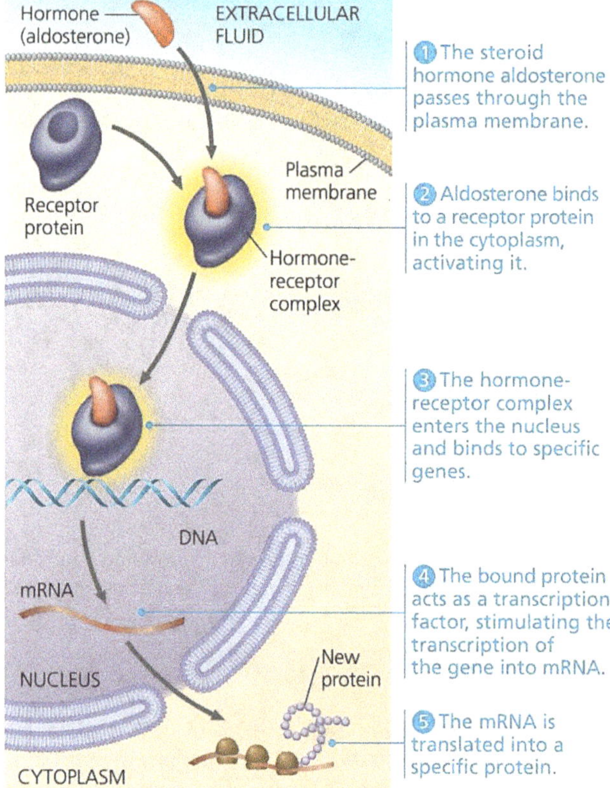

▶ **Figure 4.35** Steroid hormone interacting with an intracellular receptor.

By acting as a **transcription factor**, the aldosterone receptor itself carries out two parts of the signaling pathway, as receptor and transducer. Most other intracellular receptors function in the same way, although many of them, such as the thyroid hormone receptor, are already in the nucleus before the signaling molecule reaches them. Interestingly, many of these intracellular receptor proteins are structurally similar so the genes that code for them appear to be the evolutionary descendants of a single ancestral gene. Because of their structural similarities, they are all part of the **nuclear receptor superfamily**.

Each of these receptors has three functional domains:

1. a hormone-binding domain,

2. a DNA-binding domain, and

3. a domain that can interact with coactivators to affect the level of gene transcription

Signal transduction: Cascades of molecular interactions transmit signals from receptors to relay molecules in the cell

When receptors for signaling molecules are plasma membrane proteins, the transduction stage of cell signaling is usually a multistep pathway involving many molecules. Steps often include activation of proteins by addition or removal of phosphate groups or release of other small molecules or ions that act as signaling molecules. One benefit of using multiple steps is that a signal caused by a small number of signaling molecules can be greatly amplified. A second benefit of using multistep pathways is that they provide more opportunities for coordination and control than do simpler systems.

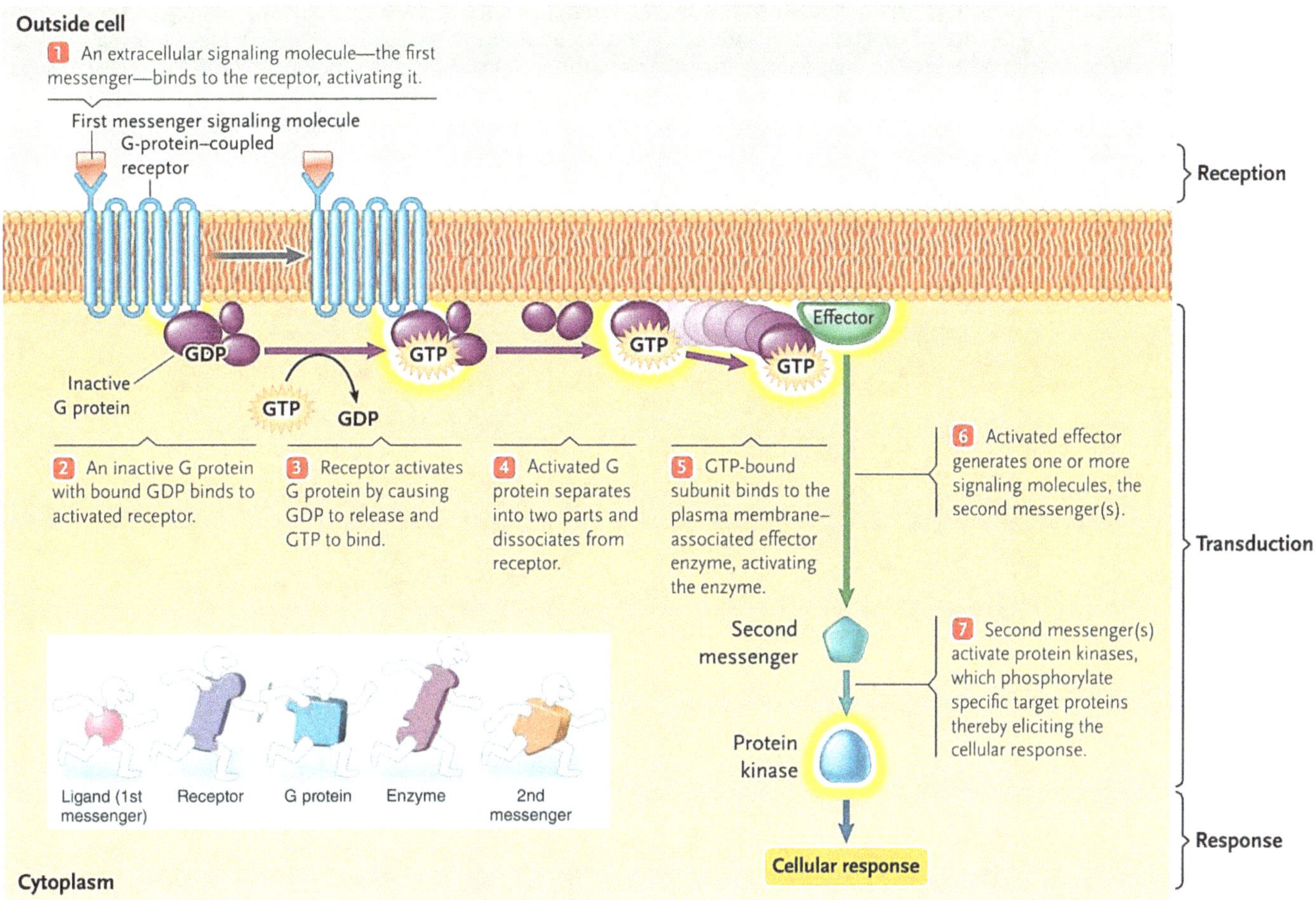

▲ **Figure 4.36** first messenger → receptor → G proteins → effector → protein kinases → target proteins. © Cengage Learning 2017

Signal transduction pathway is the chain of molecular interactions that leads to a particular response within the cell after binding of a signal to the receptor. Like falling dominoes, the signal-activated receptor activates another molecule, which activates yet another molecule, and so on, until the protein that produces the final cellular response is activated. The molecules that relay a signal from receptor to response, which we call relay molecules, are often proteins.

Protein-protein interactions are a major theme of cell signaling. For example, binding of the first messenger (the extracellular signaling molecule) to the receptor activates it. The activated receptor activates an intracellular G protein, which in turn activates a plasma membrane-associated effector molecule. The activated effector generates one or more internal, nonprotein signaling molecules called second messengers, which results in the activation of protein kinases. The activated protein kinases phosphorylate specific target proteins, thereby eliciting the cellular response.

Keep in mind that the original signaling molecule is not physically passed along a signaling pathway; in most cases, it never even enters the cell. When we say that the signal is relayed along a pathway, we mean that certain information is passed on. At each step, the signal is transduced into a different form, commonly a shape change in the next protein. Very often, the shape change is brought about by phosphorylation.

- **Protein Phosphorylation and Dephosphorylation**

Phosphorylation of proteins and its reverse, dephosphorylation, are commonly used in cells to regulate protein activity. An enzyme that transfers phosphate groups from ATP to a protein is generally known as a **protein kinase**. The receptor tyrosine kinase (RTK) is a specific kind of protein kinase that phosphorylates tyrosines on the other RTK in a dimer. Most cytoplasmic protein kinases, however, act on proteins different from themselves. Another distinction is that most cytoplasmic protein kinases phosphorylate either of two other amino acids, **serine** or **threonine**, rather than tyrosine. Serine/threonine kinases are widely involved in signaling pathways in animals, plants, and fungi.

Phosphorylation cascade is similar to many known pathways, including those triggered in yeast by mating factors and in animal cells by many growth factors. The signal is transmitted by a cascade of protein phosphorylations, each causing a shape change in the phosphorylated protein. The shape change results from the interaction of the newly added phosphate groups with charged or polar amino acids on the protein being phosphorylated. The shape change in turn alters the function of the protein, most often activating it (in some cases, though, phosphorylation instead decreases the activity of the protein).

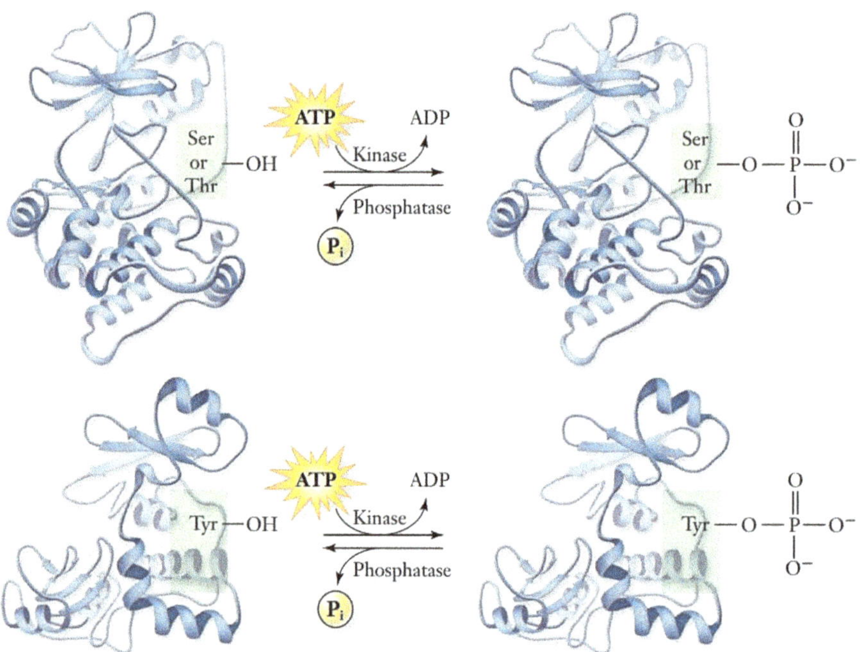

▲ **Figure 4.37** Phosphorylation of proteins. Many proteins are controlled by their phosphorylation state—that is, they are activated by phosphorylation and deactivated by dephosphorylation or the reverse. The enzymes that add phosphate groups are called kinases. These form two classes depending on the amino acid the phosphate is added to, either serine–threonine kinases or tyrosine kinases. The action of kinases is reversed by protein phosphatase enzymes.

Equally important in the phosphorylation cascade are the **protein phosphatases**, enzymes that can rapidly remove phosphate groups from proteins, a process called **dephosphorylation**. By dephosphorylating and thus inactivating protein kinases, phosphatases provide the mechanism for turning off the signal transduction pathway when the initial signal is no longer present. Phosphatases also make the protein kinases available for reuse, enabling the cell to respond again to an extracellular signal.

The phosphorylation-dephosphorylation system acts as a molecular switch in the cell, turning activities on or off, or up or down, as required. At any given moment, the activity of a protein regulated by phosphorylation depends on the balance in the cell between active kinase molecules and active phosphatase molecules.

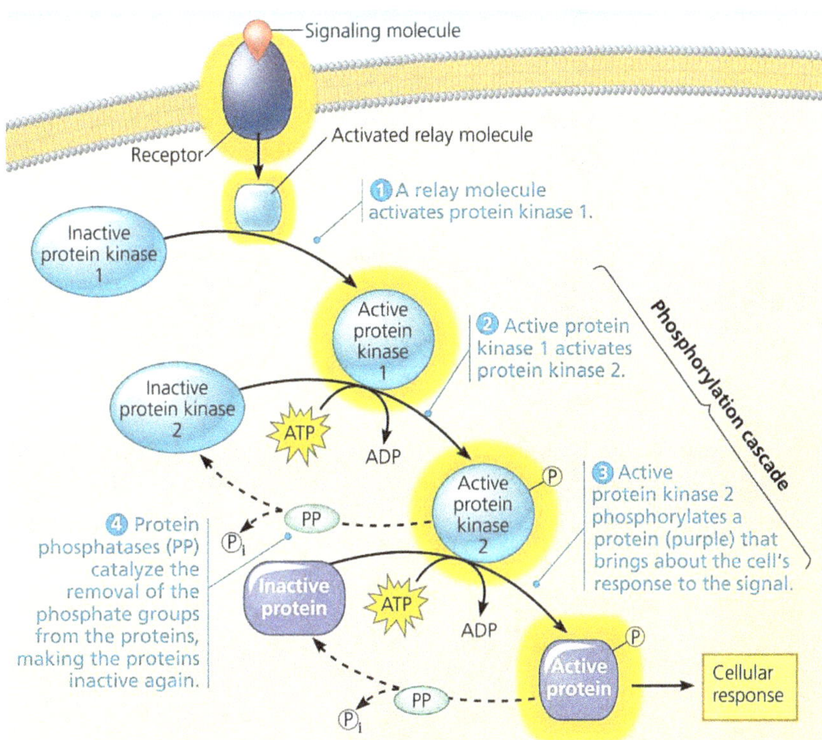

▲ **Figure 4.38** A phosphorylation cascade. Here, phosphorylation activates each protein, and dephosphorylation returns it to its inactive form.

- **Small Molecules and Ions as Second Messengers**

Not all components of signal transduction pathways are proteins. Many signaling pathways also involve small, nonprotein, water-soluble molecules or ions called **second messengers**. The pathway's "first messenger" is considered to be the extracellular signaling molecule, the ligand, that binds to the membrane receptor. Because they are small and water-soluble, they can readily spread through regions of the cell by diffusion.

Second messengers participate in pathways that are initiated by both G protein-coupled receptors and receptor tyrosine kinases. The signals that result in second messenger production often act quickly, in a matter of seconds or minutes, but their duration is usually short. Therefore, such signaling typically occurs when a cell needs a quick and short cellular response.

Cells use several different types of second messengers, and more than one type may be used at the same time. The two most common second messengers are cyclic AMP and calcium ions, Ca^{2+}. A large variety of relay proteins respond to changes in the cytosolic concentration of one or the other of these second messengers.

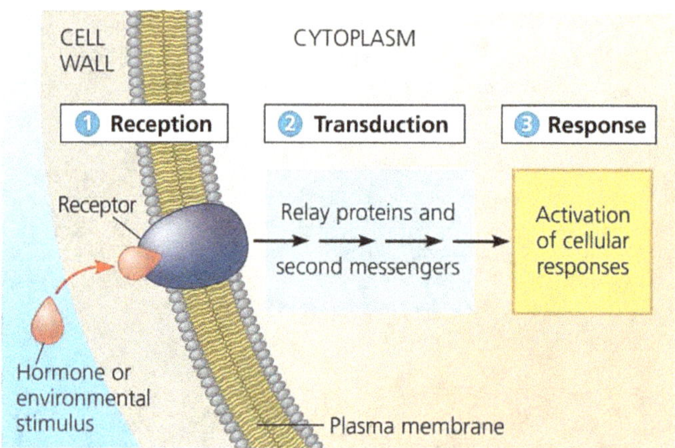

Second messengers:
1) Cyclic AMP
2) Calcium Ions
3) Inositol Trisphosphate (IP3)
4) Diacylglycerol (DAG)
5) Eicosanoids

Cyclic AMP

An enzyme embedded in the plasma membrane, **adenylyl cyclase** (also known as adenylate cyclase), converts ATP to **cAMP** (cyclic adenosine monophosphate) in response to an extracellular signal such as epinephrine. When epinephrine outside the cell binds to a GPCR, it activates a G protein that in turn activates adenylyl cyclase. Adenylyl cyclase can then catalyze the synthesis of many molecules of cAMP.

In this way, the normal cellular concentration of cAMP can be boosted 20-fold in a matter of seconds. The cAMP broadcasts the signal to the cytoplasm. It does not persist for long in the absence of the hormone because a different enzyme, called **phosphodiesterase**, converts cAMP to AMP. Another surge of epinephrine is needed to boost the cytosolic concentration of cAMP again. Peptide hormone glucagon, triggers a cAMP receptor–response pathway too, which stimulates liver cells to break down glycogen into glucose units that pass from the liver cells into the bloodstream.

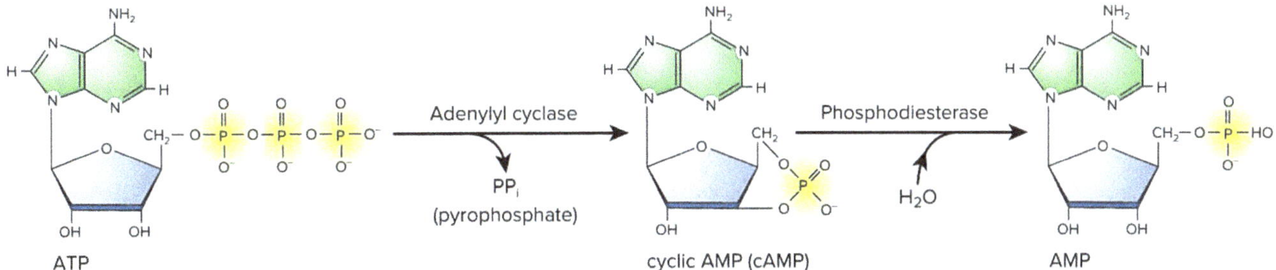

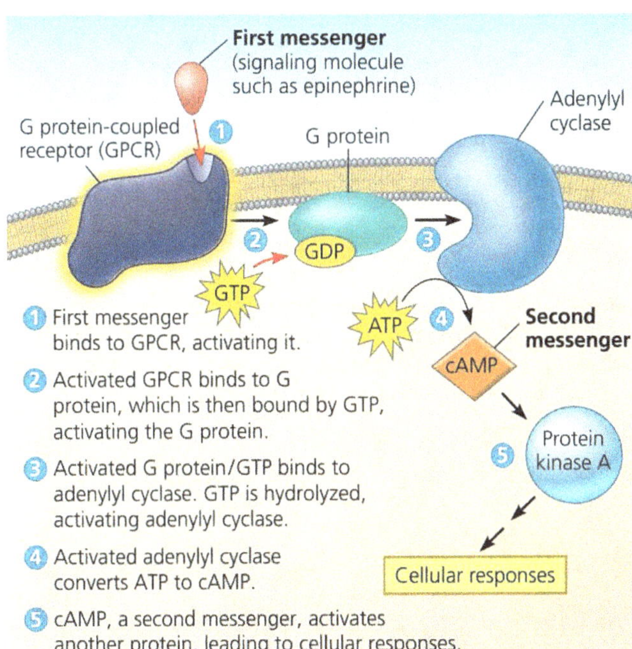

◀ **Figure 4.39** cAMP as a second messenger in a G protein signaling pathway.

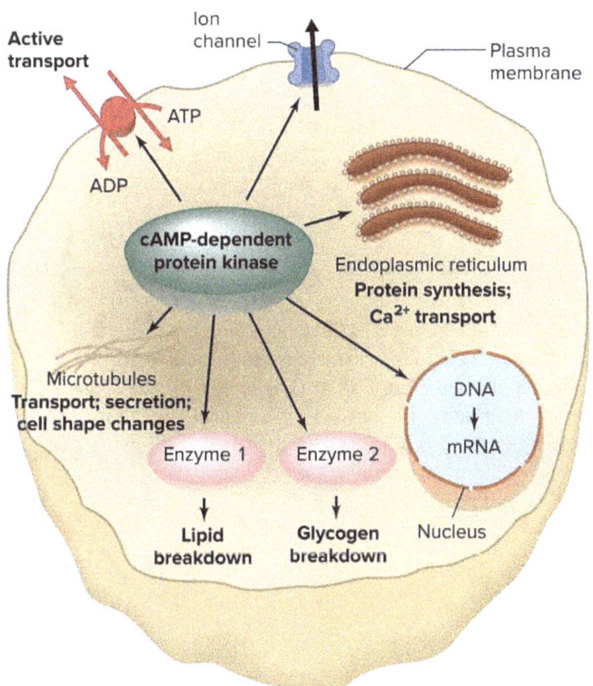

▶ **Figure 4.40** The variety of cellular responses induced by cAMP is due mainly to the fact that activated cAMP-dependent protein kinase can phosphorylate many different proteins, activating or inhibiting them.

Calcium Ions and Inositol Trisphosphate (IP3)

Many of the signaling molecules that function in animals—including neurotransmitters, growth factors, and some hormones—induce responses in their target cells through signal transduction pathways that increase the cytosolic concentration of calcium ions (Ca^{2+}). Calcium is even more widely used than cAMP as a second messenger. Increasing the local cytosolic concentration of Ca^{2+} causes many responses in animal cells, including muscle cell contraction, exocytosis of molecules (secretion), and cell division. Cells use Ca^{2+} as a second messenger in pathways triggered by both G protein-coupled receptors and receptor tyrosine kinases.

Although cells always contain some Ca^{2+}, this ion can function as a second messenger because its concentration in the cytosol is normally much lower than the concentration outside the cell. In fact, the level of Ca^{2+} in the blood and extracellular fluid of an animal is often more than 10,000 times higher than that in the cytosol. Calcium ions are actively transported out of the cell and are actively imported from the cytosol into the endoplasmic reticulum (and, under some conditions, into mitochondria and chloroplasts) by various protein pumps. As a result, the calcium concentration in the ER is usually much higher than that in the cytosol.

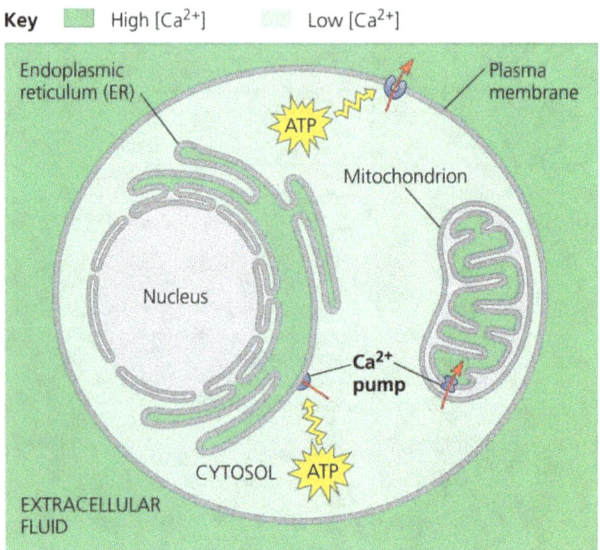

▶ **Figure 4.41** The maintenance of calcium ion concentrations in an animal cell. The Ca^{2+} concentration in the cytosol is usually much lower (light green) than in the extracellular fluid and ER (dark green). Protein pumps in the plasma membrane and the ER membrane, driven by ATP, move Ca^{2+} from the cytosol into the extracellular fluid and into the lumen of the ER. Mitochondrial pumps, driven by chemiosmosis, move Ca^{2+} into mitochondria when the calcium level in the cytosol rises significantly.

Ca^{2+} initiates some cellular responses by binding to **calmodulin**, a 148-amino-acid cytoplasmic protein that contains four binding sites for Ca^{2+}. When four Ca^{2+} ions are bound to calmodulin, the calmodulin/Ca^{2+} complex is able to bind to other proteins to activate them. These proteins include protein kinases, ion channels, receptor proteins, and cyclic nucleotide phosphodiesterases. These many uses of Ca^{2+} make it one of the most versatile second messengers in cells.

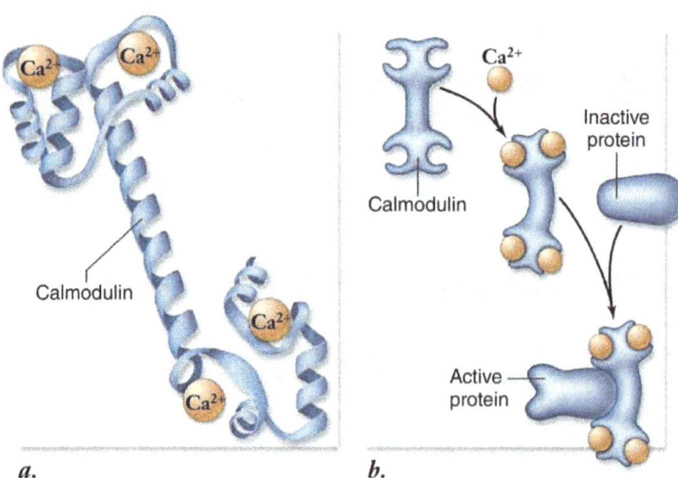

▶ **Figure 4.42** Calmodulin. a. Calmodulin is a protein containing 148 amino acid residues that mediates Ca^{2+} function. b. When four Ca^{2+} are bound to the calmodulin molecule, it undergoes a conformational change that allows it to bind to other cytoplasmic proteins and effect cellular responses.

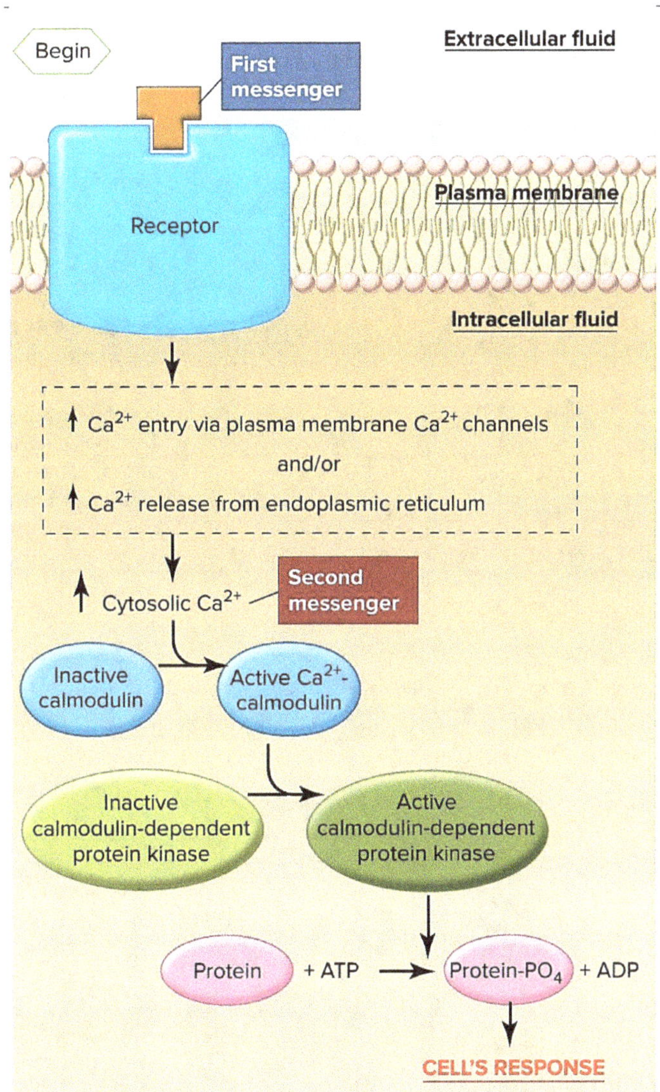

▲ **Figure 4.43** Ca²⁺, calmodulin, and the calmodulin-dependent protein kinase system. (There are multiple calmodulin-dependent protein kinases.)

In response to a signal relayed by a signal transduction pathway, the cytosolic calcium level may rise, usually by a mechanism that releases Ca^{2+} from the cell's ER. The pathways leading to calcium release involve two other second messengers, **inositol trisphosphate (IP$_3$)** and **diacylglycerol (DAG).** These two messengers are produced by cleavage of a certain kind of phospholipid in the plasma membrane (PIP$_2$ or phosphotidylinositol-4,5-bisphosphate) by **phospholipase C.** A signal causes IP$_3$ to stimulate the release of calcium from the ER. Because IP$_3$ acts before calcium in these pathways, calcium could be considered a "**thirdmessenger.**" However, scientists use the term second messenger for all small, nonprotein components of signal transduction pathways.

The IP$_3$/DAG response pathways are activated by a large number of polar hormones (including growth factors) and neurotransmitters, among the mammalian hormones that activate the pathways are antidiuretic hormone, angiotensin, and norepinephrine. Antidiuretic hormone, also known as vasopressin, helps the body conserve water by reducing the output of urine, angiotensin helps maintain blood volume and pressure and Norepinephrine (also known as noradrenaline), together with epinephrine, brings about the fight-or-flight response in threatening or stressful situations.

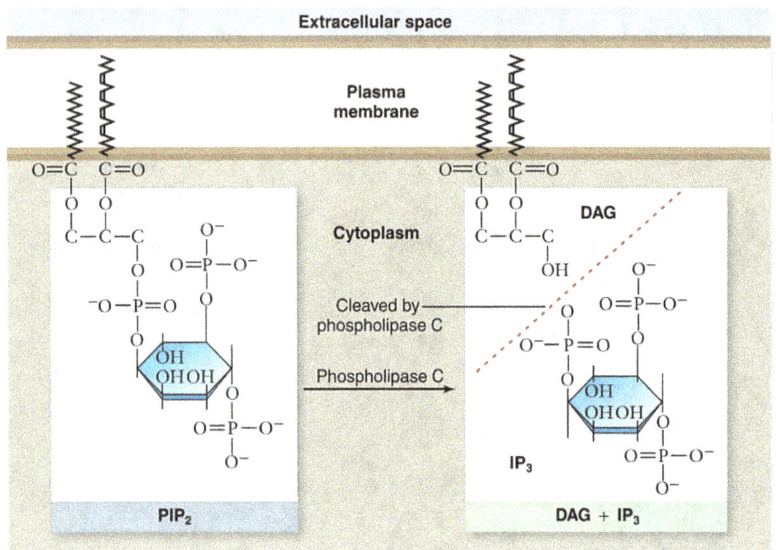

◄ **Figure 4.44** Production of second messengers. The inositol phospholipid PIP2 is composed of two lipids and a phosphate attached to glycerol. The phosphate is also attached to the sugar inositol. This molecule can be cleaved by the enzyme phospholipase C to produce two different second messengers: DAG, made up of the glycerol with the two lipids, and IP3, inositol triphosphate.

► **Figure 4.45** Calcium and IP$_3$ in signaling pathways. Calcium ions (Ca^{2+}) and inositol trisphosphate (IP$_3$) function as second messengers in many signal transduction pathways. In this figure, the process is initiated by the binding of a signaling molecule to a G protein-coupled receptor. A receptor tyrosine kinase could also initiate this pathway by activating phospholipase C.

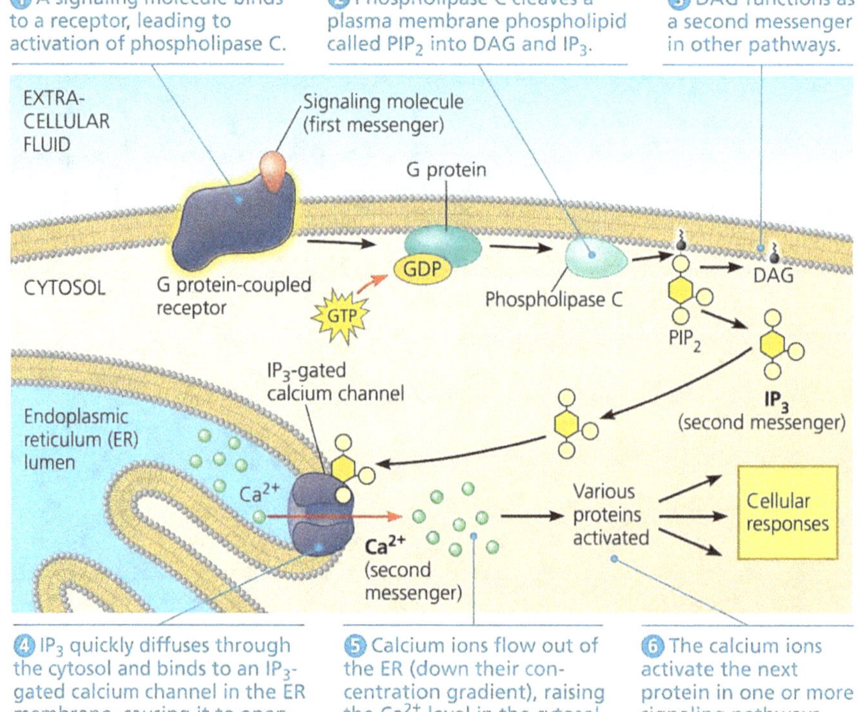

Other Messengers

The **eicosanoids** are a family of molecules produced from the polyunsaturated fatty acid arachidonic acid, which is present in plasma membrane phospholipids. The eicosanoids include the **cyclic endoperoxides**, the **prostaglandins**, the **thromboxanes**, and the **leukotrienes**. They are generated in many kinds of cells in response to different types of extracellular signals; these include a variety of growth factors, immune defense molecules, and even other eicosanoids. Thus, eicosanoids may act as both extracellular and intracellular messengers, depending on the cell type.

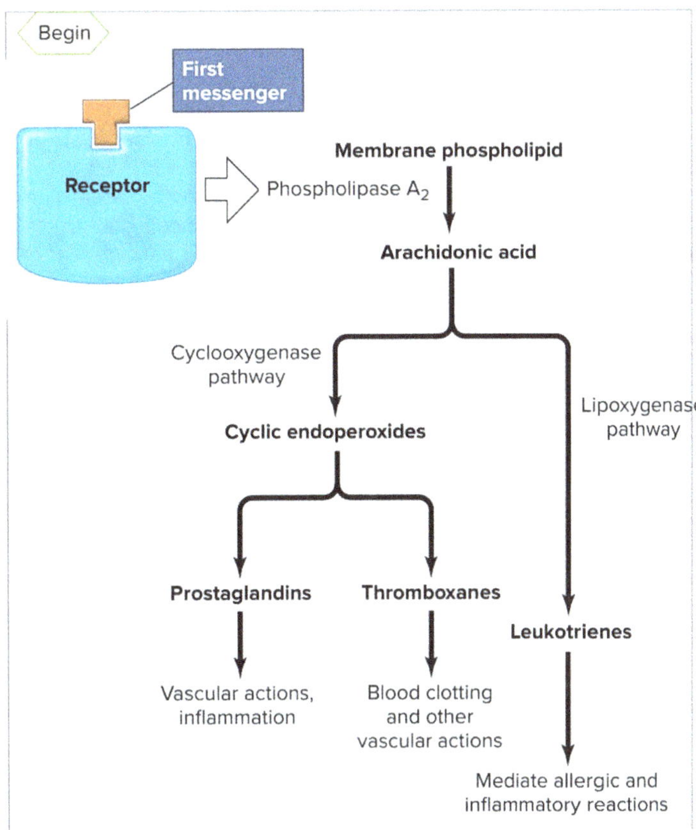

▲ **Figure 4.46** Pathways for eicosanoid synthesis and some of their major functions. Phospholipase A$_2$ is the one enzyme common to the formation of all the eicosanoids. Anti-inflammatory steroids induce expression of a protein that inhibits phospholipase A$_2$. The pathway mediated by cyclooxygenase is inhibited by aspirin and other nonsteroidal anti-inflammatory drugs (NSAIDs). There are also drugs available that inhibit the lipoxygenase enzyme, thereby blocking the formation of leukotrienes. These drugs may be helpful in controlling asthma, in which excess leukotrienes have been implicated in the allergic and inflammatory components of the disease.

Ras pathways involve tyrosine kinase receptors and G proteins

Some tyrosine kinase receptors activate G proteins. For example, when growth factors (GF) bind to tyrosine kinase receptors, **Ras proteins** are activated. Growth factors cause a rapid increase in the expression of many genes in mammals, perhaps as many as 100. Ras proteins are a group of small G proteins. Like other G proteins, Ras proteins are active when bound to GTP. Ras proteins are molecular switches that regulate signaling networks inside the cell. When activated, Ras triggers a cascade of reactions called the **Ras pathway**. In this pathway a tyrosine in specific kinase proteins is phosphorylated, leading to critical cell responses.

Ras pathways are important in gene expression, cell division, cell movement, cell differentiation, cell adhesion, embryonic development, and apoptosis. Ras genes code for Ras proteins. Certain mutations in Ras genes result in mutant Ras proteins that bind GTP but cannot hydrolyze it. The mutant Ras proteins are stuck in the "on" state, resulting in unregulated cell division. This condition is associated with several types of human cancer. In fact, mutations in Ras genes have been identified in about one-third of all human cancers. Drugs that inhibit specific tyrosine kinase receptors and Ras pathways are being developed to treat cancer.

One Ras pathway that has been extensively studied is the **MAP kinase pathway**, also known as the **ERK pathway**; MAP is an acronym for "mitogen-activated protein" (mitogens induce mitosis, the nuclear division associated with eukaryotic cell division). ERK is an acronym for "extracellular signal-regulated kinases." Several distinct groups of MAP kinases have been described. Proteins in the MAP pathway phosphorylate a nuclear protein that combines with other proteins to form a transcription factor. When specific genes are activated, proteins needed for cell growth, cell division, and cell differentiation (specialization) are synthesized. The MAP kinase cascade is the main signaling pathway for cell division and differentiation.

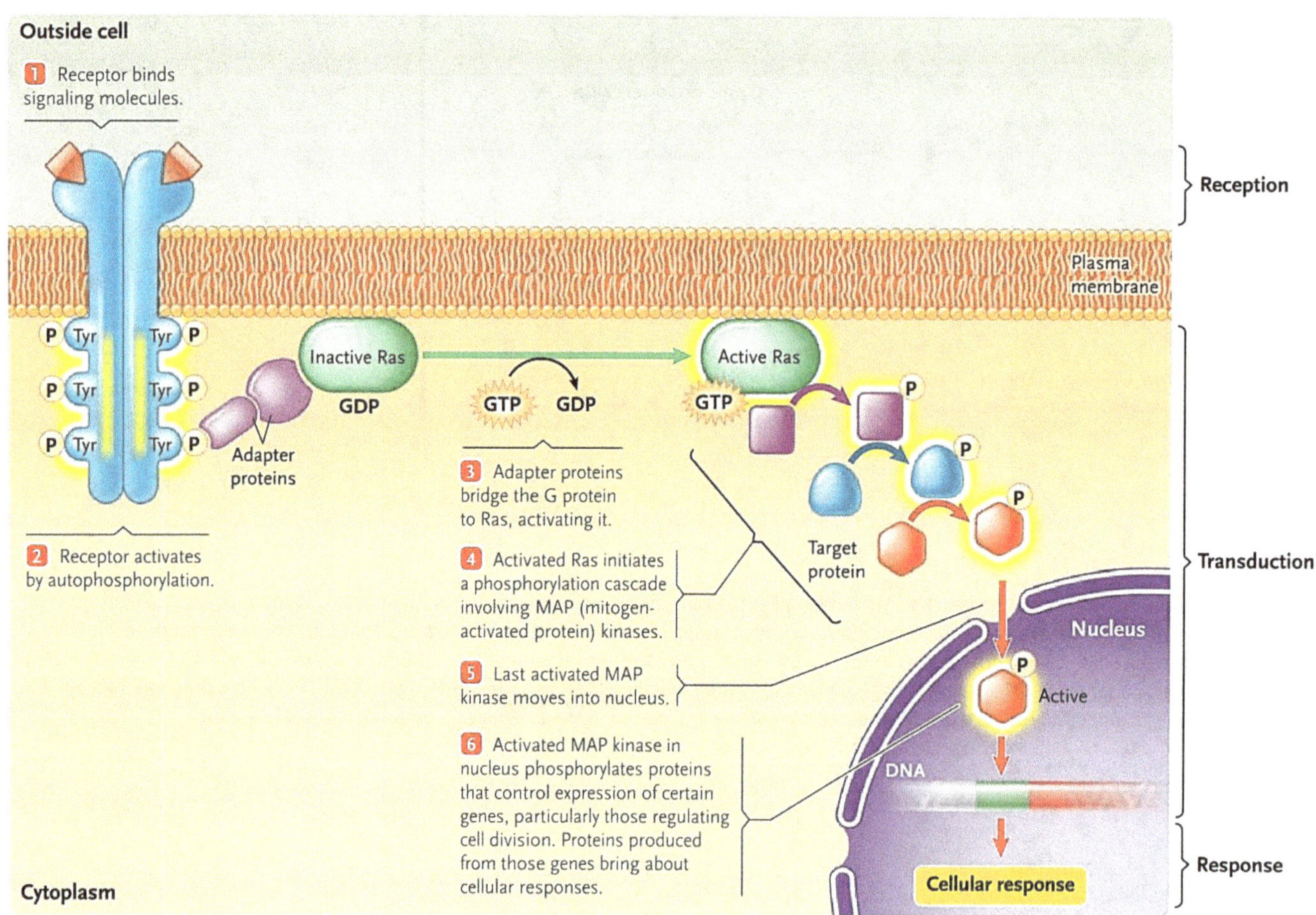

▲ **Figure 4.47** The pathway from receptor tyrosine kinases to gene regulation, including the G protein, Ras, and MAP kinase. © Cengage Learning 2017

Different receptors can produce the same second messengers

Two hormones glucagon and epinephrine can both stimulate liver cells to mobilize glucose. The reason that these different signals have the same effect is that they both act by the same signal transduction pathway to stimulate the breakdown and inhibit the synthesis of glycogen.

The binding of either hormone to its receptor activates a G protein that simulates adenylyl cyclase. The production of cAMP leads to the activation of PKA, which in turn activates another protein kinase called phosphorylase kinase. Activated phosphorylase kinase then activates glycogen phosphorylase, which cleaves off units of glucose 6-phosphate from the glycogen polymer. The action of multiple kinases again leads to amplification such that a few signaling molecules result in a large number of glucose molecules being released.

At the same time, PKA also phosphorylates the enzyme glycogen synthase, but in this case it inhibits the enzyme, thus preventing the synthesis of glycogen. In addition, PKA phosphorylates other proteins that activate the expression of genes encoding the enzymes needed to synthesize glucose. This convergence of signal transduction pathways from different receptors leads to the same result—glucose is mobilized.

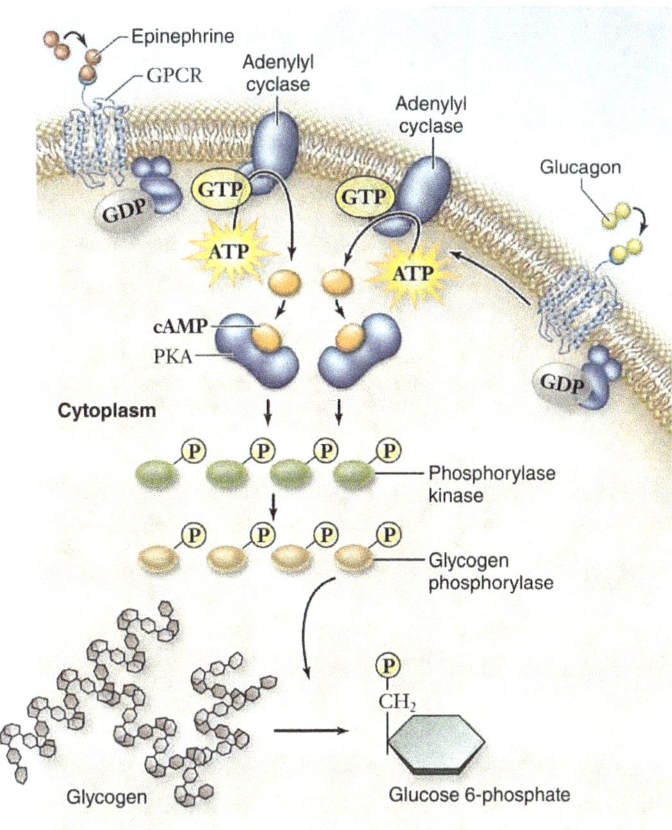

▲ **Figure 4.48** Different receptors can activate the same signaling pathway. The hormones glucagon and epinephrine both act through GPCRs. Each of these receptors acts via a G protein that activates adenylyl cyclase, producing cAMP. The activation of PKA begins a kinase cascade that leads to the breakdown of glycogen.

Receptor subtypes can lead to different effects in different cells

A single signaling molecule, epinephrine, can have different effects in different cells. One way this happens is through the existence of multiple forms of the same receptor. The receptor for epinephrine actually has nine different subtypes, or isoforms. These are encoded by different genes and are actually different receptor molecules. The sequences of these proteins are very similar, especially in the ligand-binding domain, which allows them to bind epinephrine. They differ mainly in their cytoplasmic domains, which interact with G proteins. This leads to different isoforms activating different G proteins, thereby leading to different signal transduction pathways.

Thus, in the heart, muscle cells have one isoform of the receptor that, when bound to epinephrine, activates a G protein that activates adenylyl cyclase, leading to increased cAMP. This increases the rate and force of contraction. In the intestine, smooth muscle cells have a different isoform of the receptor that, when bound to epinephrine, activates a different G protein that inhibits adenylyl cyclase, which decreases cAMP. This has the result of relaxing the muscle.

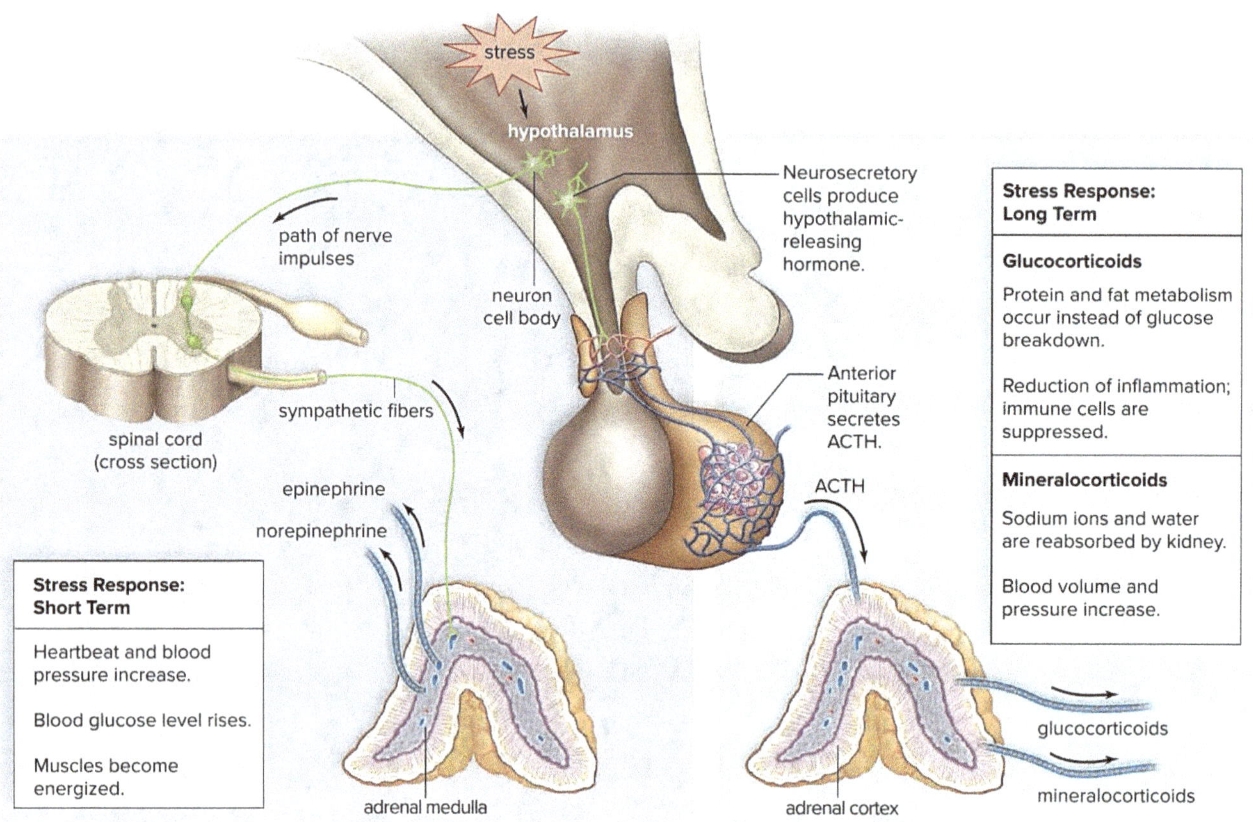

▲ **Figure 4.49** Response of the adrenal medulla and the adrenal cortex to stress.

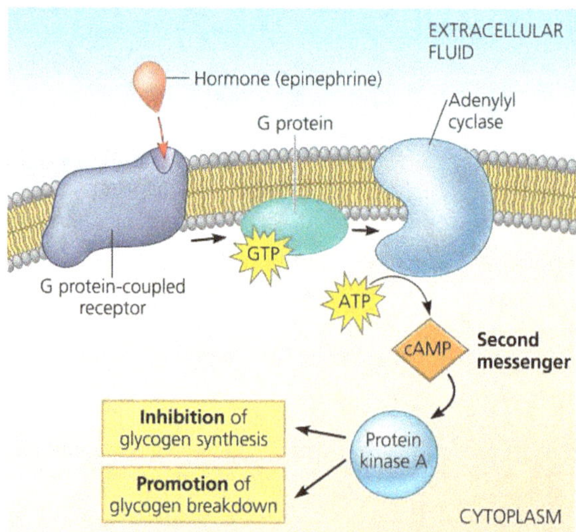

Cellular response: Cell signaling leads to regulation of transcription or cytoplasmic activities

Nuclear and Cytoplasmic Responses

Ultimately, a signal transduction pathway leads to the regulation of one or more cellular activities. The response may occur in the nucleus or in the cytoplasm of the cell. Many signaling pathways ultimately regulate protein synthesis, usually by turning specific genes on or off in the nucleus. Like an activated steroid receptor, the final activated molecule in a signaling pathway may function as a transcription factor. A signaling pathway may activates a transcription factor that turns a gene on. In other cases, the transcription factor might regulate a gene by turning it off. Often, a transcription factor regulates several different genes.

Sometimes a signaling pathway may regulate the activity of proteins rather than causing their synthesis by activating gene expression. This directly affects proteins that function outside the nucleus. For example, a signal may cause the opening or closing of an ion channel in the plasma membrane or a change in the activity of a metabolic enzyme. So the function of three common categories of proteins is controlled by cell signaling: enzymes, structural proteins, and transcription factors.

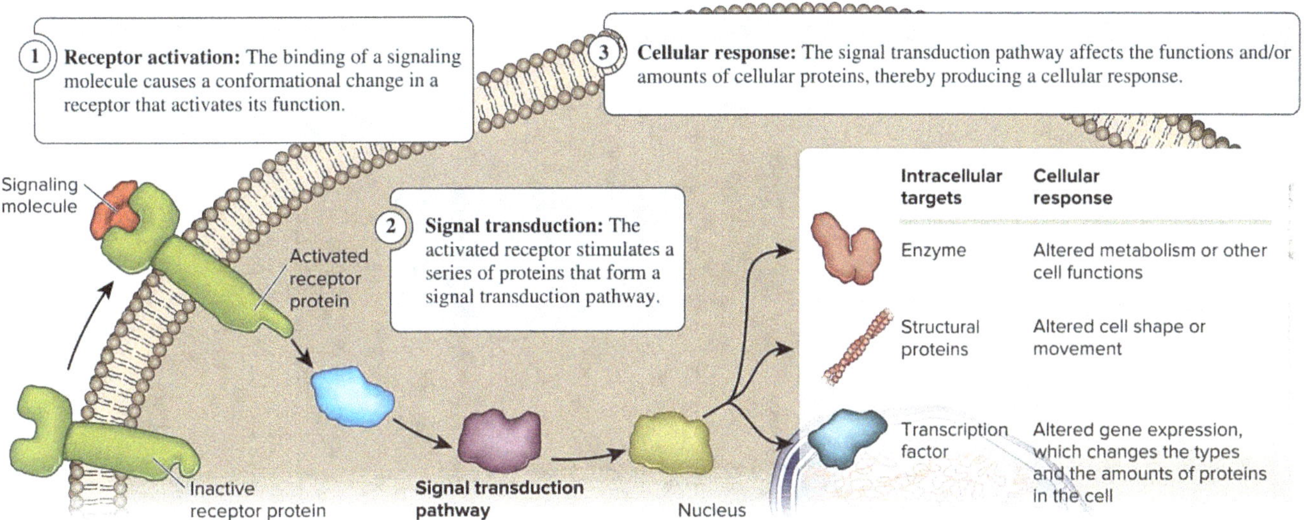

▲ **Figure 4.50** Cellular response. Transcription factors function within the nucleus, whereas enzymes and structural proteins may function in various regions throughout the cell.

Let's explore a signal transduction pathway in which a G-protein-coupled receptor recognizes the hormone epinephrine (also called adrenaline). Epinephrine is produced when an individual is confronted with a stressful situation. It promotes cellular changes, such as an increase in ATP production, that help the individual deal with a perceived threat or danger. These changes improve the individual's ability to stay and fight or to flee. For this reason, epinephrine is sometimes called the fight-or-flight hormone. The following steps occur at the cellular level.

1. Epinephrine binds to a GPCR, which promotes the dissociation of the α subunit from the G protein.

2. The α subunit then activates adenylyl cyclase, which catalyzes the production of cAMP from ATP.

3. One effect of cAMP is to activate **cAMP-dependent protein kinase** or **protein kinase A** (PKA, a serine/threonine kinase), which is composed of four subunits: two catalytic subunits that phosphorylate specific cellular proteins, and two regulatory subunits that inhibit the catalytic subunits when they are bound to those subunits. cAMP binds to the regulatory subunits of PKA. The binding of cAMP separates the regulatory and catalytic subunits, which allows each catalytic subunit to be active.

4. The catalytic subunit of PKA phosphorylates specific cellular proteins such as enzymes, structural proteins, or transcription factors, thus affecting cell structure and function.

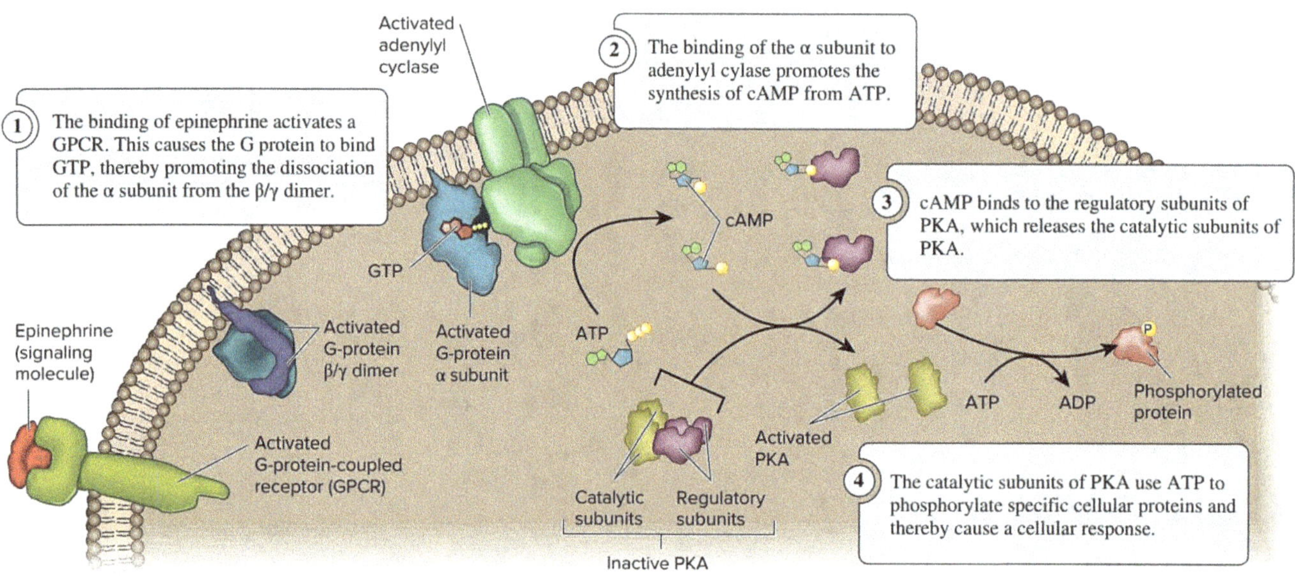

When PKA becomes active, it phosphorylates two enzymes: phosphorylase kinase and glycogen synthase. Both of these enzymes are involved with the metabolism of glycogen, which is a polymer of glucose used to store energy. When phosphorylase kinase is phosphorylated, it becomes active. The function of phosphorylase kinase is to phosphorylate another enzyme in the cell called glycogen phosphorylase, which then becomes active. This enzyme causes glycogen breakdown by phosphorylating glucose units at the ends of a glycogen polymer, which releases individual glucose-phosphate molecules from glycogen:

$$\text{Glycogen}_{n} + \text{P}_{i} \xrightarrow{\text{Glycogen phosphorylase}} \text{Glycogen}_{n-1} + \text{Glucose-phosphate}$$

where n is the number of glucose units in the glycogen polymer.

When PKA phosphorylates glycogen synthase, the function of this enzyme is inhibited rather than activated. The function of glycogen synthase is to make glycogen. Therefore, the effect of cAMP is to prevent glycogen synthesis. Taken together, the effects of epinephrine in skeletal muscle cells are to stimulate glycogen breakdown and to inhibit glycogen synthesis. This provides these cells with more glucose molecules, which they can break down to make ATP. The ATP is then used as an energy source for muscle contraction. In this way, the individual is better prepared to fight or flee.

Further regulation of cell metabolism is provided by other G protein systems that inhibit adenylyl cyclase. In these systems, a different signaling molecule activates a different receptor, which in turn activates an inhibitory G protein that blocks activation of adenylyl cyclase. Cell activities can be fine-tuned by the balance between these systems.

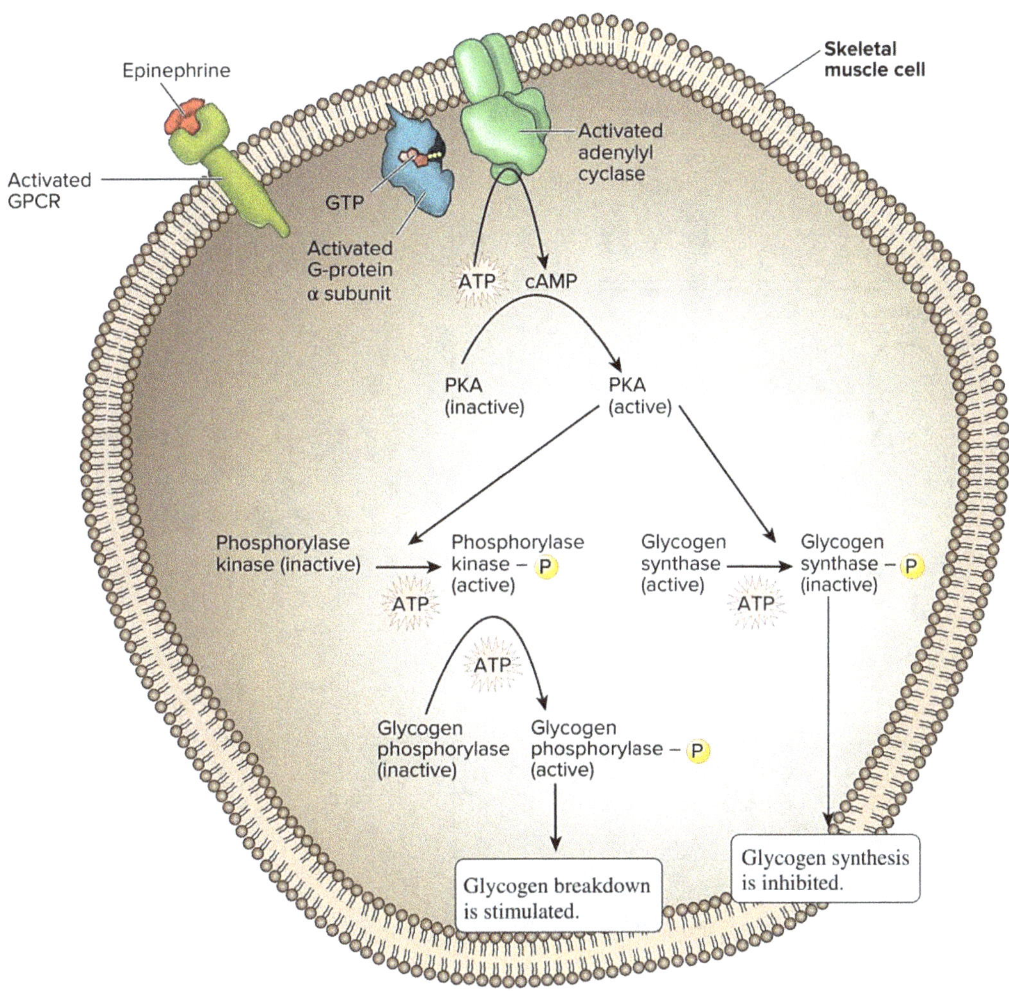

▲ **Figure 4.51** The cellular response of a skeletal muscle cell to epinephrine.

Regulation of the Response

Whether the response occurs in the nucleus or in the cytoplasm, it is not simply turned "on" or "off." Rather, the extent and specificity of the response are regulated in multiple ways. Consider four aspects of this regulation:

First, signaling pathways generally amplify the cell's response to a single signaling event. The degree of amplification depends on the function of the specific molecules in the pathway. Elaborate enzyme cascades amplify the cell's response to a signal. At each catalytic step in the cascade, the number of activated products can be much greater than in the preceding step.

As a result, just a few extracellular signal molecules binding to their receptors can produce a full internal response. For similar reasons, amplification also occurs for signal transduction pathways that involve internal receptors.

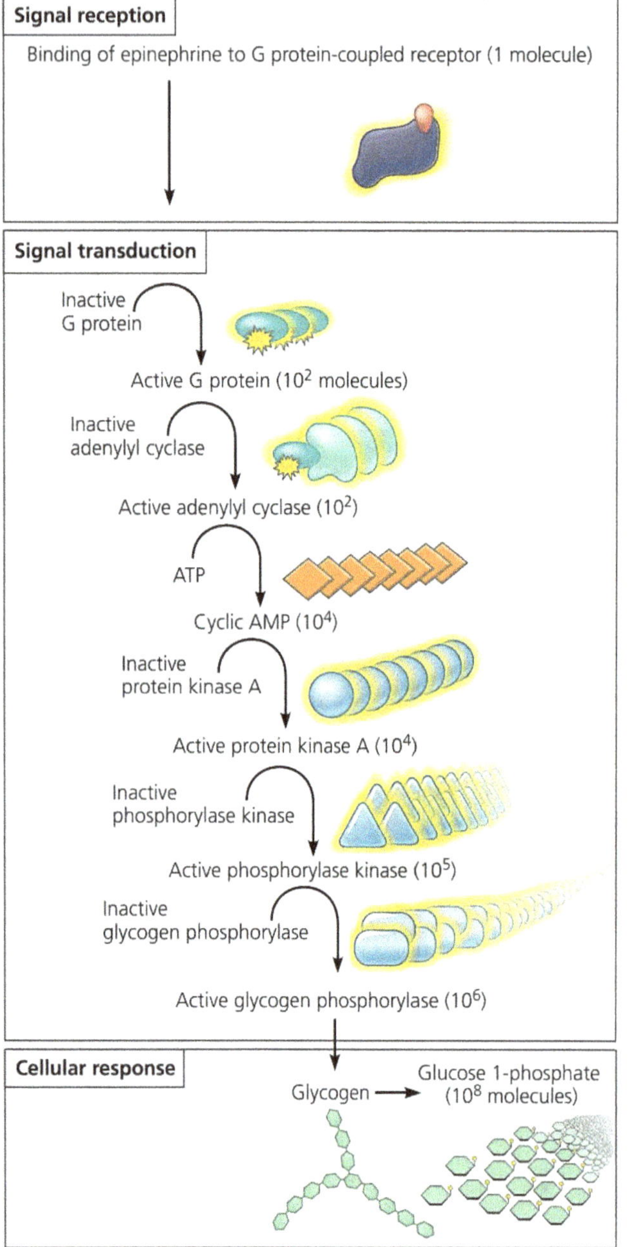

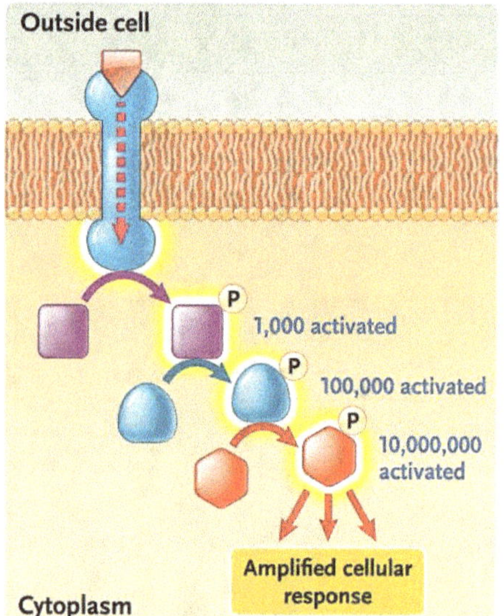

▲ **Figure 4.52** Amplification in signal transduction.

◄ **Figure 4.53** Signal amplification. The signal is amplified at each step in the pathway so that one activated receptor can give rise to thousands of final products (proteins at the end of the pathway). The response is far greater than what you might expect from a single receptor.

Second, the many steps in a multistep pathway provide control points at which the cell's response can be further regulated, contributing to the specificity of the response and allowing coordination with other signaling pathways. Consider two different cells in your body—a liver cell and a heart muscle cell, for example. Both are in contact with your bloodstream and are therefore constantly exposed to many different hormone molecules, as well as to local regulators secreted by nearby cells. Yet the liver cell responds to some signals but ignores others, and the same is true for the heart cell. And some kinds of signals trigger responses in both cells—but different responses. For instance, epinephrine stimulates the liver cell to break down glycogen, but the main response of the heart cell to epinephrine is contraction, leading to a more rapid heartbeat.

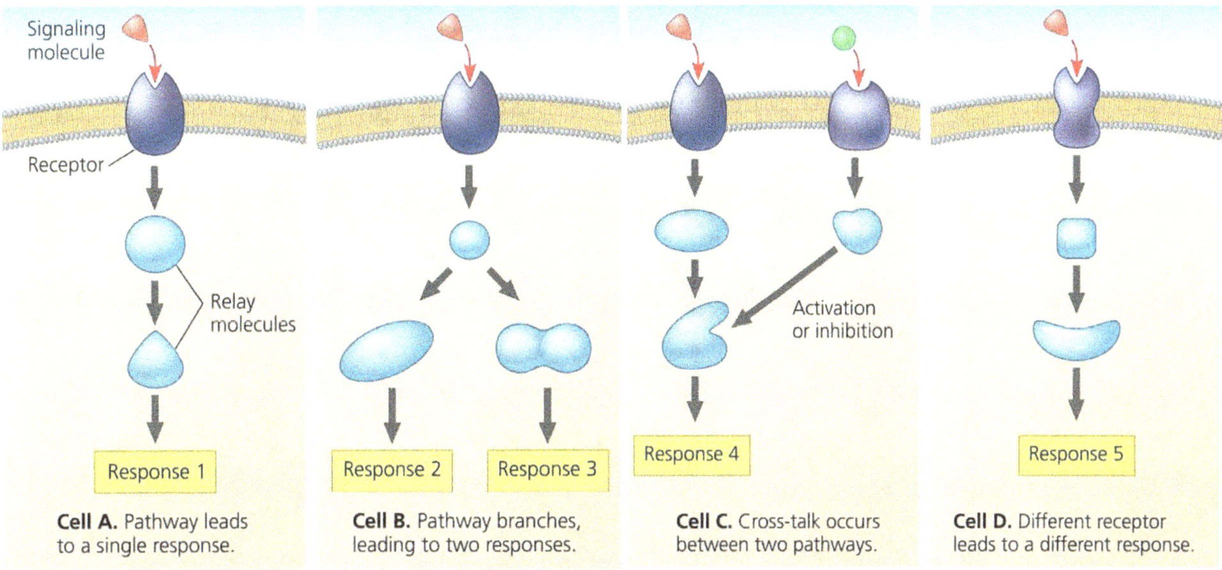

▲ **Figure 4.54** The specificity of cell signaling. The particular proteins a cell possesses determine what signaling molecules it responds to and the nature of the response. The four cells in these diagrams respond to the same signaling molecule (red) in different ways because each has a different set of proteins (purple and teal). Note, however, that the same kinds of molecules can participate in more than one pathway.

Cell A. Pathway leads to a single response.

Cell B. Pathway branches, leading to two responses.

Cell C. Cross-talk occurs between two pathways.

Cell D. Different receptor leads to a different response.

The explanation for the specificity exhibited in cellular responses to signals is the same as the basic explanation for virtually all differences between cells: Because different kinds of cells turn on different sets of genes, different kinds of cells have different collections of proteins. The response of a cell to a signal depends on its particular collection of signal receptor proteins, relay proteins, and proteins needed to carry out the response. Thus, two cells that respond differently to the same signal differ in one or more proteins that respond to the signal.

Also, within some cells, there are more complex pathways that branch or converge. Such branched pathways often involve receptor tyrosine kinases (which can activate multiple relay proteins) or second messengers (which can regulate numerous proteins). Branching of pathways and "cross-talk" (interaction) between pathways are important in regulating and coordinating a cell's responses to information coming in from different sources in the body. Moreover, the use of some of the same proteins in more than one pathway allows the cell to economize on the number of different proteins it must make.

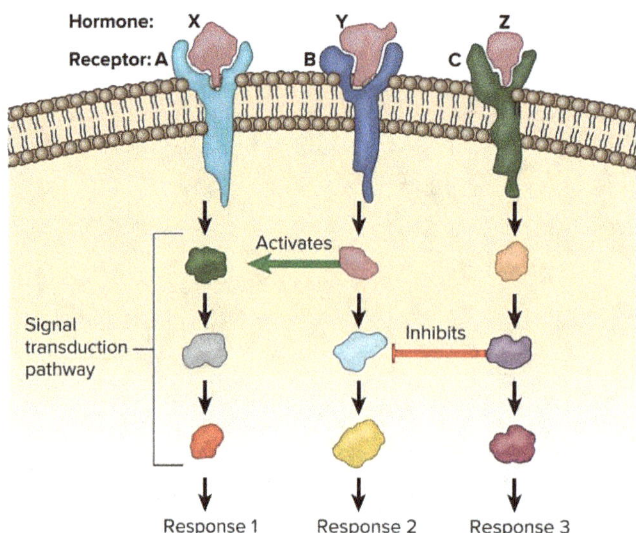

(a) Crosstalk among three different signal transduction pathways

Hormones Present	Cellular Response		
	1	2	3
None	–	–	–
X	++	–	–
Y	++	++	–
Z	–	–	++
X,Y	+++	++	–
Y,Z	++	+	++
X,Y,X	+++	+	++

(b) Effects of hormones on cellular responses

▲ **Figure 4.55** Crosstalk between different signal transduction pathways. (a) The first protein in the middle pathway activates the first protein in the pathway on the left. The second protein in the pathway on the right inhibits the second protein in the middle pathway. (b)The cellular responses depend on which hormones are present.

The designation – refers to no response, and the designations +, ++, and +++ refer to a low, moderate, or high response, respectively. The presence of hormone X promotes cellular response 1. When hormones X and Y are both present, response 1 is higher because both hormones promote this response as a result of crosstalk. In contrast, hormone Z stimulates response 3, but it inhibits response 2 via crosstalk. As seen here, crosstalk enables a cell to respond to hormones in several different ways.

Third, the overall efficiency of the response is enhanced by the presence of proteins known as scaffolding proteins. In many cases, the efficiency of signal transduction is apparently increased by the presence of scaffolding proteins, large relay proteins to which several other relay proteins are simultaneously attached. Researchers have found scaffolding proteins in brain cells that permanently hold together networks of signaling pathway proteins at synapses. This hardwiring enhances the speed and accuracy of signal transfer between cells because the rate of protein-protein interaction is not limited by diffusion. Furthermore, in some cases the scaffolding proteins themselves may directly activate relay proteins.

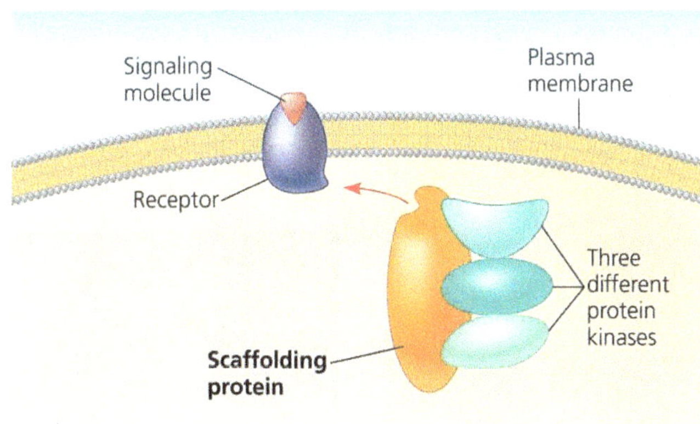

► **Figure 4.56** A scaffolding protein. The scaffolding protein shown here (orange) simultaneously binds to a specific activated membrane receptor and three different protein kinases. This physical arrangement facilitates signal transduction by these molecules.

Finally, a crucial point in regulating the response is the termination of the signal. For a cell of a multicellular organism to remain capable of responding to incoming signals, each molecular change in its signaling pathways must last only a short time. The ability of a cell to receive new signals depends on reversibility of the changes produced by prior signals. The binding of signaling molecules to receptors is reversible. As the external concentration of signaling molecules falls, fewer receptors are bound at any given moment, and the unbound receptors revert to their inactive form.

The cellular response occurs only when the concentration of receptors with bound signaling molecules is above a certain threshold. When the number of active receptors falls below that threshold, the cellular response ceases. Then, by a variety of means, the relay molecules return to their inactive forms: The GTPase activity intrinsic to a G protein hydrolyzes its bound GTP; the enzyme phosphodiesterase converts cAMP to AMP; protein phosphatases inactivate phosphorylated kinases and other proteins; and so forth. As a result, the cell is soon ready to respond to a fresh signal.

For example, when the concentration of the hormone insulin is too high for an extended period, cells decrease the number of their insulin receptors. This process is called **receptor down-regulation**. In the case of insulin, receptor down-regulation. **Receptor up-regulation** occurs in response to low hormone concentrations. In this process, a greater number of receptors are synthesized, and their increased numbers on the plasma membrane make it more likely that the signal will be received by a receptor on the cell. Receptor up-regulation thus amplifies the signaling molecule's effect on the cell. Receptor up-regulation and down-regulation are controlled in part by signals to genes that code for the receptors.

Apoptosis requires integration of multiple cell-signaling pathways

Cells that are infected, are damaged, or have reached the end of their functional life span often undergo "**programmed cell death**". The best-understood type of this controlled cell suicide is **apoptosis**. During this process, cellular agents chop up the DNA and fragment the organelles and other cytoplasmic components.

The cell shrinks and becomes lobed (a change called "blebbing"), and the cell's parts are packaged up in vesicles that are engulfed and digested by specialized scavenger cells, leaving no trace. Apoptosis protects neighboring cells from damage that they would otherwise suffer if a dying cell merely leaked out all its contents, including its many digestive enzymes.

The signal that triggers all of the complex events that occur during apoptosis can come from either outside or inside the cell. Outside the cell, signaling molecules released from other cells can initiate a signal transduction pathway that activates the genes and proteins responsible for carrying out cell death. Within a cell whose DNA has been irretrievably damaged, a series of protein-protein interactions can pass along a signal that similarly triggers cell death.

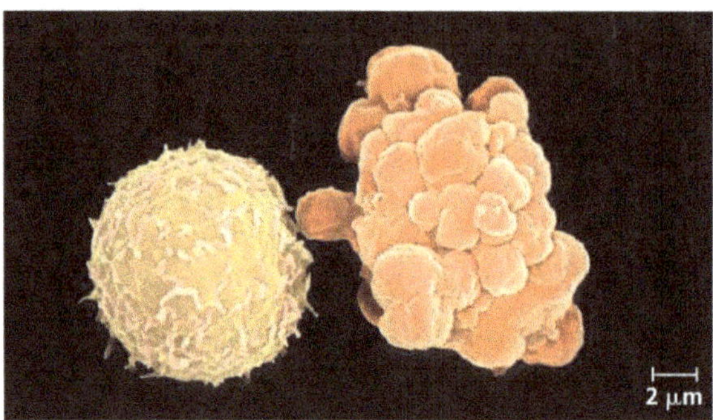

▲ **Figure 4.57** Apoptosis of a human white blood cell. On the left is a normal white blood cell, while on the right is a white blood cell undergoing apoptosis. The apoptotic cell is shrinking and forming lobes ("blebs"), which eventually are shed as membranebounded cell fragments (colorized SEMs).

Apoptotic Pathways and the Signals That Trigger Them

In humans and other mammals, several different pathways, involving about 15 different **caspases**, can carry out apoptosis. The pathway that is used depends on the type of cell and on the particular signal that initiates apoptosis. One major pathway involves certain mitochondrial proteins that are triggered to form molecular pores in the mitochondrial outer membrane, causing it to leak and release other proteins that promote apoptosis. Perhaps surprisingly, these latter include **cytochrome c**, which functions in mitochondrial electron transport in healthy cells but acts as a cell death factor when released from mitochondria. The process of mitochondrial apoptosis in mammals uses proteins similar to the nematode proteins Ced-3, Ced-4, and Ced-9. These can be thought of as relay proteins capable of transducing the apoptotic signal.

At key gateways into the apoptotic program, relay proteins integrate signals from several different sources and can send a cell down an apoptotic pathway. Often, the signal originates outside the cell which presumably was released by a neighboring cell. When a death-signaling ligand occupies a cell-surface receptor, this binding leads to activation of caspases and other enzymes that carry out apoptosis, without involving the mitochondrial pathway.

Two other types of alarm signals that can lead to apoptosis originate from inside the cell rather than from a cell-surface receptor. One signal comes from the nucleus, generated when the DNA has suffered irreparable damage, and a second comes from the endoplasmic reticulum when excessive protein misfolding occurs.

In vertebrates, apoptosis is essential for normal development of the nervous system, for normal operation of the immune system, and for normal morphogenesis of hands and feet in humans and paws in other mammals.

Significant evidence points to the involvement of apoptosis in certain degenerative diseases of the nervous system, such as Parkinson's disease and Alzheimer's disease. In Alzheimer's disease, an accumulation of aggregated proteins in neuronal cells activates an enzyme that triggers apoptosis, resulting in the loss of brain function seen in these patients. Furthermore, cancer can result from a failure of cell suicide.

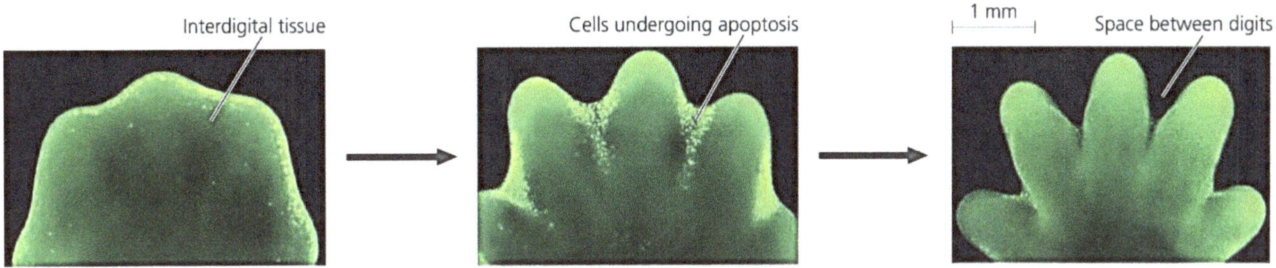

▲ **Figure 4.58** Effect of apoptosis during paw development in the mouse. In mice, humans, other mammals, and land birds, the embryonic region that develops into feet or hands initially has a solid, platelike structure. Apoptosis eliminates the cells in the interdigital regions, thus forming the digits. The embryonic mouse paws shown in these fluorescence light micrographs are stained so that cells undergoing apoptosis appear a bright yellowish green. Apoptosis of cells begins at the margin of each interdigital region (left), peaks as the tissue in these regions is reduced (middle), and is no longer visible when the interdigital tissue has been eliminated (right).

Component	Composition	Function	How It Works	Example
Phospholipid bilayer	Phospholipid molecules	Provides permeability barrier, matrix for proteins	Excludes water-soluble molecules from nonpolar interior of bilayer and cell	Bilayer of cell is impermeable to large water-soluble molecules, such as glucose
Transmembrane proteins	Carriers	Actively or passively transport molecules across membrane	Move specific molecules through the membrane in a series of conformational changes	Glycophorin carrier for sugar transport; sodium–potassium pump
	Channels	Passively transport molecules across membrane	Create a selective tunnel that acts as a passage through membrane	Sodium and potassium channels in nerve, heart, and muscle cells
	Receptors	Transmit information into cell	Signal molecules bind to cell-surface portion of the receptor protein. This alters the portion of the receptor protein within the cell, inducing activity	Specific receptors bind peptide hormones and neurotransmitters
Interior protein network	Spectrins	Determine shape of cell	Form supporting scaffold beneath membrane, anchored to both membrane and cytoskeleton	Red blood cell
	Clathrins	Anchor certain proteins to specific sites, especially on the exterior plasma membrane in receptor-mediated endocytosis	Proteins line coated pits and facilitate binding to specific molecules	Localization of low-density lipoprotein receptor within coated pits
Cell-surface markers	Glycoproteins	"Self" recognition	Create a protein/carbohydrate chain shape characteristic of individual	Major histocompatibility complex protein recognized by immune system
	Glycolipid	Tissue recognition	Create a lipid/carbohydrate chain shape characteristic of tissue	A, B, O blood group markers

▲ **Table 4.1** Components of the Cell Membrane

Process		How It Works	Example
		PASSIVE PROCESSES	
Diffusion			
Direct		Random molecular motion produces net migration of nonpolar molecules toward region of lower concentration	Movement of oxygen into cells
Facilitated Diffusion			
Protein channel		Polar molecules or ions move through a protein channel; net movement is toward region of lower concentration	Movement of ions in or out of cell
Protein carrier		Molecule binds to carrier protein in membrane and is transported across; net movement is toward region of lower concentration	Movement of glucose into cells
Osmosis			
Aquaporins		Diffusion of water across the membrane via osmosis; requires osmotic gradient	Movement of water into cells placed in a hypotonic solution
		ACTIVE PROCESSES	
Active Transport			
Protein carrier			
Na^+/K^+ pump		Carrier uses energy to move a substance across a membrane against its concentration gradient	Na^+ and K^+ against their concentration gradients
Coupled transport		Molecules are transported across a membrane against their concentration gradients by the cotransport of sodium ions or protons down their concentration gradients	Coupled uptake of glucose into cells against its concentration gradient using a Na^+ gradient
Endocytosis			
Membrane vesicle			
Phagocytosis		Particle is engulfed by membrane, which folds around it and forms a vesicle	Ingestion of bacteria by white blood cells
Pinocytosis		Fluid droplets are engulfed by membrane, which forms vesicles around them	"Nursing" of human egg cells
Receptor-mediated endocytosis		Endocytosis triggered by a specific receptor, forming clathrin-coated vesicles	Cholesterol uptake
Exocytosis			
Membrane vesicle		Vesicles fuse with plasma membrane and eject contents	Secretion of mucus; release of neurotransmitters

▲ **Table 4.2** Mechanisms for Transport Across Cell Membranes

Receptor Type	Structure	Function	Example
Intracellular Receptors	No extracellular signal-binding site	Receives signals from lipid-soluble or noncharged, nonpolar small molecules	Receptors for NO, steroid hormone, vitamin D, and thyroid hormone
Cell-Surface Receptors			
Chemically gated ion channels	Multipass transmembrane protein forming a central pore	Molecular "gates" triggered chemically to open or close	Neurons
Enzymatic receptors	Single-pass transmembrane protein	Binds signal extracellularly; catalyzes response intracellularly	Phosphorylation of protein kinases
G protein–coupled receptors	Seven-pass transmembrane protein with cytoplasmic binding site for G protein	Binding of signal to receptor causes GTP to bind a G protein; G protein, with attached GTP, detaches to deliver the signal inside the cell	Peptide hormones, rod cells in the eyes

▲ **Table 4.3** Receptors Involved in Cell Signaling

Substance	Source	Effects
Ca^{2+}	Enters cell through plasma membrane ion channels or is released into the cytosol from endoplasmic reticulum.	Activates protein kinase C, calmodulin, and other Ca^{2+}-binding proteins; Ca^{2+}–calmodulin activates calmodulin-dependent protein kinases.
Cyclic AMP (cAMP)	A G protein activates plasma membrane adenylyl cyclase, which catalyzes the formation of cAMP from ATP.	Activates cAMP-dependent protein kinase (protein kinase A).
Cyclic GMP (cGMP)	Generated from guanosine triphosphate in a reaction catalyzed by a plasma membrane receptor with guanylyl cyclase activity.	Activates cGMP-dependent protein kinase (protein kinase G).
Diacylglycerol (DAG)	A G protein activates plasma membrane phospholipase C, which catalyzes the generation of DAG and IP_3 from plasma membrane phosphatidylinositol bisphosphate (PIP_2).	Activates protein kinase C.
Inositol trisphosphate (IP_3)	See DAG above.	Releases Ca^{2+} from endoplasmic reticulum into the cytosol.

▲ **Table 4.4** Reference Table of Important Second Messengers

KEY CONCEPTS

- Membranes are sheets of phospholipid bilayers with associated proteins. Hydrophobic regions of a membrane are oriented inward and hydrophilic regions oriented outward. In the fluid mosaic model, proteins float on or in the lipid bilayer.

- In eukaryotic cells membranes have four components: a phospholipid bilayer, transmembrane proteins (integral membrane proteins), interior proteins which are associated with the membrane but are not an integral part, and cell-surface markers.

- Cholesterol and sphingolipid can associate to form microdomains. The two leaflets of the plasma membrane are also not identical.

- Transmission electron microscopy (TEM) and scanning electron microscopy (SEM) have confirmed the structure predicted by the fluid mosaic model.

- Phospholipids are composed of two fatty acids and a phosphate group linked to a three-carbon glycerol molecule.

- The phosphate group of a phospholipid is polar and hydrophilic; the fatty acids are nonpolar and hydrophobic, and they orient away from the polar head of the phospholipids. The nonpolar interior of the lipid bilayer impedes the passage of water and water-soluble substances.

- Hydrogen bonding of water keeps the membrane in its bilayer configuration; however, phospholipids and unanchored proteins in the membrane are loosely associated and can diffuse laterally.

- Membrane fluidity depends on the fatty acid composition of the membrane. Unsaturated fats tend to make the membrane more fluid because of the "kinks" of double bonds in the fatty acid tails. Temperature also affects fluidity.

- Different membrane compartments have different phospholipid composition. This affects both structure and function of different membranes, and persists despite membrane traffic.

- Transporters are integral membrane proteins that carry specific substances through the membrane. Enzymes often occur on the interior surface of the membrane. Cell-surface receptors respond to external chemical messages and change conditions inside the cell; cell identity markers on the surface allow recognition of the body's cells as "self." Cell-to-cell adhesion proteins glue cells together; surface proteins that interact with other cells anchor to the cytoskeleton.

- Surface proteins are attached to the surface by nonpolar regions that associate with polar regions of phospholipids. Transmembrane proteins may cross the bilayer a number of times, and each membrane-spanning region is called a transmembrane domain. Such a domain is composed of hydrophobic amino acids usually arranged in α helices. In certain proteins, β-pleated sheets in the nonpolar region form a pipelike passageway having a polar environment. An example is the porin class of proteins.

- Simple diffusion is the passive movement of a substance along a chemical or electrical gradient. Biological membranes pose a barrier to hydrophilic polar molecules, while they allow hydrophobic substances to diffuse freely.

- Ions and large hydrophilic molecules cannot cross the phospholipid bilayer. Diffusion can still occur via channel or carrier proteins by facilitated diffusion. Channels allow the diffusion of specific ions, by forming an aqueous pore in the membrane. Carrier proteins bind to the molecules they transport, much like an enzyme. The rate of transport by a carrier is limited by the number of carriers in the membrane.

- The direction of movement due to osmosis depends on the solute concentration on either side of the membrane. Solutions can be isotonic, hypotonic, or hypertonic. Cells in an isotonic solution are in osmotic balance; cells in a hypotonic solution will gain water; and cells in a hypertonic solution will lose water. Aquaporins are water channels that facilitate the diffusion of water.

- Active transport uses specialized protein carriers that couple a source of energy to transport. They are classified based on the number of molecules and direction of transport. Uniporters transport a specific molecule in one direction; symporters transport two molecules in the same direction; and antiporters transport two molecules in opposite directions.

- The sodium–potassium pump moves Na^+ out of the cell and K^+ into the cell against their concentration gradients using ATP. In every cycle of the pump, three Na^+ leave the cell and two K^+ enter it. This pump appears to be almost universal in animal cells.

- Coupled transport uses a concentration gradient of one molecule to move another against a gradient in the same direction. Counter transport is similar, but the two molecules move in opposite directions.

- Bulk transport moves large quantities of substances that cannot pass through the cell membrane.

- In endocytosis, the cell membrane surrounds material and pinches off to form a vesicle. In receptor-mediated endocytosis, specific molecules bind to receptors on the cell membrane.

- In exocytosis, material in a vesicle is discharged when the vesicle fuses with the membrane.

CHAPTER 5

Metabolism, Energy and Enzyme

Chapter Contents:

- An organism's metabolism transforms matter and energy
- Forms of Energy
- The Laws of Energy Transformation
- ATP powers cellular work
- The Regeneration of ATP
- Redox Reactions
- Enzymes are organized into teams in metabolic pathways
- Catalysis in the Enzyme's Active Site
- Effect of Temperature on Enzyme Activity
- Effect of pH on Enzyme Activity
- Effect of substrate or enzyme concentration on Enzyme Activity
- Regulation of enzyme activity helps control metabolism
- Enzyme Inhibitors
- Comparing enzyme affinities
- RNA-Based Biological Catalysts: Ribozymes
- Metabolic Pathways Are Regulated in Three General Ways
- KEY CONCEPTS

An organism's metabolism transforms matter and energy

Metabolism as a whole manages the material and energy resources of the cell. Metabolism, which occurs only in living organisms, comprises thousands of biochemical reactions that accomplish the special activities we associate with life, such as growth, reproduction, movement, and the ability to respond to stimuli. In other words, metabolism underlies essentially all life activities. In living cells, chemical reactions are coordinated with each other and often occur in a series of steps called a **metabolic pathway**, with each step catalyzed by a different **enzyme**, a macromolecule that speeds up a chemical reaction.

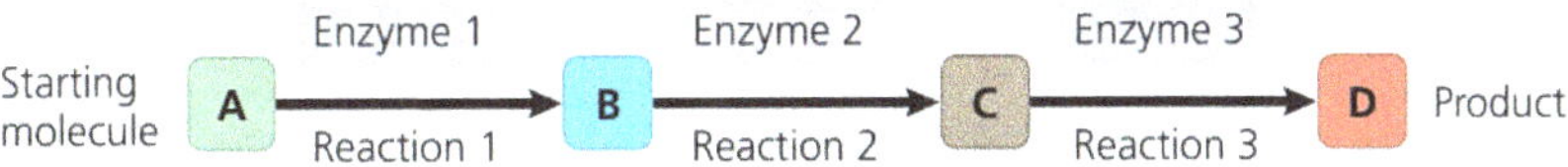

Some metabolic pathways release energy by breaking down complex molecules to simpler compounds. These degradative processes are called **catabolic pathways**, or **breakdown pathways**. One major catabolic pathway is **cellular respiration**, which breaks down glucose and other organic fuels in the presence of oxygen to carbon dioxide and water. Energy stored in the organic molecules becomes available to do cellular work, such as ciliary beating or membrane transport.

Anabolic pathways, in contrast, consume energy to build complicated molecules from simpler ones; they are sometimes called **biosynthetic pathways**. Examples of anabolism are synthesis of an amino acid from simpler molecules and synthesis of a protein from amino acids.

Bioenergetics, the study of how energy flows through living organisms.

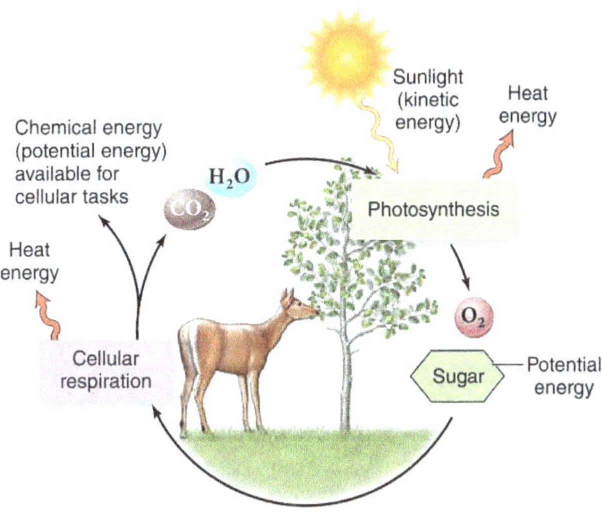

▲ **Figure 5.1** Energy Can Take Many Forms. In photosynthesis, plants transform the kinetic energy in sunlight into potential energy stored in the chemical bonds of sugars and other organic molecules. Respiration, in turn, releases this potential energy. Heat energy is lost at every step along the way.

Forms of Energy

Energy is the capacity to cause change. In everyday life, energy is important because some forms of energy can be used to do work. Energy exists in various forms, and the work of life depends on the ability of cells to transform energy from one form to another. Even when you are asleep, cells of your muscles, brain, and other parts of your body are at work and using energy. Energy can exist in many different forms, including heat, chemical, electrical, mechanical, and radiant energy. Visible light, infrared and ultraviolet light, gamma rays, and X-rays are all types of radiant energy. Although the forms of energy are different, energy can be converted readily from one form to another.

All forms of energy can exist in one of two states: **kinetic** and **potential. Thermal energy** is kinetic energy associated with the random movement of atoms or molecules; thermal energy in transfer from one object to another is called **heat**. Light is also a type of energy that can be harnessed to perform work, such as powering photosynthesis in green plants. An object not presently moving may still possess energy. Energy that is not kinetic is called **potential energy**; it is energy that matter possesses because of its location or structure. Molecules possess energy because of the arrangement of electrons in the bonds between their atoms. **Chemical energy** is a term used by biologists to refer to the potential energy available for release in a chemical reaction.

Catabolic pathways release energy by breaking down complex molecules. Biologists say that these complex molecules, such as glucose, are high in chemical energy. During a catabolic reaction, some bonds are broken and others are formed, releasing energy and resulting in lower-energy breakdown products. The organic food molecules provide the necessary chemical energy for working. This chemical energy is itself derived from light energy absorbed by plants during photosynthesis. Organisms are energy transformers.

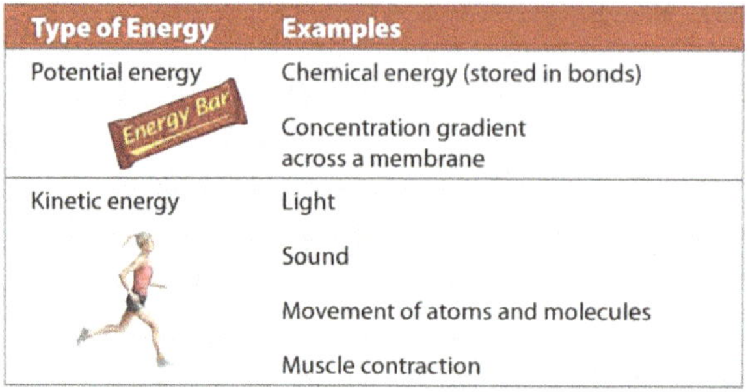

Type of Energy	Examples
Potential energy	Chemical energy (stored in bonds)
	Concentration gradient across a membrane
Kinetic energy	Light
	Sound
	Movement of atoms and molecules
	Muscle contraction

▲ **Figure 5.2** Examples of Energy in Biology

The Laws of Energy Transformation

Thermodynamics is the branch of chemistry concerned with energy changes. Cells are governed by the laws of physics and chemistry, so we must understand these laws in order to understand how cells function. When discussing thermodynamics, scientists refer to a system, which is the object under study. A system is whatever we define it to be—a single molecule, a cell, a planet. Everything outside a system is its surroundings. The universe, in this context, is the total of the system and the surroundings. There are three types of systems:

- An isolated system does not exchange matter or energy with its surroundings. A perfectly insulated Thermos flask is an example of an isolated system.

- A closed system can exchange energy but not matter with its surroundings. Earth is a closed system. It takes in a great amount of energy from the Sun and releases heat, but essentially no matter is exchanged between Earth and the rest of the universe (barring the odd space probe).

- An open system can exchange both energy and matter with its surroundings. All living organisms are open systems.

Organisms are **open systems**. They absorb energy—for instance, light energy or chemical energy in the form of organic molecules—and release heat and metabolic waste products, such as carbon dioxide, to the surroundings. Two laws of thermodynamics govern energy transformations in organisms and all other collections of matter. Two laws about energy—the first and second laws of thermodynamics—apply to all things in the universe.

First law of thermodynamics

According to the **first law of thermodynamics**, the energy of the universe is constant: Energy can be transferred and transformed, but it cannot be created or destroyed. The first law is also known as the principle of conservation of energy. The energy of any system plus its surroundings is constant. A system may absorb energy from its surroundings or may give up some energy to its surroundings, but the total energy content of that system plus its surroundings is always the same.

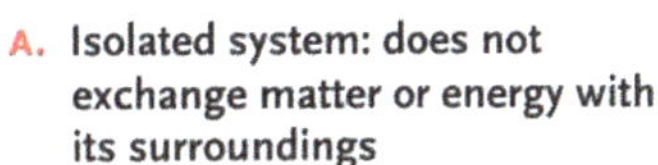

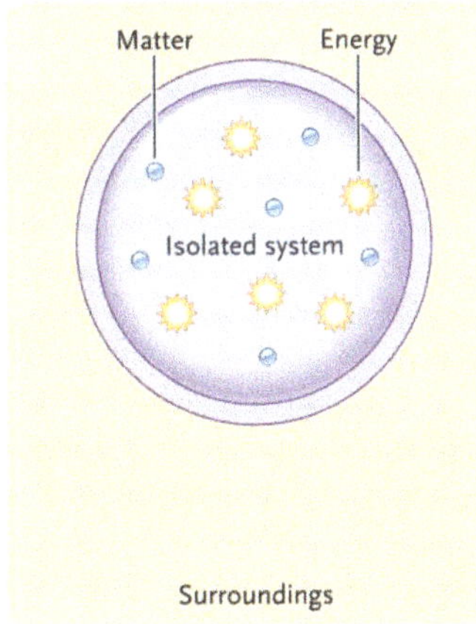

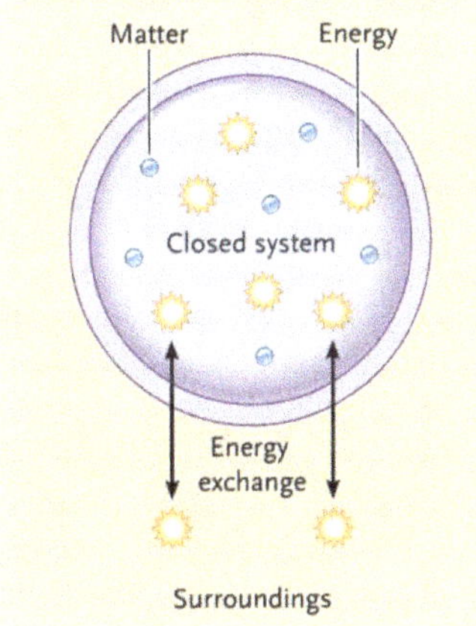

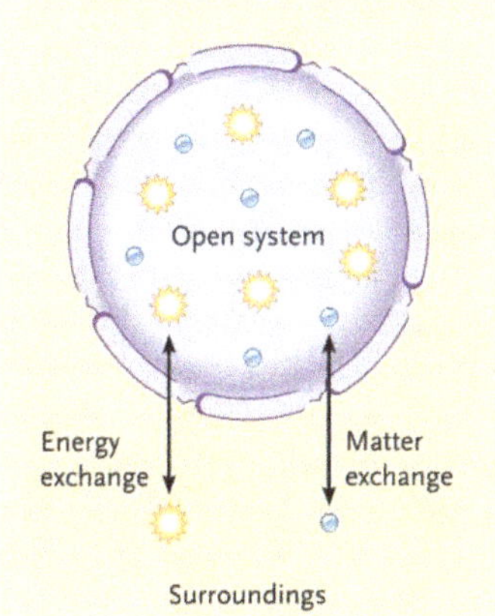

▲ **Figure 5.3** Isolated, closed, and open systems in thermodynamics.

The second law of thermodynamics:

The second law of thermodynamics states that when energy is converted from one form to another, some usable energy—that is, energy available to do work—is converted into heat that disperses into the surroundings. Heat is the kinetic energy of randomly moving particles. So every energy transfer or transformation increases the **entropy** of the universe. Scientists use a quantity called entropy as a measure of molecular disorder, or randomness. The more randomly arranged a collection of matter is, the greater its entropy. As a result, the amount of usable energy available to do work in the universe decreases over time.

The free energy is denoted by the symbol G (for "Gibbs free energy"). G is equal to the energy contained in a molecule's chemical bonds (called **enthalpy** and designated H) .

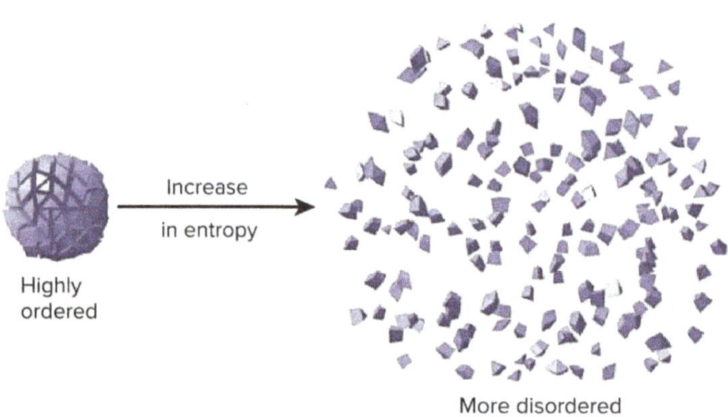

▲ **Figure 5.4** Entropy, a measure of the disorder of a system. An increase in entropy means an increase in disorder.

The free-energy change (ΔG) of a reaction tells us whether or not the reaction occurs spontaneously (the word spontaneous implies that it is energetically favorable, not that it will occur rapidly). Once we know the value of ΔG for a process, we can use it to predict whether the process will be spontaneous (that is, whether it is energetically favorable and will occur without an input of energy) so that only processes with a negative ΔG are spontaneous. In other words, every spontaneous process decreases the system's free energy, and processes that have a positive or zero ΔG are never spontaneous.

For a reaction to have a negative ΔG, the system must lose free energy during the change from initial state to final state. Because it has less free energy, the system in its final state is less likely to change and is therefore more **stable** than it was previously. Another term that describes a state of maximum stability is **equilibrium**. At equilibrium, the forward and reverse reactions occur at the same rate, and there is no further net change in the relative concentration of products and reactants. As a system moves toward equilibrium, its free energy becomes progressively lower and reaches its lowest point when the system achieves equilibrium ($\Delta G = 0$). A process is spontaneous and can perform work only when it is moving toward equilibrium.

Based on their free-energy changes, chemical reactions can be classified as either **exergonic** ("energy outward") or **endergonic** ("energy inward"). An exergonic or exothermic reaction proceeds with a net release of free energy so the products contain less energy than the reactants. Because the chemical mixture loses free energy (G decreases), ΔG is negative for an exergonic reaction. Using ΔG as a standard for spontaneity, exergonic reactions are those that occur spontaneously. Such reactions break large, complex molecules into their smaller, simpler components. Cellular respiration, the breakdown of glucose to carbon dioxide and water, is an example.

An endergonic or endothermic reaction is one that absorbs free energy from its surroundings so, the products contain more energy than the reactants. Because this kind of reaction essentially stores free energy in molecules (G increases), ΔG is positive. Such reactions are non-spontaneous. Typically, endergonic reactions build complex molecules from simpler components. Photosynthesis for example, is an endergonic reaction.

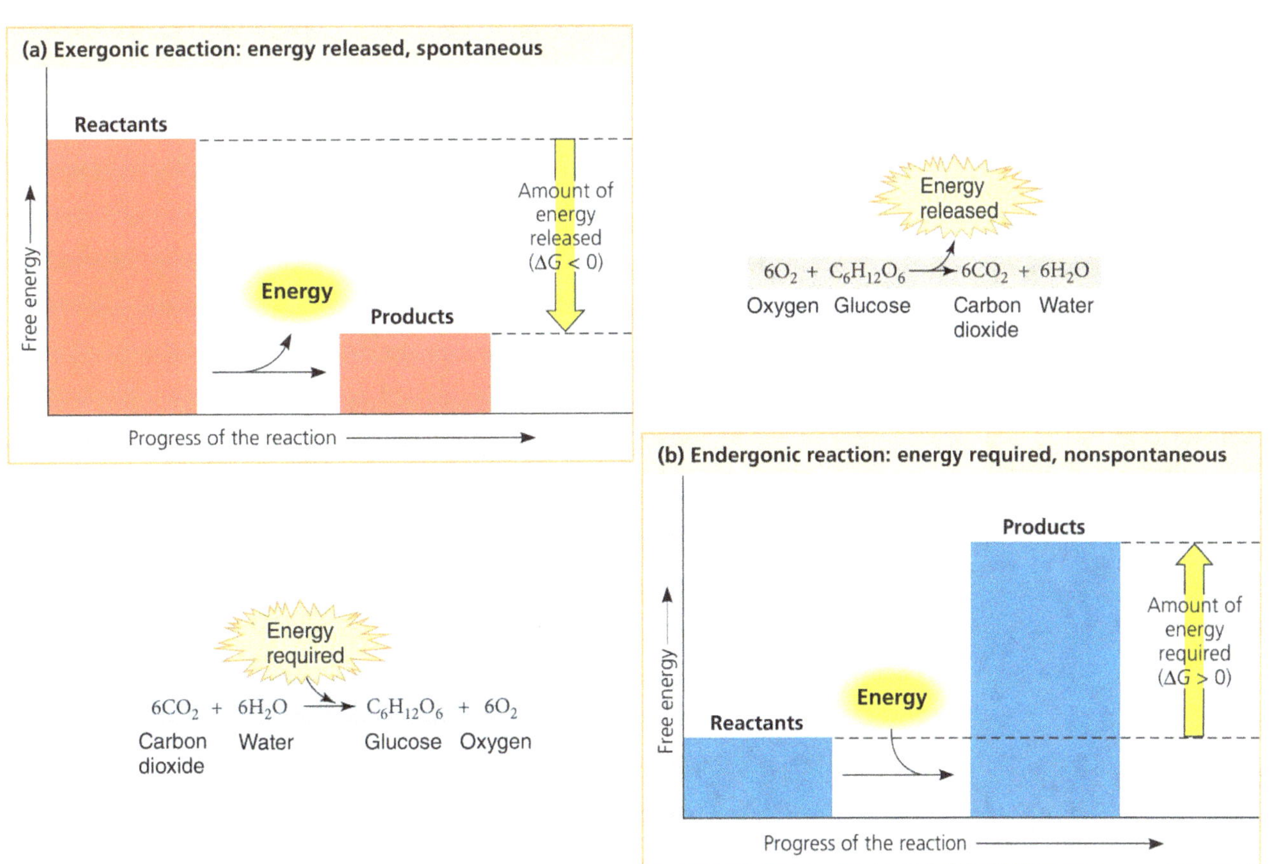

▲ **Figure 5.5** Free energy changes (ΔG) in exergonic and endergonic reactions.

ATP powers cellular work by coupling exergonic reactions to endergonic reactions

A cell does three main kinds of work:

- Chemical work, the pushing of endergonic reactions that would not occur spontaneously, such as the synthesis of polymers from monomers.

- Transport work, the pumping of substances across membranes against the direction of spontaneous movement.

- Mechanical work, such as the beating of cilia, the contraction of muscle cells, and the movement of chromosomes during cellular reproduction.

A key feature in the way cells manage their energy resources to do this work is **energy coupling**, the use of an exergonic process to drive an endergonic one. ATP is responsible for mediating most energy coupling in cells, and in most cases it acts as the immediate source of energy that powers cellular work.

ATP is composed of three smaller components. The first component is a 5-carbon sugar, ribose, which serves as the framework to which the other two subunits are attached. The second component is adenine, an organic molecule composed of two carbon–nitrogen rings. Each of the nitrogen atoms in the ring has an unshared pair of electrons and weakly attracts hydrogen ions, making adenine chemically a weak base. The third component of ATP is a chain of three phosphates, thus adenosine triphosphate.

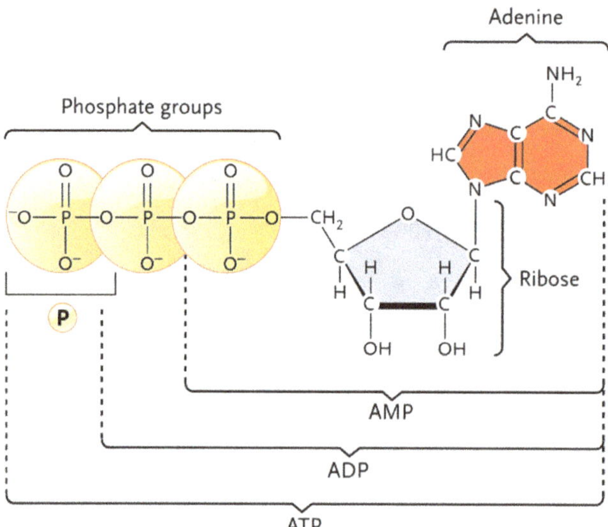

In addition to its role in energy coupling, ATP is also one of the nucleoside triphosphates used to make RNA. The bonds between the phosphate groups of ATP can be broken by hydrolysis. When the terminal phosphate bond is broken by addition of a water molecule, a molecule of inorganic phosphate ($HOPO_3^{2-}$, abbreviated as Pi) leaves the ATP, which becomes adenosine diphosphate, or ADP. The reaction is exergonic and releases 7.3 kcal of energy per mole of ATP hydrolyzed:

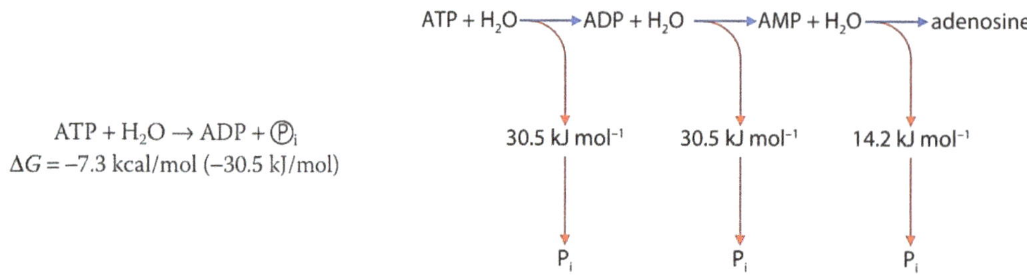

Both of the two terminal phosphates can be hydrolyzed to release energy, leaving adenosine monophosphate (AMP), but the third phosphate is not attached by a high-energy bond. With only one phosphate group, AMP has no other phosphates to provide the electrostatic repulsion that makes the bonds holding the two terminal phosphate groups high-energy bonds.

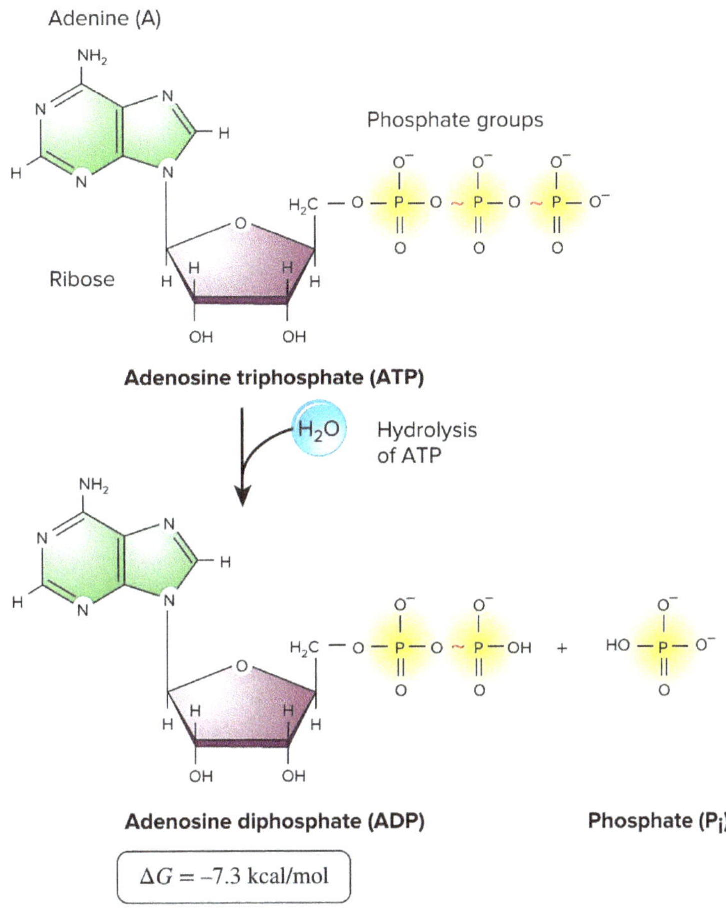

Adenine (A)

Phosphate groups

Ribose

Adenosine triphosphate (ATP)

H_2O Hydrolysis of ATP

Adenosine diphosphate (ADP)   **Phosphate (P_i)**

$\Delta G = -7.3 \text{ kcal/mol}$

At any time, a typical cell contains more than 10 ATP molecules for each ADP molecule. Because the cell maintains the ATP concentration at such a high level (relative to the concentration of ADP), its hydrolysis reaction is even more strongly exergonic and more able to drive the endergonic reactions to which it is coupled. In eukaryotic cells, organelles called mitochondria produce most of a cell's ATP. A mitochondrion uses the potential energy in the bonds of one glucose molecule to generate dozens of ATP molecules in cellular respiration. Not surprisingly, the most energy-hungry cells, such as those in the muscles and brain, also contain the most mitochondria. ATP is made when a phosphate group is combined with ADP. This is done in two main ways:

- using energy provided directly by another chemical reaction – this is called a **substrate-linked reaction**

- by **chemiosmosis**, a process that takes place across the inner membranes of mitochondria, using energy released by the movement of hydrogen ions down their concentration gradient (respiration). In plants ATP is made also by photosynthesis.

In humans, all of the ATP is made in respiration, both by substrate-linked reactions and by chemiosmosis.

If the ΔG of an endergonic reaction is less than the amount of energy released by ATP hydrolysis, then the two reactions can be coupled so that, overall, the coupled reactions are exergonic. This usually involves phosphorylation, the transfer of a phosphate group from ATP to some other molecule, such as the reactant. The recipient molecule with the phosphate group covalently bonded to it is then called a **phosphorylated intermediate**. The key to coupling exergonic and endergonic reactions is the formation of this phosphorylated intermediate, which is more reactive (less stable, with more free energy) than the original unphosphorylated molecule.

Organisms require huge amounts of ATP. A typical human cell uses the equivalent of 2 billion ATP molecules a minute just to stay alive. Even though ATP is essential to life, cells do not stockpile it in large quantities. ATP's high-energy phosphate bonds make the molecule too unstable for long-term storage. Instead, cells store energy-rich molecules such as fats, starch, and glycogen. When ATP supplies run low, cells divert some of their lipid and carbohydrate reserves to the metabolic pathways of cellular respiration. This process soon produces additional ATP.

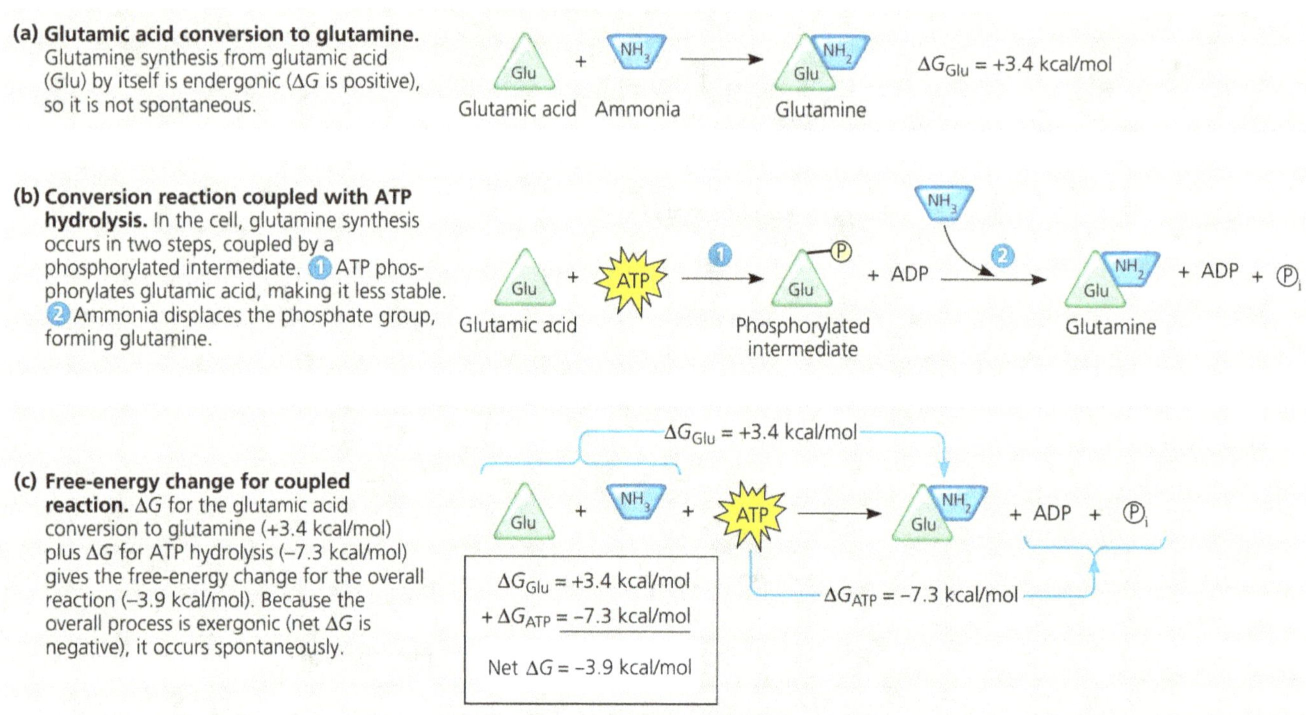

▲ **Figure 5.6** How ATP drives chemical work: energy coupling using ATP hydrolysis. In this example, the exergonic process of ATP hydrolysis drives an endergonic process—synthesis of the amino acid glutamine

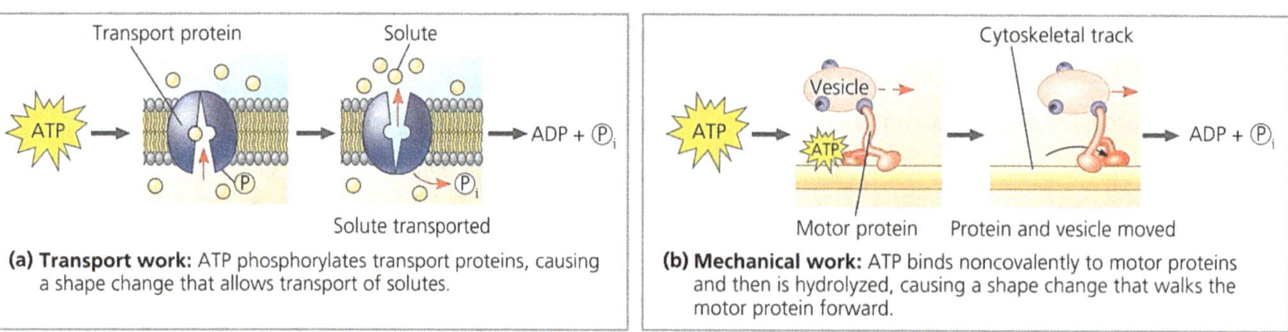

(a) **Transport work:** ATP phosphorylates transport proteins, causing a shape change that allows transport of solutes.

(b) **Mechanical work:** ATP binds noncovalently to motor proteins and then is hydrolyzed, causing a shape change that walks the motor protein forward.

▲ **Figure 5.7** How ATP drives transport and mechanical work. ATP hydrolysis causes changes in the shapes and binding affinities of proteins. This can occur either (a) directly, by phosphorylation, as shown for a membrane protein carrying out active transport of a solute or (b) indirectly, via noncovalent binding of ATP and its hydrolytic products, as is the case for motor proteins that move vesicles (and other organelles) along cytoskeletal "tracks" in the cell.

The Regeneration of ATP

An organism at work uses ATP continuously, but ATP is a renewable resource that can be regenerated by the addition of phosphate to ADP. The free energy required to phosphorylate ADP comes from exergonic breakdown reactions (catabolism) in the cell. This shuttling of inorganic phosphate and energy is called the ATP cycle, and it couples the cell's energy-yielding (exergonic) processes to the energy consuming (endergonic) ones. The ATP cycle proceeds at an astonishing pace. For example, a working muscle cell recycles its entire pool of ATP in less than a minute. That turnover represents 10 million molecules of ATP consumed and regenerated per second per cell. If ATP could not be regenerated by the phosphorylation of ADP, humans would use up nearly their body weight in ATP each day. Because both directions of a reversible process cannot be downhill, the regeneration of ATP from ADP and Pi is necessarily endergonic:

$$ADP + \textcircled{P}_i \rightarrow ATP + H_2O$$
$$\Delta G = +7.3 \text{ kcal/mol } (+30.5 \text{ kJ/mol}) \text{ (standard conditions)}$$

Since ATP formation from ADP and Pi is not spontaneous, free energy must be spent to make it occur. Catabolic (exergonic) pathways, especially cellular respiration, provide the energy for the endergonic process of making ATP. Plants also use light energy to produce ATP. Thus, the ATP cycle is a key player in bioenergetics, functioning as a revolving door through which energy passes during its transfer from catabolic to anabolic pathways.

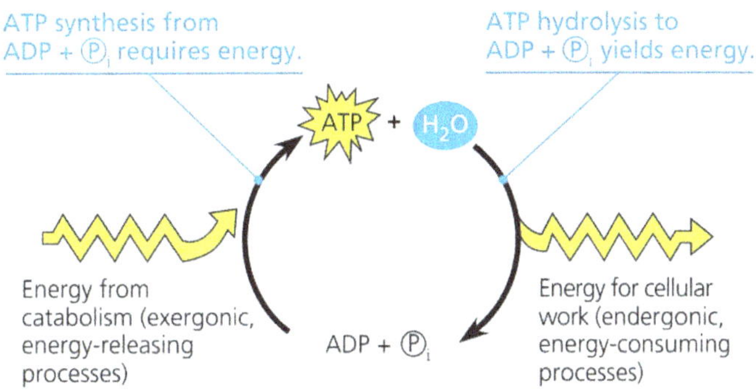

▲ **Figure 5.8** The ATP cycle. Energy released by breakdown reactions (catabolism) in the cell is used to phosphorylate ADP, regenerating ATP. Chemical potential energy stored in ATP drives most cellular work.

If the ΔG of an endergonic reaction is less than the amount of energy released by ATP hydrolysis, then the two reactions can be coupled so that, overall, the coupled reactions are exergonic. This usually involves phosphorylation, the transfer of a phosphate group from ATP to some other molecule, such as the reactant. The recipient molecule with the phosphate group covalently bonded to it is then called a **phosphorylated intermediate**. The key to coupling exergonic and endergonic reactions is the formation of this phosphorylated intermediate, which is more reactive (less stable, with more free energy) than the original unphosphorylated molecule .

Redox Reactions

You have seen that cells transfer energy through the transfer of a phosphate group from ATP. Energy is also transferred through the transfer of electrons. Oxidation is the chemical process in which a substance loses electrons, whereas reduction is the complementary process in which a substance gains electrons. Because electrons released during an oxidation reaction cannot exist in the free state in living cells, every oxidation reaction must be accompanied by a reduction reaction in which the electrons are accepted by another atom, ion, or molecule. Oxidation and reduction reactions are often called **redox reactions** because they occur simultaneously so that electrons removed from one molecule during oxidation must join another molecule and reduce it. The substance that becomes oxidized gives up energy as it releases electrons, and the substance that becomes reduced receives energy as it gains electrons.

Redox reactions often occur in a series as electrons are transferred from one molecule to another. These electron transfers, which are equivalent to energy transfers, are an essential part of cellular respiration, photosynthesis, and many other chemical processes. Redox reactions, for example, release the energy stored in food molecules so that ATP can be synthesized using that energy.

Generally, it is not easy to remove one or more electrons from a covalent compound; it is much easier to remove a whole atom. For this reason, redox reactions in cells usually involve the transfer of a hydrogen atom rather than just an electron. A hydrogen atom contains an electron, plus a proton that does not participate in the oxidation–reduction reaction. When an electron, either singly or as part of a hydrogen atom, is removed from an organic compound, it takes with it some of the energy stored in the chemical bond of which it was a part. That electron, along with its energy, is transferred to an acceptor molecule. An electron progressively loses free energy as it is transferred from one acceptor to another.

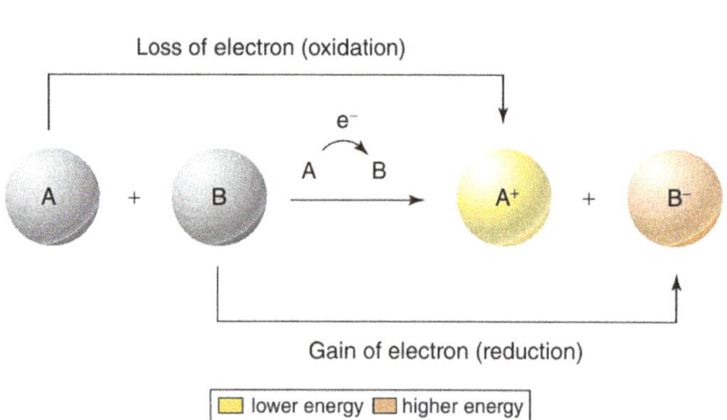

▲ **Figure 5.9** Redox reactions. Oxidation is the loss of an electron; reduction is the gain of an electron. In this example, the charges of molecules A and B appear as superscripts in each molecule. Molecule A loses energy as it loses an electron, and molecule B gains that energy as it gains an electron.

One of the most common acceptor molecules in cellular processes is **nicotin amide adenine dinucleotide** (NAD⁺). When NAD⁺ becomes reduced, it temporarily stores large amounts of free energy. Here is a generalized equation showing the transfer of hydrogen from a compound, which we call X, to NAD⁺. Note that the NAD⁺ becomes reduced when it combines with hydrogen. NAD⁺ is an ion with a net charge of +1. When 2 electrons and 1 proton are added, the charge is neutralized and the reduced form of the compound, NADH, is produced (Although the correct way to write the reduced form of NAD⁺ is NADH + H⁺, for simplicity we present the reduced form as NADH.) Some energy stored in the bonds holding the hydrogen atoms to molecule X has been transferred by this redox reaction and is temporarily held by NADH. When NADH transfers the electrons to some other molecule, some of their energy is transferred. This energy is usually then transferred through a series of reactions that ultimately result in the formation of ATP.

$$XH_2 + NAD^+ \longrightarrow X + NADH + H^+$$

Oxidized → Reduced

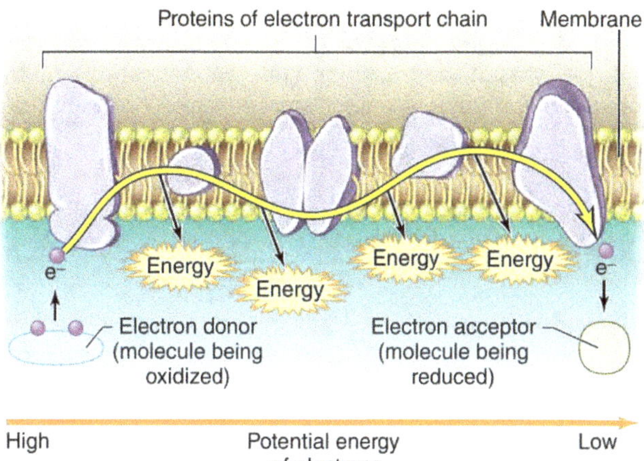

▶ **Figure 5.10** Electron Transport Chain.

Nicotinamide adenine dinucleotide phosphate (NADP⁺) is a hydrogen acceptor that is chemically similar to NAD⁺ but has an extra phosphate group. Unlike NADH, the reduced form of NADP⁺, abbreviated NADPH, is not involved in ATP synthesis. Instead, the electrons of NADPH are used more directly to provide energy for certain reactions, including certain essential reactions of photosynthesis. Other important hydrogen acceptors or electron acceptors are FAD and the **cytochromes**. **Flavin adenine dinucleotide** (FAD) is a nucleotide that accepts hydrogen atoms and their electrons; its reduced form is FADH₂. The cytochromes are proteins that contain iron; the iron component accepts electrons from hydrogen atoms and then transfers these electrons to some other compound. Like NAD⁺ and NADP⁺, FAD and the cytochromes are electron transfer agents. Each exists in a reduced state, in which it has more free energy, or in an oxidized state, in which it has less. Each is an essential component of many redox reaction sequences in cells.

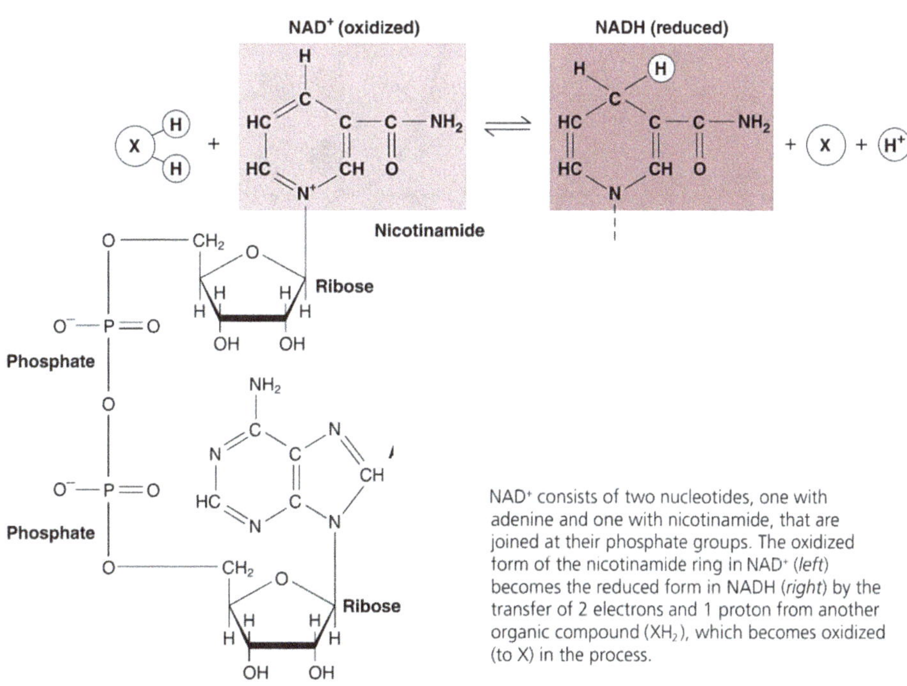

NAD⁺ consists of two nucleotides, one with adenine and one with nicotinamide, that are joined at their phosphate groups. The oxidized form of the nicotinamide ring in NAD⁺ (*left*) becomes the reduced form in NADH (*right*) by the transfer of 2 electrons and 1 proton from another organic compound (XH₂), which becomes oxidized (to X) in the process.

Enzymes

Every chemical reaction between molecules involves both bond breaking and bond forming. Changing one molecule into another generally involves contorting the starting molecule into a highly unstable state before the reaction can proceed. To reach the contorted state where bonds can change, reactant molecules must absorb energy from their surroundings. When the new bonds of the product molecules form, energy is released as heat, and the molecules return to stable shapes with lower energy than the contorted state. The initial investment of energy for starting a reaction—the energy required to contort the reactant molecules so the bonds can break—is known as the free energy of activation, or **activation energy (E_A)**.

Activation energy is often supplied by heat in the form of thermal energy that the reactant molecules absorb from the surroundings. The absorption of thermal energy accelerates the reactant molecules, so they collide more often and more forcefully. It also agitates the atoms within the molecules, making the breakage of bonds more likely. When the molecules have absorbed enough energy for the bonds to break, the reactants are in an unstable condition known as the **transition state**. The reactants in the transition state are activated, and their bonds can be broken. As the atoms then settle into their new, more stable bonding arrangements, energy is released to the surroundings.

$$\text{Enzyme} + \text{Substrate(s)} \rightleftharpoons \text{Enzyme-substrate complex} \rightleftharpoons \text{Enzyme} + \text{Product(s)}$$

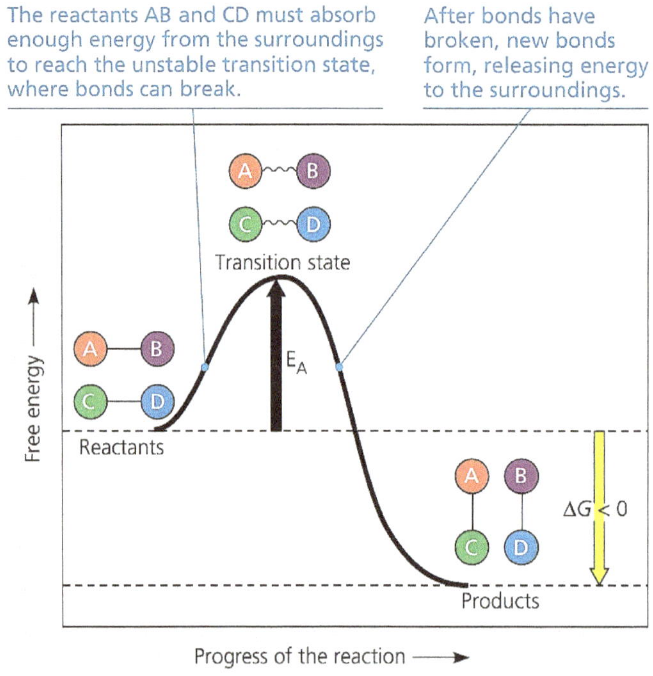

The reactants AB and CD must absorb enough energy from the surroundings to reach the unstable transition state, where bonds can break.

After bonds have broken, new bonds form, releasing energy to the surroundings.

Transition state

Free energy

E_A

Reactants

Products

$\Delta G < 0$

Progress of the reaction

▲ **Figure 5.11** Energy profile of an exergonic reaction. The "molecules" are hypothetical, with A, B, C, and D representing portions of the molecules. Thermodynamically, this is an exergonic reaction, with a negative ΔG, and the reaction occurs spontaneously. However, the activation energy (EA) provides a barrier that determines the rate of the reaction.

Enzymes are among the most important of all biological molecules. An enzyme is an organic molecule that catalyzes (speeds up) a chemical reaction without being consumed. Most enzymes are proteins, although some are made of RNA and are called **ribozymes.** Most of the known ribozymes speed the cutting and splicing reactions that remove surplus segments from RNA molecules as part of their conversion into finished form. Some have other functions, however. For example, ribosomes, the cell structures that assemble amino acids into proteins.

Many of the cell's organelles, including mitochondria, chloroplasts, lysosomes, and peroxisomes, are specialized sacs of enzymes. Enzymes copy DNA, build proteins, digest food, recycle a cell's worn-out parts, and catalyze oxidation–reduction reactions, just to name a few of their jobs. Without enzymes, all of these reactions would proceed far too slowly to support life.

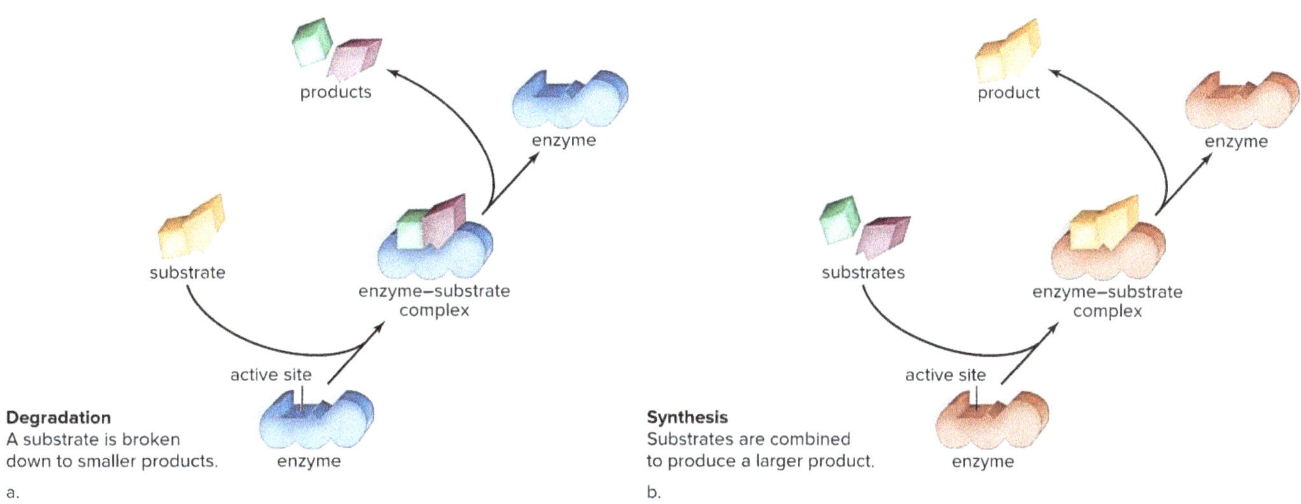

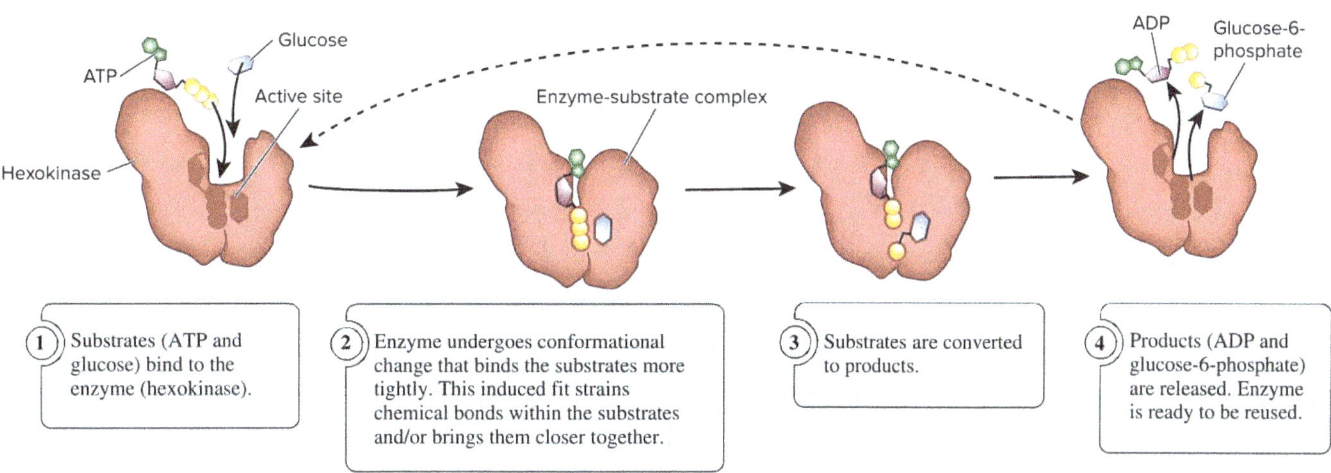

▲ **Figure 5.12** The steps of an enzyme-catalyzed reaction. The example shown here involves the enzyme hexokinase, which binds glucose and ATP. The products are glucose-6-phosphate and ADP, which are released from the enzyme.

An enzyme catalyzes a reaction by **lowering the activation energy barrier**, enabling the reactant molecules to absorb enough energy to reach the transition state even at moderate temperatures. The enzyme brings reactants (also called substrates) into contact with one another, so that less energy is required for the reaction to proceed. The catalytic ability of some enzymes is truly impressive. For example, hydrogen peroxide (H_2O_2) breaks down extremely slowly if the reaction is uncatalyzed, but a single molecule of the enzyme catalase brings about the decomposition of 40 million molecules of hydrogen peroxide per second! Catalase has the highest catalytic rate known for any enzyme. It protects cells by destroying hydrogen peroxide, a poisonous substance produced as a byproduct of some cell reactions.

The reactant an enzyme acts on is referred to as the enzyme's **substrate**. The enzyme binds to its substrate (or substrates, when there are two or more reactants), forming an **enzymesubstrate complex**. While enzyme and substrate are joined, the catalytic action of the enzyme converts the substrate to the **product** (or products) of the reaction.

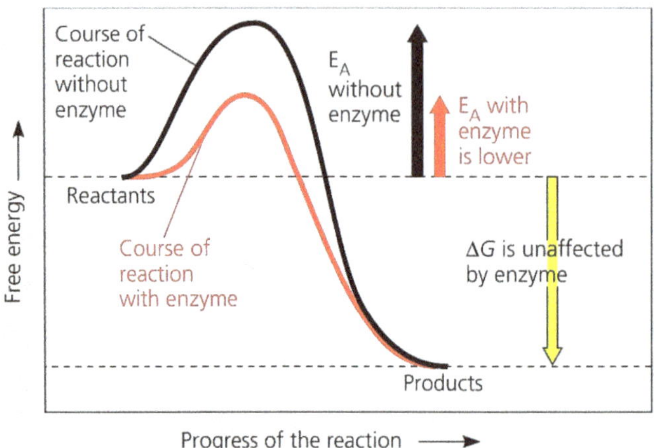

▲ **Figure 5.13** The effect of an enzyme on activation energy. Without affecting the free-energy change (ΔG) for a reaction, an enzyme speeds the reaction by reducing its activation energy (E_A)

Scientists usually name enzymes by adding the suffix –ase to the name of the substrate. The enzyme sucrase, for example, splits sucrose into glucose and fructose. A few enzymes retain traditional names that do not end in -ase; some of these end -zyme. For example, lysozyme is an enzyme found in tears and saliva; it breaks down bacterial cell walls. Other examples of enzymes with traditional names are pepsin and trypsin, which break peptide bonds in proteins.

Enzyme Class	Function
Oxidoreductases	Catalyze oxidation–reduction reactions
Transferases	Catalyze the transfer of a functional group from a donor molecule to an acceptor molecule
Hydrolases	Catalyze hydrolysis reactions
Isomerases	Catalyze conversion of a molecule from one isomeric form to another
Ligases	Catalyze certain reactions in which two molecules become joined in a process coupled to the hydrolysis of ATP
Lyases	Catalyze certain reactions in which double bonds form or break

Enzymes are organized into teams in metabolic pathways:

Enzymes play an essential role in reaction coupling because they usually work in sequence, with the product of one enzyme-controlled reaction serving as the substrate for the next. Each enzyme carries out one step, such as changing molecule A into molecule B. Then molecule B is passed along to the next enzyme, which converts it into molecule C, and so on. Such a series of reactions is called a **metabolic pathway**. Some metabolic pathways are linear, meaning that the reactions run straight from reactant to product. Other metabolic pathways are cyclic. In a cyclic pathway, the last step regenerates a reactant for the first step.

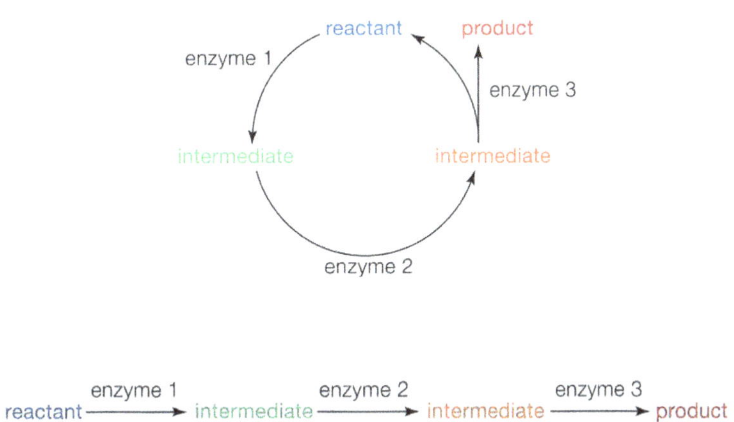

Assume that reactant A is continually supplied and that its concentration remains constant. Enzyme 1 converts reactant A to product B.

The concentration of B is always lower than the concentration of A because B is removed as it is converted to C in the reaction catalyzed by enzyme 2. If C is removed as quickly as it is formed (perhaps by leaving the cell), the entire reaction pathway is "pulled" toward C.

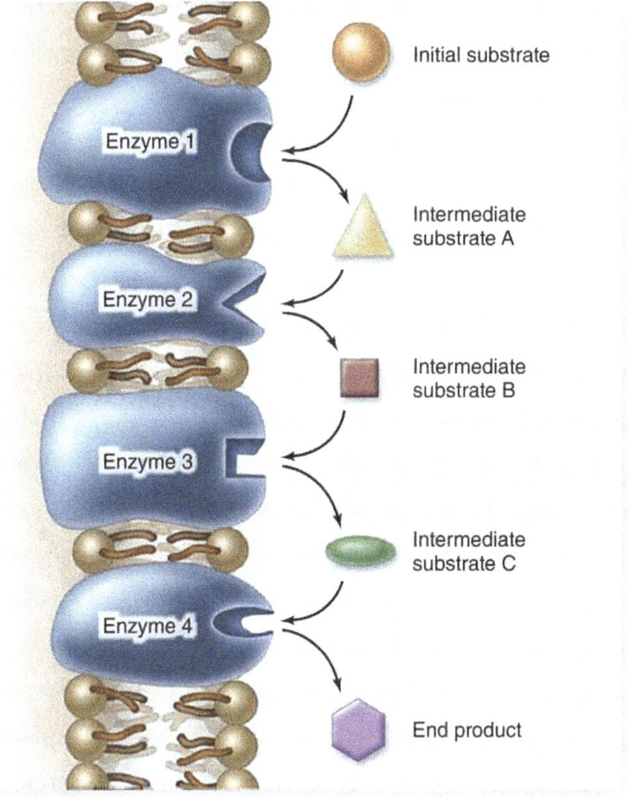

▶ **Figure 5.14** A biochemical pathway

The reaction catalyzed by each enzyme is very specific; an enzyme can recognize its specific substrate even among closely related compounds. The specificity of an enzyme results from its shape, which is a consequence of its amino acid sequence. Only a restricted region of the enzyme molecule actually binds to the substrate. This region, called the **active site**, is typically a pocket or groove on the surface of the enzyme where catalysis occurs. The specificity of an enzyme is attributed to a complementary fit between the shape of its active site and the shape of the substrate.

The lock-and-key hypothesis: The shape of the active site allows the substrate to fit perfectly. The idea that the enzyme has a particular shape into which the substrate fits exactly is known as the lock-and-key hypothesis. The substrate is the key whose shape fits the lock of the enzyme. The substrate is held in place by temporary bonds which form between the substrate and some of the R groups of the enzyme's amino acids. This combined structure is termed the enzyme–substrate complex. Each enzyme will act on only one type of substrate molecule. This is because the shape of the active site will only allow one shape of molecule to fit. The enzyme is said to be specific for this substrate.

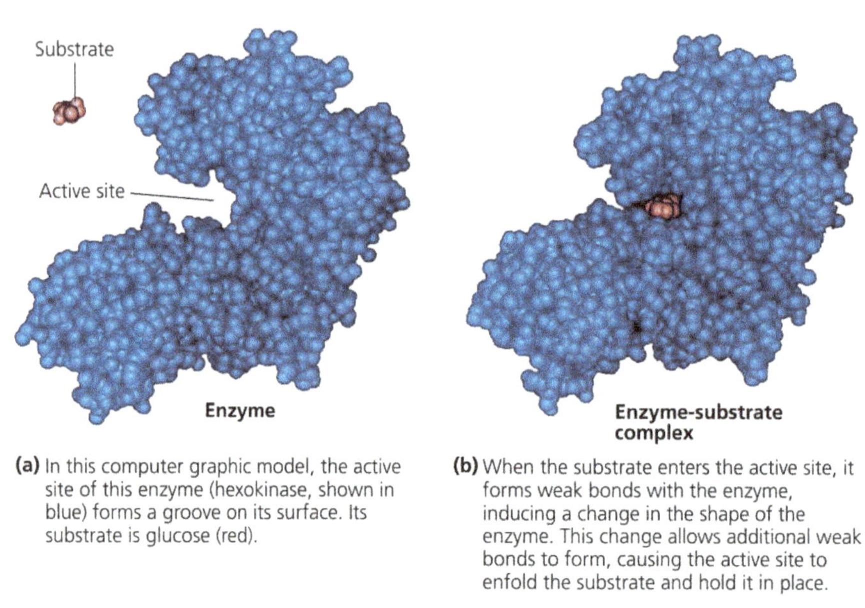

(a) In this computer graphic model, the active site of this enzyme (hexokinase, shown in blue) forms a groove on its surface. Its substrate is glucose (red).

(b) When the substrate enters the active site, it forms weak bonds with the enzyme, inducing a change in the shape of the enzyme. This change allows additional weak bonds to form, causing the active site to enfold the substrate and hold it in place.

▲ **Figure 5.15** Induced fit between an enzyme and its substrate.

The induced-fit hypothesis: It is basically the same as the lock-and-key hypothesis, but adds the idea that the enzyme, and sometimes the substrate, can change shape slightly as the substrate molecule enters the enzyme, in order to ensure a perfect fit. This makes the catalysis even more efficient. The active site itself is also not a rigid receptacle for the substrate.

When the substrate enters the active site, the enzyme changes shape slightly due to interactions between the substrate's chemical groups and chemical groups on the side chains of the amino acids that form the active site. This shape change makes the active site fit even more snugly around the substrate. The tightening of the binding after initial contact—called **induced fit**—is like a clasping handshake. Induced fit brings chemical groups of the active site into positions that enhance their ability to catalyze the chemical reaction.

Catalysis in the Enzyme's Active Site

In most enzymatic reactions, the substrate is held in the active site by so-called weak interactions, such as hydrogen bonds and ionic bonds. The R groups of a few of the amino acids that make up the active site catalyze the conversion of substrate to product, and the product departs from the active site. The enzyme is then free to take another substrate molecule into its active site. The entire cycle happens so fast that a single enzyme molecule typically acts on about 1,000 substrate molecules per second, and some enzymes are even faster. Enzymes, like other catalysts, emerge from the reaction in their original form. Therefore, very small amounts of enzyme can have a huge metabolic impact by functioning over and over again in catalytic cycles.

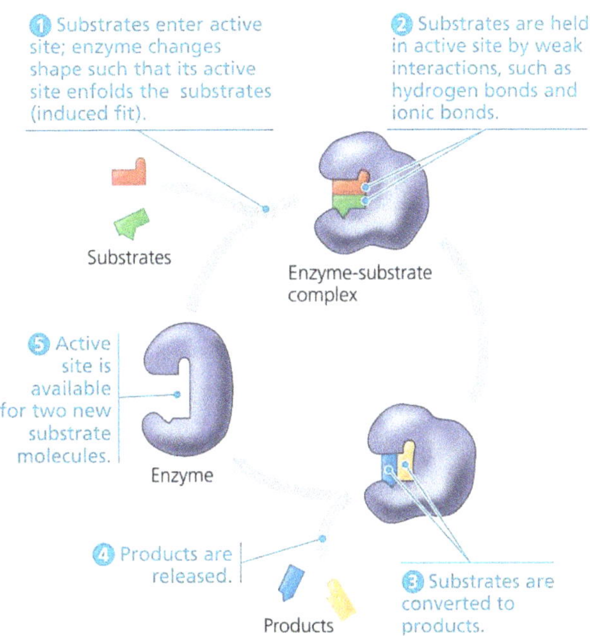

▲ **Figure 5.16** The active site and catalytic cycle of an enzyme. An enzyme can convert one or more reactant molecules to one or more product molecules. The enzyme shown here converts two substrate molecules to two product molecules.

Effect of Temperature on Enzyme Activity

The 3-D structures of proteins are sensitive to their environment. As a consequence, each enzyme works better under some conditions than others, because these **optimal conditions** favor the most active shape for the enzyme.

Temperature and pH are environmental factors important in the activity of an enzyme. Up to a point, the rate of an enzymatic reaction increases with increasing temperature, partly because substrates collide with active sites more frequently when the molecules move rapidly. Above that temperature, however, the speed of the enzymatic reaction drops sharply. The thermal agitation of the enzyme molecule disrupts the hydrogen bonds, ionic bonds, and other weak interactions that stabilize the active shape of the enzyme, and the protein molecule eventually denatures.

Each enzyme has an **optimal temperature** at which its reaction rate is greatest. Without denaturing the enzyme, this temperature allows the greatest number of molecular collisions and the fastest conversion of the reactants to product molecules. Most human enzymes have optimal temperatures of about 35-40°C (close to human body temperature). The thermophilic bacteria that live in hot springs contain enzymes with optimal temperatures of 70°C or higher.

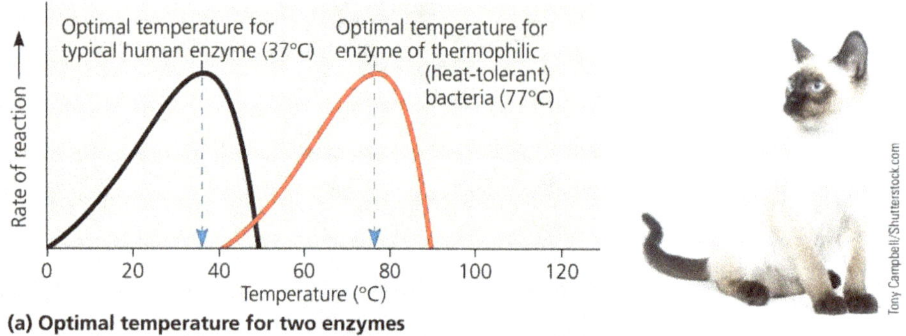

(a) Optimal temperature for two enzymes

▲ **Figure 5.17** Visible effects of environmental temperature on enzyme activity in Siamese cats. The fur on the extremities— ears, nose, paws, and tail—contains more dark brown pigment (melanin) than the rest of the body. A heat-sensitive enzyme controlling melanin production is denatured in warmer body regions, so dark pigment is not produced and fur color is lighter. © Cengage Learning 2017

Effect of pH on Enzyme Activity

Just as each enzyme has an optimal temperature, it also has a pH at which it is most active. The **optimal pH** values for most enzymes fall in the range of pH 6–8, but there are exceptions. For example, pepsin, a digestive enzyme in the human stomach, works best at a very low pH. Such an acidic environment denatures most enzymes, but pepsin is well-adapted evolutionarily to maintain its functional 3-D structure in the acidic environment of the stomach. In contrast, trypsin, a digestive enzyme residing in the more alkaline environment of the human intestine would be denatured in the stomach.

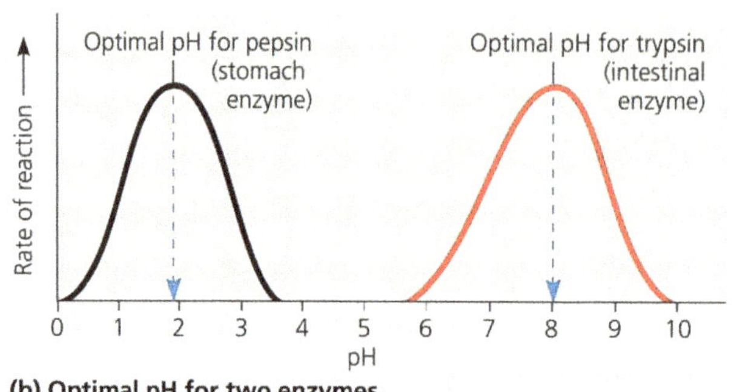

(b) Optimal pH for two enzymes

Effect of substrate concentration or enzyme concentration on Enzyme Activity

If the pH and temperature are kept constant, the rate of the reaction can be affected by the substrate concentration or by the enzyme concentration. If an excess of substrate is present, the enzyme concentration is the rate-limiting factor. The initial rate of the reaction is then directly proportional to the enzyme concentration.

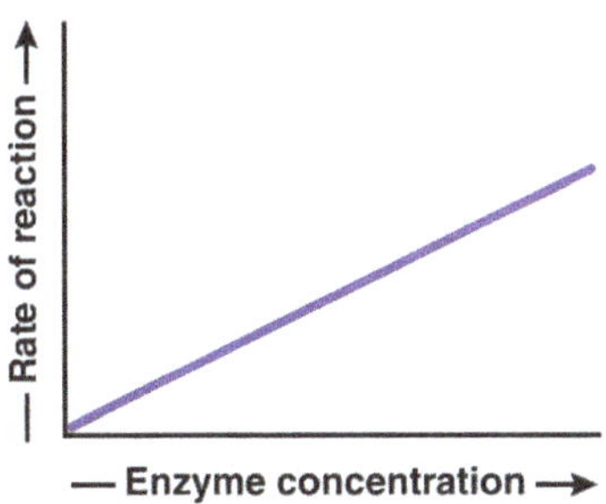

(a) In this example the rate of reaction is measured at different enzyme concentrations, with an excess of substrate present. (Temperature and pH are constant.) The rate of the reaction is directly proportional to the enzyme concentration.

If the enzyme concentration is kept constant, the rate of an enzymatic reaction is proportional to the concentration of substrate present. Substrate concentration is the rate limiting factor at lower concentrations; the rate of the reaction is therefore directly proportional to the substrate concentration. However, at higher substrate concentrations, the enzyme molecules become saturated with substrate; that is, substrate molecules are bound to all available active sites of enzyme molecules. In this situation, increasing the substrate concentration does not increase the net reaction rate.

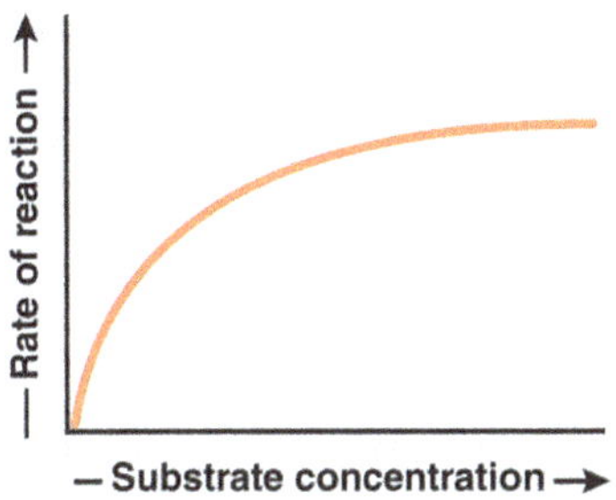

(b) In this example the rate of the reaction is measured at different substrate concentrations, and enzyme concentration, temperature, and pH are constant. If the substrate concentration is relatively low, the reaction rate is directly proportional to substrate concentration. However, higher substrate concentrations do not increase the reaction rate because the enzymes become saturated with substrate.

Regulation of enzyme activity helps control metabolism:

Chemical chaos would result if all of a cell's metabolic pathways were operating simultaneously. Intrinsic to life's processes is a cell's ability to tightly regulate its metabolic pathways by controlling when and where its various enzymes are active. It does this either by switching on and off the genes that encode specific enzymes or by regulating the activity of enzymes once they are made.

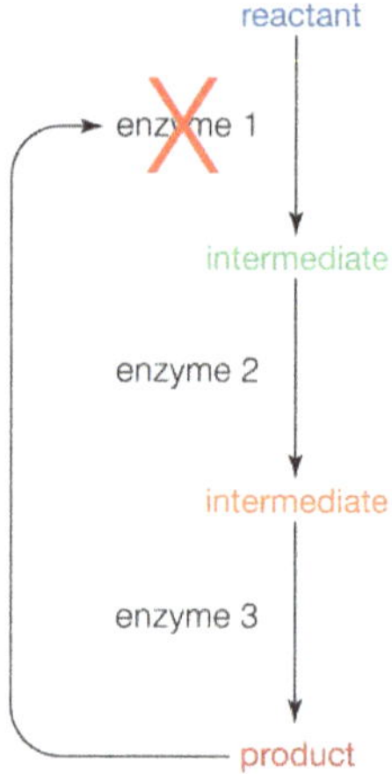

1. Allosteric Regulation of Enzymes

In many cases, the molecules that naturally regulate enzyme activity in a cell behave something like reversible noncompetitive inhibitors. Most noncompetitive inhibitors bind to a specific portion of the enzyme called an **allosteric site**. These sites serve as chemical on/off switches; the binding of a substance to the site can switch the enzyme between its active and inactive configurations. A substance that binds to an allosteric site and reduces enzyme activity is called an **allosteric inhibitor.** An **allosteric activator** binds to allosteric sites to keep an enzyme in its active configuration, thereby increasing enzyme activity.

These regulatory molecules change an enzyme's shape and the functioning of its active site by binding to a site elsewhere on the molecule, via noncovalent interactions. **Allosteric regulation** is the term used to describe any case in which a protein's function at one site is affected by the binding of a regulatory molecule to a separate site. It may result in either inhibition or stimulation of an enzyme's activity. Most enzymes known to be allosterically regulated are constructed from two or more subunits, each composed of a polypeptide chain with its own active site. The entire complex oscillates between two different shapes, one catalytically active and the other inactive.

→ In the simplest kind of allosteric regulation, an activating or inhibiting regulatory molecule binds to a regulatory site (sometimes called an allosteric site), often located where subunits join. The binding of an activator to a regulatory site stabilizes the shape that has functional active sites, whereas the binding of an inhibitor stabilizes the inactive form of the enzyme. The subunits of an allosteric enzyme fit together in such a way that a shape change in one subunit is transmitted to all others. Through this interaction of subunits, a single activator or inhibitor molecule that binds to one regulatory site will affect the active sites of all subunits.

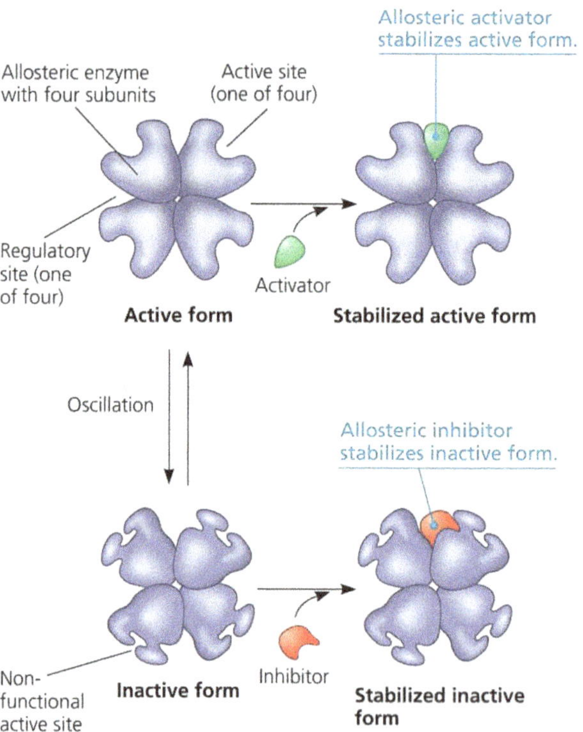

At low concentrations, activators and inhibitors dissociate from the enzyme. The enzyme can then oscillate again.

→ In another kind of allosteric activation, a substrate molecule binding to one active site in a multisubunit enzyme triggers a shape change in all the subunits, thereby increasing catalytic activity at the other active sites. Called **cooperativity**, this mechanism amplifies the response of enzymes to substrates: One substrate molecule primes an enzyme to act on additional substrate molecules more readily. Cooperativity is considered allosteric regulation because even though substrate is binding to an active site, its binding affects catalysis in another active site.

The inactive form shown on the left oscillates with the active form when the active form is not stabilized by substrate.

2. Feedback regulation:

- **Negative feedback:** A metabolic pathway is halted by the inhibitory binding of its end product to an enzyme that acts early in the pathway.

- **Positive feedback**: in which a product activates the pathway leading to its own production. Blood clotting, for example, involves a biochemical pathway with multiple steps. Damage to a blood vessel wall initiates these reactions. As a clot begins to form, the reactions accelerate, which further stimulates clotting, speeding up the reactions even more, and so on. Once the clot is large enough to stem the flow of blood, however, the pathway shuts down—an example of negative feedback. Positive feedback is much rarer than negative feedback in organisms.

▶ **Figure 5.18** Feedback inhibition in isoleucine synthesis.

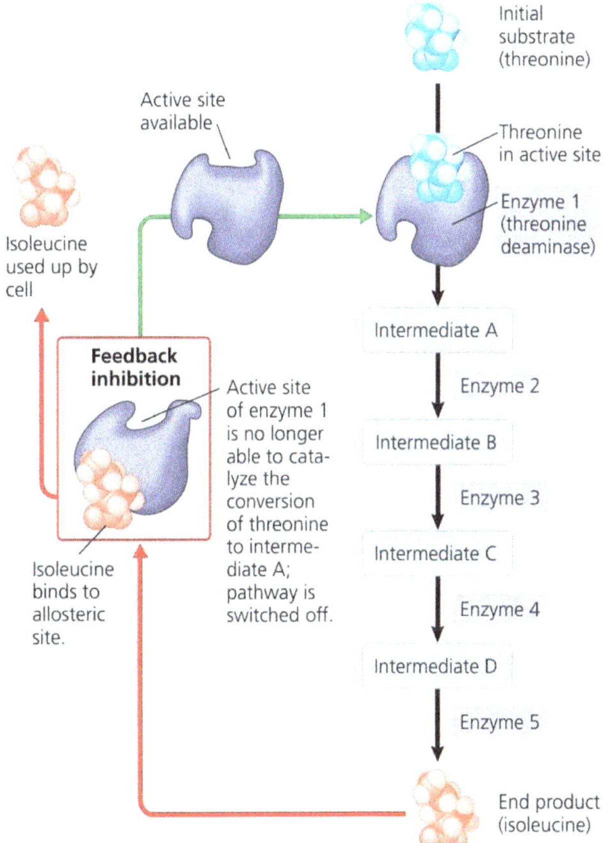

Enzyme Inhibitors:

Certain chemicals selectively inhibit the action of specific enzymes. Sometimes the inhibitor attaches to the enzyme by covalent bonds, in which case the inhibition is usually irreversible. Toxins and poisons are often irreversible enzyme inhibitors. Many enzyme inhibitors, however, bind to the enzyme by weak interactions, and when this occurs the inhibition is reversible. Some reversible inhibitors resemble the normal substrate molecule and compete for admission into the active site and called **competitive inhibitors**. This kind of inhibition can be overcome by increasing the concentration of substrate so that as active sites become available, more substrate molecules than inhibitor molecules are around to gain entry to the sites.

▶ **Figure 5.19**

How enzymes can be inhibited.

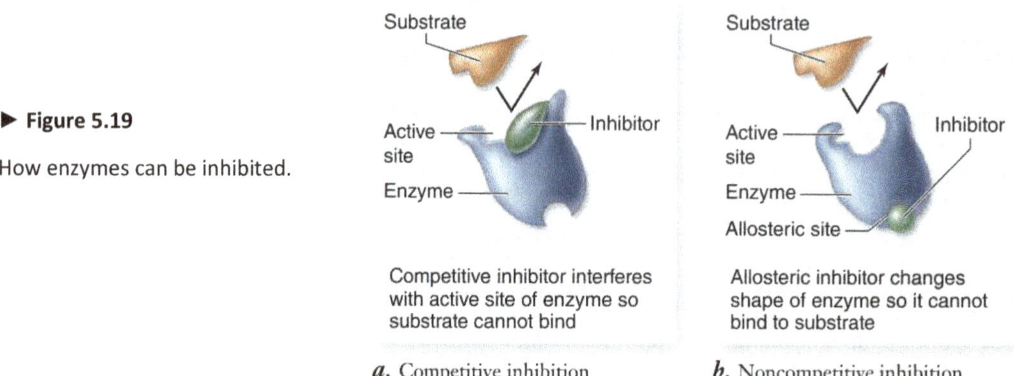

a. Competitive inhibition

b. Noncompetitive inhibition

In contrast, **noncompetitive inhibitors** do not directly compete with the substrate to bind to the enzyme at the active site. Instead, they impede enzymatic reactions by binding to another part of the enzyme. This interaction causes the enzyme molecule to change its shape in such a way that the active site becomes much less effective at catalyzing the conversion of substrate to product. Enzyme inhibitors have many practical applications. Aspirin relieves pain by binding to an enzyme that cells use to produce pain-related molecules called prostaglandins.

An irreversible inhibitor permanently inactivates or destroys an enzyme when the inhibitor combines with one of the enzyme's functional groups, either at the active site or elsewhere. Many poisons are irreversible enzyme inhibitors. For example, heavy metals such as mercury and lead bind irreversibly to and denature many proteins, including enzymes. Physicians treat many bacterial infections with drugs that directly or indirectly inhibit bacterial enzyme activity. For example, sulfa drugs have a chemical structure similar to that of the nutrient para-aminobenzoic acid (PABA).

When PABA is available, microorganisms can synthesize the vitamin folic acid, which is necessary for growth. Humans do not synthesize folic acid from PABA. For this reason, sulfa drugs selectively affect bacteria. When a sulfa drug is present, the drug competes with PABA for the active site of the bacterial enzyme. When bacteria use the sulfa drug instead of PABA, they synthesize a compound that cannot be used to make folic acid. Therefore, the bacterial cells are unable to grow.

Penicillin and related antibiotics irreversibly inhibit a bacterial enzyme called transpeptidase. This enzyme establishes some of the chemical linkages in the bacterial cell wall. Bacteria susceptible to these antibiotics cannot produce properly constructed cell walls and are prevented from multiplying effectively. Human cells do not have cell walls and therefore do not use this enzyme. Thus, except for individuals allergic to it, penicillin is harmless to humans. Unfortunately, during the years since it was introduced, resistance to penicillin has evolved in many bacterial strains. The resistant bacteria fight back with an enzyme of their own, penicillinase, which breaks down the penicillin and renders it ineffective. Because bacteria evolve at such a rapid rate, drug resistance is a growing problem in medical practice.

Comparing enzyme affinities

Affinity is a measure of the strength of attraction between two things. A high affinity means there is a strong attraction. When applied to enzymes, affinity is a measure of the strength of attraction between the enzyme and its substrate. The greater the affinity of an enzyme for its substrate, the faster it works. Another way of thinking about this is to say that the higher the affinity, the more likely it is that the product will be formed when a substrate molecule enters the active site. If the affinity is low, the substrate may leave the active site before a reaction takes place. At V_{max} all the enzyme molecules are bound to substrate molecules – the enzyme is saturated with substrate. All the active sites are full. As substrate concentration is increased, reaction rate rises until the reaction reaches its maximum rate, V_{max}.

$\frac{1}{2}V_{max}$ is exactly half the maximum velocity, This is the substrate concentration at which half the enzyme's active sites are occupied by substrate. The substrate concentration which causes $\frac{1}{2}V_{max}$ is labelled Km that is known as the Michaelis–Menten constant. The Michaelis–Menten constant of an enzyme is the substrate concentration at which the enzyme works at half its maximum rate. It is used as a measure of the affinity of the enzyme for its substrate, the lower the value of Km, the more efficient the enzyme so the higher the affinity of an enzyme for its substrate, the lower its Km will be.

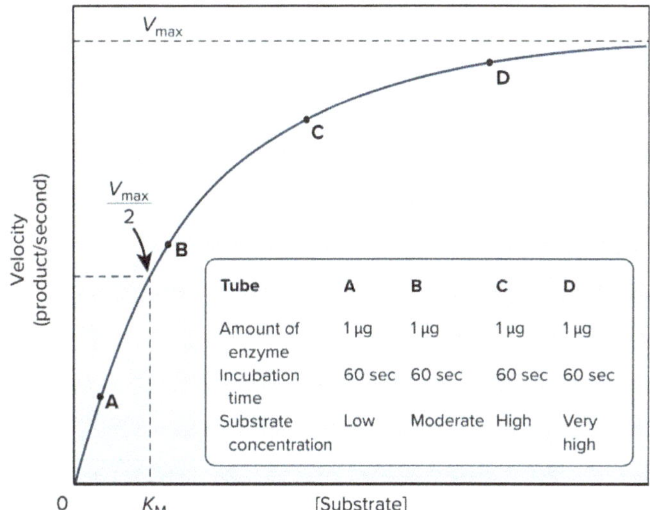

Tube	A	B	C	D
Amount of enzyme	1 μg	1 μg	1 μg	1 μg
Incubation time	60 sec	60 sec	60 sec	60 sec
Substrate concentration	Low	Moderate	High	Very high

(a) Reaction velocity in the absence of inhibitors

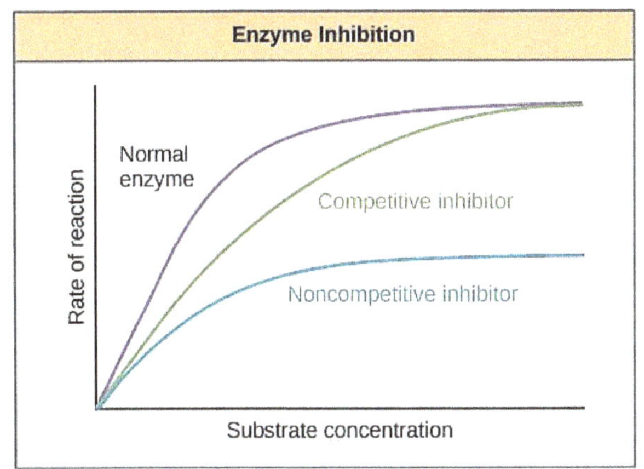

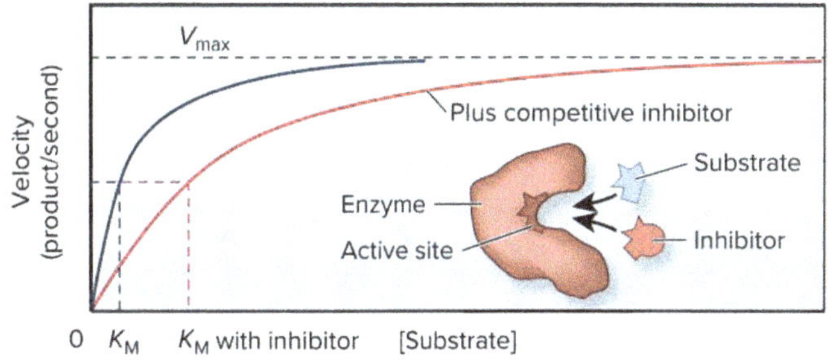

(b) Competitive inhibition

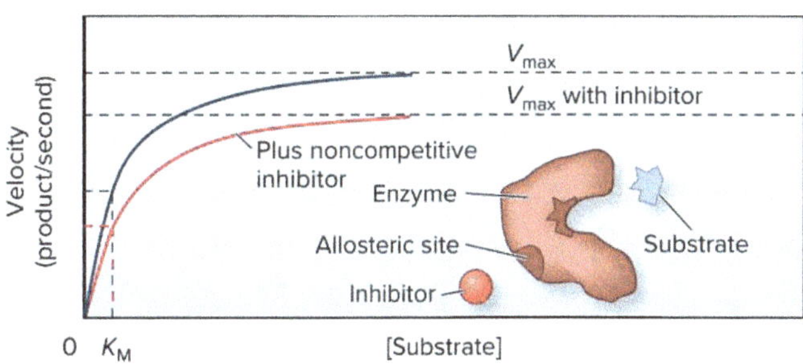

(c) Noncompetitive inhibition

▲ **Figure 5.20** The relationship between velocity and substrate concentration in an enzyme-catalyzed reaction, and the effects of inhibitors. (a) In the absence of an inhibitor, the maximal velocity (V_{max}) of an enzyme catalyzed reaction is achieved when the substrate concentration is high enough to be saturating. The K_M value for an enzyme is the substrate concentration at which the velocity of the reaction is half the maximal velocity. (b) A competitive inhibitor binds to the active site of an enzyme and raises the K_M. (c) A noncompetitive inhibitor binds to an allosteric site outside the active site and lowers the V_{max}

Enzymes, which are proteins, sometimes require the following nonprotein molecules or ions to carry out their functions:

- **Prosthetic groups** are small molecules that are permanently attached to the surface of an enzyme and aid in enzyme function.

- **Cofactors** are usually inorganic ions, such as Fe^{3+} or Zn^{2+}, that temporarily bind to the surface of an enzyme and promote a chemical reaction. Cofactors are often oxidized or reduced during the reaction, but, like enzymes, they are not consumed. Instead, they return to their original state when the reaction is complete. Neither the **apoenzyme** nor the cofactor alone has catalytic activity; only when the two are combined (**holoenzyme**) does the enzyme function.

- **Coenzymes** are organic molecules that temporarily bind to an enzyme and participate in the chemical reaction that the enzyme catalyzes but are left unchanged when the reaction is completed. Some of these coenzymes are synthesized by cells, but many of them are taken in as dietary vitamins by animal cells. The cell uses many water soluble vitamins, including B1, B2, B6, B12, niacin, and folic acid, to produce coenzymes; vitamin C is a coenzyme itself. Diets lacking in vitamins can lead to reduced enzyme function and, eventually, serious illness or even death. Modified nucleotides are also used as coenzymes such as NADH, NADPH, and FADH2; they transfer electrons.

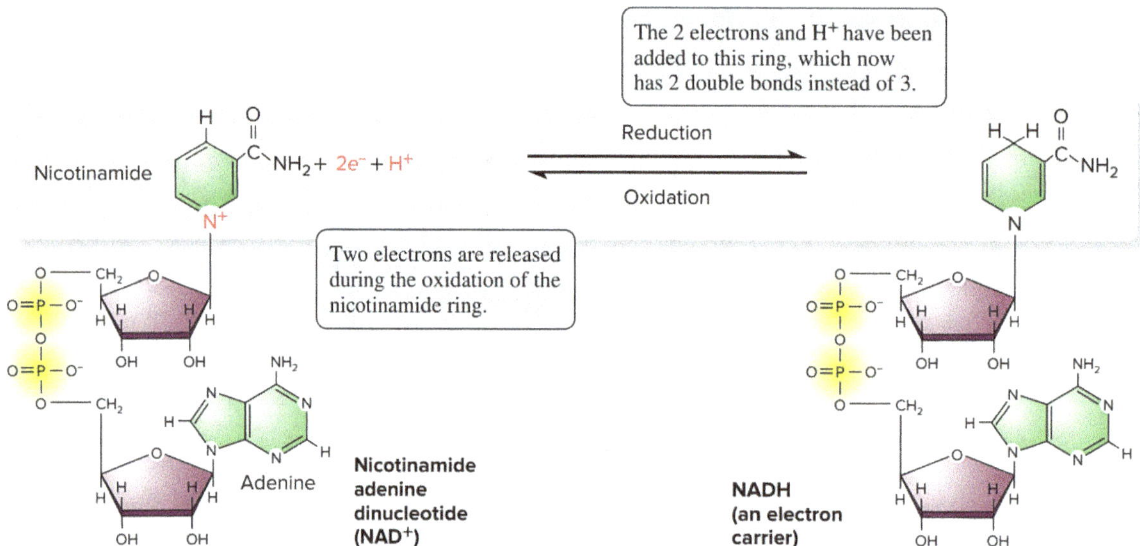

▲ **Figure 5.21** The reduction of NAD+ to produce NADH. NAD+ is composed of two nucleotides: one with an adenine base and one with a nicotinamide base. The oxidation of organic molecules releases electrons. These electrons and an H+ bind to NAD+, which results in the formation of NADH. The two electrons and H+ are incorporated into the nicotinamide ring. Note: The actual net charges of NAD+ and NADH are -1 and -2, respectively. They are designated NAD+ and NADH to emphasize the net charge of the nicotinamide ring, which is involved in redox reactions.

RNA-Based Biological Catalysts: Ribozymes

RNA-based catalysts, called ribozymes, are part of the biochemical machinery of all cells. Most of the known ribozymes speed the cutting and splicing reactions that remove surplus segments from RNA molecules as part of their conversion into finished form. Some have other functions, however. For example, ribosomes, the cell structures that assemble amino acids into proteins, can still link amino acids together even if their proteins are removed. After the proteins are extracted, only RNA molecules are left in the ribosomes, indicating that a ribozyme catalyzes this central reaction of protein synthesis.

Ribozymes provide a possible solution to a long-standing "chicken-or-egg" paradox about the evolution of life: Did proteins or nucleic acids come first in evolution? It is difficult to understand how DNA could exist without the enzymatic proteins required for its duplication. At the same time, it is difficult to understand how enzymes could exist without nucleic acids, which contain the information required to make them. Ribozymes offer a way around this dilemma because they could have acted as both enzymes and informational molecules when cellular life first appeared. The earliest forms of life therefore might have inhabited an "RNA world" in which neither DNA nor proteins played critical roles.

Metabolic Pathways Are Regulated in Three General Ways

The regulation of metabolic pathways is important for a variety of reasons. Catabolic pathways are regulated so that organic molecules are broken down only when they are no longer needed or when the cell requires energy. During anabolic reactions, regulation ensures that a cell synthesizes molecules only when they are needed. The regulation of catabolic and anabolic pathways occurs at the genetic, cellular, and biochemical levels.

Gene Regulation: Enzymes are proteins that are coded by genes. One way that cells control metabolic pathways is through gene regulation. For example, if a bacterial cell is not exposed to a particular sugar in its environment, it will turn off the genes that code the enzymes that are needed to break down that sugar. Then, if the sugar becomes available, the genes are switched back on.

Regulation via Cell-Signaling Pathways: Metabolic reactions are also coordinated at the cellular level. Cells integrate signals from their environment and adjust their metabolic pathways to adapt to those signals. Cell-signaling pathways often lead to the activation of protein kinases—enzymes that covalently attach a phosphate group to a target protein. For example, when people are frightened, they secrete a hormone called epinephrine into the bloodstream. This hormone binds to the surface of muscle cells and stimulates an intracellular pathway that leads to the phosphorylation of specific enzymes involved in carbohydrate metabolism. These activated enzymes promote the breakdown of carbohydrates, an event that supplies the frightened individual with more energy. Epinephrine is sometimes called the fight or flight hormone because the added energy prepares an individual to either stay and fight or run away quickly.

Biochemical Regulation: Many metabolic pathways use feedback inhibition (an example of noncompetitive inhibition) as a form of biochemical regulation. In such cases, the enzyme being inhibited has two binding sites. One site is the active site, where the reactants are converted to products. In addition, enzymes controlled by feedback inhibition also have an allosteric site, where a molecule can bind noncovalently and affect the enzyme's function.

The binding of a molecule to an allosteric site causes a conformational change in the enzyme that inhibits its catalytic function. For example, the conformational change may alter the structure of the active site so that reactants cannot bind. Allosteric sites are often found in the enzymes that catalyze the early steps in a metabolic pathway. Such allosteric sites typically bind molecules that are the products of the metabolic pathway. When the products bind to these sites, they inhibit the function of these enzymes, thereby preventing the formation of too much product.

Cell-signaling and biochemical regulation are important, rapid ways to control chemical reactions in a cell. For a metabolic pathway composed of several enzyme-catalyzed reactions, which enzyme should be controlled? In many cases, a metabolic pathway has a rate-limiting ced, such changes will have the greatest influence on the formation of the final product of the metabolic pathway. Rather than affecting all of the enzymes in a metabolic pathway, cell-signaling and biochemical regulation are often directed at the enzyme that catalyzes the rate-limiting step. This is an efficient, rapid way to control the amount of product of a pathway.

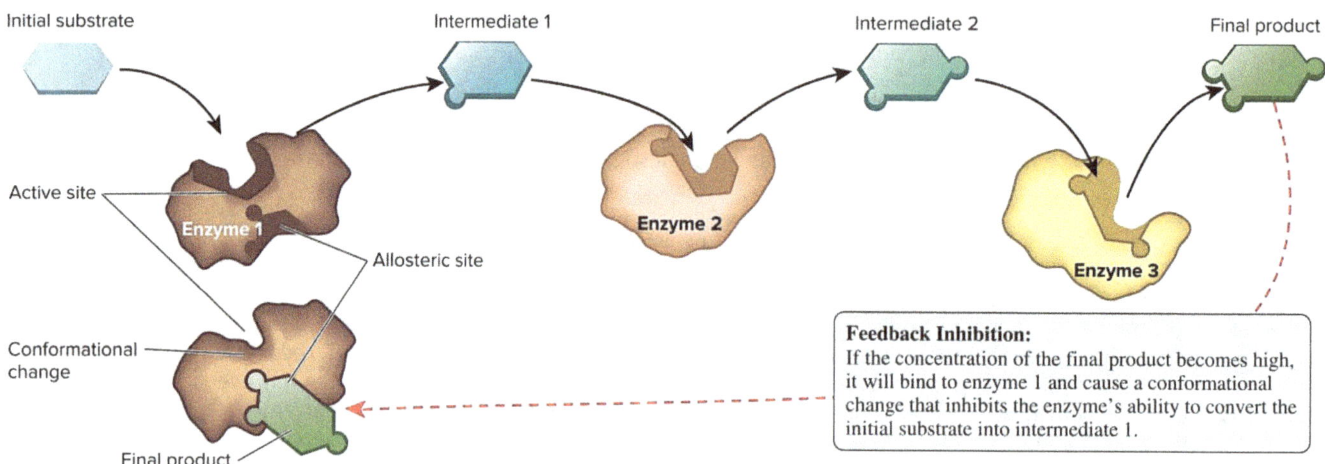

▲ **Figure 5.22** Feedback inhibition.

- Thermodynamics is the study of energy changes.

- Energy is the capacity to do work. Potential energy is stored energy, and kinetic energy is the energy of motion. Energy can take many forms: mechanical, heat, sound, electric current, light, or radioactive radiation. Energy is measured in units of heat known as kilocalories.

- Photosynthesis stores light energy from the Sun as potential energy in the covalent bonds of sugar molecules. Breaking these bonds in living cells releases energy for use in other reactions.

- Oxidation is a reaction involving the loss of electrons. Reduction is the gain of electrons. These two reactions take place together and are therefore termed redox reactions.

- Virtually all activities of living organisms require energy. Energy changes form as it moves through organisms and their biochemical systems, but it is not created or destroyed.

- The disorder, or entropy, of the universe is continuously increasing. In an open system like the Earth, which is receiving energy from the Sun, this may not be the case. To increase order, however, energy must be expended. In energy conversions, some energy is always lost as heat.

- Free energy (G) is the energy available to do work in any system. Changes in free energy (ΔG) predict the direction of reactions. Reactions with a negative ΔG are spontaneous (exergonic) reactions, and reactions with a positive ΔG are not spontaneous (endergonic). Endergonic chemical reactions absorb energy from the surroundings, whereas exergonic reactions release energy to the surroundings.

- Activation energy is the energy required to destabilize chemical bonds and initiate chemical reactions. Even exergonic reactions require this activation energy. Catalysts speed up chemical reactions by lowering the activation energy.

- Adenosine triphosphate (ATP) is the molecular currency used for cellular energy transactions.

- The energy of ATP is stored in the bonds between its terminal phosphate groups. These groups repel each other due to their negative charge and therefore the covalent bonds joining these phosphates are unstable.

- Enzymes hydrolyze the terminal phosphate group of ATP to release energy for reactions. If ATP hydrolysis is coupled to an endergonic reaction with a positive ΔG with magnitude less than that for ATP hydrolysis, the two reactions together will be exergonic.

- ATP hydrolysis releases energy to drive endergonic reactions, and it is synthesized with energy from exergonic reactions.

- Enzymes lower the activation energy needed to initiate a chemical reaction.

- Substrates bind to the active site of an enzyme. Enzymes adjust their shape to the substrate so there is a better fit.

- Enzymes can be free in the cytosol or exist as components bound to membranes and organelles. Enzymes involved in a biochemical pathway can form multienzyme complexes. Although most enzymes are proteins, some are actually RNA molecules, called ribozymes.

- An enzyme's functionality depends on its ability to maintain its three dimensional shape, which can be affected by temperature and pH. The activity of enzymes can be affected by inhibitors. Competitive inhibitors compete for the enzyme's active site, which leads to decreased enzyme activity. Enzyme activity can be controlled by effectors. Allosteric enzymes have a second site, located away from the active site, that binds effectors to activate or inhibit the enzyme. Noncompetitive inhibitors and activators bind to the allosteric site, changing the structure of the enzyme to inhibit or activate it. Cofactors are nonorganic metals necessary for enzyme function. Coenzymes are nonprotein organic molecules, such as certain vitamins, needed for enzyme function. Often coenzymes serve as electron acceptors.

- Metabolism is the sum of all biochemical reactions in a cell. Anabolic reactions require energy to build up molecules, and catabolic reactions break down molecules and release energy.

- Chemical reactions in biochemical pathways use the product of one reaction as the substrate for the next. In the primordial "soup" of the early oceans, many reactions were probably single-step reactions combining two molecules. As one of the substrate molecules was depleted, organisms having an enzyme that could synthesize the substrate would have a selective advantage. In this manner, biochemical pathways are thought to have evolved "backward" with new reactions producing limiting substrates for existing reactions.

- Biosynthetic pathways are often regulated by the end product of the pathway. Feedback inhibition occurs when the end-product of a reaction combines with an enzyme's allosteric site to shut down the enzyme's activity.

CHAPTER 6

Cellular Respiration and Fermentation

Chapter Contents:

- Catabolic pathways yield energy by oxidizing organic fuel
- Catabolic Pathways and Production of ATP
- Oxidation of Organic Fuel Molecules During Cellular Respiration
- The Stages of Cellular Respiration
- Glycolysis harvests chemical energy by oxidizing glucose to pyruvate
- Oxidation of Pyruvate to Acetyl CoA
- The Citric Acid Cycle
- The Pathway of Electron Transport
- Chemiosmosis: The Energy-Coupling Mechanism
- An Accounting of ATP Production by Cellular Respiration
- Fermentation and anaerobic respiration
- Types of Fermentation
- Comparing Fermentation with Anaerobic and Aerobic Respiration
- Glycolysis and the citric acid cycle connect to many other metabolic pathways
- Biosynthesis (Anabolic Pathways)
- Glycolysis and Citric Acid Cycle Stages of Cellular Respiration Are Regulated by Feedback Mechanisms
- Photosynthesis and Respiration Are Ancient Pathways
- KEY CONCEPTS

Catabolic pathways yield energy by oxidizing organic fuels

Plants, algae, and some bacteria harvest the energy of sunlight through **photosynthesis**, converting radiant energy into chemical energy. These organisms, along with a few others that use chemical energy in a similar way, are called **autotrophs**. All other organisms live on the organic compounds autotrophs produce, using them as food, and are called **heterotrophs**. At least 95% of the kinds of organisms on Earth—all animals and fungi, and most protists and prokaryotes—are heterotrophs. Autotrophs also extract energy from organic compounds—they just have the additional capacity to use the energy from sunlight to synthesize these compounds. The process by which energy is harvested is **cellular respiration**—the oxidation of organic compounds to extract energy from chemical bonds. Cellular respiration is the complete oxidation of glucose.

Living cells require transfusions of energy from outside sources to perform their many tasks—for example, assembling polymers, pumping substances across membranes, moving, and reproducing. The job of extracting energy from the complex organic mixture in most foods is tackled in stages. First, enzymes break down the large molecules into smaller ones, a process called digestion, then, other enzymes dismantle these fragments a bit at a time, harvesting energy from C—H and other chemical bonds at each stage.

Respiration breaks food down, using oxygen (O_2) and generating ATP. The waste products of this type of respiration, carbon dioxide (CO_2) and water (H_2O), are the raw materials for photosynthesis. Metabolic pathways that release stored energy by breaking down complex molecules are called **catabolic pathways**.

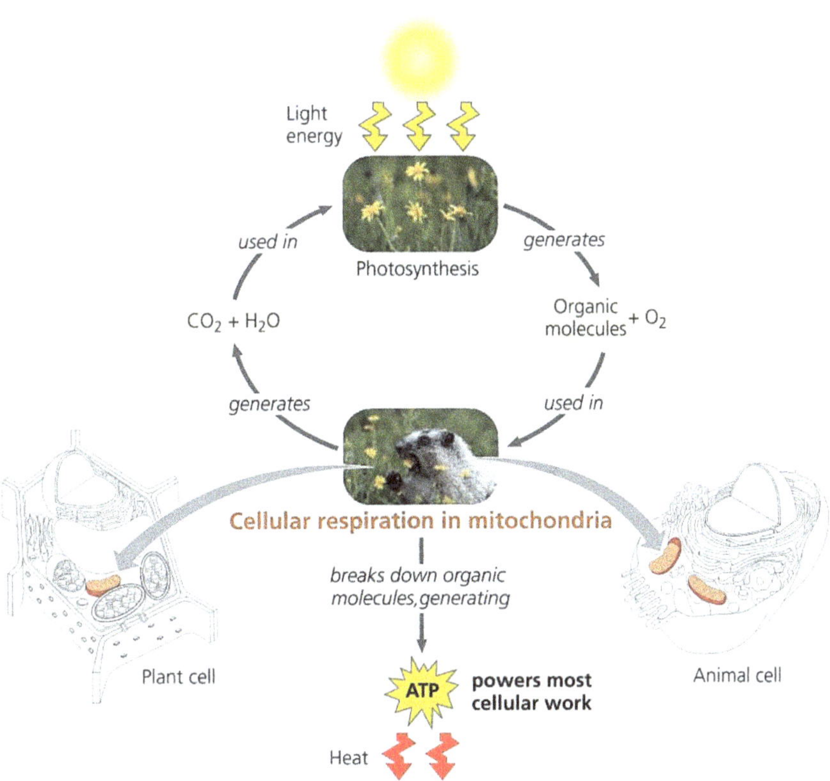

Catabolic Pathways and Production of ATP

Organic compounds possess potential energy as a result of the arrangement of electrons in the bonds between their atoms. Compounds that can participate in exergonic reactions can act as fuels. Through the activity of enzymes, a cell systematically degrades complex organic molecules that are rich in potential energy to simpler waste products that have less energy. Some of the energy taken out of chemical storage can be used to do work; the rest is dissipated as heat.

One catabolic process, **fermentation**, is a partial degradation of sugars or other organic fuel that occurs without the use of oxygen. However, the most efficient catabolic pathway is **aerobic respiration**, in which oxygen is consumed as a reactant along with the organic fuel. The cells of most eukaryotic and many prokaryotic organisms can carry out aerobic respiration. In humans and many other animals, O_2 from inhaled air diffuses into the bloodstream across the walls of microscopic air sacs in the lungs. The circulatory system carries the inhaled O_2 to cells, where gas exchange occurs. O_2 diffuses into the cell's mitochondria, the sites of respiration. Meanwhile, CO_2 diffuses out of the cells and into the bloodstream. After moving from the blood into the lungs, the CO_2 is exhaled.

Plants use O_2 to respire about half of the glucose they produce. Why do plants have a reputation for producing O_2, if they also consume it? The reason is that plants incorporate much of the remaining glucose into cellulose, starch, and other stored organic molecules. Therefore, they absorb much more CO_2 in photosynthesis than they release in respiration, and they release more O_2 than they consume. Some prokaryotes use substances other than oxygen as reactants in a similar process that harvests chemical energy without oxygen; this process is called **anaerobic respiration**. Technically, the term cellular respiration includes both aerobic and anaerobic processes. Fermentation is not respiration because it generates ATP from glycolysis only.

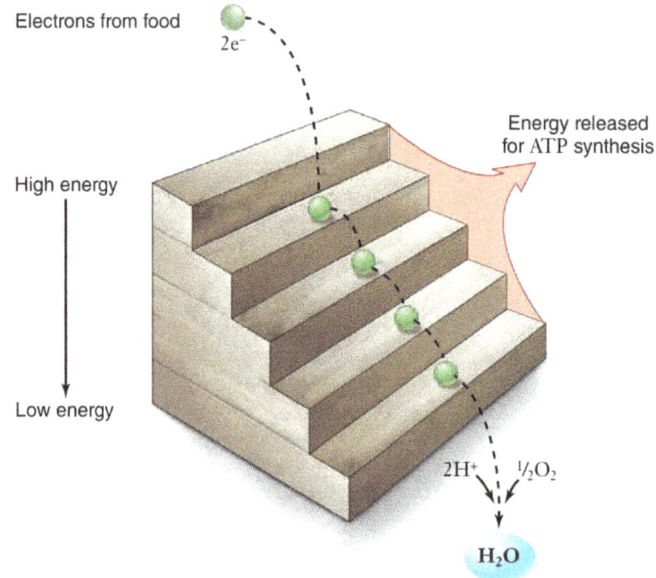

▶ **Figure 6.1** How electron transport works. This diagram shows how ATP is generated when electrons transfer from one energy level to another. Rather than releasing a single explosive burst of energy, electrons "fall" to lower and lower energy levels in steps, releasing stored energy with each fall as they tumble to the lowest (most electronegative) electron acceptor, O_2.

Aerobic respiration is in principle similar to the combustion of gasoline in an automobile engine after oxygen is mixed with the fuel (hydrocarbons). Food provides the fuel for respiration, and the exhaust is carbon dioxide and water. The overall process can be summarized as follows, glucose becomes oxidized and oxygen becomes reduced:

$$\text{Organic compounds} + \text{Oxygen} \rightarrow \text{Carbon dioxide} + \text{Water} + \text{Energy}$$

Carbohydrates, fats, and proteins from food can all be processed and consumed as fuel. In animal diets, a major source of carbohydrates is starch, a storage polysaccharide that can be broken down into glucose ($C_6H_{12}O_6$) subunits.

$$C_6H_{12}O_6 + 6\,O_2 \rightarrow 6\,CO_2 + 6\,H_2O + \text{Energy (ATP + heat)}$$

This breakdown of glucose is exergonic, having a free energy change of -686 kcal (-2,870 kJ) per mole of glucose decomposed (ΔG= -686 kcal/mol). A negative ΔG indicates that the products of the chemical process store less energy than the reactants and that the reaction can happen spontaneously—in other words, without an input of energy.

Catabolic pathways do not directly move flagella, pump solutes across membranes, polymerize monomers, or perform other cellular work. Catabolism is linked to work by a chemical drive shaft, **ATP**. To keep working, the cell must regenerate its supply of ATP from ADP and Pi.

Based on the transfer of electrons during the chemical reactions, catabolic pathways that decompose glucose and other organic fuels yield energy. The relocation of electrons releases energy stored in organic molecules, and this energy ultimately is used to synthesize ATP. This reaction does not happen all at once. If a cell released all the potential energy in glucose's chemical bonds in one uncontrolled step, the sudden release of heat would destroy the cell; in effect, it would act like a tiny bomb. Rather, the chemical bonds and atoms in glucose are rearranged one step at a time, releasing a tiny bit of energy with each transformation. According to the second law of thermodynamics, some of this energy is lost as heat. But much of it is stored in the chemical bonds of ATP.

In many chemical reactions, there is a transfer of one or more electrons (e^-) from one reactant to another. These electron transfers are called oxidation-reduction reactions, or redox reactions for short. In a redox reaction, the loss of electrons from one substance is called **oxidation**, and the addition of electrons to another substance is known as **reduction**, the electron donor is called the reducing agent and the electron acceptor, is the oxidizing agent. Because an electron transfer requires both an electron donor and an acceptor, oxidation and reduction always go hand in hand.

We could generalize a redox reaction this way:

$$
\begin{array}{ccccccc}
\multicolumn{7}{c}{\overset{\longrightarrow \text{becomes oxidized} \longrightarrow}{}} \\
Xe^- & + & Y & \longrightarrow & X & + & Ye^- \\
\multicolumn{7}{c}{\underset{\longleftarrow \text{becomes reduced} \longleftarrow}{}}
\end{array}
$$

Oxidation of Organic Fuel Molecules During Cellular Respiration

In general, organic molecules that have an abundance of hydrogen are excellent fuels because their bonds are a source of "hilltop" electrons, whose energy may be released as these electrons "fall" down an energy gradient during their transfer to oxygen. The summary equation for respiration indicates that hydrogen is transferred from glucose to the O atoms in O_2. But the important point, not visible in the summary equation, is that the energy state of the electron changes as hydrogen (with its electron) is transferred to oxygen. In respiration, the oxidation of glucose transfers electrons to a lower energy state, liberating energy that becomes available for ATP synthesis. So, in general, we see fuels with multiple C–H bonds oxidized into products with multiple C–O bonds.

The main energy-yielding foods, carbohydrates and fats, are reservoirs of electrons associated with hydrogen, often in the form of C–H bonds. Only the barrier of activation energy holds back the flood of electrons to a lower energy state. Without this barrier, a food substance like glucose would combine almost instantaneously with O_2. If we supply the activation energy by igniting glucose, it burns in air, releasing 686 kcal (2,870 kJ) of heat per mole of glucose. Body temperature is not high enough to initiate burning, of course. Instead, if you swallow some glucose, enzymes in your cells will lower the barrier of activation energy, allowing the sugar to be oxidized in a series of steps.

If energy is released from a fuel all at once, it cannot be harnessed efficiently for constructive work. Cellular respiration does not oxidize glucose (or any other organic fuel) in a single explosive step either. Rather, glucose is broken down in a series of steps, each one catalyzed by an enzyme. At key steps, electrons are stripped from the glucose. As is often the case in oxidation reactions, each electron travels with a proton—thus, as a hydrogen atom.

The hydrogen atoms are not transferred directly to O_2, but instead are usually passed first to an electron carrier, a coenzyme called **nicotinamide adenine dinucleotide**, a derivative of the vitamin niacin. The two nucleotides that make up NAD^+, nicotinamide monophosphate (NMP) and adenosine monophosphate (AMP), are joined by their phosphate groups. The two nucleotides serve different functions in the NAD^+ molecule: AMP acts as the core, providing a shape recognized by many enzymes; NMP is the active part of the molecule, because it is readily reduced—that is, it easily accepts electrons.

This coenzyme is well suited as an electron carrier because it can cycle easily between its oxidized form, NAD^+, and its reduced form, NADH. As an electron acceptor, NAD^+ functions as an oxidizing agent during respiration.

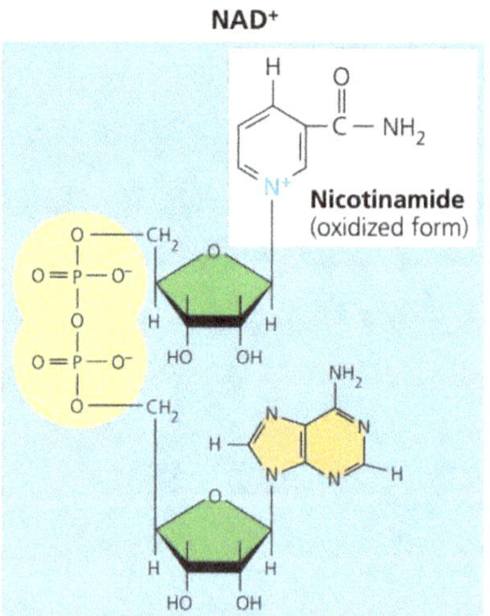

▶ **Figure 6.2** NAD^+ as an electron shuttle. The full name for NAD^+, nicotinamide adenine dinucleotide, describes its structure: The molecule consists of two nucleotides joined together at their phosphate groups (shown in yellow).

Nicotinamide is a nitrogenous base, although not one that is present in DNA or RNA. The enzymatic transfer of 2 electrons and 1 proton (H^+) from an organic molecule in food to NAD^+ reduces the NAD^+ to NADH; the second proton (H^+) is released. Most of the electrons removed from food are transferred initially to NAD^+.

Enzymes called **dehydrogenases** remove a pair of hydrogen atoms (2 electrons and 2 protons) from the substrate (glucose, in the preceding example), thereby oxidizing it. The enzyme delivers the 2 electrons along with 1 proton to its coenzyme, NAD⁺, forming NADH. The other proton is released as a hydrogen ion (H⁺) into the surrounding solution:

$$\text{H}-\overset{|}{\underset{|}{\text{C}}}-\text{OH} + \text{NAD}^+ \xrightarrow{\text{Dehydrogenase}} \overset{|}{\underset{|}{\text{C}}}=\text{O} + \text{NADH} + \text{H}^+$$

Each NADH molecule formed during respiration represents stored energy that can be tapped to make ATP when the electrons complete their "fall" down an energy gradient from NADH to O_2. An **electron transport chain** consists of a number of molecules, mostly proteins, built into the inner membrane of the mitochondria of eukaryotic cells (and the plasma membrane of respiring prokaryotes). Electrons removed from glucose are shuttled by NADH to the "top," higher energy end of the chain. At the "bottom," lower-energy end, O_2 captures these electrons along with hydrogen nuclei (H⁺), forming water. (Anaerobically respiring prokaryotes have an electron acceptor at the end of the chain that is different from O_2).

In summary, during cellular respiration, most electrons travel the following "downhill" route: glucose → NADH → electron transport chain → oxygen.

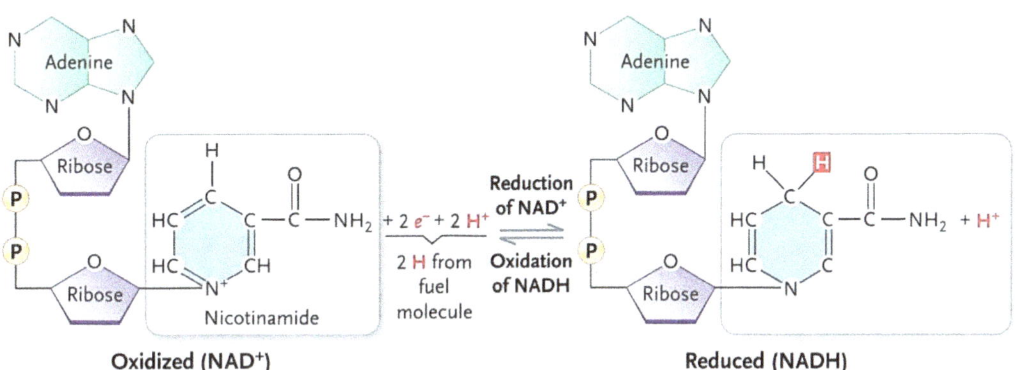

▲ **Figure 6.3** The emergent properties of a compound. The metal sodium combines with the poisonous gas chlorine, forming the edible compound sodium chloride, or table salt. Electron carrier NAD⁺. When a fuel molecule is oxidized, releasing two hydrogen atoms, NAD⁺, the oxidized form of the carrier, accepts a proton (H⁺) and two electrons and is transformed into NADH, the reduced form of the carrier. The nitrogenous base (blue) of NAD that adds and releases electrons and protons is nicotinamide, which is derived from the vitamin niacin (nicotinic acid). © Cengage Learning 2017

The Stages of Cellular Respiration

Most reactions involved in aerobic respiration are one of three types: dehydrogenations, decarboxylations, and those we informally categorize as preparation reactions. **Dehydrogenations** are reactions in which two hydrogen atoms (actually, 2 electrons plus 1 or 2 protons) are removed from the substrate and transferred to NAD^+ or FAD. **Decarboxylations** are reactions in which part of a carboxyl group (—COOH) is removed from the substrate as a molecule of CO_2.

The carbon dioxide you exhale with each breath is derived from decarboxylations that occur in your cells. The rest of the reactions are preparation reactions in which molecules undergo rearrangements and other changes so that they can undergo further dehydrogenations or decarboxylations. The harvesting of energy from glucose by cellular respiration is a cumulative function of three metabolic stages:

- **Glycolysis**

- **Pyruvate oxidation (transition step or preparatory reaction or link reaction) and the citric acid cycle**

- **Oxidative phosphorylation or chemiosmotic phosphorylation: electron transport and chemiosmosis**

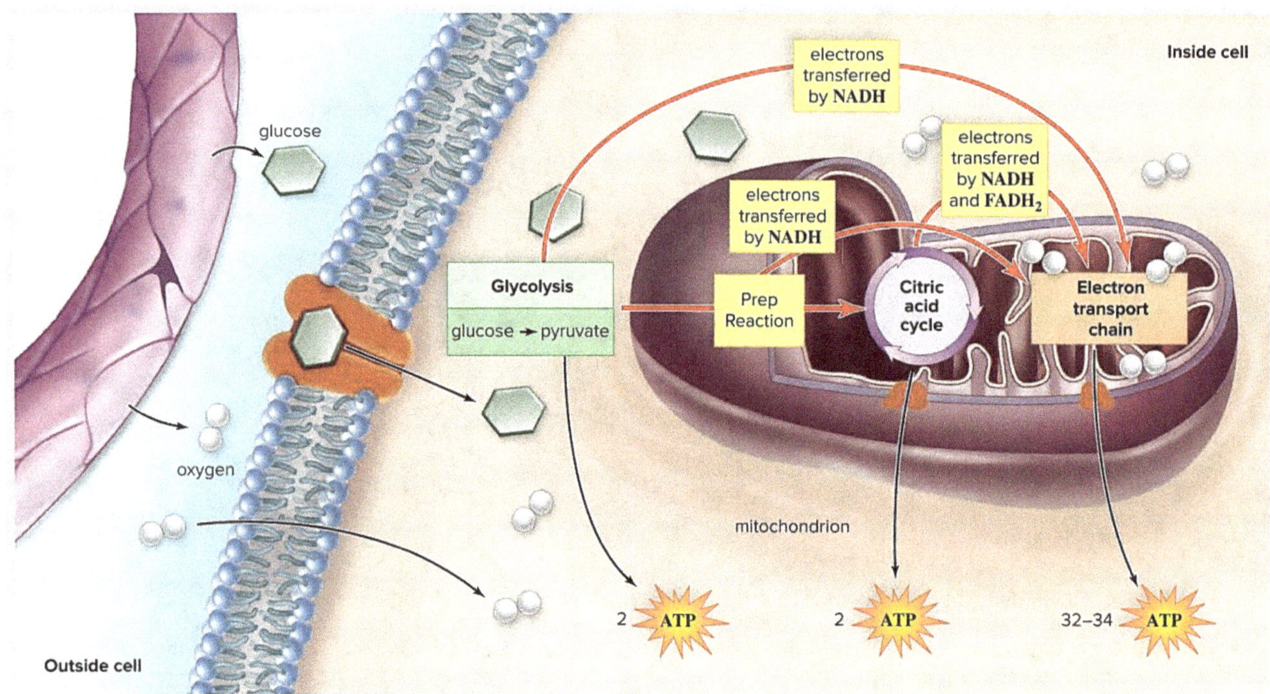

▲ **Figure 6.4** The four stages of aerobic respiration

Glycolysis and then pyruvate oxidation and the citric acid cycle are the catabolic pathways that break down glucose and other organic fuels. Glycolysis, which occurs in the cytosol, begins the degradation process by breaking glucose into two molecules of a compound called pyruvate. In eukaryotes, pyruvate enters the mitochondrion and is oxidized to a compound called acetyl CoA, which enters the citric acid cycle. There, the breakdown of glucose to carbon dioxide is completed (In prokaryotes, these processes take place in the cytosol). Thus, the carbon dioxide produced by respiration represents fragments of oxidized organic molecules.

Some of the steps of glycolysis and the citric acid cycle are redox reactions in which dehydrogenases transfer electrons from substrates to NAD^+ or the related electron carrier FAD, forming NADH or $FADH_2$. In the third stage of respiration, the electron transport chain accepts electrons from NADH or $FADH_2$ generated during the first two stages and passes these electrons down the chain. At the end of the chain, the electrons are combined with molecular oxygen (O_2) and hydrogen ions (H^+), forming water. The energy released at each step of the chain is stored in a form the mitochondrion (or prokaryotic cell) can use to make ATP from ADP. This mode of ATP synthesis is called oxidative phosphorylation because it is powered by the redox reactions of the electron transport chain.

▶ **Figure 6.5** An overview of cellular respiration. During glycolysis, each glucose molecule is broken down into two molecules of the compound pyruvate. In eukaryotic cells, as shown here, the pyruvate enters the mitochondrion. There it is oxidized to acetyl CoA, which is further oxidized to CO_2 in the citric acid cycle. NADH and a similar electron carrier, a coenzyme called $FADH_2$, transfer electrons derived from glucose to electron transport chains, which are built into the inner mitochondrial membrane. (In prokaryotes, the electron transport chains are located in the plasma membrane.) During oxidative phosphorylation, electron transport chains convert the chemical energy to a form used for ATP synthesis in the process called chemiosmosis.

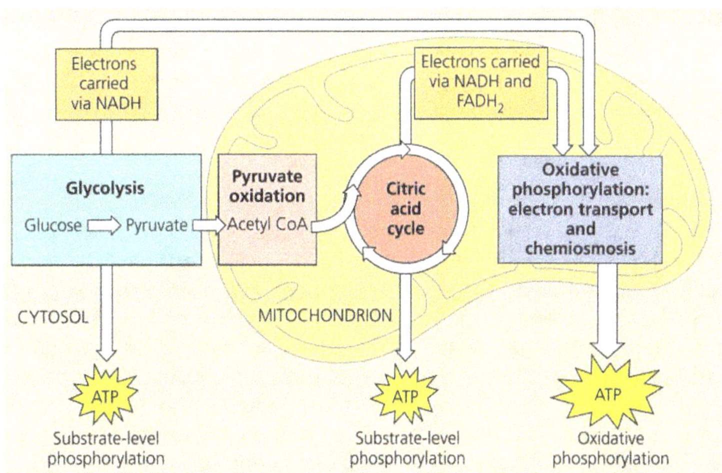

In eukaryotic cells, the inner membrane of the mitochondrion is the site of electron transport and another process called chemiosmosis, together making up **oxidative phosphorylation** (In prokaryotes, these processes take place in the plasma membrane). Oxidative phosphorylation accounts for almost 90% of the ATP generated by respiration. A smaller amount of ATP is formed directly in a few reactions of glycolysis and the citric acid cycle by a mechanism called **substrate-level phosphorylation**. This mode of ATP synthesis occurs when an enzyme transfers a phosphate group from a substrate molecule to ADP, rather than adding an inorganic phosphate to ADP as in oxidative phosphorylation. "Substrate molecule" here refers to an organic molecule generated as an intermediate during the catabolism of glucose.

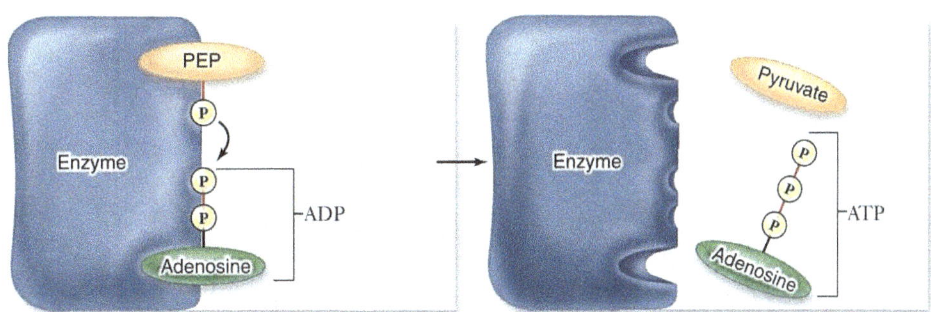

▲ **Figure 6.6** Substrate-level phosphorylation. Some molecules, such as phosphoenolpyruvate (PEP), possess a high-energy phosphate (P) bond similar to the bonds in ATP. When PEP's phosphate group is transferred enzymatically to ADP, the energy in the bond is conserved, and ATP is created.

Glycolysis harvests chemical energy by oxidizing glucose to pyruvate

The word glycolysis means "sugar splitting," and that is exactly what happens during this pathway. Glucose, a six carbon sugar, is split into two three-carbon sugars. These smaller sugars are then oxidized and their remaining atoms rearranged to form two molecules of pyruvate (the ionized form of pyruvic acid). Glycolysis can be divided into two phases: the energy investment phase and the energy payoff phase. During the energy investment phase, the cell actually spends ATP. This investment is repaid with interest during the energy payoff phase, when ATP is produced by substrate-level phosphorylation and NAD^+ is reduced to NADH by electrons released from the oxidation of glucose. The net energy yield from glycolysis, per glucose molecule, is 2 ATP plus 2 NADH.

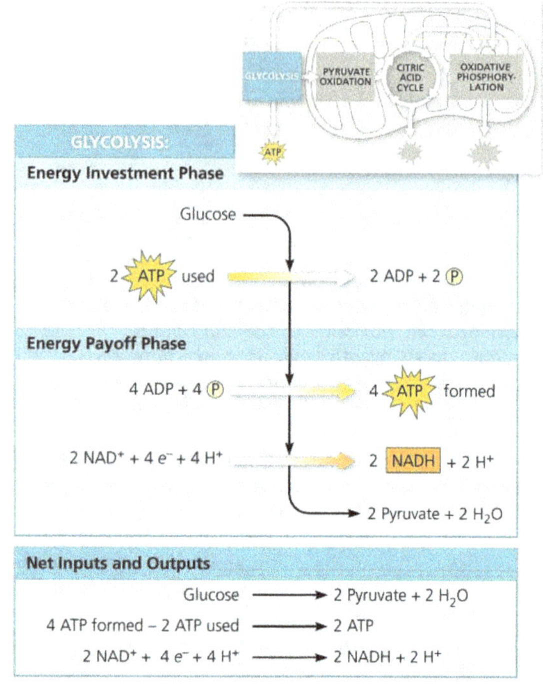

▶ **Figure 6.7** The inputs and outputs of glycolysis.

All of the carbon originally present in glucose is accounted for in the two molecules of pyruvate; no carbon is released as CO_2 during glycolysis. Glycolysis occurs whether or not O_2 is present. However, if O_2 is present, the chemical energy stored in pyruvate and NADH can be extracted by pyruvate oxidation, the citric acid cycle, and oxidative phosphorylation. All the intermediates between glucose and the end product pyruvate contain one or more ionized phosphate groups. Plasma membranes are impermeable to such highly ionized molecules; therefore, these molecules remain trapped within the cell.

▼ **Figure 6.8** The steps of glycolysis.

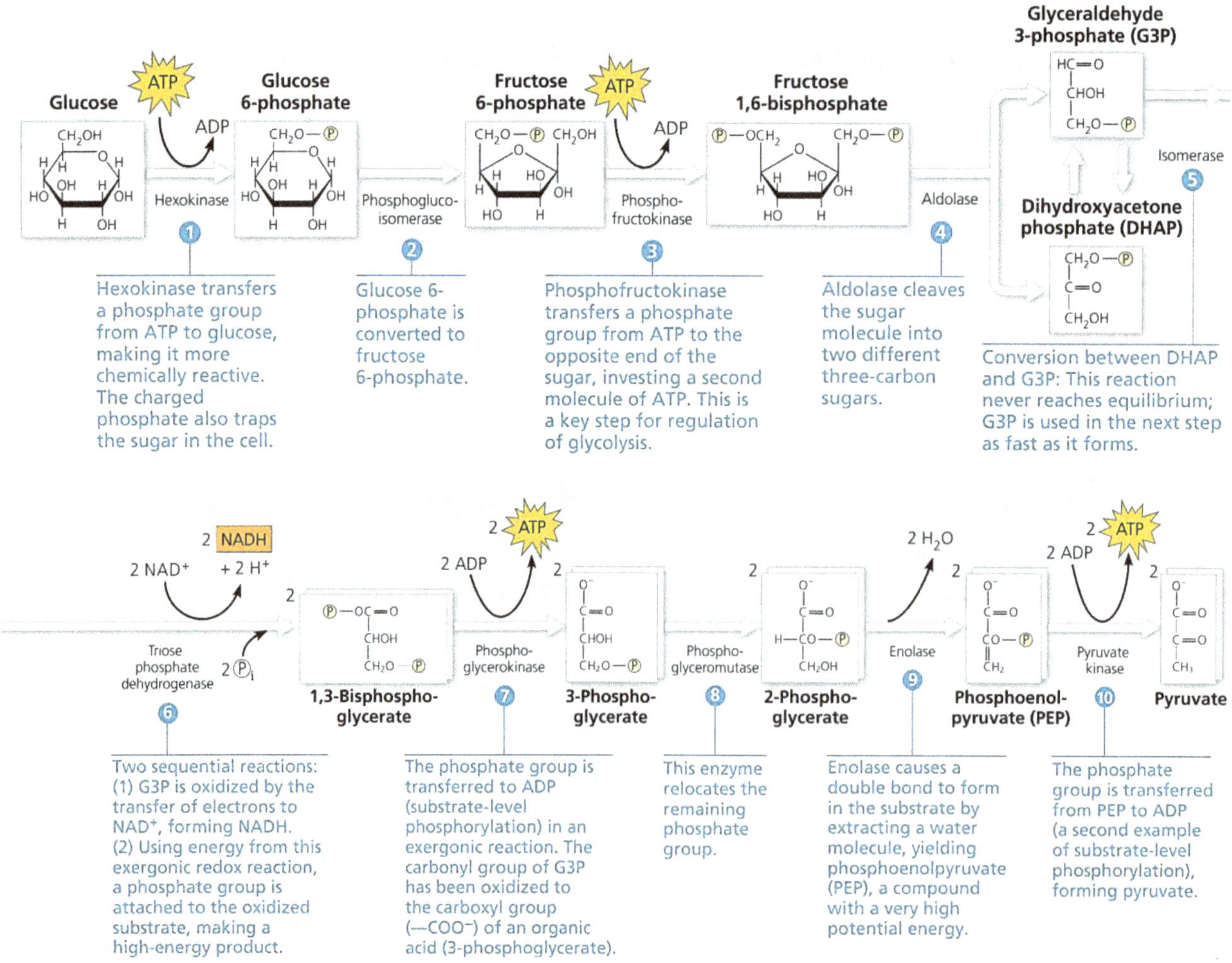

Hexokinase transfers a phosphate group from ATP to glucose, making it more chemically reactive. The charged phosphate also traps the sugar in the cell.

Glucose 6-phosphate is converted to fructose 6-phosphate.

Phosphofructokinase transfers a phosphate group from ATP to the opposite end of the sugar, investing a second molecule of ATP. This is a key step for regulation of glycolysis.

Aldolase cleaves the sugar molecule into two different three-carbon sugars.

Conversion between DHAP and G3P: This reaction never reaches equilibrium; G3P is used in the next step as fast as it forms.

Two sequential reactions: (1) G3P is oxidized by the transfer of electrons to NAD^+, forming NADH. (2) Using energy from this exergonic redox reaction, a phosphate group is attached to the oxidized substrate, making a high-energy product.

The phosphate group is transferred to ADP (substrate-level phosphorylation) in an exergonic reaction. The carbonyl group of G3P has been oxidized to the carboxyl group (—COO⁻) of an organic acid (3-phosphoglycerate).

This enzyme relocates the remaining phosphate group.

Enolase causes a double bond to form in the substrate by extracting a water molecule, yielding phosphoenolpyruvate (PEP), a compound with a very high potential energy.

The phosphate group is transferred from PEP to ADP (a second example of substrate-level phosphorylation), forming pyruvate.

Oxidation of Pyruvate to Acetyl CoA

This step, linking glycolysis and the citric acid cycle. When O_2 is present, the pyruvate in eukaryotic cells enters a mitochondrion, where the oxidation of glucose is completed. In aerobically respiring prokaryotic cells, this process occurs in the cytosol.

Upon entering the mitochondrion via active transport, pyruvate is first converted to a compound called **acetyl coenzyme A**, or **acetyl CoA**. In this series of reactions, pyruvate undergoes a process known as **oxidative decarboxylation**. First, a carboxyl group is removed as carbon dioxide, which diffuses out of the cell. Then the remaining 2-carbon fragment becomes oxidized, and NAD^+ accepts the electrons removed during the oxidation.

Finally, the oxidized 2-carbon fragment, an acetyl group, becomes attached to coenzyme A, yielding acetyl CoA. **Pyruvate dehydrogenase**, the enzyme that catalyzes these reactions, is an enormous multienzyme complex consisting of 72 polypeptide chains. Coenzyme A transfers groups derived from organic acids. In this case, coenzyme A transfers an acetyl group, which is related to acetic acid. Coenzyme A is manufactured in the cell from one of the B vitamins, pantothenic acid.

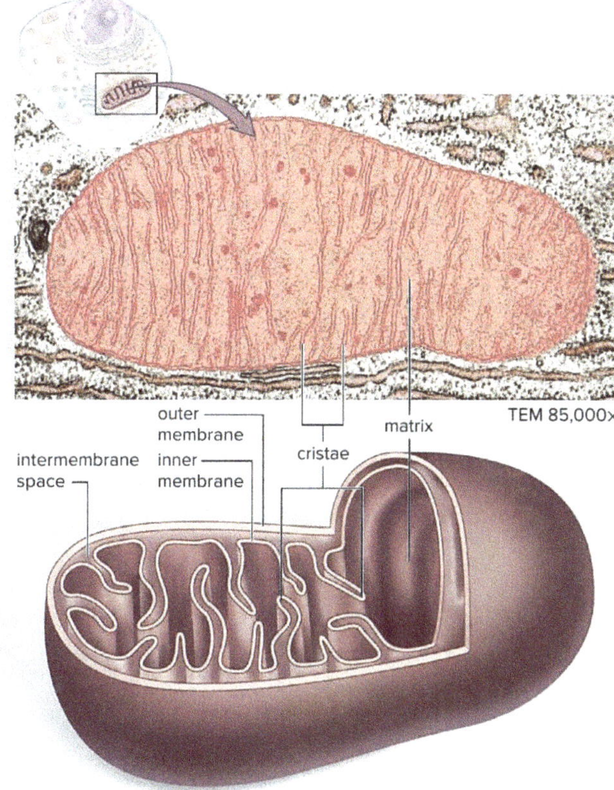

1. Pyruvate's carboxyl group ($-$ COO⁻), already somewhat oxidized and thus carrying little chemical energy, is now fully oxidized and given off as a molecule of CO_2. This is the first step in which CO_2 is released during respiration (**Decarboxylation**).

2. Next, the remaining two-carbon fragment is oxidized and the electrons transferred to NAD^+, storing energy in the form of NADH (**Dehydrogenation**).

3. Finally, the acetyl group (a two-carbon intermediate) is attached to CoA (a sulfur-containing compound derived from a B vitamin) via a covalent bond to a sulfur atom. The hydrolysis of this bond releases a large amount of free energy, making it possible for the acetyl group to be transferred to other organic molecules.

$$2 \text{ pyruvate} + 2 \text{ } NAD^+ + 2 \text{ CoA} \longrightarrow 2 \text{ acetyl CoA} + 2 \text{ NADH} + 2 \text{ } CO_2$$

► **Figure 6.9** Oxidation of pyruvate to acetyl CoA, the step before the citric acid cycle. Pyruvate enters the mitochondrion through a transport protein and is processed by a complex of several enzymes known as pyruvate dehydrogenase. This complex catalyzes the three numbered steps. The CO_2 molecule will diffuse out of the cell. The NADH will be used in oxidative phosphorylation. The acetyl group of acetyl CoA will enter the citric acid cycle. (Coenzyme A is abbreviated S-CoA when it is attached to a molecule, emphasizing its sulfur atom, S.)

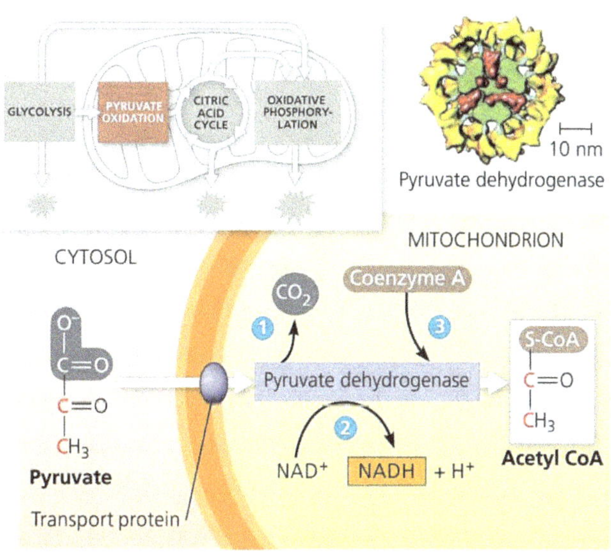

The Citric Acid Cycle

The **citric acid cycle** functions as a metabolic furnace that further oxidizes organic fuel derived from pyruvate. The citric acid cycle is also called the **tricarboxylic acid cycle** (TCA) or the **Krebs cycle**. The cycle has eight steps, each catalyzed by a specific enzyme. Recall that each glucose gives rise to two molecules of acetyl CoA that enter the cycle. Because the numbers noted earlier are obtained from a single acetyl group entering the pathway, the total yield per glucose from the citric acid cycle turns out to be doubled, or $4CO_2$, 6 NADH, 2 $FADH_2$, and the equivalent of 2 ATP.

$$\text{Acetyl CoA} + 3\,\text{NAD}^+ + \text{FAD} + \text{GDP} + \text{P}_i + 2\,\text{H}_2\text{O} \longrightarrow$$
$$2\,\text{CO}_2 + \text{CoA} + 3\,\text{NADH} + 3\,\text{H}^+ + \text{FADH}_2 + \text{GTP}$$

Most of the ATP produced by respiration is generated later, from oxidative phosphorylation, when the NADH and $FADH_2$ produced by the citric acid cycle and earlier steps relay the electrons extracted from food to the electron transport chain. In the process, they supply the necessary energy for the phosphorylation of ADP to ATP. In eukaryotic cells, all the citric acid cycle enzymes are located in the mitochondrial matrix except for the enzyme that catalyzes step 6, which resides in the inner mitochondrial membrane.

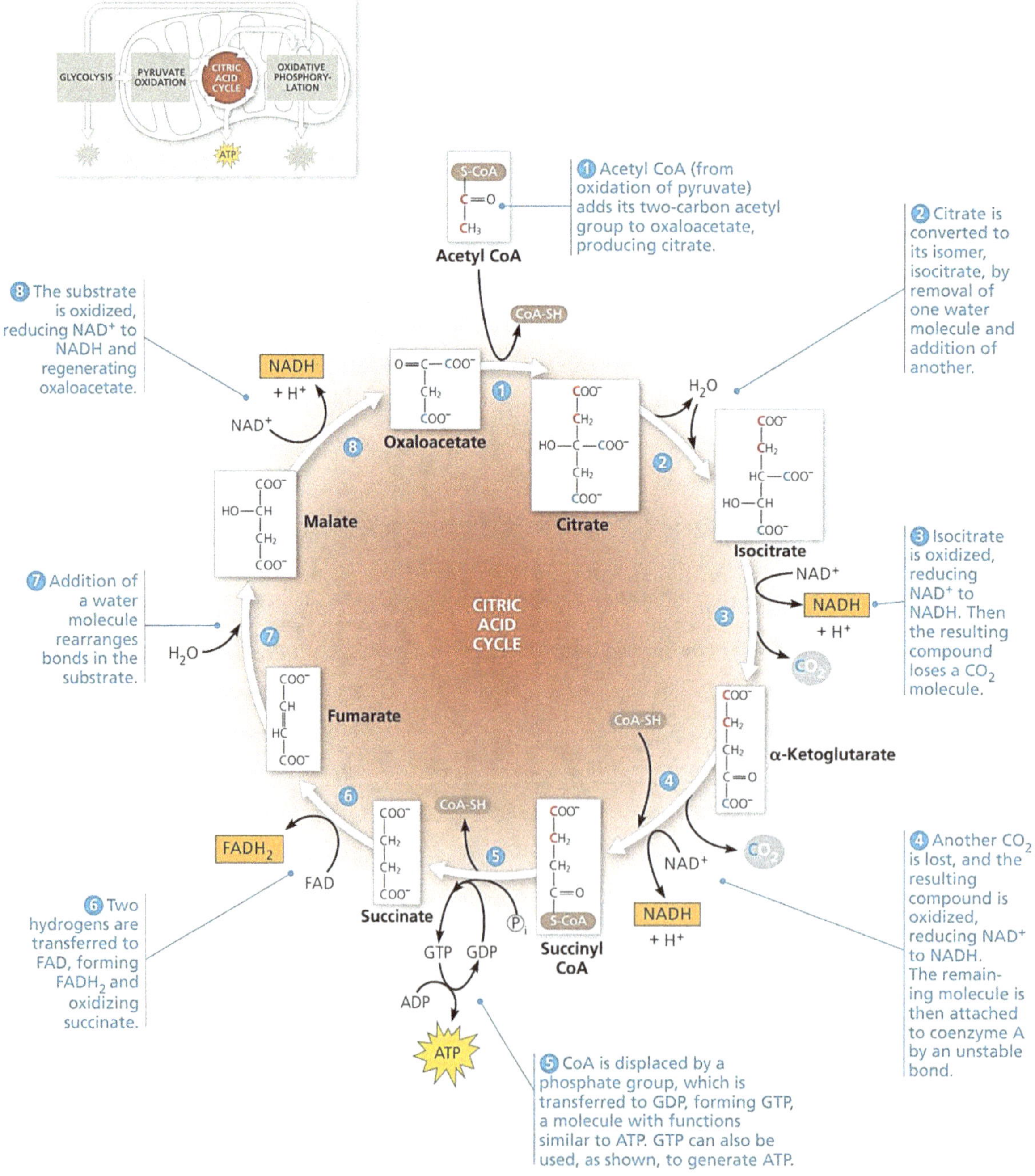

Reaction 1: Condensation Citrate is formed from acetyl-CoA and oxaloacetate. This condensation reaction is irreversible, committing the 2-carbon acetyl group to the Krebs cycle. The reaction is inhibited when the cell's ATP concentration is high and stimulated when it is low. The result is that when the cell possesses ample amounts of ATP, the Krebs cycle shuts down, and acetyl-CoA is channeled into fat synthesis.

Reactions 2 and 3: Isomerization Before the oxidation reactions can begin, the hydroxyl (–OH) group of citrate must be repositioned. This rearrangement is done in two steps: First, a water molecule is removed from one carbon; then water is added to a different carbon. As a result, an –H group and an –OH group change positions. The product is an isomer of citrate called isocitrate. This rearrangement facilitates the subsequent reactions.

Reaction 4: The First Oxidation In the first energy-yielding step of the cycle, isocitrate undergoes an oxidative decarboxylation reaction. First, isocitrate is oxidized, yielding a pair of electrons that reduce a molecule of NAD^+ to NADH. Then the oxidized intermediate is decarboxylated; the central carboxyl group splits off to form CO_2, yielding a 5- carbon molecule called α--ketoglutarate.

Reaction 5: The Second Oxidation Next, α-ketoglutarate is decarboxylated by a multienzyme complex similar to pyruvate dehydrogenase. The succinyl group left after the removal of CO_2 joins to coenzyme A, forming succinyl-CoA. In the process, two electrons are extracted, and they reduce another molecule of NAD^+ to NADH.

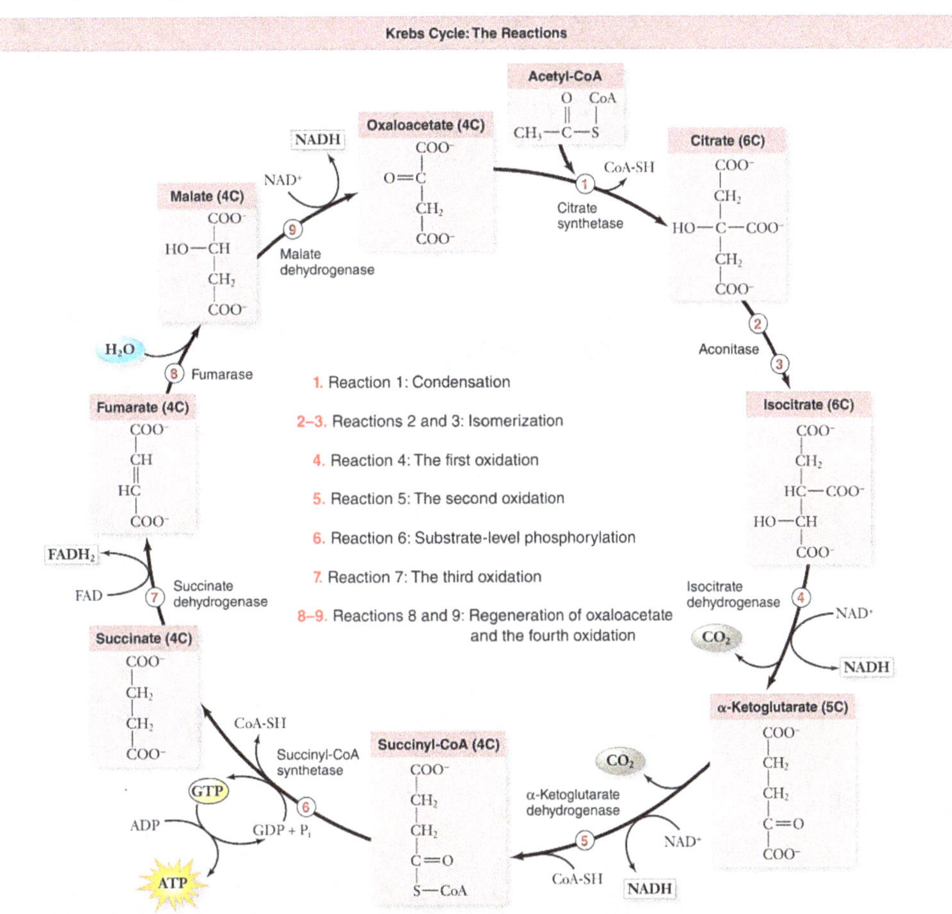

Reaction 6: Substrate-Level Phosphorylation. The   linkage between the 4-carbon succinyl group and CoA is a high-energy bond. In a coupled reaction similar to those that take place in glycolysis, this bond is cleaved, and the energy released drives the phosphorylation of guanosine diphosphate (GDP), forming guanosine triphosphate (GTP). GTP can transfer a phosphate to ADP converting it into ATP. The 4-carbon molecule that remains is called succinate.

Reaction 7: The Third Oxidation Next, succinate is oxidized to fumarate by an enzyme located in the inner mitochondrial membrane. The free-energy change in this reaction is not large enough to reduce NAD^+. Instead, FAD is the electron acceptor. Unlike NAD^+, FAD is not free to diffuse within the mitochondrion; it is tightly associated with its enzyme in the inner mitochondrial membrane. Its reduced form, $FADH_2$, can only contribute electrons to the electron transport chain in the membrane.

Reactions 8 and 9: Regeneration of Oxaloacetate In the final two reactions of the cycle, a water molecule is added to fumarate, forming malate. Malate is then oxidized, yielding a 4-carbon molecule of oxaloacetate and two electrons that reduce a molecule of NAD^+ to NADH. Oxaloacetate, the molecule that began the cycle, is now free to combine with another 2-carbon acetyl group from acetyl-CoA and begin the cycle again.

The Pathway of Electron Transport

Glycolysis and the citric acid cycle, produce only 4 ATP molecules per glucose molecule, all by substrate level phosphorylation: 2 net ATP from glycolysis and 2 ATP from the citric acid cycle. At this point, molecules of NADH (and $FADH_2$) account for most of the energy extracted from each glucose molecule. These electron escorts link glycolysis and the citric acid cycle to the machinery of oxidative phosphorylation, which uses energy released by the **electron transport chain (ETC)** to power ATP synthesis. During oxidative phosphorylation, chemiosmosis couples electron transport to ATP synthesis.

The electron transport chain is a collection of molecules embedded in the inner membrane of the mitochondrion in eukaryotic cells. (In prokaryotes, these molecules reside in the plasma membrane.) The folding of the inner membrane to form cristae increases its surface area, providing space for thousands of copies of each component of the electron transport chain in a mitochondrion. The infolded membrane with its concentration of electron carrier molecules is well-suited for the series of sequential redox reactions that take place along the electron transport chain. Most components of the chain are proteins, which exist in multiprotein complexes numbered I through IV. Tightly bound to these proteins are prosthetic groups, nonprotein components such as cofactors and coenzymes essential for the catalytic functions of certain enzymes.

These protein complexes, numbered I, III, and IV, are integral membrane proteins located in the inner mitochondrial membrane. In addition, a smaller complex, complex II, is bound to the inner mitochondrial membrane on the matrix side. Associated with the system are two small, highly mobile electron carriers, cytochrome c and ubiquinone (also known as coenzyme Q, or CoQ), which shuttle electrons between the major complexes. During this electron transport, electron carriers alternate between reduced and oxidized states as they accept and then donate electrons. Each component of the chain becomes reduced when it accepts electrons from its "uphill" neighbor, which has a lower affinity for electrons. It then returns to its oxidized form as it passes electrons to its "downhill" neighbor, which has a higher affinity for electrons.

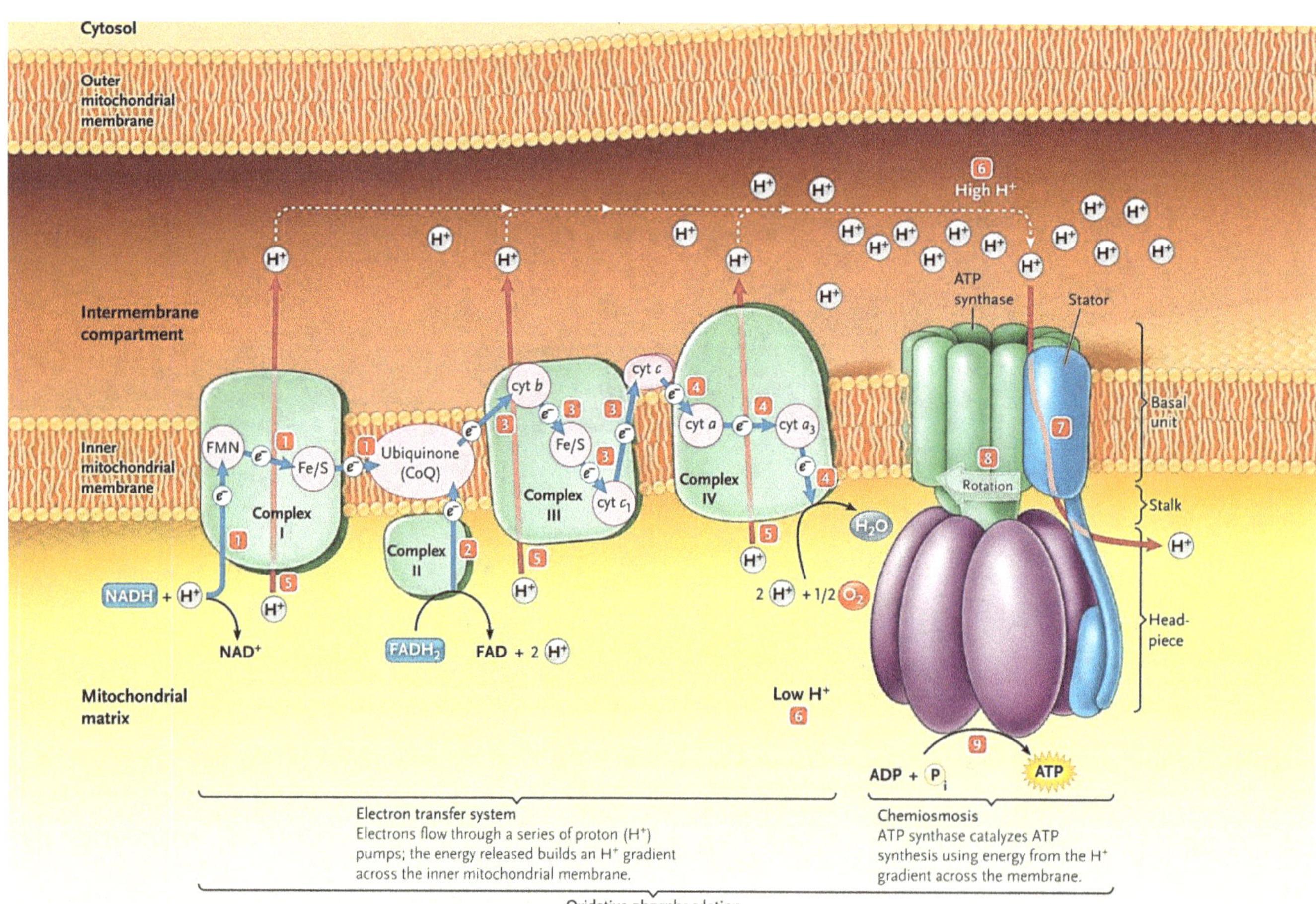

▲ **Figure 6.10** Oxidative phosphorylation involves the electron transfer system (steps 1–6), and chemiosmosis by ATP synthase (steps 7–9). Blue arrows indicate electron flow; red arrows indicate H⁺ movement.

Electrons acquired from glucose by NAD+ during glycolysis and the citric acid cycle are transferred from NADH to the first molecule of the electron transport chain in complex I. This molecule is a flavoprotein, so named because it has a prosthetic group called flavin mononucleotide (FMN). In the next redox reaction, the flavoprotein returns to its oxidized form as it passes electrons to an iron-sulfur protein (Fe.S in complex I), one of a family of proteins with both iron and sulfur tightly bound. The iron-sulfur protein then passes the electrons to a compound called ubiquinone. This electron carrier is a small hydrophobic molecule, the only member of the electron transport chain that is not a protein.

The lipid ubiquinone is individually mobile within the membrane rather than residing in a particular complex. (Another name for ubiquinone is coenzyme Q, or CoQ).Most of the remaining electron carriers between ubiquinone and oxygen are proteins called cytochromes. Their prosthetic group, called a heme group, has an iron atom that accepts and donates electrons. (The heme group in a cytochrome is similar to the heme group in hemoglobin, the protein of red blood cells, except that the iron in hemoglobin carries oxygen, not electrons.)

The electron transport chain has several types of cytochromes, each named "cyt" with a letter and number to distinguish it as a different protein with a slightly different electron-carrying heme group. The last cytochrome of the chain, cyt a_3, passes its electrons to oxygen (in O_2), which is very electronegative. Each O also picks up a pair of hydrogen ions (protons) from the aqueous solution, neutralizing the -2 charge of the added electrons and forming water. The ETC is also called the **respiratory chain** because the oxygen we breathe is used in this process. However, some microorganisms can use molecules other than O2 as their final electron acceptor.

Another source of electrons for the electron transport chain is $FADH_2$, the other reduced product of the citric acid cycle. Notice that $FADH_2$ adds its electrons from within complex II, at a lower energy level than NADH does. Consequently, although NADH and $FADH_2$ each donate an equivalent number of electrons (2) for oxygen reduction, the electron transport chain provides about one-third less energy for ATP synthesis when the electron donor is $FADH_2$ rather than NADH.

▶ **Figure 6.11** Free-energy change during electron transport. The overall energy drop (ΔG) for electrons traveling from NADH to oxygen is 53 kcal/mol, but this "fall" is broken up into a series of smaller steps by the electron transport chain. (An oxygen atom is represented here as $1/2O_2$ to show that O_2 is reduced, not individual oxygen atoms.)

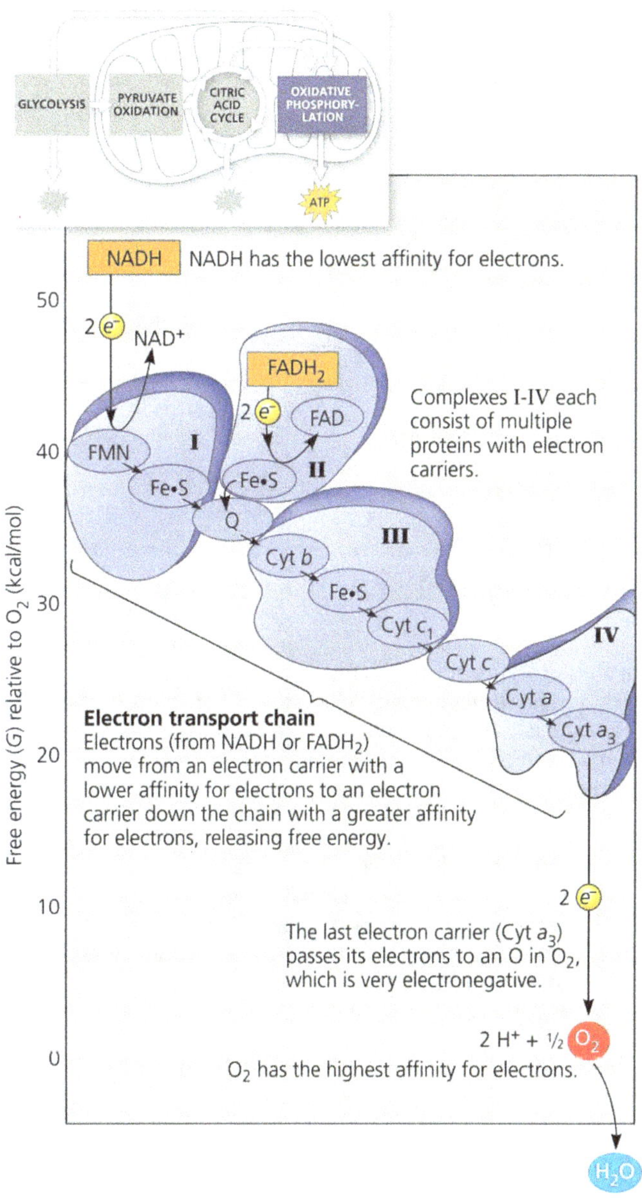

Complex I (NADH–ubiquinone oxidoreductase) accepts electrons from NADH molecules that were produced during glycolysis, the formation of acetyl CoA, and the citric acid cycle. Complex II (succinate–ubiquinone reductase) accepts electrons from $FADH_2$ molecules that were produced during the citric acid cycle. Complexes I and II both produce the same product, reduced ubiquinone, which is the substrate of complex III (ubiquinone–cytochrome c oxidoreductase).

That is, complex III accepts electrons from reduced ubiquinone and passes them on to cytochrome c. Complex IV (cytochrome c oxidase)accepts electrons from cytochrome c and uses these electrons to reduce molecular oxygen, forming water in the process. The electrons simultaneously unite with protons from the surrounding medium to form hydrogen, and the chemical reaction between hydrogen and oxygen produces water.

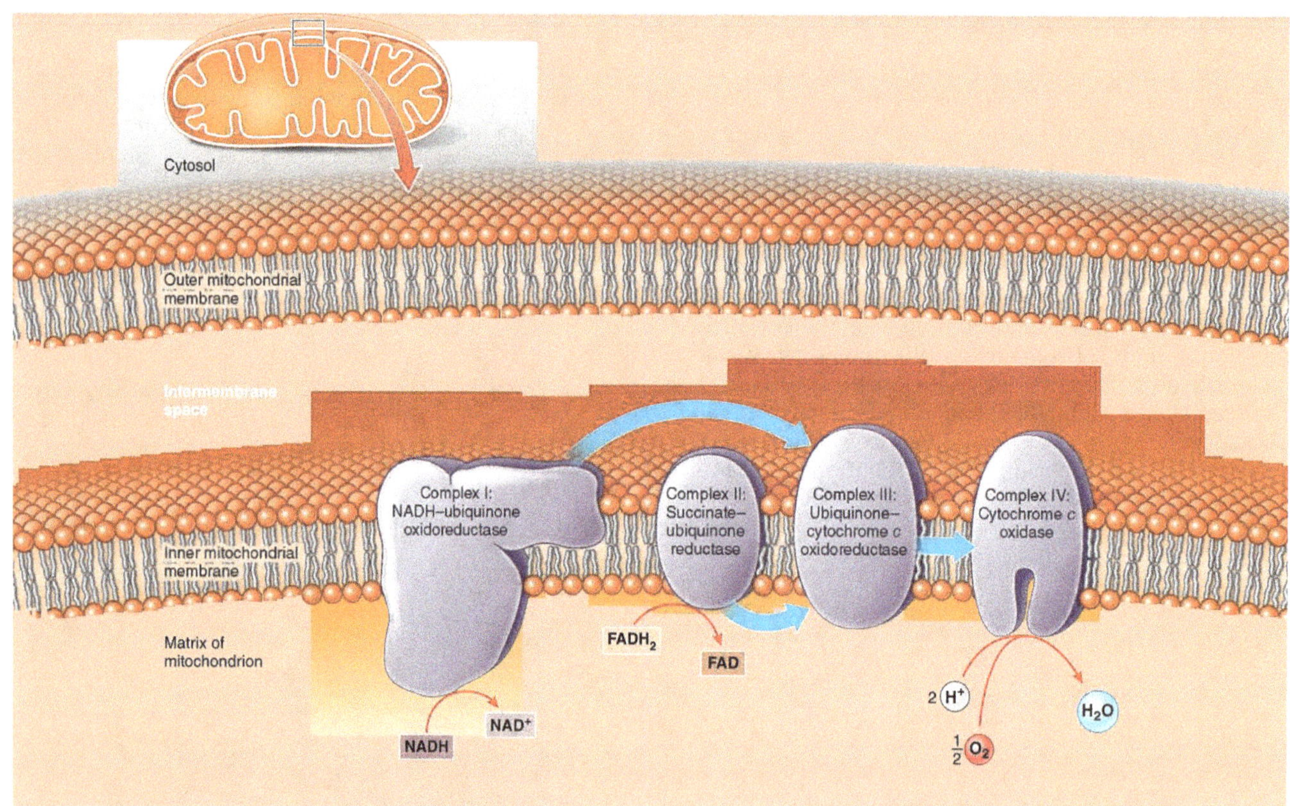

▲ **Figure 6.12** Electron carriers in the mitochondrial inner membrane transfer electrons from NADH and FADH2 to oxygen.

Thirteen proteins of the electron transfer system complexes are encoded by the 13 protein-coding genes of the mitochondrial genome. The remaining approximately 77 proteins of the electron transfer system are encoded by nuclear genes, synthesized in the cytosol, and imported into the mitochondria. Similarly, all of the proteins involved in the transcription, translation, and assembly of the 13 mitochondrial-encoded proteins, and proteins for pyruvate oxidation and the citric acid cycle, are nuclear-encoded.

The poison cyanide does its deadly work by blocking the transfer of electrons from complex IV to oxygen. When that occurs, each acceptor molecule in the chain retains its electrons (each remains in its reduced state), and the entire chain is blocked all the way back to NADH. Because oxidative phosphorylation is coupled to electron transport, no additional ATP is produced by way of the electron transport chain. The gas carbon monoxide inhibits complex IV activity, leading to abnormalities in mitochondrial function. In this way, the carbon monoxide in tobacco smoke contributes to the development of diseases associated with smoking.

Chemiosmosis: The Energy-Coupling Mechanism

Populating the inner membrane of the mitochondrion or the prokaryotic plasma membrane are many copies of a protein complex called **ATP synthase**, the enzyme that makes ATP from ADP and inorganic phosphate. ATP synthase works like an ion pump running in reverse. Ion pumps usually use ATP as an energy source to transport ions against their gradients. Enzymes can catalyze a reaction in either direction, depending on the ΔG for the reaction, which is affected by the local concentrations of reactants and products.

Under the conditions of cellular respiration, rather than hydrolyzing ATP to pump protons against their concentration gradient, ATP synthase uses the energy of an existing ion gradient to power ATP synthesis. The power source for ATP synthase is a difference in the concentration of H^+ (a pH difference) on opposite sides of the inner mitochondrial membrane. This process, in which energy stored in the form of a hydrogen ion gradient across a membrane is used to drive cellular work such as the synthesis of ATP, is called chemiosmosis.

ATP synthase is a multi subunit complex with four main parts, each made up of multiple polypeptides. It consists of a basal unit embedded in the inner mitochondrial membrane (the F0 membrane-bound complex) connected by a stalk to a headpiece located in the matrix (the F1 complex). A peripheral stalk called a stator bridges the basal unit and headpiece. Protons move one by one into binding sites on the rotor, causing it to spin in a way that catalyzes ATP production from ADP and Pi.

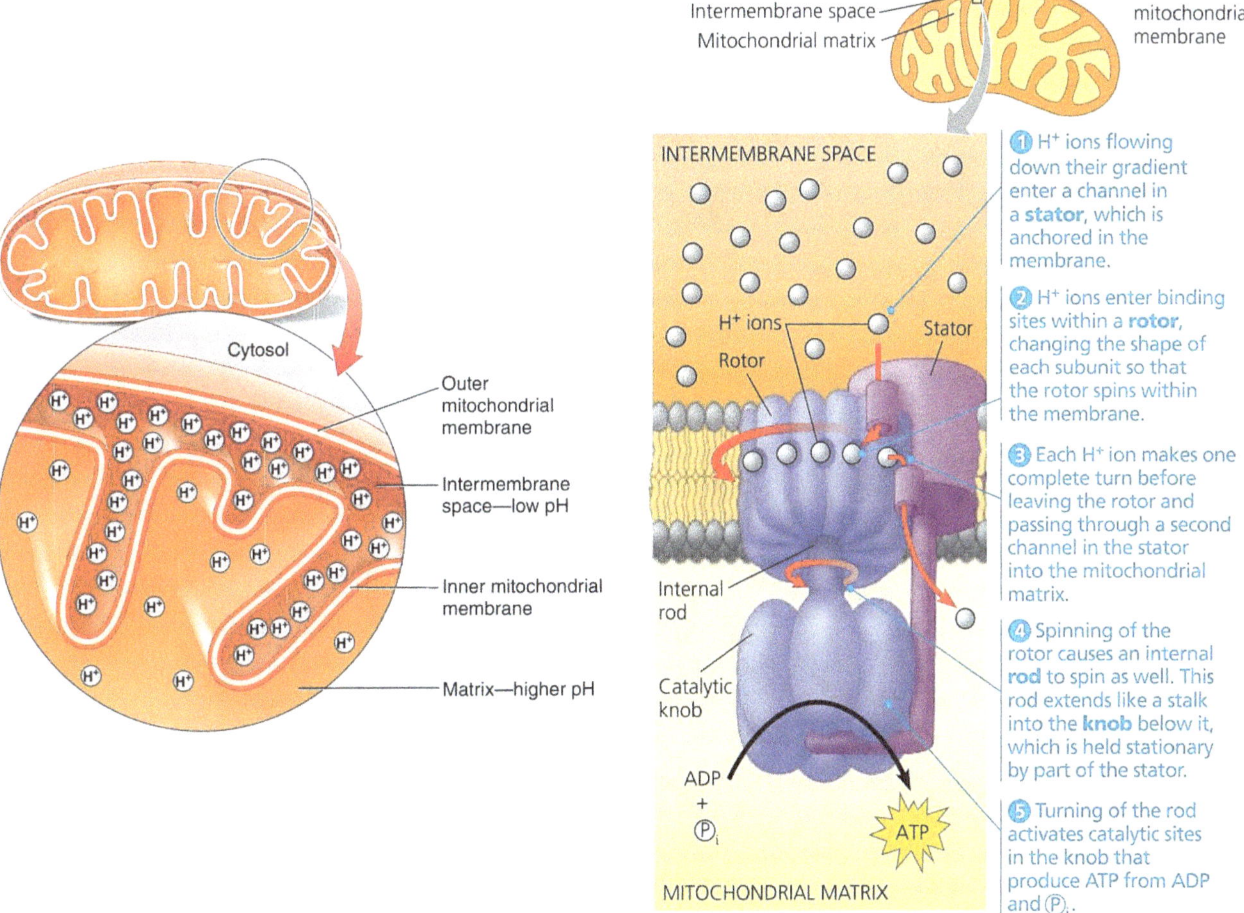

▲ **Figure 6.13** The ATP rotary engine. Protons move across the membrane down their concentration gradient. The energy released causes the rotor and stalk structures to rotate. This mechanical energy alters the conformation of the ATP synthase enzyme to catalyze the formation of ATP.

The nonmembrane-embedded portion consists of 1 ε, 1 γ, 1 δ, 3 α, and 3 β subunits. Movement of H⁺ from the intermembrane space to the matrix causes the ring of c subunits to rotate clockwise, which, in turn, causes the γ subunit to rotate. The rotation, in 120° increments, causes the β subunits to progress through a series of 3 conformational changes that lead to the synthesis of ATP from ADP and P_i.

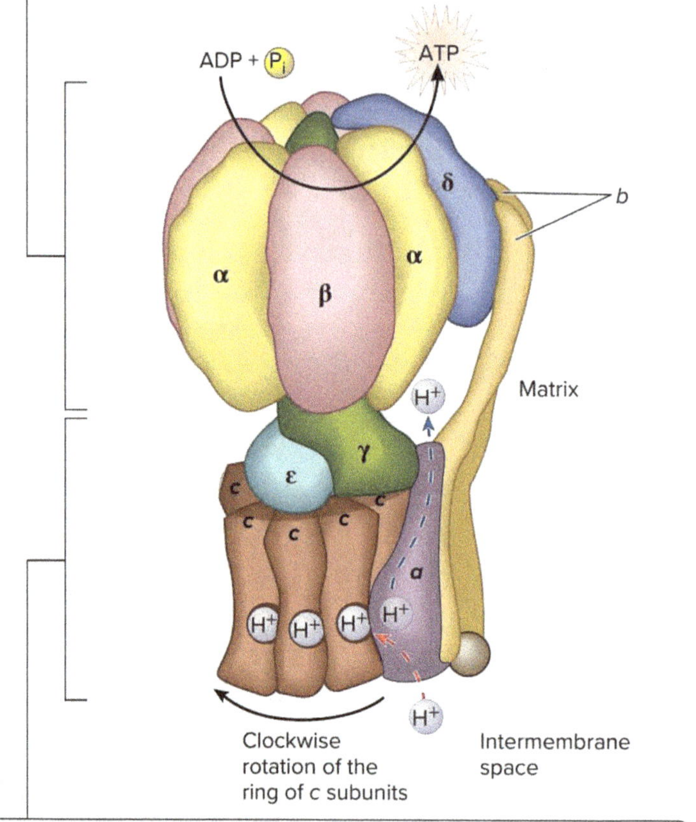

An H⁺ in the intermembrane space moves through a half-channel in the a subunit and binds to a c subunit. An H⁺ is released from an adjacent c subunit and moves through a half-channel in the a subunit to the matrix. These events promote the turning of the ring of c subunits in the clockwise direction.

▲ **Figure 6.14** The subunit structure and function of ATP synthase, a Rotary Machine.

ATP synthase is a rotary machine. The term machine refers to the observations that it has moving parts and is powered by an energy source. The region embedded in the membrane is composed of three types of subunits called a, b, and c. Approximately 10–14 c subunits form a ring in the membrane. One a subunit is bound to this ring, and two b subunits are attached to the a subunit and protrude from the membrane. The nonmembrane-embedded subunits are designated with Greek letters. One ε and one γ subunit bind to the ring of c subunits. The γ subunit forms a long stalk that extends upward into the center of another ring of three α and three β subunits. Each β subunit contains a catalytic site where ATP is made. Finally, the δ subunit forms a connection between the ring of α and β subunits and the two b subunits.

Establishing the H⁺ gradient is a major function of the electron transport chain. The chain is an energy converter that uses the exergonic flow of electrons from NADH and $FADH_2$ to pump H⁺ across the membrane, from the mitochondrial matrix into the intermembrane space. The H⁺ has a tendency to move back across the membrane, diffusing down its gradient. And the ATP synthases are the only sites that provide a route through the membrane for H⁺. The passage of H⁺ through ATP synthase uses the exergonic flow of H⁺ to drive the phosphorylation of ADP. Thus, the energy stored in an **H⁺ electrochemical gradient** across a membrane couples the redox reactions of the electron transport chain to ATP synthesis.

Certain members of the electron transport chain accept and release protons (H⁺) along with electrons. (The aqueous solutions inside and surrounding the cell are a ready source of H⁺.) At certain steps along the chain, electron transfers cause H⁺ to be taken up and released into the surrounding solution. In eukaryotic cells, the electron carriers are spatially arranged in the inner mitochondrial membrane in such a way that H⁺ is accepted from the mitochondrial matrix and deposited in the intermembrane space. The H⁺ gradient that results is referred to as a **proton-motive force**, emphasizing the capacity of the gradient to perform work. The force drives H⁺ back across the membrane through the H⁺ channels provided by ATP synthases.

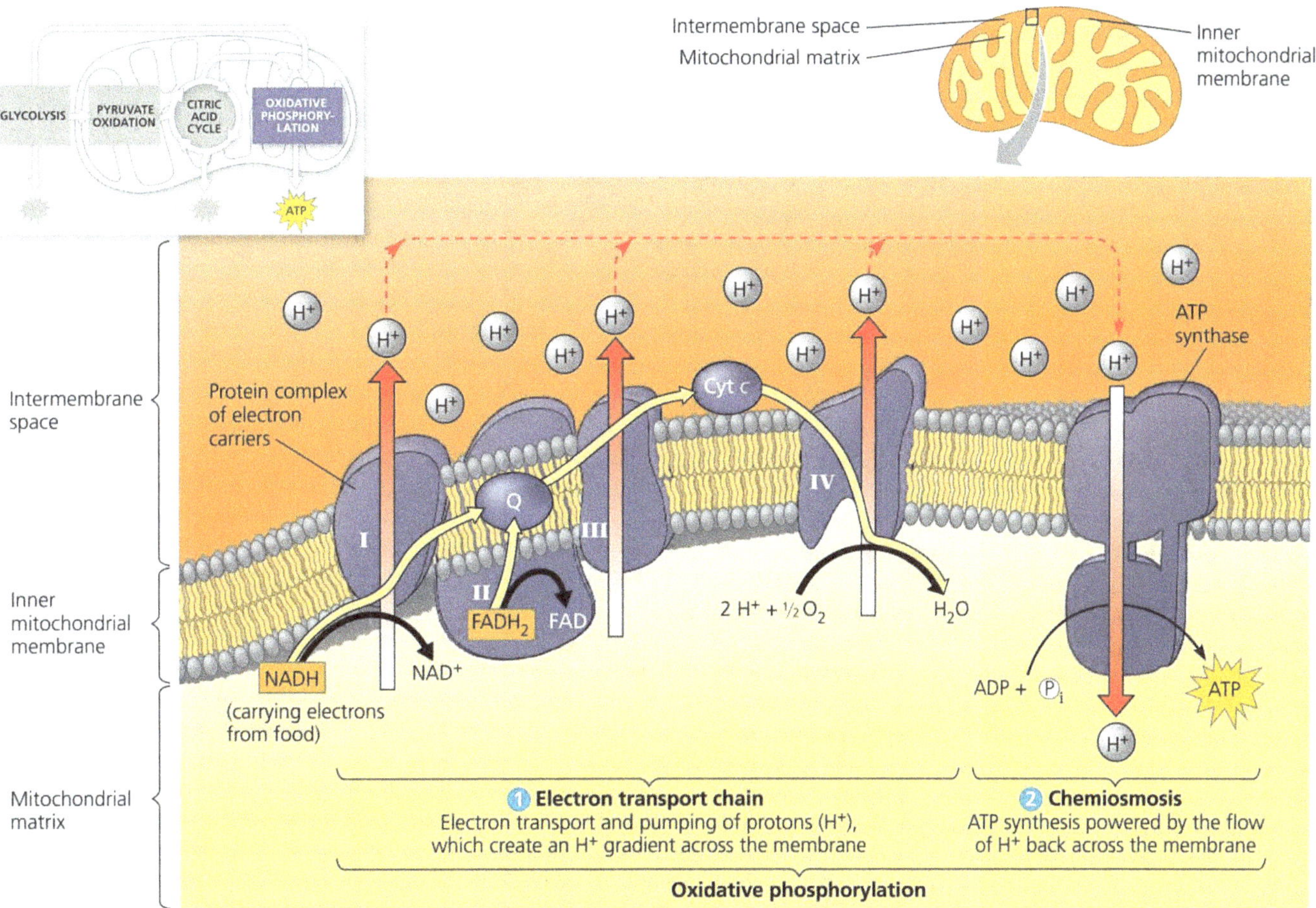

▲ **Figure 6.15** Chemiosmosis couples the electron transport chain to ATP synthesis. NADH and $FADH_2$ shuttle high-energy electrons extracted from food during glycolysis and the citric acid cycle into an electron transport chain built into the inner mitochondrial membrane. Most of the electron carriers of the chain are grouped into four complexes (I–IV). Two mobile carriers, ubiquinone (Q) and cytochrome c (Cyt c), move rapidly, ferrying electrons between the large complexes. As the complexes shuttle electrons, they pump protons from the mitochondrial matrix into the intermembrane space. $FADH_2$ deposits its electrons via complex II and so results in fewer protons being pumped into the intermembrane space than occurs with NADH. Chemical energy originally harvested from food is transformed into a proton-motive force, a gradient of H⁺ across the membrane. During chemiosmosis, the protons flow back down their gradient via ATP synthase, which is built into the membrane nearby. The ATP synthase harnesses the proton-motive force to phosphorylate ADP, forming ATP. Together, electron transport and chemiosmosis make up oxidative phosphorylation.

In general terms, chemiosmosis is an energy-coupling mechanism that uses energy stored in the form of an H⁺ gradient across a membrane to drive cellular work. In mitochondria, the energy for gradient formation comes from exergonic redox reactions along the electron transport chain, and ATP synthesis is the work performed. But chemiosmosis also occurs elsewhere and in other variations. Chloroplasts use chemiosmosis to generate ATP during photosynthesis; in these organelles, light (rather than chemical energy) drives both electron flow down an electron transport chain and the resulting H⁺ gradient formation.

Prokaryotes generate H⁺ gradients across their plasma membranes. They then tap the proton-motive force not only to make ATP inside the cell but also to rotate their flagella and to pump nutrients and waste products across the membrane.

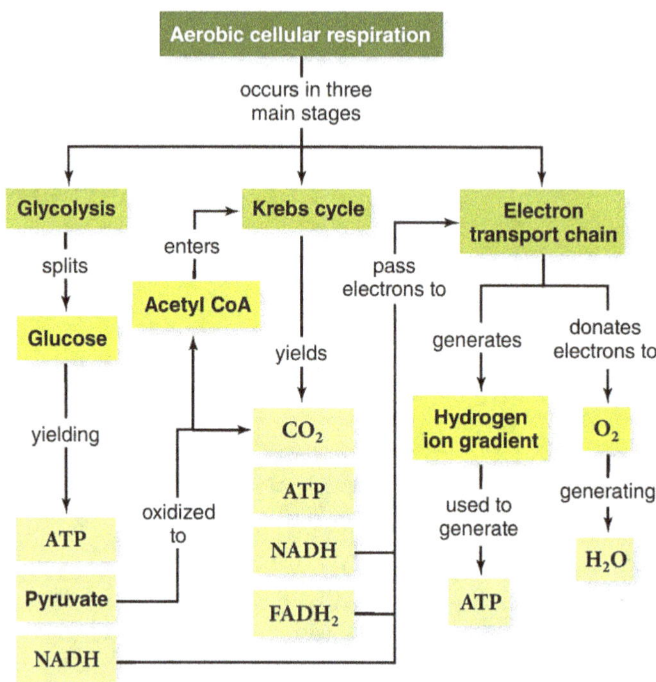

▲ **Figure 6.16** How Cells Release Energy.

Stage	Summary	Some Starting Materials	Some End Products
1. Glycolysis (in cytosol)	Series of reactions in which glucose is degraded to pyruvate; net profit of 2 ATPs; electrons are transferred to carriers; can proceed anaerobically	Glucose, ATP, NAD⁺, ADP, P₍ᵢ₎	Pyruvate, ATP, NADH
2. Formation of acetyl CoA (in mitochondria)	Pyruvate is degraded and combined with coenzyme A to form acetyl CoA; electrons are transferred to carriers; CO₂ is released	Pyruvate, coenzyme A, NAD⁺	Acetyl CoA, CO₂, NADH
3. Citric acid cycle (in mitochondria)	Series of reactions in which the acetyl portion of acetyl CoA is degraded to CO₂; electrons are transferred to carriers; ATP is synthesized	Acetyl CoA, H₂O, NAD⁺, FAD, ADP, P₍ᵢ₎	CO₂, NADH, FADH₂, ATP
4. Electron transport and chemiosmosis (in mitochondria)	Chain of several electron transport molecules; electrons are passed along chain; released energy is used to form a proton gradient; ATP is synthesized as protons diffuse down the gradient; oxygen is final electron acceptor	NADH, FADH₂, O₂, ADP, P₍ᵢ₎	ATP, H₂O, NAD⁺, FAD

▲ **Table 6.1** Summary of Aerobic Respiration

An Accounting of ATP Production by Cellular Respiration

During respiration, most energy flows in this sequence: glucose → NADH → electron transport chain → proton-motive force → ATP.

Figure gives a detailed accounting of the ATP yield for each glucose molecule that is oxidized. The tally adds the 4 ATP produced directly by substrate-level phosphorylation during glycolysis and the citric acid cycle to the many more molecules of ATP generated by oxidative phosphorylation. Each NADH that transfers a pair of electrons from glucose to the electron transport chain contributes enough to the proton motive force to generate 2.5 ATP (We know that 1 NADH results in 10 H^+ being transported out across the inner mitochondrial membrane, the exact number of H^+ that must reenter the mitochondrial matrix via ATP synthase to generate 1 ATP is 4 H^+). Since $FADH_2$ electrons enter later in the chain, each molecule of this electron carrier is responsible for transport of only enough H^+ for the synthesis of 1.5 ATP.

ATP yield varies slightly depending on **first**, the slight energetic cost of moving the ATP formed in the mitochondrion out into the cytosol, **second**, the type of shuttle used to transport electrons from the cytosol into the mitochondrion. The mitochondrial inner membrane is not permeable to NADH, so NADH in the cytosol is segregated from the machinery of oxidative phosphorylation. The 2 electrons of NADH captured in glycolysis must be conveyed into the mitochondrion by one of several electron shuttle systems. Depending on the kind of shuttle in a particular cell type, the electrons are passed either to NAD^+ or to FAD in the mitochondrial matrix.

If the electrons are passed to FAD, as in skeletal muscle and brain cells, only about 1.5 ATP can result from each NADH that was originally generated in the cytosol. If the electrons are passed to mitochondrial NAD^+, as in heart, liver, and kidney cells, the yield is about 2.5 ATP per NADH and **third** the use of the proton-motive force generated by the redox reactions of respiration to drive other kinds of work. For example, the proton-motive force powers the mitochondrion's uptake of pyruvate from the cytosol via an H^+/pyruvate symporter. These "expenses" reduce the actual ATP yield to about 30 per glucose.

Stage	ATP used	ATP made	Net gain in ATP
glycolysis	2	4	2
link reaction	0	0	0
Krebs cycle	0	2	2
oxidative phosphorylation	0	28	28
Total	2	34	32

If all the proton-motive force generated by the electron transport chain were used to drive ATP synthesis, one glucose molecule could generate a maximum of 28 ATP produced by oxidative phosphorylation plus 4 ATP (net) from substrate-level phosphorylation to give a total yield of about **32 ATP** (or only about 30 ATP if the less efficient shuttle were functioning).

However, about 34% of the potential chemical energy in glucose has been transferred to ATP, the rest of the energy stored in glucose is lost as heat. We humans use some of this heat to maintain our relatively high body temperature (37°C), and we dissipate the rest through sweating and other cooling mechanisms.

Surprisingly, perhaps, it may be beneficial under certain conditions to reduce the efficiency of cellular respiration. A remarkable adaptation is shown by hibernating mammals, which overwinter in a state of inactivity and lowered metabolism. Although their internal body temperature is lower than normal, it still must be kept significantly higher than the external air temperature. One type of tissue, called brown fat, is made up of cells packed full of mitochondria. The inner mitochondrial membrane contains a channel protein called the uncoupling protein that allows protons to flow back down their concentration gradient without generating ATP. Activation of these proteins in hibernating mammals results in ongoing oxidation of stored fuel (fats), generating heat without any ATP production.

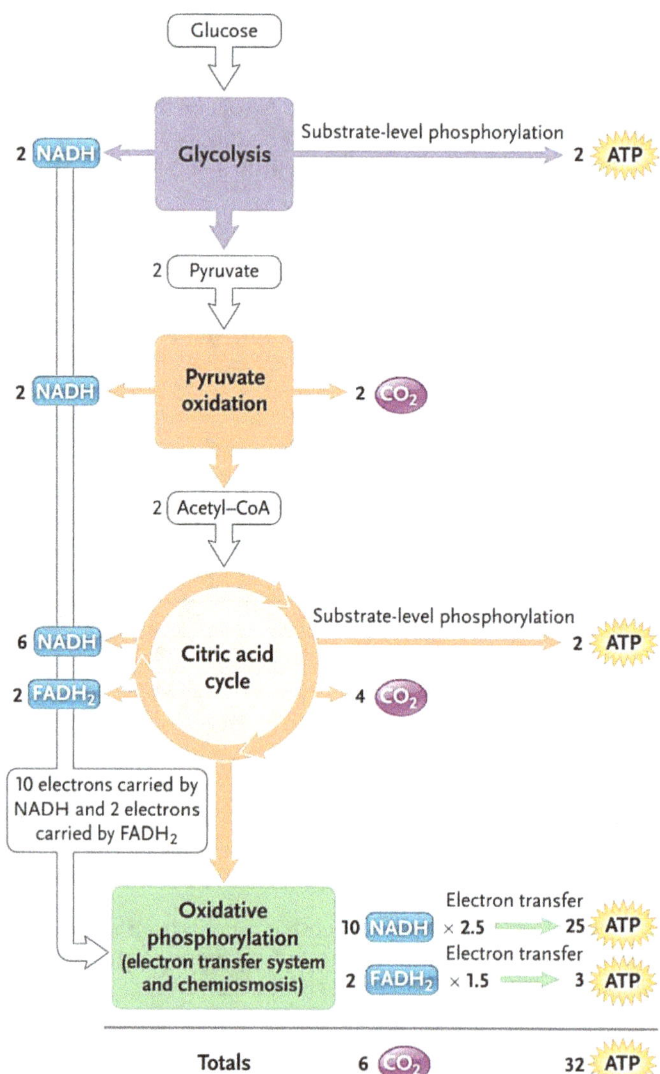

▲ **Figure 6.17** Summary of ATP production from the complete oxidation of a molecule of glucose. The total of 32 ATP assumes that electrons carried from glycolysis by NADH are transferred to NAD^+ inside mitochondria. If the electrons from glycolysis are instead transferred to FAD inside mitochondria, total production will be 30 ATP. © Cengage Learning 2017

Fermentation and anaerobic respiration enable cells to produce ATP without the use of oxygen

There are two general mechanisms by which certain cells can oxidize organic fuel and generate ATP without the use of O_2: anaerobic respiration and fermentation. The distinction between these two is that an electron transport chain is used in anaerobic respiration but not in fermentation (The electron transport chain is also called the respiratory chain because of its role in both types of cellular respiration). Anaerobic respiration, takes place in certain prokaryotic organisms that live in environments without O_2. These organisms have an electron transport chain but do not use O_2 as a final electron acceptor at the end of the chain. O_2 performs this function very well because it consists of two extremely electronegative atoms, but other substances can also serve as final electron acceptors. Alternative electron acceptors include NO_3^- (nitrate), SO_4^{2-} (sulfate), and CO_2. The number of ATPs generated per molecule of glucose depends on the electron acceptor, but it is always lower than the ATP yield for aerobic respiration.

Many bacteria and archaea generate ATP by anaerobic respiration, and they play starring roles in nutrient cycles wherever O_2 is scarce. For example, in waterlogged, oxygen-poor soils, bacteria that use NO_3^- as an electron acceptor begin a chain reaction that ends with the production of nitrogen gas (N_2). This gas drifts into the atmosphere, leaving the soil less fertile for plant growth. Under anaerobic conditions Escherichia coli, a bacterial species found in your intestinal tract, produces an enzyme called nitrate reductase. This enzyme uses nitrate (NO^{3-}) as the final electron acceptor of the electron transport chain. At the end of the chain, NO^{3-} is converted to nitrite (NO^{2-}). Sulfur bacteria that live in wetlands may use SO_4^{2-}, producing smelly hydrogen sulfide (H_2S) as a byproduct. Methanogen archaea living inside the intestines of cattle use CO_2 as an electron acceptor, generating methane gas (CH_4). Other inorganic electron acceptors used for anaerobic respiration include CO_2, Fe^{3+}, and Mn^{4+}.

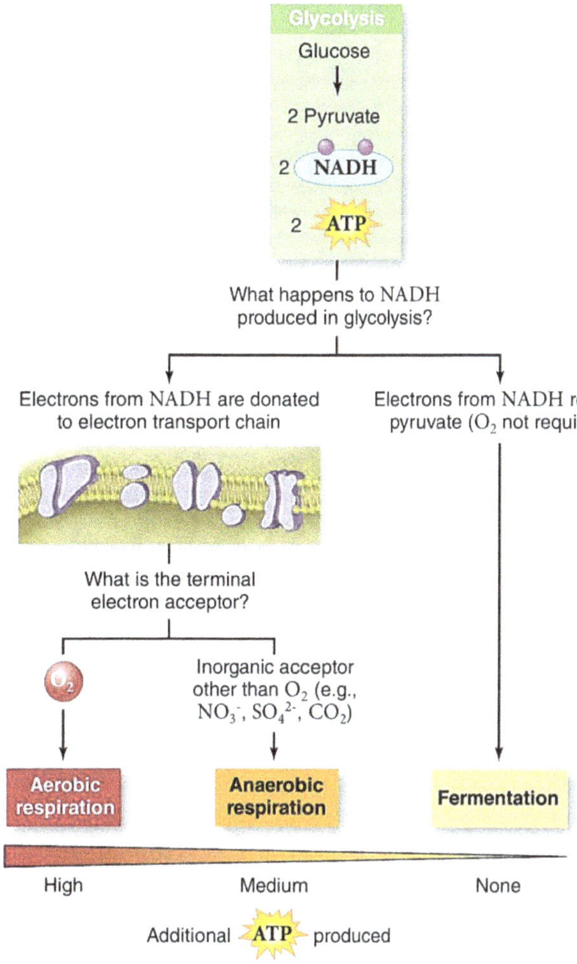

► **Figure 6.18** Alternative Metabolic Pathways. If O_2 is available, most organisms generate ATP in aerobic respiration. Two other pathways, anaerobic respiration and fermentation, can occur in the absence of O_2. Both alternatives yield less ATP than does aerobic respiration.

Fermentation uses organic compounds as electron acceptors and is a way of harvesting chemical energy without using either O_2 or any electron transport chain—in other words, without cellular respiration. Oxidation simply refers to the loss of electrons to an electron acceptor, so it does not need to involve O_2. Glycolysis oxidizes glucose to two molecules of pyruvate. The oxidizing agent of glycolysis is NAD^+, and neither O_2 nor any electron transfer chain is involved. So in fermentation, electrons are transferred to an organic molecule rather than to an electron transfer system.

Overall, glycolysis is exergonic, and some of the energy made available is used to produce 2 ATP (net) by substrate-level phosphorylation. If O_2 is present, then additional ATP is made by oxidative phosphorylation when NADH passes electrons removed from glucose to the electron transport chain. But glycolysis generates 2 ATP whether oxygen is present or not—that is, whether conditions are aerobic or anaerobic.

As an alternative to respiratory oxidation of organic nutrients, fermentation is an extension of glycolysis that allows continuous generation of ATP by the substrate-level phosphorylation of glycolysis. For this to occur, there must be a sufficient supply of NAD^+ to accept electrons during the oxidation step of glycolysis. Without some mechanism to recycle NAD^+ from NADH, glycolysis would soon deplete the cell's pool of NAD^+ by reducing it all to NADH and would shut itself down for lack of an oxidizing agent. Under aerobic conditions, NAD^+ is recycled from NADH by the transfer of electrons to the electron transport chain. An anaerobic alternative is to transfer electrons from NADH to pyruvate, the end product of glycolysis.

	Aerobic Respiration	Anaerobic Respiration	Fermentation
Immediate fate of electrons in NADH	Transferred to electron transport chain	Transferred to electron transport chain	Transferred to organic molecule
Terminal electron acceptor of electron transport chain	O_2	Inorganic substances such as NO_3^- or SO_4^{2-}	No electron transport chain
Reduced product(s) formed	Water	Relatively reduced inorganic substances	Relatively reduced organic compounds (commonly, alcohol or lactate)
Mechanism of ATP synthesis	Oxidative phosphorylation/ chemiosmosis; also substrate-level phosphorylation	Oxidative phosphorylation/ chemiosmosis; also substrate-level phosphorylation	Substrate-level phosphorylation only (during glycolysis)

▲ **Table 6.2** A Comparison of Aerobic Respiration, Anaerobic Respiration, and Fermentation

Types of Fermentation

Fermentation consists of glycolysis plus reactions that regenerate NAD^+ by transferring electrons from NADH to pyruvate or derivatives of pyruvate. NAD^+ is required for glycolysis to continue. If virtually all NAD^+ becomes reduced to NADH during glycolysis, glycolysis stops and no more ATP is produced. The NAD^+ can then be reused to oxidize sugar by glycolysis, which nets two molecules of ATP by substrate-level phosphorylation. There are many types of fermentation, differing in the end products formed from pyruvate. Two types are alcohol fermentation and lactic acid fermentation, and both are harnessed by humans for food and industrial production.

- **Alcohol fermentation**: In alcohol fermentation, pyruvate is converted to ethanol (ethyl alcohol) in two steps. The first step releases CO_2 from the pyruvate, which is converted to the two-carbon compound acetaldehyde. In the second step, acetaldehyde is reduced by NADH to ethanol. This regenerates the supply of NAD^+ needed for the continuation of glycolysis. Many bacteria carry out alcohol fermentation under anaerobic conditions. Yeast (a fungus), in addition to aerobic respiration, also carries out alcohol fermentation. For thousands of years, humans have used yeast in baking. The CO_2 bubbles generated by baker's yeast (Saccharomyces cerevisiae) during alcohol fermentation allow bread to rise.

- **Lactic acid fermentation**: During lactic acid fermentation, pyruvate is reduced directly by NADH to form lactate as an end product, regenerating NAD^+ with no release of CO_2 (Lactate is the ionized form of lactic acid). Lactic acid fermentation by certain fungi and bacteria (Lactobacillus) is used in the dairy industry to make cheese and yogurt.

There are two types of skeletal muscle fibers. One (red muscle) preferentially oxidizes glucose completely to CO_2; Red fibers have many mitochondria and produce ATP by aerobic respiration. These fibers sustain prolonged activity such as marathon runs. They are red because they contain an abundance of myoglobin, a protein that stores oxygen for aerobic respiration in muscle tissue. The other (white muscle) produces significant amounts of lactate from the pyruvate made during glycolysis, even under aerobic conditions, offering fast but energetically inefficient ATP production.

White muscle fibers contain few mitochondria and no myoglobin, so they do not carry out a lot of aerobic respiration. The low ATP yield does not support prolonged activity. That is one reason why chickens cannot fly very far: Their flight muscles consist mostly of white fibers (thus, the "white" breast meat). Chickens fly only in short bursts. More often, they walk or run. Their leg muscles consist mostly of red muscle fibers, the "dark meat." The lactate product is then mostly oxidized by red muscle cells in the vicinity, with the remainder exported to liver or kidney cells for glucose formation. During strenuous exercise, when carbohydrate catabolism outpaces the supply of O_2 from the blood to the muscle, lactate can't be oxidized to pyruvate. The lactate that accumulates was once thought to cause muscle fatigue during intense exercise and pain a day or so later.

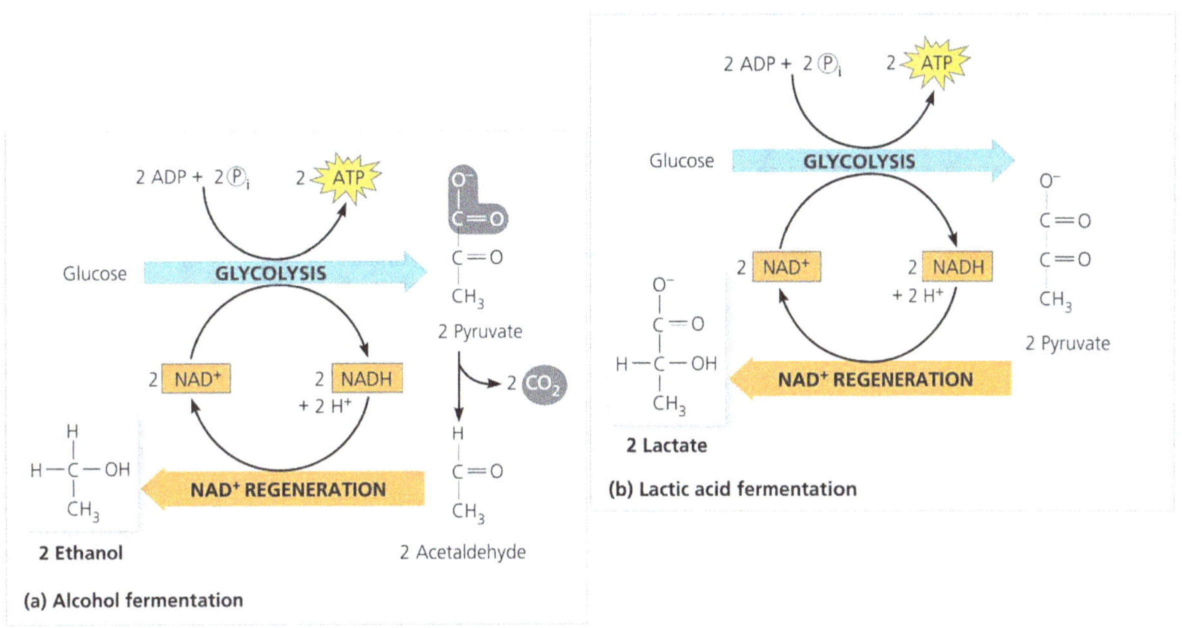

▲ **Figure 6.19 Fermentation.** In the absence of oxygen, many cells use fermentation to produce ATP by substrate-level phosphorylation. NAD^+ is regenerated for use in glycolysis when pyruvate, the end product of glycolysis, serves as an electron acceptor for oxidizing NADH. Two of the common end products formed from fermentation are (a)ethanol and (b)lactate, the ionized form of lactic acid.

Comparing Fermentation with Anaerobic and Aerobic Respiration

Fermentation, anaerobic respiration, and aerobic respiration are three alternative cellular pathways for producing ATP by harvesting the chemical energy of food. All three use glycolysis to oxidize glucose and other organic fuels to pyruvate, with a net production of 2 ATP by substrate-level phosphorylation. And in all three pathways, NAD^+ is the oxidizing agent that accepts electrons from food during glycolysis.

A key difference is the contrasting mechanisms for oxidizing NADH back to NAD^+, which is required to sustain glycolysis. In fermentation, the final electron acceptor is an organic molecule such as pyruvate (lactic acid fermentation) or acetaldehyde (alcohol fermentation). In cellular respiration, by contrast, electrons carried by NADH are transferred to an electron transport chain, which regenerates the NAD^+ required for glycolysis.

Another major difference is the amount of ATP produced. Fermentation yields two molecules of ATP, produced by substrate-level phosphorylation. In the absence of an electron transport chain, the energy stored in pyruvate is unavailable. In cellular respiration, however, pyruvate is completely oxidized in the mitochondrion. Most of the chemical energy from this process is shuttled by NADH and $FADH_2$ in the form of electrons to the electron transport chain. There, the electrons move stepwise down a series of redox reactions to a final electron acceptor.

In aerobic respiration, the final electron acceptor is O_2; in anaerobic respiration, the final acceptor is another molecule with a high affinity for electrons, although less so than O_2. Stepwise electron transport drives oxidative phosphorylation, yielding ATP. Thus, cellular respiration harvests much more energy from each sugar molecule than fermentation can. In fact, aerobic respiration yields up to 32 molecules of ATP per glucose molecule, up to 16 times as much as does fermentation.

Some organisms, called **obligate anaerobes**, carry out only fermentation or anaerobic respiration. In fact, these organisms cannot survive in the presence of oxygen, some forms of which can actually be toxic if protective systems are not present in the cell. A few cell types, such as cells of the vertebrate brain, can carry out only aerobic oxidation of pyruvate, and need O_2 to survive. Other organisms, including yeasts and many bacteria, can make enough ATP to survive using either fermentation or respiration. Such species are called **facultative anaerobes**.

In yeast cells, for example, pyruvate is a fork in the metabolic road that leads to two alternative catabolic routes. Under aerobic conditions, pyruvate can be converted to acetyl CoA, and oxidation continues in the citric acid cycle via aerobic respiration. Under anaerobic conditions, lactic acid fermentation occurs. To make the same amount of ATP, a facultative anaerobe has to consume sugar at a much faster rate when fermenting than when respiring. Some prokaryotic and eukaryotic cells are **strict aerobes**, meaning that they have an absolute requirement for oxygen to survive and are unable to live solely by fermentations. Vertebrate brain cells are key examples of strict aerobes.

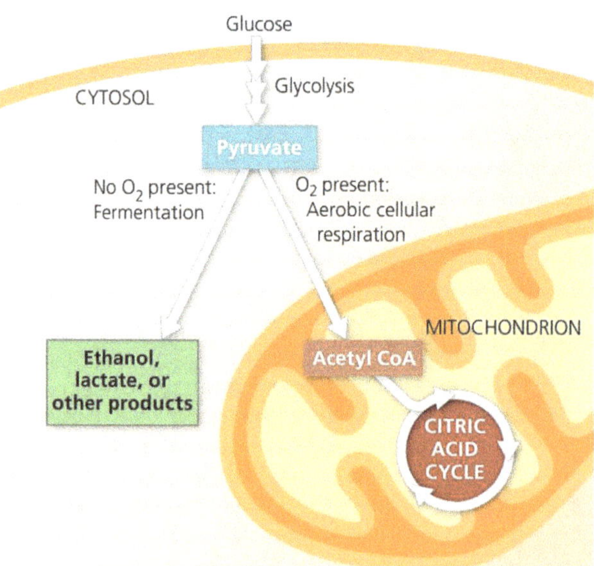

▲ **Figure 6.20** Pyruvate as a key juncture in catabolism. Glycolysis is common to fermentation and cellular respiration. The end product of glycolysis, pyruvate, represents a fork in the catabolic pathways of glucose oxidation. In a facultative anaerobe, capable of both aerobic cellular respiration and fermentation, pyruvate is committed to one of those two pathways, usually depending on whether or not oxygen is present.

Glycolysis and the citric acid cycle connect to many other metabolic pathways

We obtain most of our calories in the form of fats, proteins, and carbohydrates such as sucrose and other disaccharides, and starch, a polysaccharide. All these organic molecules in food can be used by cellular respiration to make ATP.

Glycolysis can accept a wide range of carbohydrates for catabolism. In the digestive tract, starch is hydrolyzed to glucose, which is broken down in cells by glycolysis and the citric acid cycle. Glycogen, the polysaccharide that humans and many other animals store in their liver and muscle cells, can be hydrolyzed to glucose between meals as fuel for respiration. Digestion of disaccharides, including sucrose, provides glucose and other monosaccharides as fuel for respiration.

Proteins can also be used for fuel, but first they must be digested to their constituent amino acids. Many of the amino acids are used by the organism to build new proteins. Amino acids present in excess are converted by enzymes to intermediates of glycolysis and the citric acid cycle. For example, alanine is converted into pyruvate, glutamate into α-ketoglutarate and aspartate into oxaloacetate. Before amino acids can feed into glycolysis or the citric acid cycle, their amino groups must be removed, a process called **deamination**. The nitrogenous waste is excreted from the animal in the form of ammonia (NH_3), urea, or other waste products.

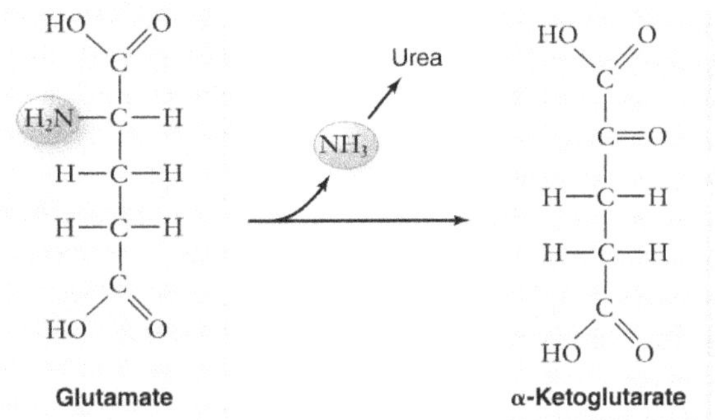

▲ **Figure 6.21** Deamination. After proteins are broken down into their amino acid constituents, the amino groups are removed from the amino acids to form molecules that participate in glycolysis and the Krebs cycle.

Catabolism can also harvest energy stored in fats obtained either from food or from fat cells in the body. After fats are digested to glycerol and fatty acids, the glycerol is converted to glyceraldehyde 3-phosphate, an intermediate of glycolysis. Most of the energy of a fat is stored in the fatty acids. A metabolic sequence called **beta oxidation** breaks the fatty acids down to two-carbon fragments, which enter the citric acid cycle as acetyl CoA.

NADH and $FADH_2$ are also generated during beta oxidation; they can enter the electron transport chain, leading to further ATP production. Fats make excellent fuels, in large part due to their chemical structure and the high energy level of their electrons (present in many C–H bonds, equally shared between C and H) compared to those of carbohydrates. A gram of fat oxidized by respiration produces more than twice as much ATP as a gram of carbohydrate.

Although most cells store small amounts of fat, most of the body's fat is stored in specialized cells known as adipocytes. Almost the entire cytoplasm of each of these cells is filled with a single, large fat droplet. Clusters of adipocytes form adipose tissue, most of which is in deposits underlying the skin or surrounding internal organs. The function of adipocytes is to synthesize and store triglycerides during periods of food uptake and then, when food is not being absorbed from the small intestine, to release fatty acids and glycerol into the blood for uptake and use by other cells to provide the energy needed for ATP formation.

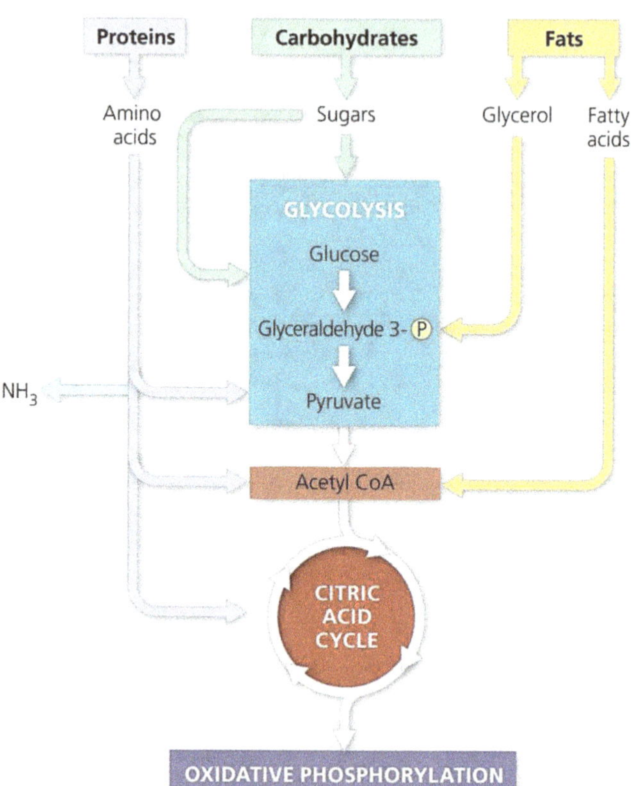

▶ **Figure 6.22** The catabolism of various molecules from food. Carbohydrates, fats, and proteins can all be used as fuel for cellular respiration. Monomers of these molecules enter glycolysis or the citric acid cycle at various points. Glycolysis and the citric acid cycle are catabolic funnels through which electrons from all kinds of organic molecules flow on their exergonic fall to oxygen.

The ratio of carbon dioxide produced to oxygen used is called the **respiratory quotient** (RQ). Measuring this ratio is necessary to work out which substrate is being used in respiration. It can also show whether or not anaerobic respiration is occurring.

$$RQ = \frac{\text{volume of carbon dioxide given out in unit time}}{\text{volume of oxygen taken in unit time}}$$

Or, from an equation:

$$RQ = \frac{\text{moles or molecules of carbon dioxide given out}}{\text{moles or molecules of oxygen taken in}}$$

Respiratory substrate	Respiratory quotient (RQ)
carbohydrate	1.0
lipid	0.7
protein	0.9

Biosynthesis (Anabolic Pathways)

Cells need substance as well as energy. Not all the organic molecules of food are destined to be oxidized as fuel to make ATP. In addition to calories, food must also provide the carbon skeletons that cells require to make their own molecules. Some organic monomers obtained from digestion can be used directly. For example, amino acids from the hydrolysis of proteins in food can be incorporated into the organism's own proteins.

Often, however, the body needs specific molecules that are not present as such in food. Compounds formed as intermediates of glycolysis and the citric acid cycle can be diverted into anabolic pathways as precursors from which the cell can synthesize the molecules it requires. For example, humans can make about half of the 20 amino acids in proteins by modifying compounds siphoned away from the citric acid cycle; the rest are "essential amino acids" that must be obtained in the diet.

Also, glucose can be made from pyruvate, and fatty acids can be synthesized from acetyl CoA. When ATP levels are high, the oxidative pathway is inhibited, and acetyl-CoA is channeled into fatty acid synthesis. This explains why many animals (humans included) develop fat reserves when they consume more food than their activities require. Of course, these anabolic, or biosynthetic, pathways do not generate ATP, but instead consume it. Alternatively, when ATP levels are low, the oxidative pathway is stimulated, and acetyl-CoA flows into energy- producing oxidative metabolism.

In addition, glycolysis and the citric acid cycle function as metabolic interchanges that enable our cells to convert some kinds of molecules to others as we need them. For example, an intermediate compound generated during glycolysis, dihydroxyacetone phosphate, can be converted to one of the major precursors of fats. If we eat more food than we need, we store fat even if our diet is fat free. Metabolism is remarkably versatile and adaptable. Therefore, glucose can readily be metabolized and used to synthesize fat, but the fatty acid portion of fat cannot be used to synthesize glucose.

In addition, when energy is not needed by the body, glucose can be synthesized from intermediates of the glycolysis and citric acid cycle pathways, as well as from molecules derived from those pathways. Examples are pyruvate, lactate, malate, oxaloacetate, and several amino acids. The glucose biosynthesis process is called **gluconeogenesis**. Its reactions are the reverse of those of glycolysis, involving many of the same enzymes. However, four enzymes are unique to gluconeogenesis, making the pathway distinct from that of glycolysis. Gluconeogenesis consumes ATP, rather than produces it. As you would expect, the regulation of glycolysis and gluconeogenesis is carefully controlled according to the energy needs of the body.

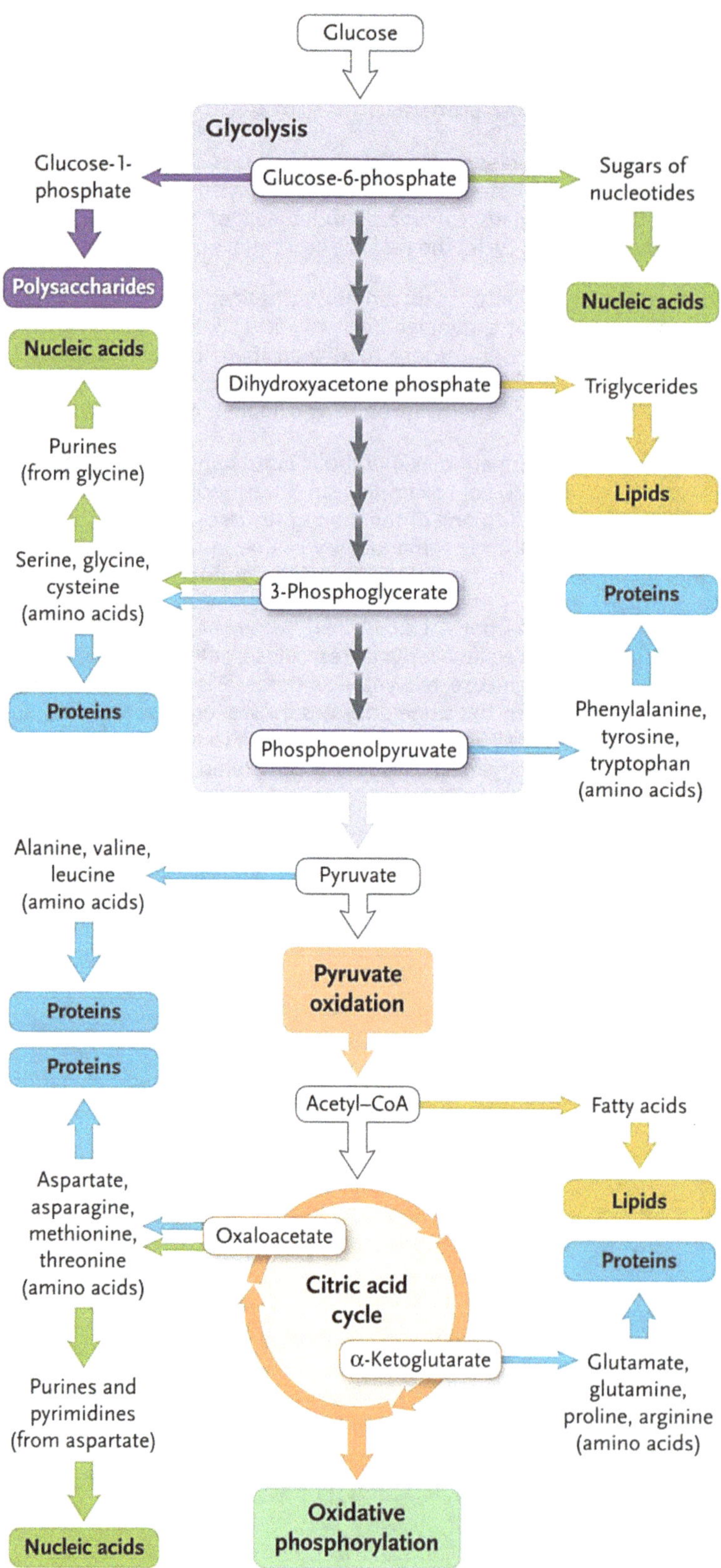

▲ **Figure 6.23** Molecules synthesized from compounds in glycolysis, pyruvate oxidation, and the citric acid cycle. The products are used in the biosynthesis of many of the cell's important molecules, such as nucleic acids, proteins, polysaccharides, and lipids. © Cengage Learning 2017

Glycolysis and Citric Acid Cycle Stages of Cellular Respiration Are Regulated by Feedback Mechanisms

The cell does not waste energy making more of a particular substance than it needs. If there is a surplus of a certain amino acid, for example, the anabolic pathway that synthesizes that amino acid from an intermediate of the citric acid cycle is switched off. The most common mechanism for this control is feedback inhibition: The end product of the anabolic pathway inhibits the enzyme that catalyzes an early step of the pathway.

The cell also controls its catabolism. If the cell is working hard and its ATP concentration begins to drop, cellular respiration speeds up. When there is plenty of ATP to meet demand, respiration slows down, sparing valuable organic molecules for other functions. Again, control is based mainly on regulating the activity of enzymes at strategic points in the catabolic pathway. One important switch is phosphofructokinase, the enzyme that catalyzes step 3 of glycolysis. That is the first step that commits the substrate irreversibly to the glycolytic pathway. By controlling the rate of this step, the cell can speed up or slow down the entire catabolic process. Phosphofructokinase can thus be considered the pacemaker of cellular respiration.

Phosphofructokinase is an allosteric enzyme with receptor sites for specific inhibitors and activators. It is inhibited by ATP and stimulated by AMP (adenosine monophosphate), which the cell derives from ADP. As ATP accumulates, inhibition of the enzyme slows down glycolysis. The enzyme becomes active again as cellular work converts ATP to ADP (and AMP) faster than ATP is being regenerated. Phosphofructokinase is also sensitive to citrate, the first product of the citric acid cycle.

If citrate accumulates in mitochondria, some of it passes into the cytosol and inhibits phosphofructokinase. This mechanism helps synchronize the rates of glycolysis and the citric acid cycle. As citrate accumulates, glycolysis slows down, and the supply of pyruvate and thus acetyl groups to the citric acid cycle decreases. If citrate consumption increases, either because of a demand for more ATP or because anabolic pathways are draining off intermediates of the citric acid cycle, glycolysis accelerates and meets the demand. Metabolic balance is augmented by the control of enzymes that catalyze other key steps of glycolysis and the citric acid cycle.

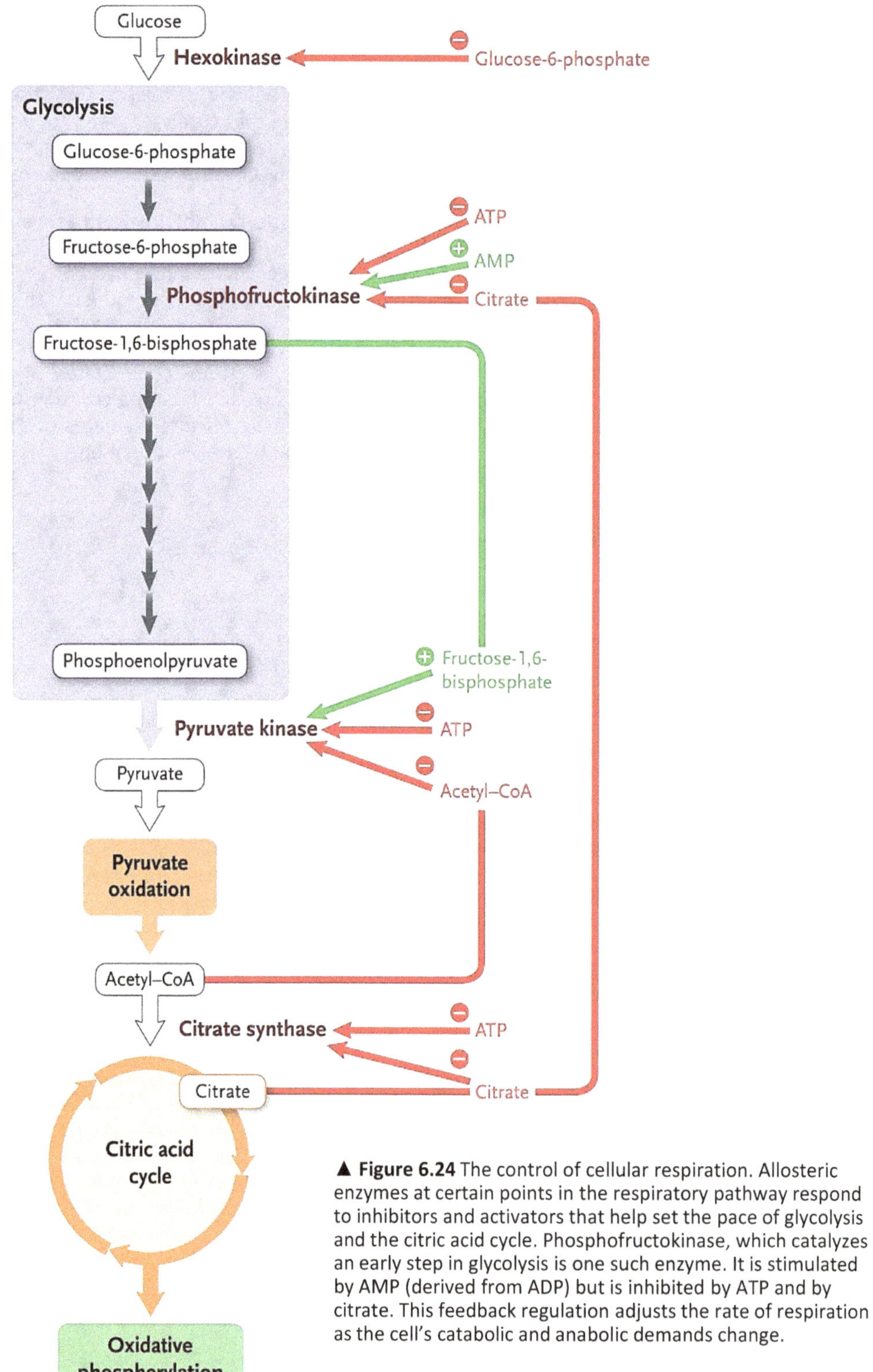

▲ **Figure 6.24 The control of cellular respiration.** Allosteric enzymes at certain points in the respiratory pathway respond to inhibitors and activators that help set the pace of glycolysis and the citric acid cycle. Phosphofructokinase, which catalyzes an early step in glycolysis is one such enzyme. It is stimulated by AMP (derived from ADP) but is inhibited by ATP and by citrate. This feedback regulation adjusts the rate of respiration as the cell's catabolic and anabolic demands change.

Photosynthesis and Respiration Are Ancient Pathways

Photosynthesis, glycolysis, and cellular respiration are intimately related. The carbohydrate product of photosynthesis, glucose, is the starting material for glycolysis. The O_2 released in photosynthesis becomes the final electron acceptor in aerobic respiration. CO_2 generated in respiration enters the carbon reactions in chloroplasts. Finally, photosynthesis splits water produced by aerobic respiration. Together, these energy reactions sustain life.

Glycolysis is probably the most ancient of the energy pathways because it occurs in virtually all cells. Glycolysis evolved when the atmosphere lacked or had very little O_2. These reactions enabled the earliest organisms to extract energy from simple organic compounds in the nonliving environment. Photosynthesis, in turn, may have evolved from glycolysis; some of the reactions of the Calvin cycle are the reverse of some of those of glycolysis.

The first photosynthetic organisms could not have been plants, because such complex organisms were not present on the early Earth. Rather, photosynthesis may have originated in an anaerobic cell that used hydrogen sulfide (H_2S) instead of water as an electron donor. These first photosynthetic microorganisms would have released sulfur, rather than O_2, into the environment. Eventually, changes in pigment molecules enabled some of these organisms to use water instead of H_2S as an electron source.

Fossil evidence of cyanobacteria shows that oxygen-generating photosynthesis arose at least 3.5 billion years ago. Once this pathway started, the accumulation of O_2 in the primitive atmosphere altered life on Earth forever. Later, in a process called endosymbiosis, a large "host" cell engulfed one of those ancient cyanobacteria and thereby transformed itself into a eukaryotic-like cell, complete with chloroplasts. Mitochondria evolved in a similar way, when larger cells engulfed bacteria capable of using O_2.

The double membranes of both chloroplasts and mitochondria are a consequence of endosymbiosis. The engulfed bacterium's cell membrane developed into each organelle's inner membrane, and the vesicle membrane remained as the outer membrane. Endosymbiosis therefore explains why the electron transport chain is in the bacterial cell membrane but in the inner mitochondrial membrane of a eukaryotic cell. The observation that both mitochondria and chloroplasts contain DNA and ribosomes lends additional support to the endosymbiosis theory.

As time went on, different types of complex cells probably diverged, leading to the evolution of a great variety of eukaryotic organisms. Today, the interrelationships among photosynthesis, glycolysis, and aerobic respiration, along with the great similarities of these reactions in diverse species, demonstrate a unifying theme of biology: All types of organisms are related at the biochemical level.

	Photosynthesis	Respiration
Food	Produced	Consumed
Energy	Stored as glucose	Released from glucose
Light	Required	Not required
H_2O	Consumed	Released
CO_2	Consumed	Released
O_2	Released	Consumed

▲ **Table 6.3** Photosynthesis and Respiration Compared

KEY CONCEPTS

- Aerobic respiration uses oxygen as the final electron acceptor for redox reactions. Anaerobic respiration utilizes inorganic molecules as acceptors, and fermentation uses organic molecules.

- Electron carriers can be reversibly oxidized and reduced. For example, NAD^+ is reduced to NADH by acquiring two electrons; NADH supplies these electrons to other molecules to reduce them.

- Mitochondria of eukaryotic cells move electrons in steps via the electron transport chain to capture energy efficiently.

- The ultimate goal of cellular respiration is synthesis of ATP, which is used to power most of the cell's activities.

- Substrate-level phosphorylation transfers a phosphate directly to ADP Oxidative phosphorylation generates ATP via the enzyme ATP synthase, powered by a proton gradient.

- Priming reactions add two phosphates to glucose; this is cleaved into two 3-carbon molecules of glyceraldehyde 3-phosphate (G3P). Oxidation of G3P transfers electrons to NAD^+, yielding NADH. After four more reactions, the final product is two molecules of pyruvate. Glycolysis produces 2 net ATP, 2 NADH, and 2 pyruvate.

- Glycolysis is an ancient process with a low energy yield, but it can be efficient with up to 40% of available energy trapped as ATP. Glycolysis was probably the first catabolic reaction to evolve.

- In the presence of oxygen, pyruvate is oxidized to acetyl-CoA, which can be oxidized by the Krebs cycle. This process leads to a large amount of ATP. In the absence of oxygen, a fermentation reaction uses all or part of pyruvate to oxidize NADH.

- In the presence of oxygen, NADH passes electrons to the electron transport chain. In the absence of oxygen, NADH passes the electrons to an organic molecule such as acetaldehyde (fermentation).

- Pyruvate is oxidized to yield 1 CO_2, 1 NADH, and 1 acetyl-CoA. Acetyl-CoA enters the Krebs cycle as 2-carbon acetyl units.

- The first reaction is an irreversible condensation that produces citrate; it is inhibited when ATP is plentiful. The second and third reactions rearrange citrate to isocitrate. The fourth and fifth reactions are oxidations; in each reaction, one NAD^+ is reduced to NADH. The sixth reaction is a substrate-level phosphorylation producing GTP, and from that ATP. The seventh reaction is another oxidation that reduces FAD to $FADH_2$. Reactions eight and nine regenerate oxaloacetate, including one final oxidation that reduces NAD^+ to NADH.

- As a glucose molecule is broken down to CO_2, some of its energy is preserved in 4 ATP, 10 NADH, and 2 $FADH_2$.

- In the inner mitochondrial membrane, NADH is oxidized to NAD^+ by NADH dehydrogenase. Electrons move through ubiquinone and the bc_1 complex to cytochrome oxidase, where they join with H^+ and O_2 to form H_2O. This results in three protons being pumped into the intermembrane space. For $FADH_2$, electrons are passed directly to ubiquinone. Thus only two protons are pumped into the intermembrane space.

- Electron transport powers proton pumps in the inner membrane.

- Chemiosmosis utilizes the electrochemical gradient to produce ATP.

- Protons diffuse back into the mitochondrial matrix via the ATP synthase channel. The enzyme uses this energy to synthesize ATP.

- Electron transport powers proton pumps in the inner membrane.

- Chemiosmosis utilizes the electrochemical gradient to produce ATP.

- Protons diffuse back into the mitochondrial matrix via the ATP synthase channel. The enzyme uses this energy to synthesize ATP.

- The theoretical yield for eukaryotes is 30 molecules of ATP per glucose molecule.

- Glucose catabolism is controlled by the concentration of ATP molecules and intermediates in the Krebs cycle.

- In the absence of oxygen other final electron acceptors can be used for respiration.

- Fermentation is the regeneration of NAD^+ by oxidation of NADH and reduction of an organic molecule. In yeast, pyruvate is decarboxylated, then reduced to ethanol. In animals, pyruvate is reduced directly to lactate.

- Fatty acids are converted to acetyl groups by successive rounds of βoxidation. These acetyl groups feed into the Krebs cycle to be oxidized and generate NADH for electron transport.

- A small number of key intermediates connect metabolic pathways.

- Major milestones are recognized in the evolution of metabolism; the order of events is hypothetical.

Photosynthesis

Chapter Contents:

- Photosynthesis feeds the biosphere
- Chloroplasts: The Sites of Photosynthesis in Plants
- The Two Stages of Photosynthesis
- The light reactions
- Photosynthetic Pigments: The Light Receptors
- Excitation of Chlorophyll by Light
- Photosystem
- Linear Electron Flow
- Cyclic Electron Flow
- A Comparison of Chemiosmosis in Chloroplasts and Mitochondria
- The Calvin cycle
- Alternative mechanisms of carbon fixation
- C_4 Plants have evolved to minimize photorespiration
- CAM Plants
- Limiting factors in photosynthesis
- KEY CONCEPTS

Photosynthesis feeds the biosphere

The conversion process that transforms the energy of sunlight into chemical energy stored in sugars and other organic molecules is called **photosynthesis**. It is an oxidation–reduction (redox) process. "Oxidation" means that electrons are removed from an atom or molecule; "reduction" means electrons are added. Photosynthesis strips electrons from the oxygen atoms in H_2O. Photosynthesis nourishes almost the entire living world directly or indirectly. An organism acquires the organic compounds it uses for energy and carbon skeletons by one of two major modes: autotrophic nutrition or heterotrophic nutrition.

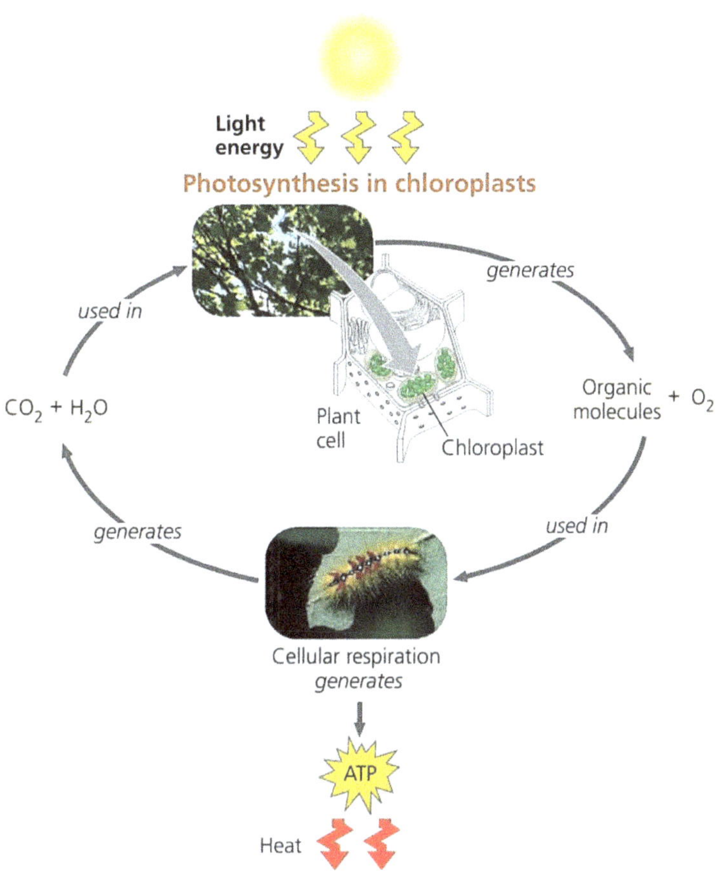

Autotrophs are "self-feeders"; they sustain themselves without eating anything derived from other living beings. Photosynthesizers and other organisms that make all of their required organic molecules from CO_2 and other inorganic sources such as water are called autotrophs. They are the ultimate sources of organic compounds for all nonautotrophic organisms, and for this reason, biologists refer to autotrophs as the **producers** of the biosphere.

Almost all plants are autotrophs; the only nutrients they require are water and minerals from the soil and CO_2 from the air. Specifically, plants are **photoautotrophs**, organisms that use light as a source of energy to synthesize organic substances. Photosynthesis also occurs in algae, certain other protists, and some prokaryotes (multicellular algae, some non-algal unicellular protists such as Euglena, the prokaryotes called cyanobacteria and other photosynthetic prokaryotes such as purple sulfur bacteria which produce sulfur and some archaea). Some other prokaryotes are **chemoautotrophs**, which obtain their energy from the oxidation of reduced inorganic molecules such as hydrogen sulfide (H_2S), nitrite (NO_2^-), or ammonia (NH_3). Some of this captured energy is subsequently used to carry out carbon fixation.

Heterotrophs are unable to make their own food; they live on compounds produced by other organisms. Heterotrophs are the biosphere's **consumers**. Almost all heterotrophs, including humans, are completely dependent, either directly or indirectly, on photoautotrophs for food—and also for oxygen, a by-product of photosynthesis.

Chloroplasts: The Sites of Photosynthesis in Plants

Chloroplast, is the eukaryotic organelle that absorbs energy from sunlight and uses it to drive the synthesis of organic compounds from carbon dioxide (CO_2) and water (H_2O). All green parts of a plant, including green stems and unripened fruit, have chloroplasts, but the leaves are the major sites of photosynthesis in most plants. There are about half a million chloroplasts in a chunk of leaf with a top surface area of 1 mm^2. Chloroplasts are found mainly in the cells of the **mesophyll**, the tissue in the interior of the leaf. CO_2 enters the leaf, and O_2 exits, by way of microscopic pores called **stomata** (singular, stoma; from the Greek, meaning "mouth"). Water absorbed by the roots is delivered to the leaves in veins. Leaves also use veins to export sugar to roots and other nonphotosynthetic parts of the plant.

A chloroplast has two membranes surrounding a dense fluid called the **stroma**. The stroma of a chloroplast contains small ribosomes and small circles of DNA, used to synthesise proteins. The stroma also contains starch grains, which store some of the carbohydrate made, in an insoluble form. Suspended within the stroma is a third membrane system, made up of sacs called **thylakoids**. The thylakoid membrane encloses a fluid-filled interior space, the **thylakoid lumen**. In some places, thylakoid sacs are stacked in columns called **grana** (singular, granum). Connections between grana are termed **stroma lamella**. **Chlorophyll**, the green pigment that gives leaves their color, resides in the thylakoid membranes of the chloroplast. It is the light energy absorbed by chlorophyll that drives the synthesis of organic molecules in the chloroplast.

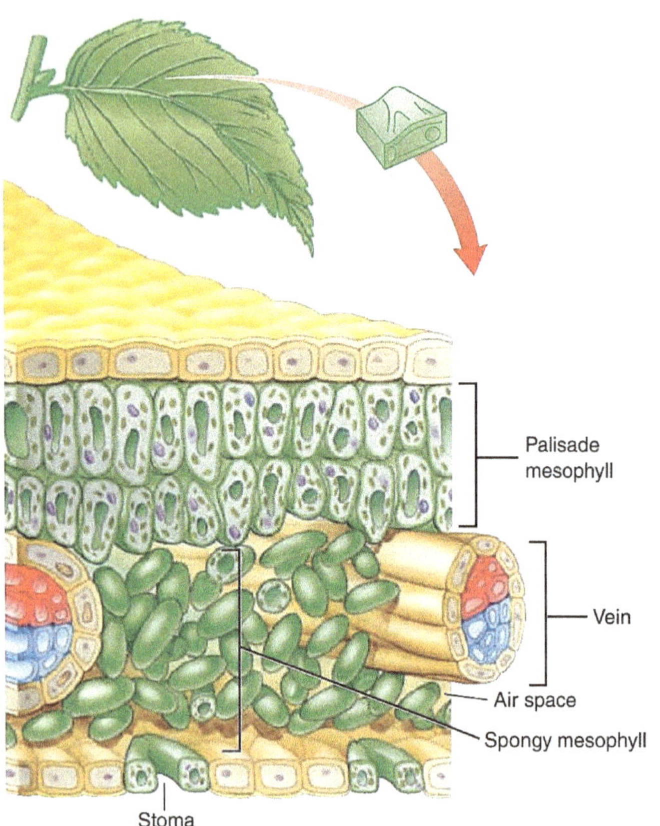

◄ **Figure 7.1** This leaf cross section reveals that the mesophyll is the photosynthetic tissue. CO_2 enters the leaf through tiny pores or stomata, and H_2O is carried to the mesophyll in veins.

► **Figure 7.2** In the chloroplast, pigments necessary for the light-capturing reactions of photosynthesis are part of thylakoid membranes, whereas the enzymes for the synthesis of carbohydrate molecules are in the stroma.

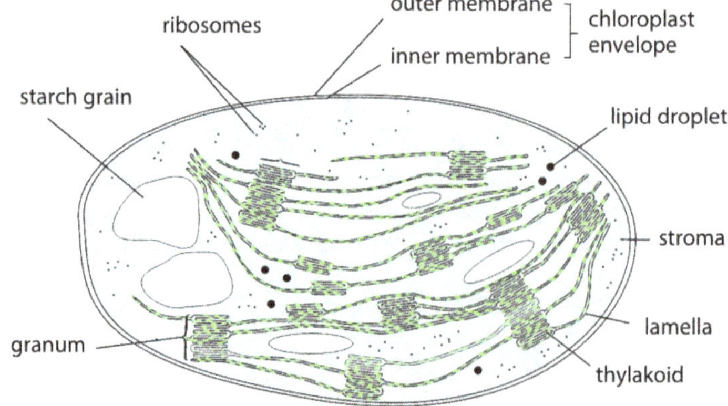

The overall photosynthetic equation has been known since the 1800s: In the presence of light, the green parts of plants produce organic compounds and O_2 from CO_2 and H_2O. We can summarize the complex series of chemical reactions in photosynthesis with this chemical equation:

$$6CO_2 + 12H_2O + light \longrightarrow C_6H_{12}O_6 + 6H_2O + 6O_2$$

carbon   water   glucose   water   oxygen
dioxide

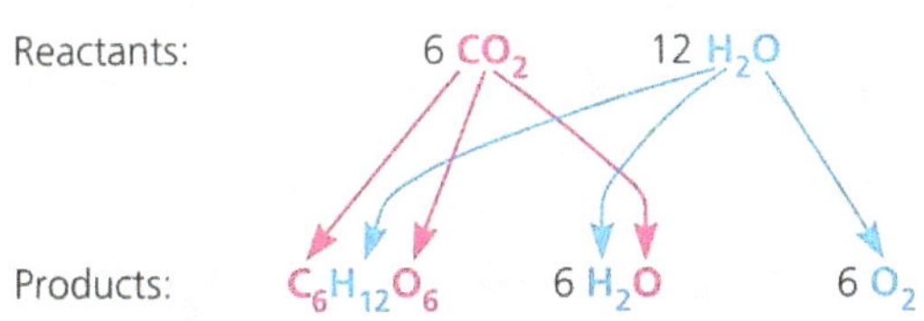

$$\text{Energy} + 6\,CO_2 + 6\,H_2O \longrightarrow C_6H_{12}O_6 + 6\,O_2$$

becomes reduced
becomes oxidized

Water is split, and its electrons are transferred along with hydrogen ions (H+) from the water to carbon dioxide, reducing it to sugar. Because the electrons increase in potential energy as they move from water to sugar, this process requires energy—in other words, it is endergonic. This energy boost that occurs during photosynthesis is provided by light.

▲ **Figure 7.3** Oxygen produced by photosynthesis. On sunny days, the oxygen released by aquatic plants is sometimes visible as bubbles in the water. This plant (Elodea) is actively carrying on photosynthesis.

Nigel Cattlin/Visuals Unlimited Inc.

The Two Stages of Photosynthesis

Actually, photosynthesis is not a single process, but two processes, each of which has multiple steps. These two stages of photosynthesis are known as the **light reactions** (the photopart of photosynthesis) and the **Calvin cycle** (the synthesis part).

The light reactions are the steps of photosynthesis that convert solar energy to chemical energy. Water is split, providing a source of electrons and protons (hydrogen ions, H^+) and giving off O_2 as a by-product. Light absorbed by chlorophyll drives a transfer of the electrons and hydrogen ions from water to an acceptor called **$NADP^+$ (nicotinamide adenine dinucleotide phosphate)**, where they are temporarily stored. The light reactions use solar energy to reduce $NADP^+$ to **NADPH** by adding a pair of electrons along with an H^+ (Although the correct way to write the reduced form of $NADP^+$ is NADPH + H^+, for simplicity it is presented as NADPH).

The light reactions also generate ATP, using chemiosmosis to power the addition of a phosphate group to ADP, a process called **photophosphorylation**. Thus, light energy is initially converted to chemical energy in the form of two compounds: NADPH and ATP. NADPH, a source of electrons, acts as "reducing power" that can be passed along to an electron acceptor, reducing it, while ATP is the versatile energy currency of cells. Notice that the light reactions produce no sugar; that happens in the second stage of photosynthesis, the Calvin cycle.

The cycle begins by incorporating CO_2 from the air into organic molecules already present in the chloroplast. This initial incorporation of carbon into organic compounds is known as **carbon fixation**. The Calvin cycle then reduces the fixed carbon to carbohydrate by the addition of electrons. The reducing power is provided by NADPH, which acquired its cargo of electrons in the light reactions. To convert CO_2 to carbohydrate, the Calvin cycle also requires chemical energy in the form of ATP, which is also generated by the light reactions. Thus, it is the Calvin cycle that makes sugar, but it can do so only with the help of the NADPH and ATP produced by the light reactions. The metabolic steps of the Calvin cycle are sometimes referred to as the **dark reactions**, or **light-independent reactions**, because none of the steps requires light directly.

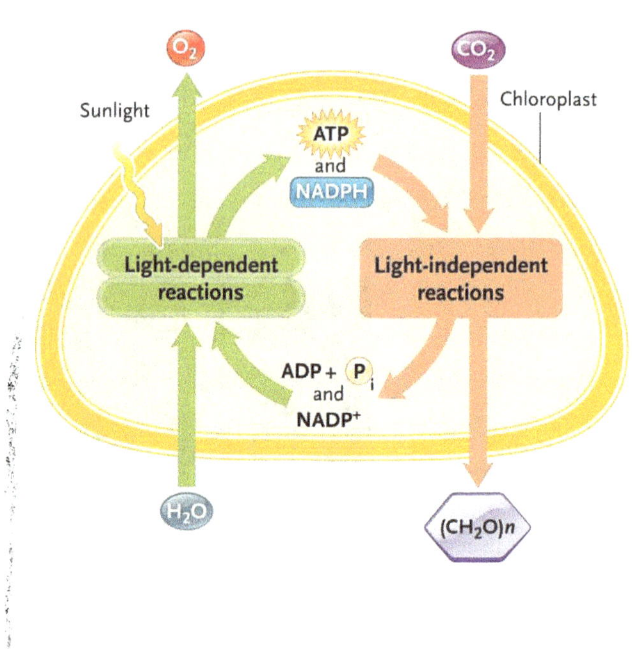

► **Figure 7.4** The light-dependent and light-independent reactions of photosynthesis, and their interlinking reactants and products. Both series of reactions occur in the chloroplasts of plants and algae. © Cengage Learning 2017

The thylakoids of the chloroplast are the sites of the light reactions, while the Calvin cycle occurs in the stroma. On the outside of the thylakoids, during the light reactions, molecules of $NADP^+$ and ADP pick up electrons and phosphate, respectively, and NADPH and ATP are then released to the stroma, where they play crucial roles in the Calvin cycle.

The light reactions convert solar energy to the chemical energy of ATP and NADPH

Light is a form of energy known as electromagnetic energy, also called electromagnetic radiation. Electromagnetic energy travels in rhythmic waves analogous to those created by dropping a pebble into a pond. Electromagnetic waves, however, are disturbances of electric and magnetic fields rather than disturbances of a material medium such as water. The distance between the crests of electromagnetic waves is called the **wavelength**. Wavelengths range from less than a nanometer (for gamma rays) to more than a kilometer (for radio waves). This entire range of radiation is known as the **electromagnetic spectrum.** The segment most important to life is the narrow band from about 380 nm to 740 nm in wavelength. This radiation is known as **visible light** because it can be detected as various colors by the human eye.

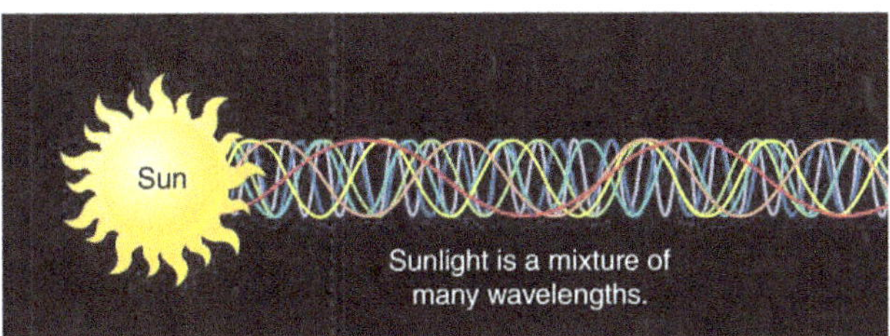

The model of light as waves explains many of light's properties, but in certain respects light behaves as though it consists of discrete particles, called **photons**. Photons are not tangible objects, but they act like objects in that each of them has a fixed quantity of energy. The sunlight that reaches Earth's surface consists of three main components of the electromagnetic spectrum: **ultraviolet radiation**, **visible light**, and **infrared radiation**. The amount of energy is inversely related to the wavelength of the light: The shorter the wavelength, the greater the energy of each photon of that light. Thus, a photon of violet light packs nearly twice as much energy as a photon of red light. Its high- energy photons damage DNA, causing sunburn and skin cancer.

In the middle range of wavelengths is visible light, which provides the energy that powers photosynthesis; we perceive visible light of different wavelengths as distinct colors. Infrared radiation, with its longer wavelengths, contains too little energy per photon to be useful to organisms. Most of its energy is converted immediately to heat. Although the sun radiates the full spectrum of electromagnetic energy, the atmosphere acts like a selective window, allowing visible light to pass through while screening out a substantial fraction of other radiation.

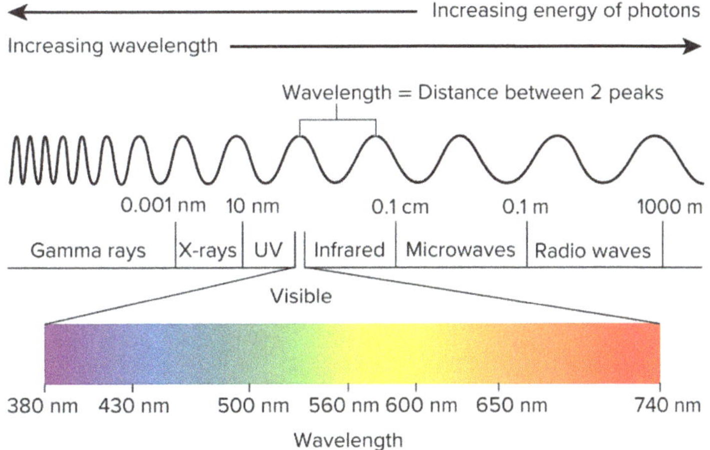

▲ **Figure 7.5** The electromagnetic spectrum. Light is a form of electromagnetic energy conveniently thought of as a wave. The shorter the wavelength of light, the greater its energy. Visible light represents only a small part of the electromagnetic spectrum between 400 and 740 nm.

Photosynthetic Pigments: The Light Receptors

When light meets matter, it may be reflected, transmitted, or absorbed. Substances that absorb visible light are known as **pigments**. A pigment is an organic molecule that selectively absorbs light of certain wavelengths. Different pigments absorb light of different wavelengths, and the wavelengths that are absorbed disappear. If a pigment is illuminated with white light, the color we see is the color most reflected or transmitted by the pigment. (If a pigment absorbs all wavelengths, it appears black.) We see green when we look at a leaf because chlorophyll absorbs violet-blue and red light while transmitting and reflecting green light.

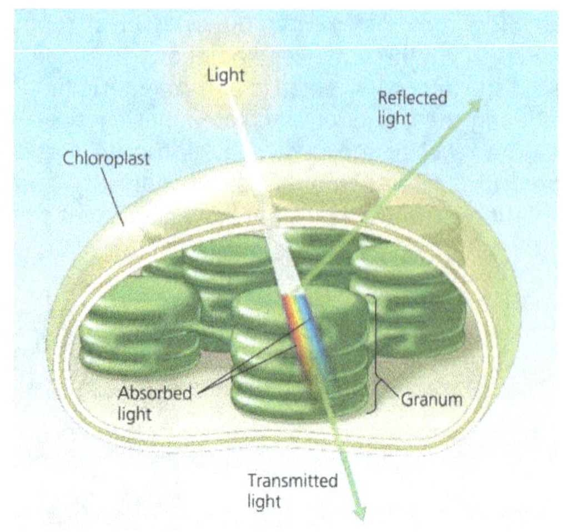

▶ **Figure 7.6** Why leaves are green: interaction of light with chloroplasts. The chlorophyll molecules of chloroplasts absorb violet-blue and red light (the colors most effective in driving photosynthesis) and reflect or transmit green light. This is why leaves appear green.

A graph plotting a pigment's light absorption versus wavelength is called an **absorption spectrum.** There are three types of pigments in chloroplasts: **chlorophyll a**, the key light-capturing pigment that participates directly in the light reactions; the accessory pigment **chlorophyll b**; and a separate group of accessory pigments called **carotenoids**. A slight structural difference between chlorophyll a and chlorophyll b is enough to cause the two pigments to absorb at slightly different wavelengths in the red and blue parts of the spectrum. As a result, chlorophyll a appears blue green and chlorophyll b olive green under visible light.

Carotenoids absorb light in the blue and blue-green regions of the visible spectrum, reflecting yellow, orange, and red. **Action spectrum**, plots the rate of photosynthesis as a function of the wavelength of light. The highest rates of photosynthesis in green plants correlate with the wavelengths that are strongly absorbed by the chlorophylls and carotenoids. Photosynthesis is poor in the green region of the spectrum.

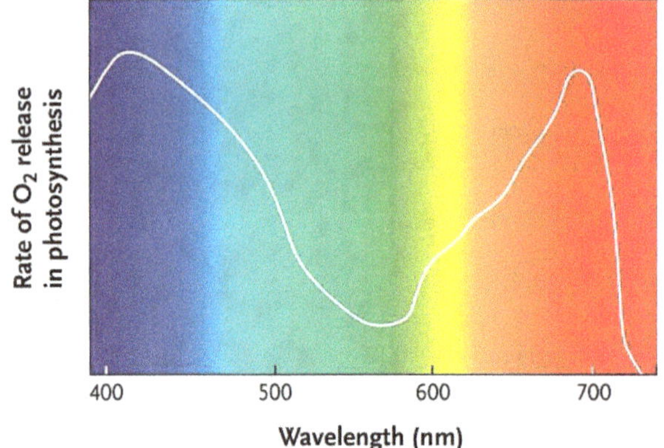

◀ **Table 7.8** The action spectrum in higher plants

▶ **Table 7.7** The absorption spectra of chlorophylls a and b.

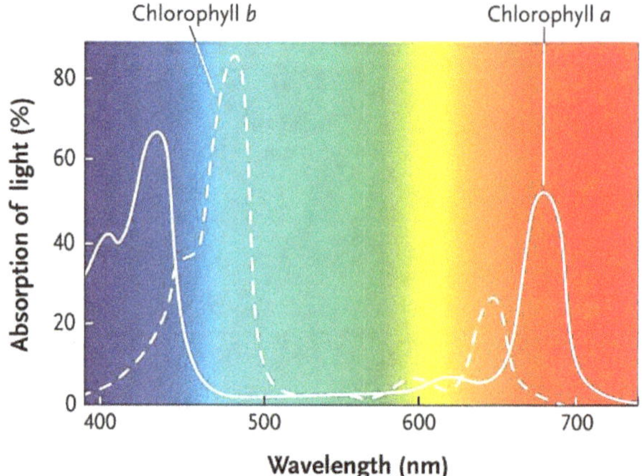

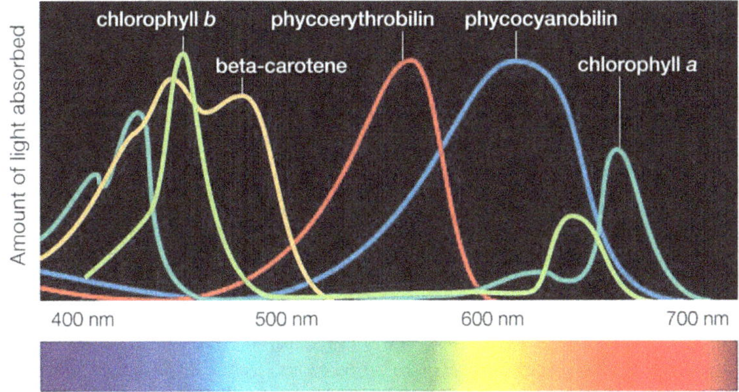

◄ **Figure 7.9** A few photosynthetic pigments. The curves in this graph show the efficiency at which each pigment absorbs the different wavelengths of visible light. Line color indicates the pigment's characteristic color. Using a combination of pigments allows photosynthetic organisms to maximize the range of wavelengths they can capture for photosynthesis. Credits: top, © Photobac/ Shutterstock; bottom, © Cengage Learning.

Chlorophylls: Two types of chlorophyll pigments, termed chlorophyll a and chlorophyll b, are found in green plants and green algae. In the chloroplast, both chlorophylls a and b are bound to integral membrane proteins in the thylakoid membrane, which traverse the membrane.

The chlorophylls contain a **porphyrin** ring and a hydrocarbon tail, also called a **phytol** tail. A magnesium ion ($Mg2+$) is bound to the porphyrin ring. An electron in the porphyrin ring is able to hop from one atom to another within this ring (the green-shaded region). Because this electron isn't restricted to a single atom, it is called a delocalized electron.

The delocalized electron can absorb light energy. The phytol tail in chlorophyll is a long hydrocarbon that is hydrophobic. Its function is to anchor the pigment to the surface of hydrophobic proteins within the thylakoid membrane of chloroplasts. All chlorophyll molecules in the thylakoid membrane are associated with specific chlorophyll-binding proteins; biologists have identified about 15 different kinds. Each thylakoid membrane is filled with precisely oriented chlorophyll molecules and chlorophyll-binding proteins that facilitate the transfer of energy from one molecule to another.

► **Figure 7.10** The structure of chlorophyll. Chlorophyll consists of a porphyrin ring and a hydrocarbon side chain. The porphyrin ring, with a magnesium atom in its center, contains a system of alternating double and single bonds; these bonds are commonly found in molecules that strongly absorb certain wavelengths of visible light and reflect others (chlorophyll reflects green). Note that at the top right corner of the diagram, the methyl group (CH_3) distinguishes chlorophyll a from chlorophyll b, which has a carbonyl group (—CHO) in this position. The hydrophobic hydrocarbon side chain anchors chlorophyll to the thylakoid membrane.

Carotenoids: Other accessory pigments include **carotenoids**, hydrocarbons that are various shades of yellow and orange because they absorb violet and blue-green light. Carotenoids may broaden the spectrum of colors that can drive photosynthesis. However, a more important function of at least some carotenoids seems to be **photoprotection**. These compounds absorb and dissipate excessive light energy that would otherwise damage chlorophyll or interact with oxygen, forming reactive oxidative molecules that are dangerous to the cell.

Interestingly, carotenoids similar to the photoprotective ones in chloroplasts have a photoprotective role in the human eye. You may have heard that eating carrots can enhance vision. If this effect is real, it is probably due to the high content of **β-carotene** in carrots. This carotenoid consists of two molecules of vitamin A joined together. The oxidation of vitamin A produces retinal, the pigment used in vertebrate vision.

Phycobiliproteins are accessory pigments found in cyanobacteria and some algae. These pigments contain a system of alternating double bonds similar to those found in other pigments and molecules that transfer electrons. Phycobiliproteins can be organized to form another light-harvesting complex that can absorb green light, which is reflected by chlorophyll. These complexes are probably ecologically important to cyanobacteria, helping them to exist in low-light situations in oceans. In this habitat, green light remains because red and blue light has been absorbed by green algae closer to the surface.

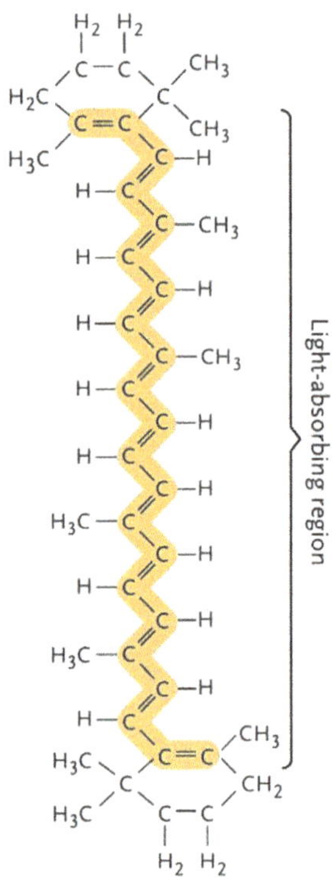

▲ **Figure 7.11** Carotenoid structure

▲ **Figure 7.12** Fall colors are produced by carotenoids and other accessory pigments. During the spring and summer, chlorophyll in leaves masks the presence of carotenoids and other accessory pigments. When cool fall temperatures cause leaves to cease manufacturing chlorophyll, the chlorophyll is no longer present to reflect green light, and the leaves reflect the orange and yellow light that carotenoids and other pigments do not absorb.

Pigment	Color(s)	Organisms
Major pigment		
Chlorophyll *a*	Blue-green	Plants, algae, cyanobacteria
Accessory pigments		
Chlorophyll *b*	Yellow-green	Plants, green algae
Carotenoids (carotenes and xanthophylls)	Red, orange, yellow	Plants, algae, bacteria

▲ **Table 7.1** Pigments of Photosynthesis.

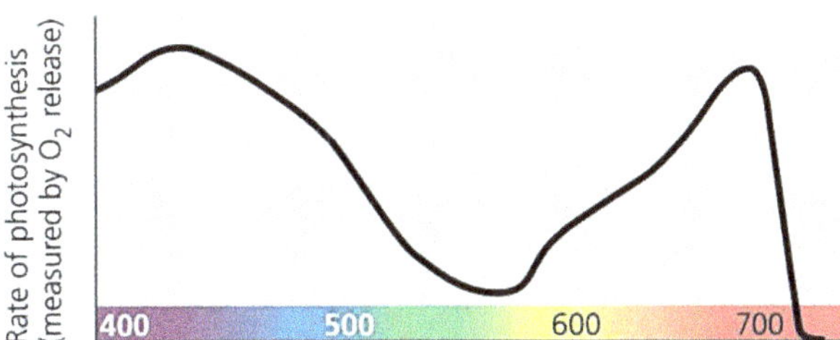

(a) **Absorption spectra.** The three curves show the wavelengths of light best absorbed by three types of chloroplast pigments.

(b) **Action spectrum.** This graph plots the rate of photosynthesis versus wavelength. The resulting action spectrum resembles the absorption spectrum for chlorophyll *a* but does not match exactly (see part a). This is partly due to the absorption of light by accessory pigments such as chlorophyll *b* and carotenoids.

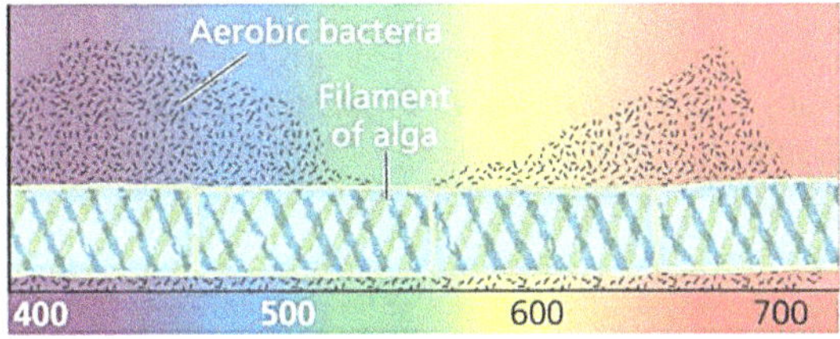

(c) **Engelmann's experiment.** In 1883, Theodor W. Engelmann illuminated a filamentous alga with light that had been passed through a prism, exposing different segments of the alga to different wavelengths. He used aerobic bacteria, which concentrate near an oxygen source, to determine which segments of the alga were releasing the most O_2 and thus photosynthesizing most. Bacteria congregated in greatest numbers around the parts of the alga illuminated with violet-blue or red light.

Conclusion Light in the violet-blue and red portions of the spectrum is most effective in driving photosynthesis.

Source T. W. Engelmann, *Bacterium photometricum*. Ein Beitrag zur vergleichenden Physiologie des Licht-und Farbensinnes, *Archiv. für Physiologie* 30:95–124 (1883).

▲ **Figure 7.13** Which wavelengths of light are most effective in driving photosynthesis?

Excitation of Chlorophyll by Light

What exactly happens when chlorophyll and other pigments absorb light? When a molecule absorbs a photon of light, one of the molecule's electrons is elevated to an orbital where it has more potential energy. When the electron is in its normal orbital, the pigment molecule is said to be in its **ground state**. Absorption of a photon boosts an electron to an orbital of higher energy, and the pigment molecule is then said to be in an **excited state**. The only photons absorbed are those whose energy is exactly equal to the energy difference between the ground state and an excited state, and this energy difference varies from one kind of molecule to another. Thus, a particular compound absorbs only photons corresponding to specific wavelengths, which is why each pigment has a unique absorption spectrum.

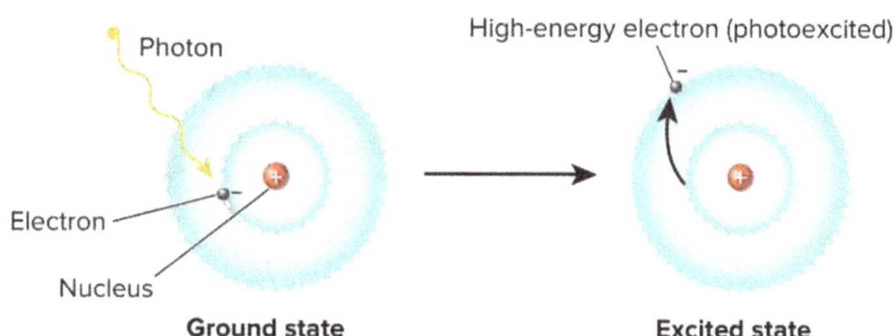

▶ **Figure 7.14** Absorption of light energy by an electron. When a photon of light having the correct amount of energy strikes an electron, the electron is boosted from the ground (unexcited) state to a higher energy level (an excited state). When this occurs, the electron occupies an orbital in which it is farther away from the nucleus of the atom. At this farther distance, the electron is held less firmly and is considered unstable.

After an electron absorbs energy, it is said to be in an excited state. Usually, this is an unstable condition. The electron can release the energy in different ways:

- An excited electron can release **heat**. For example, on a sunny day, the sidewalk heats up because it absorbs light energy that is released as heat.

- An electron can release energy in the form of light. Certain organisms, such as jellyfish, possess molecules that make them glow. This glowing is due to the release of light when electrons drop down to lower energy levels, a phenomenon called **fluorescence**.

- An excited electron can transfer its extra energy to an electron in a nearby molecule, a process called **resonance energy transfer**.

- In the case of certain photosynthetic pigments, another event can happen that is critical for the process of photosynthesis. Rather than releasing energy or transferring it to another molecule, an excited electron in reaction center is removed from that molecule and transferred to another molecule where the electron is stable. When this occurs, the energy in the electron is said to be "**captured**" because the electron does not readily drop down to a lower energy level and release heat or light.

▶ **Figure 7.15** Excitation of isolated chlorophyll by light. Absorption of a photon causes a transition of the chlorophyll molecule from its ground state to its excited state. The photon boosts an electron to an orbital where it has more potential energy. If the illuminated molecule exists in isolation, its excited electron immediately drops back down to the ground-state orbital, and its excess energy is given off as heat and fluorescence (light).

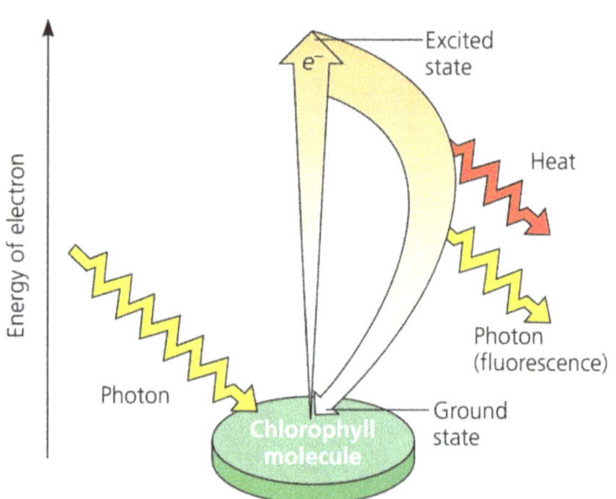

A Photosystem: A Reaction-Center Complex with Light-Harvesting Complexes

In their native environment of the thylakoid membrane, chlorophyll molecules are organized along with other small organic molecules and proteins into complexes called photosystems. A **photosystem** is composed of a reaction-center complex surrounded by several light-harvesting complexes. Each photosystem includes some 300 chlorophyll a molecules and 50 accessory pigments. Each **light-harvesting** or **antenna complex** consists of various pigment molecules (which may include chlorophyll a, chlorophyll b, and multiple carotenoids) bound to proteins.

The pigments and associated proteins are arranged as highly ordered groups that include about 250 chlorophyll molecules associated with specific enzymes and other proteins. Each antenna complex absorbs light energy and transfers it to its reaction center. The **reaction-center complex** is an organized association of proteins holding a special pair of chlorophyll a molecules and a primary electron acceptor. Energy derived from light is converted to chemical energy in the reaction centers by a series of electron transfer reactions.

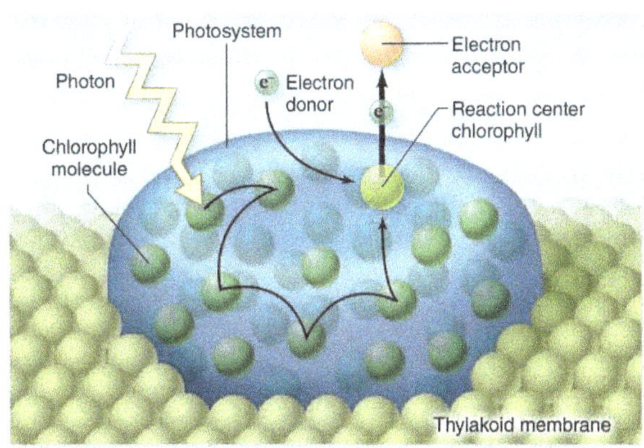

▶ **Figure 7.16** How the antenna complex works. When light of the proper wavelength strikes any pigment molecule within a photosystem, the light is absorbed by that pigment molecule. The excitation energy is then transferred from one molecule to another within the cluster of pigment molecules until it encounters the reaction center chlorophyll a. When excitation energy reaches the reaction center chlorophyll, electron transfer is initiated.

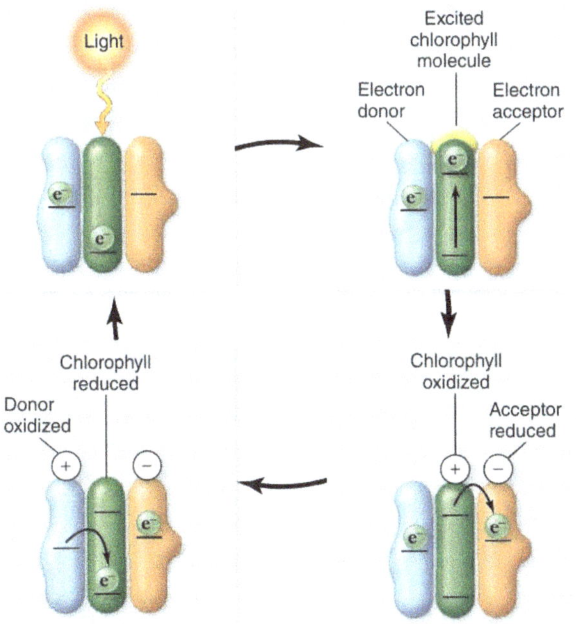

◀ **Figure 7.17** Converting light to chemical energy. When a chlorophyll in the reaction center absorbs a photon of light, an electron is excited to a higher energy level. This light-energized electron can be transferred to the primary electron acceptor, reducing it. The oxidized chlorophyll then fills its electron "hole" by oxidizing a donor molecule. The source of this donor varies with the photosystem

The number and variety of pigment molecules enable a photosystem to harvest light over a larger surface area and a larger portion of the spectrum than could any single pigment molecule alone. Together, these light-harvesting complexes act as an antenna for the reaction-center complex. When a pigment molecule absorbs a photon, the energy is transferred from pigment molecule to pigment molecule within a light-harvesting complex, until it is passed to the pair of chlorophyll a molecules in the reaction-center complex. This pair of chlorophyll a molecules is special because their molecular environment—their location and the other molecules with which they are associated—enables them to use the energy from light not only to boost one of their electrons to a higher energy level, but also to transfer it to a different molecule—the **primary electron acceptor**, which is a molecule capable of accepting electrons and becoming reduced.

The solar-powered transfer of an electron from the reaction-center chlorophyll a pair to the primary electron acceptor is the first step of the light reactions. As soon as the chlorophyll electron is excited to a higher energy level, the primary electron acceptor captures it; this is a redox reaction. Thus, each photosystem—a reaction-center complex surrounded by light-harvesting complexes— functions in the chloroplast as a unit. It converts light energy to chemical energy, which will ultimately be used for the synthesis of sugar.

The thylakoid membrane is populated by two types of photosystems that cooperate in the light reactions of photosynthesis: **photosystem II** (PS II) and **photosystem I** (PS I). Each has a characteristic reaction-center complex—a particular kind of primary electron acceptor next to a special pair of chlorophyll a molecules associated with specific proteins. The reaction-center chlorophyll a of photosystem II is known as **P680** because this pigment is best at absorbing light having a wavelength of 680 nm (in the red part of the spectrum).

The chlorophyll a at the reaction-center complex of photosystem I is called **P700** because it most effectively absorbs light of wavelength 700 nm (in the far-red part of the spectrum). These two pigments, P680 and P700, are nearly identical chlorophyll a molecules. However, their association with different proteins in the thylakoid membrane affects the electron distribution in the two pigments and accounts for the slight differences in their light-absorbing properties. An electron transport chain connects the two photosystems. A second electron transport chain extending from photosystem I ends in the production of NADPH.

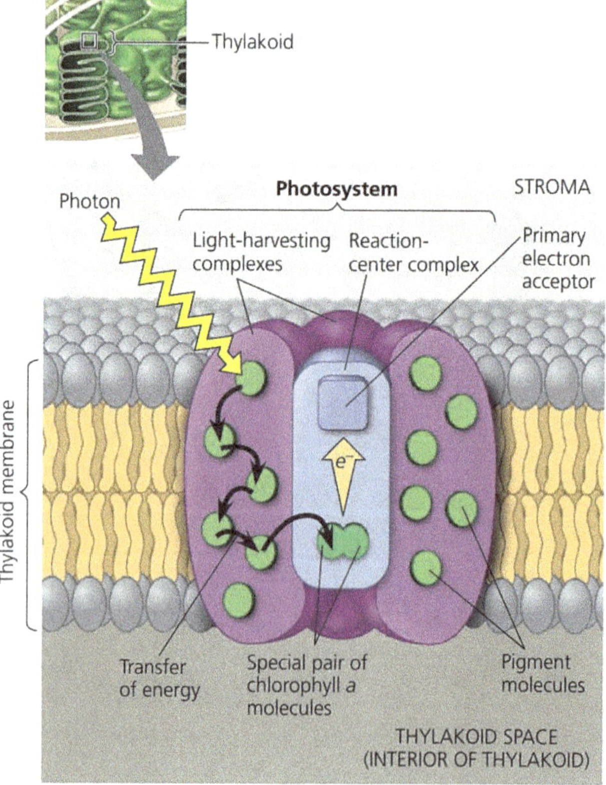

▶ **Figure 7.18** How a photosystem harvests light. When a photon strikes a pigment molecule in a light-harvesting complex, the energy is passed from molecule to molecule until it reaches the reaction-center complex. Here, an excited electron from the special pair of chlorophyll a molecules is transferred to the primary electron acceptor.

Linear Electron Flow

Light drives the synthesis of ATP and NADPH by energizing the two types of photosystems embedded in the thylakoid membranes of chloroplasts. The key to this energy transformation is a flow of electrons through the photosystems and other molecular components built into the thylakoid membrane. This is called **linear electron flow: 1.** A photon of light strikes one of the pigment molecules in a light-harvesting complex of PS II, boosting one of its electrons to a higher energy level. As this electron falls back to its ground state, an electron in a nearby pigment molecule is simultaneously raised to an excited state.

The process continues through a process known as resonance, with the energy being relayed to other pigment molecules until it reaches the P680 pair of chlorophyll a molecules in the PS II reaction-center complex. It excites an electron in this pair of chlorophylls to a higher energy state. **2.** This electron is transferred from the excited P680 to the primary electron acceptor (a highly modified chlorophyll molecule known as **pheophytin**). We can refer to the resulting form of P680, missing the negative charge of an electron, as P680$^+$.

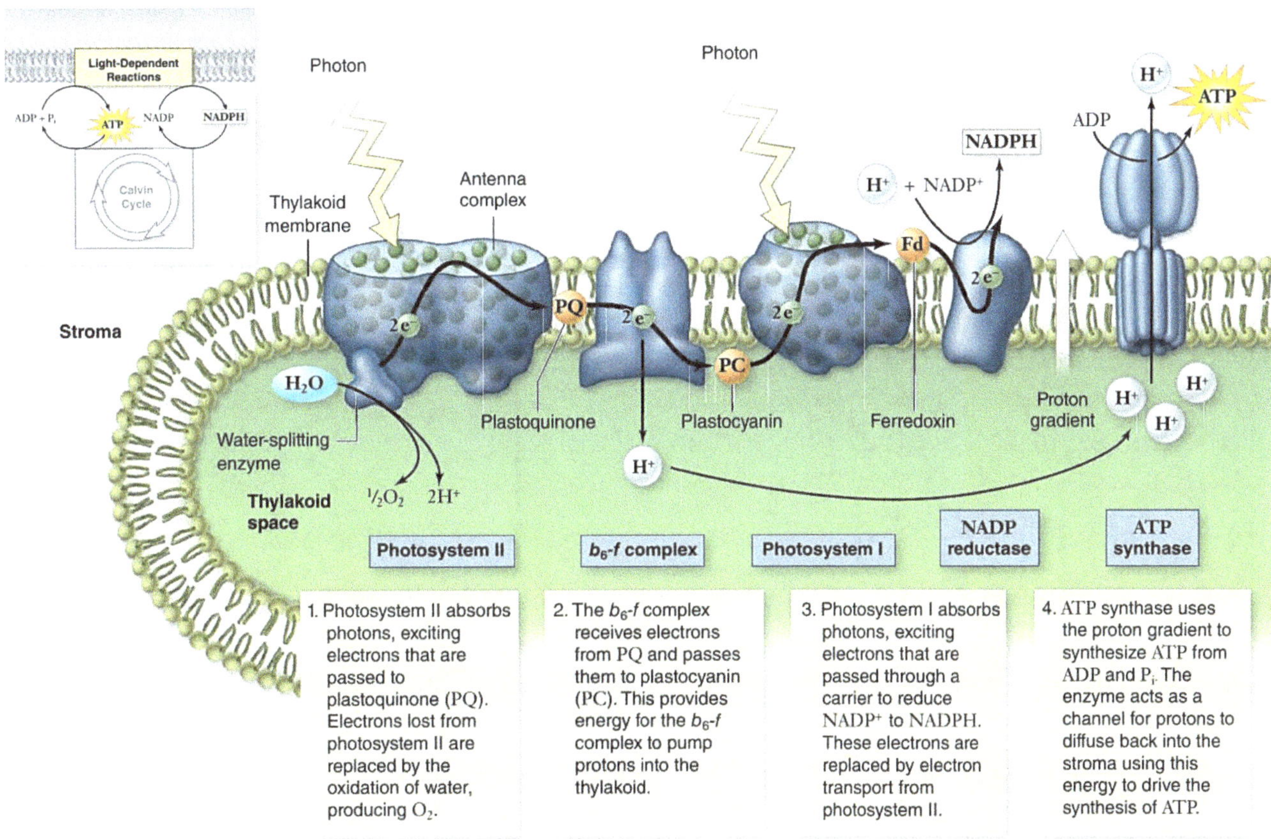

3. An unique, manganese containing enzyme that is sometimes known as the **oxygen-evolving complex**, or the **water-splitting complex**, catalyzes the splitting of a water molecule through **photolysis** (light splitting), into two electrons, two hydrogen ions (H$^+$), and an oxygen atom. The electrons are supplied one by one to the P680$^+$ pair, each electron replacing one transferred to the primary electron acceptor. (P680$^+$ is the strongest biological oxidizing agent known; its electron "hole" must be filled. This greatly facilitates the transfer of electrons from the split water molecule). The H$^+$ are released into the thylakoid space (interior of the thylakoid). The oxygen atom immediately combines with an oxygen atom generated by the splitting of another water molecule, forming O$_2$.

4. Each photo excited electron passes from the primary electron acceptor of PS II to PS I via an electron transport chain, the components of which are similar to those of the electron transport chain that functions in cellular respiration. The electron transport chain between PS II and PS I is made up of the electron carrier **plastoquinone (Pq**, the reduced quinone that results from accepting a pair of electrons), a **cytochrome b$_6$-f complex**, and a copper containing protein called **plastocyanin (Pc)**. Each component carries out redox reactions as electrons flow down the electron transport chain, releasing free energy that is used to pump protons (H$^+$) into the thylakoid space, contributing to a proton gradient across the thylakoid membrane.

5. The potential energy stored in the proton gradient is used to make ATP in a process called **chemiosmosis**. **6.** Meanwhile, light energy has been transferred via light- harvesting complex pigments to the PS I reaction-center complex, exciting an electron of the P700 pair of chlorophyll a molecules located there. The photo excited electron is then transferred to PS I's primary electron acceptor, creating an electron "hole" in the P700—which we now can call P700$^+$. In other words, P700$^+$ can now act as an electron acceptor, accepting an electron that reaches the bottom of the electron transport chain from PS II.

7. Photo excited electrons are passed in a series of redox reactions from the primary electron acceptor of PS I down a second electron transport chain through the protein **ferredoxin** (**Fd**, an iron–sulfur protein). This chain does not create a proton gradient and thus does not produce ATP. **8.** The enzyme **ferredoxin–NADP$^+$ reductase** catalyzes the transfer of electrons from Fd to NADP$^+$. Two electrons are required for its reduction to NADPH. Electrons in NADPH are at a higher energy level than they are in water (where they started), so they are more readily available for the reactions of the Calvin cycle. This process also removes an H$^+$ from the stroma.

The light reactions use solar power to generate ATP and NADPH, which provide chemical energy and reducing power, respectively, to the carbohydrate-synthesizing reactions of the Calvin cycle. Notice that NADPH, like ATP, is produced on the side of the membrane facing the stroma, where the Calvin cycle reactions take place. This is the **Z scheme** because of the zigzag-like changes in electron energy level: An electron on a nonexcited pigment molecule in the light harvesting complex of photosystem II has the lowest energy. In photosystem II, light boosts such an electron to a much higher energy level. As the electron travels from photosystem II to photosystem I, some of the energy is released. The input of light in photosystem I boosts the electron to an even higher energy than it attained in photosystem II. The electron releases a little energy before it is eventually transferred to NADP$^+$.

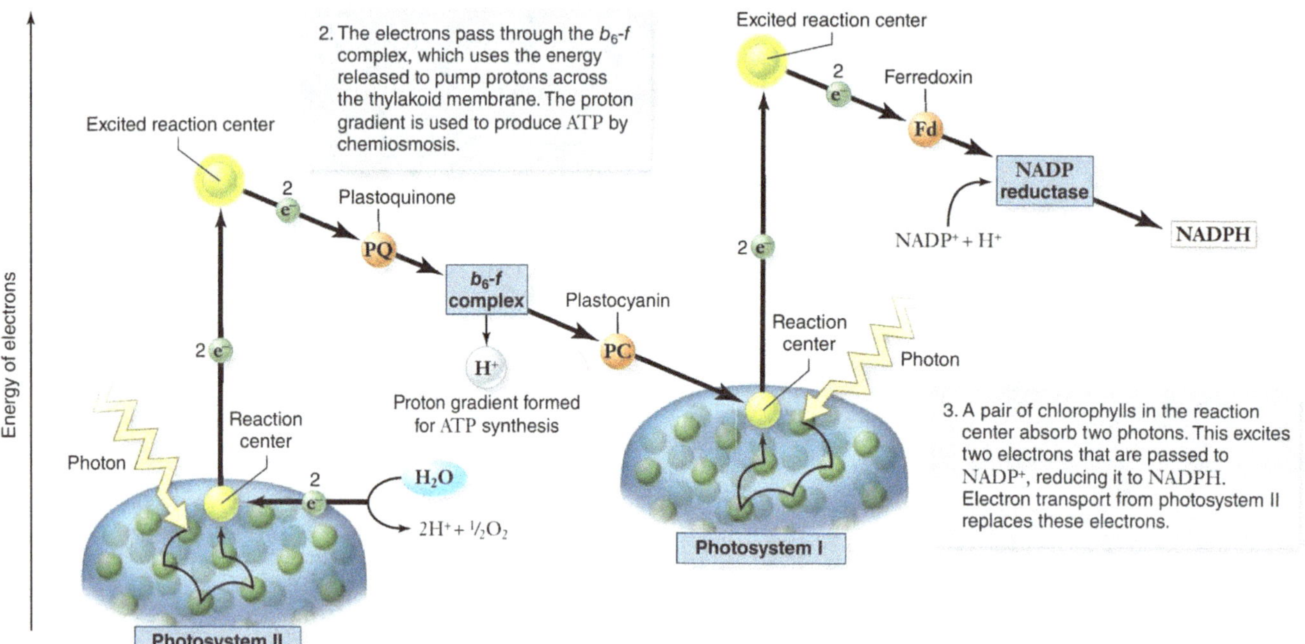

▲ **Figure 7.19** Z diagram of photosystems I and II. Two photosystems work sequentially and have different roles. Photosystem II passes energetic electrons to photosystem I via an electron transport chain. The electrons lost are replaced by oxidizing water. Photosystem I uses energetic electrons to reduce NADP$^+$ to NADPH.

Cyclic Electron Flow

In certain cases, photoexcited electrons can take an alternative path called **cyclic electron flow**, which uses photosystem I but not photosystem II. The electrons cycle back from ferredoxin (Fd) to the cytochrome complex, then via a plastocyanin molecule (Pc) to a P700 chlorophyll in the PS I reaction-center complex. There is no production of NADPH and no release of oxygen that results from this process. On the other hand, cyclic flow does generate ATP.

Rather than having both PS II and PS I, several of the currently existing groups of photosynthetic bacteria are known to have a single photosystem related to either PS II or PS I. For these species, which include the purple sulfur bacteria and the green sulfur bacteria, cyclic electron flow is the one and only means of generating ATP during the process of photosynthesis.

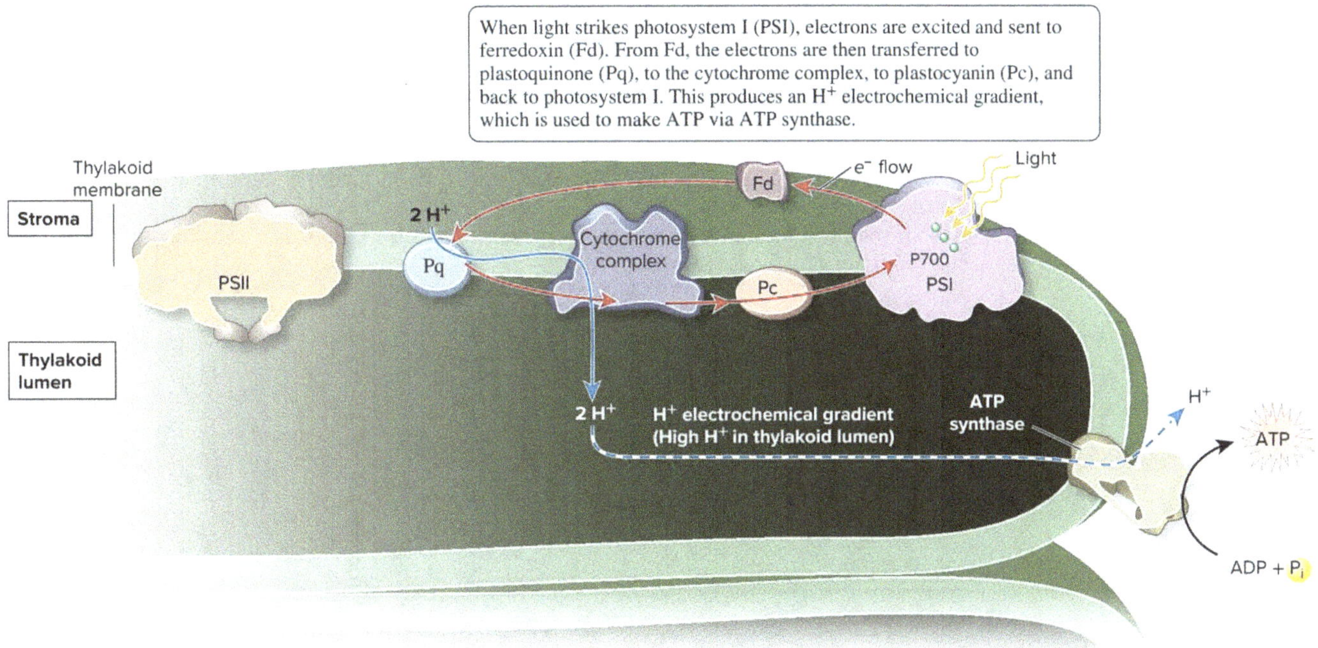

▲ **Figure 7.20** The synthesis of only ATP by cyclic photophosphorylation. In this process, electrons follow a cyclic path that is powered by photosystem I (PSI). This produces an H+ electrochemical gradient, which is then used by ATP synthase to make ATP.

	Noncyclic Electron Transport	Cyclic Electron Transport
Electron source	H_2O	None—electrons cycle through the system
Oxygen released?	Yes (from H_2O)	No
Terminal electron acceptor	$NADP^+$	None—electrons cycle through the system
Form in which energy is temporarily captured	ATP (by chemiosmosis); NADPH	ATP (by chemiosmosis)
Photosystem(s) required	PS I (P700) and PS II (P680)	PS I (P700) only

▲ **Table 7.2** A Comparison of Noncyclic and Cyclic Electron Transport

A Comparison of Chemiosmosis in Chloroplasts and Mitochondria

Chloroplasts and mitochondria generate ATP by the same basic mechanism: **chemiosmosis**. An electron transport chain pumps protons (H^+) across a membrane as electrons are passed through a series of carriers that have progressively more affinity for electrons. Thus, electron transport chains transform redox energy to a **proton-motive force**, potential energy stored in the form of an H^+ gradient across a membrane. An ATP synthase complex in the same membrane couples the diffusion of hydrogen ions down their gradient to the phosphorylation of ADP, forming ATP.

Some of the electron carriers, including the iron-containing proteins called cytochromes, are very similar in mitochondria and chloroplasts. The **ATP synthase complexes** of the two organelles are also quite similar. But there are noteworthy differences between **photophosphorylation** in chloroplasts and oxidative phosphorylation in mitochondria. Both work by way of chemiosmosis, but in chloroplasts, the high-energy electrons dropped down the transport chain come from water, while in mitochondria, they are extracted from organic molecules (which are thus oxidized). The movement of protons through ATP synthase is thought to induce changes in the conformation of the enzyme that are necessary for the synthesis of ATP. It is estimated that for every 4 protons that move through ATP synthase, 1 ATP molecule is synthesized.

Chloroplasts do not need molecules from food to make ATP; their photosystems capture light energy and use it to drive the electrons from water to the top of the transport chain. In other words, mitochondria use chemiosmosis to transfer chemical energy from food molecules to ATP, whereas chloroplasts use it to transform light energy into chemical energy in ATP. Although the spatial organization of chemiosmosis differs slightly between chloroplasts and mitochondria, it is easy to see similarities in the two. Electron transport chain proteins in the inner membrane of the mitochondrion pump protons from the mitochondrial matrix out to the intermembrane space, which then serves as a reservoir of hydrogen ions. Similarly, electron transport chain protein in the thylakoid membrane of the chloroplast, **cytochrome/b6-f complex**, pump protons from the stroma into the thylakoid space, which functions as the H^+ reservoir.

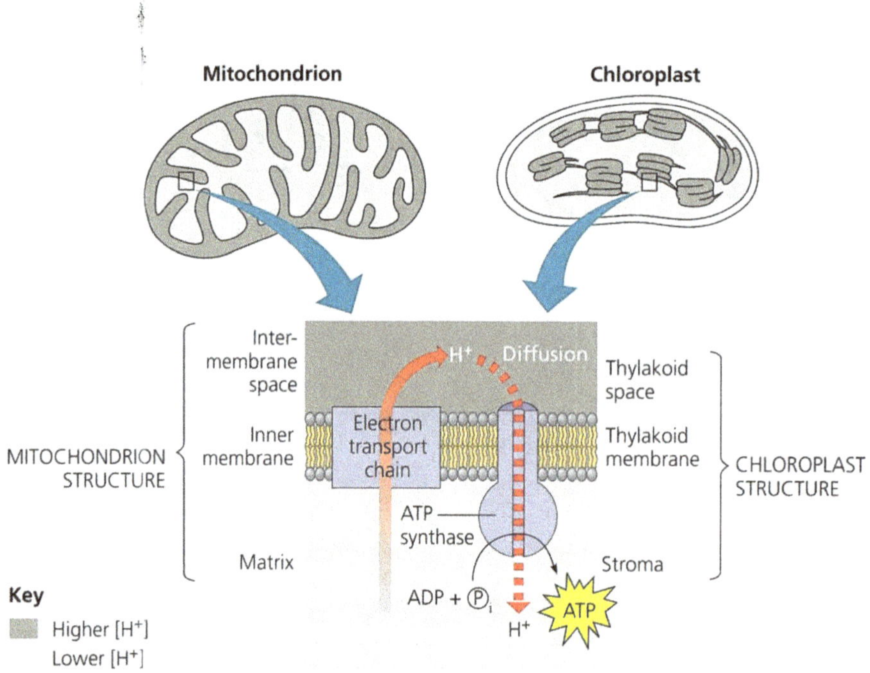

▲ **Figure 7.21** Comparison of chemiosmosis in mitochondria and chloroplasts.

The thylakoid membrane is impermeable to H⁺. Protons also accumulate in the thylakoid lumen as water is split during noncyclic electron transport. Because protons are actually hydrogen ions (H⁺), the accumulation of protons causes the pH of the thylakoid interior to fall to a pH of about 5 in the thylakoid lumen, compared with a pH of about 8 in the stroma. This difference of about 3 pH units across the thylakoid membrane means that there is an approximately thousand-fold difference in hydrogen ion concentration. In the mitochondrion, protons diffuse down their concentration gradient from the intermembrane space through ATP synthase to the matrix, driving ATP synthesis. In the chloroplast, ATP is synthesized as the hydrogen ions diffuse from the thylakoid space back to the stroma through ATP synthase complexes whose catalytic knobs are on the stroma side of the membrane. Thus, ATP forms in the stroma, where it is used to help drive sugar synthesis during the Calvin cycle.

An H⁺ electrochemical gradient is generated in three ways:

- The splitting of water, which places H⁺ in the thylakoid lumen.

- The movement of high-energy electrons along the ETC from photosystem II to photosystem I, which pumps H⁺ into the thylakoid lumen.

- The formation of NADPH, which consumes H⁺ in the stroma.

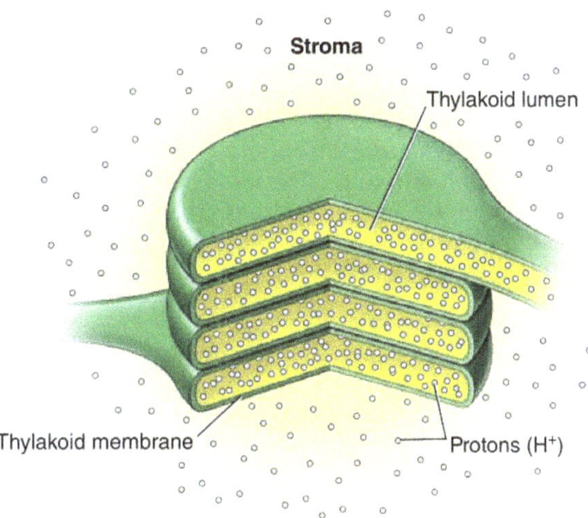

The four complexes responsible for the light-dependent reactions—namely photosystems I and II, cytochrome b₆-f, and ATP synthase—are not randomly arranged in the thylakoid. Researchers are beginning to image these complexes with the atomic force microscope, which can resolve nanometer scale structures, and a picture is emerging in which photosystem II is found primarily in the grana, whereas photosystem I and ATP synthase are found primarily in the stroma lamella. Photosystem I and ATP synthase may also be found in the edges of the grana that are not stacked. The cytochrome b₆-f complex is found in the borders between grana and stroma lamella.

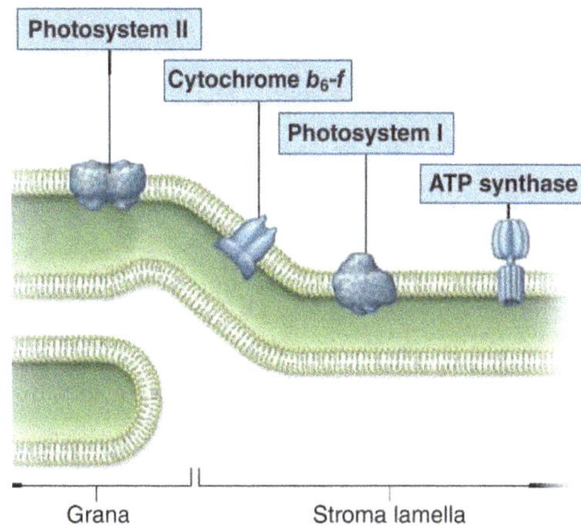

▶ **Figure 7.22** Model for the arrangement of complexes within the thylakoid. The arrangement of the two kinds of photosystems and the other complexes involved in photosynthesis is not random. Photosystem II is concentrated within grana, especially in stacked areas.

Photosystem I and ATP synthase are concentrated in stroma lamella and the edges of grana. The cytochrome b₆-f complex is in the margins between grana and stroma lamella. This is one possible model for this arrangement.

The Calvin cycle

The **Calvin cycle** takes place in the stroma; it is similar to the citric acid cycle in that a starting material is regenerated after some molecules enter and others exit the cycle. However, the citric acid cycle is catabolic, oxidizing acetyl CoA and using the energy to synthesize ATP, while the Calvin cycle is anabolic, building carbohydrates from smaller molecules and consuming energy. Carbon enters the Calvin cycle in the form of CO_2 and leaves in the form of sugar. The cycle spends ATP as an energy source and consumes NADPH as reducing power for adding high-energy electrons to make the sugar. The carbohydrate produced directly from the Calvin cycle is not glucose.

It is actually a three-carbon sugar called **glyceraldehyde 3- phosphate (G3P)**. For the net synthesis of one molecule of G3P, the cycle must take place three times, fixing three molecules of CO_2—one per turn of the cycle (the term carbon fixation refers to the initial incorporation of CO_2 into organic material). The Calvin cycle is divided into three phases through a sequence of 13 reactions: carbon fixation, reduction, and regeneration of the CO_2 acceptor. All 13 enzymes that catalyze steps in the Calvin cycle are located in the stroma of the chloroplast. Ten of the enzymes also participate in glycolysis. These enzymes catalyze reversible reactions, degrading carbohydrate molecules in cellular respiration and synthesizing carbohydrate molecules in photosynthesis.

Phase 1: Carbon fixation. The Calvin cycle incorporates each CO_2 molecule, one at a time, by attaching it to a five-carbon sugar named **ribulose 1,5–bisphosphate** (which is abbreviated **RuBP**). The enzyme that catalyzes this first step is **ribulose bisphosphate carboxylase/oxygenase**, or **rubisco**, a large, 16-subunit enzyme found in the chloroplast stroma and the most abundant protein in chloroplasts and is also thought to be the most abundant protein on Earth. The product of the reaction is a six-carbon intermediate that is short-lived because it is so energetically unstable that it immediately splits in half, forming two molecules of **3-phosphoglycerate**, **PGA** or **GP** (for each CO_2 fixed). GP can be converted to glycerol and fatty acids to produce lipids for cellular membranes.

Rubisco has essentially the same overall structure in almost all photosynthetic organisms: eight copies each of a large and a small polypeptide, joined together in a 16-subunit structure. The large subunit contains all of the known active sites where substrates, including CO_2 and RuBP, can bind (Rubisco has a ten-times greater affinity for CO_2 than it does for O_2, however, at higher concentrations of O_2 within the leaf, oxygen acts as a competitive inhibitor of the enzyme, and this favors the reaction of RuBP with O_2 rather than with CO_2: **photorespiration** occurs).

Although the small subunit has no active sites, it has a very significant effect on rubisco's rate of catalysis. That is, 99% of the catalytic activity of the enzyme is lost if the small subunit is removed. Rubisco is also the key regulatory site of the Calvin cycle. The enzyme is stimulated by both NADPH and ATP. During the daytime, when sunlight powers the light-dependent reactions, the abundant NADPH and ATP supplies keep the Calvin cycle running. In darkness, when NADPH and ATP become unavailable, the enzyme is inhibited and the Calvin cycle slows or stops.

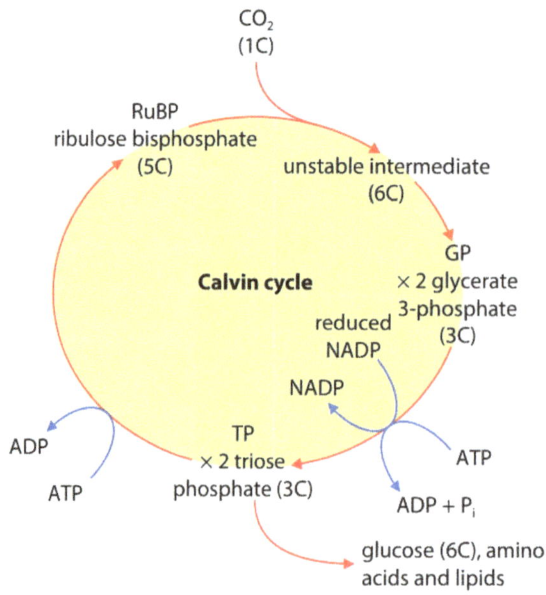

Phase 2: Reduction. Each molecule of 3- phosphoglycerate receives an additional phosphate group from ATP, becoming 1,3-bisphosphoglycerate. Next, a pair of electrons donated from NADPH reduces 1,3-bisphosphoglycerate, which also loses a phosphate group in the process, becoming **glyceraldehyde 3- phosphate** (**G3P** or **phosphoglyceraldehyde, PGAL**). Specifically, the electrons from NADPH reduce a carboxyl group on 1,3-bisphosphoglycerate to the aldehyde group of G3P, which stores more potential energy. G3P or **triose phosphate** (TP) is a sugar—the same three-carbon sugar formed in glycolysis by the splitting of glucose.

For every three molecules of CO_2 that enter the cycle, there are six molecules of G3P formed. But only one molecule of this three-carbon sugar can be counted as a net gain of carbohydrate because the rest are required to complete the cycle. The cycle began with 15 carbons' worth of carbohydrate in the form of three molecules of the five-carbon sugar RuBP. Now there are 18 carbons' worth of carbohydrate in the form of six molecules of G3P. One molecule exits the cycle to be used by the plant cell, but the other five molecules must be recycled to regenerate the three molecules of RuBP. Some of triose phosphates condense to become hexose phosphates. These are used to produce starch for storage, sucrose for translocation around the plant, or cellulose for making cell walls. Plants can also produce all 20 of the naturally occurring amino acids that they need for protein synthesis, using ammonium ions absorbed from the soil and carbohydrates produced in the light independent reactions.

Phase 3: Regeneration of the CO_2 acceptor (RuBP). In a complex series of reactions, the carbon skeletons of five molecules of G3P are rearranged by the last steps of the Calvin cycle into three molecules of RuBP. To accomplish this, the cycle spends three more molecules of ATP. The RuBP is now prepared to receive CO_2 again, and the cycle continues.

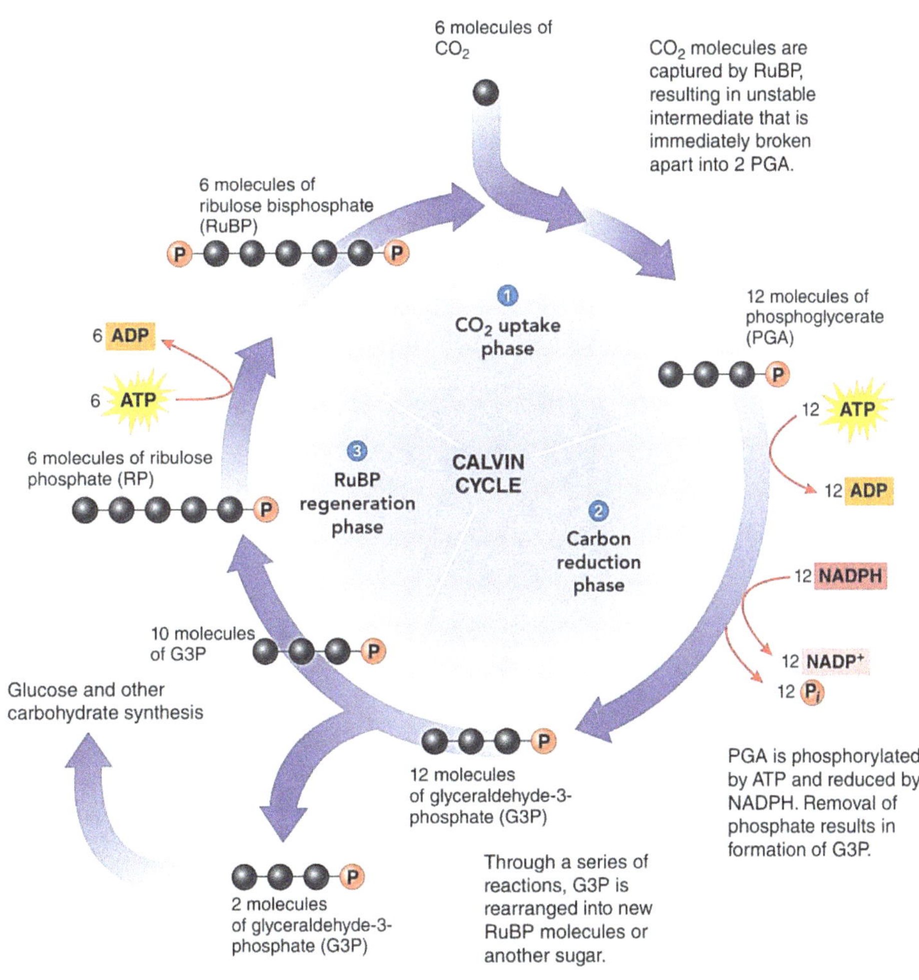

▲ **Figure 7.23** A detailed look at the Calvin cycle

For the net synthesis of one G3P molecule, the Calvin cycle consumes a total of nine molecules of ATP and six molecules of NADPH. The light reactions regenerate the ATP and NADPH. The G3P spun off from the Calvin cycle becomes the starting material for metabolic pathways that synthesize other organic compounds, including glucose (from two molecules of G3P), sucrose, and other carbohydrates. Neither the light reactions nor the Calvin cycle alone can make sugar from CO_2. As for the fates of photosynthetic products, enzymes in the chloroplast and cytosol convert the G3P made in the Calvin cycle to many other organic compounds.

Although glucose is the major product of photosynthesis, other monosaccharides, disaccharides, polysaccharides, lipids, and amino acids are also produced indirectly. In fact, all the organic molecules of plants are assembled as direct or indirect products of photosynthesis. The sugar made in the chloroplasts supplies the entire plant with chemical energy and carbon skeletons for the synthesis of all the major organic molecules of plant cells. About 50% of the organic material made by photosynthesis is consumed as fuel for cellular respiration in plant cell mitochondria.

Technically, green cells are the only autotrophic parts of the plant. The rest of the plant depends on organic molecules exported from leaves through veins. In most plants, carbohydrate is transported out of the leaves to the rest of the plant in the form of sucrose, a disaccharide. After arriving at nonphotosynthetic cells, the sucrose provides raw material for cellular respiration and a multitude of anabolic pathways that synthesize proteins, lipids, and other products. A considerable amount of sugar in the form of glucose is linked together to make the polysaccharide cellulose, especially in plant cells that are still growing and maturing. Cellulose, the main ingredient of cell walls, is the most abundant organic molecule in the plant—and probably on the surface of the planet.

Carbon Atoms in Carbohydrates Are Electron-Rich Compared to the Carbon Atom in CO_2

As we have just seen, the Calvin cycle begins by using carbon from an inorganic source (that is, CO_2) and ends with organic molecules that will be used by the plant to make other molecules. You may be wondering why CO_2 molecules cannot be directly linked to each other to form these larger molecules. The answer lies in the electrons that are located around carbon atoms. Figure below shows the relative locations of electrons around a carbon atom in CO_2, compared to the electrons around one of the carbon atoms in G3P. In CO_2, the carbon atom is considered electron poor. Oxygen is a very electronegative atom that monopolizes the electrons it shares with other atoms.

In a double covalent bond between carbon and oxygen, the shared electrons are closer to the oxygen atom. By comparison, in a carbohydate such as glyceraldehyde-3-phosphate (G3P), the carbon atom is electron rich. During the Calvin cycle, ATP provides energy and NADPH donates high energy electrons, so the carbon originally in CO_2 has been reduced. The Calvin cycle combines less electronegative atoms with carbon atoms so that C—H and C—C bonds are formed. In C—H and C—C bonds, the electrons are closer to the carbon atom compared to C—O bonds. This allows the eventual synthesis of larger organic molecules, including glucose, amino acids, and so on. In addition, the covalent bonds within these molecules store large amounts of energy.

▶ **Figure 7.24** The valence electrons around a carbon atom in carbon dioxide and around one of the carbon atoms in glyceraldehyde-3-phosphate (G3P).

Carbon dioxide

G3P

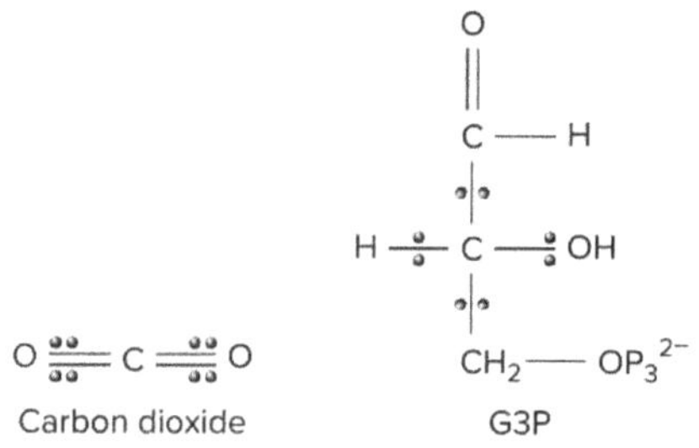

Summary Of Photosynthesis

$$6\,CO_2 + 12\,H_2O \rightarrow C_6H_{12}O_6 + 6\,O_2 + 6\,H_2O$$

$$18\,ATP + 18\,H_2O \rightarrow 18\,ADP + 18\,P_i$$

$$12\,NADPH \rightarrow 12\,NADP^+ + 12\,H^+ + 24\,e^-$$

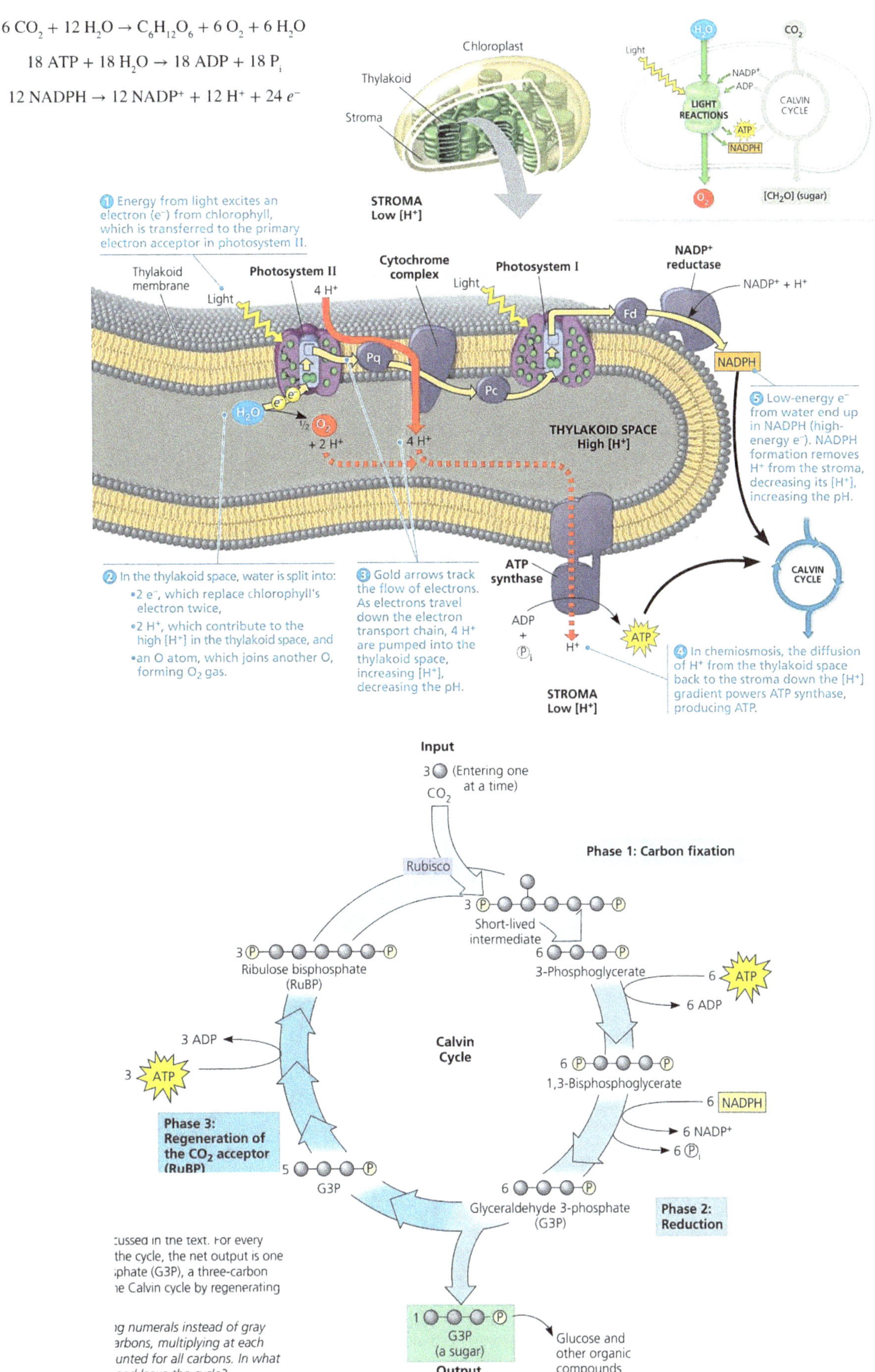

Alternative mechanisms of carbon fixation have evolved in hot, arid climates

The CO_2 required for photosynthesis enters a leaf (and the resulting O_2 exits) via stomata, the pores on the leaf surface. However, stomata are also the main avenues of **transpiration**, the evaporative loss of water from leaves. On a hot, dry day, most plants close their stomata, a response that conserves water but also reduces CO_2 levels. With stomata even partially closed, CO_2 concentrations begin to decrease in the air spaces within the leaf, and the concentration of O_2 released from the light reactions begins to increase. These conditions within the leaf favor an apparently wasteful process called photorespiration. In most plants (about 95% of plant), initial fixation of carbon occurs via rubisco, the Calvin cycle enzyme that adds CO_2 to ribulose bisphosphate. Such plants are called **C_3 plants** because the first organic product of carbon fixation is a three-carbon compound, 3- phosphoglycerate.

$$RuBP + CO_2 \rightarrow 2\ 3PG$$

C_3 plants include important agricultural plants such as rice, wheat, and soybeans. When their stomata close on hot, dry days, C_3 plants produce less sugar because the declining level of CO_2 in the leaf starves the Calvin cycle. In addition, rubisco is capable of binding O_2 in place of CO_2. As CO_2 becomes scarce within the air spaces of the leaf and O_2 builds up, rubisco adds O_2 to the Calvin cycle instead of CO_2. The product splits, and a two-carbon compound (**glycolate**) leaves the chloroplast. Peroxisomes and mitochondria within the plant cell rearrange and split this compound, releasing CO_2. The process is called **photorespiration** because it occurs in the light (photo) and consumes O_2 while producing CO_2 (respiration).

$$\text{Phosphoglycolate} \rightarrow \text{Glycolate} \rightarrow \rightarrow \text{Organic molecule} + CO_2$$

However, unlike normal cellular respiration, photorespiration uses ATP rather than generating it. And unlike photosynthesis, photorespiration produces no sugar. In fact, photorespiration decreases photosynthetic output by siphoning organic material from the Calvin cycle and releasing CO_2 that would otherwise be fixed. This CO_2 can eventually be fixed if it is still in the leaf once the CO_2 concentration builds up to a high enough level. In the meantime, though, the process is energetically costly, much like a hamster running on its wheel. In some plant species, alternate modes of carbon fixation have evolved that minimize photorespiration and optimize the Calvin cycle—even in hot, arid climates. The two most important of these photosynthetic adaptations are **C_4 photosynthesis** and **crassulacean acid metabolism (CAM)**.

$$RuBP + O_2 \rightarrow 3PG + \text{Phosphoglycolate}$$

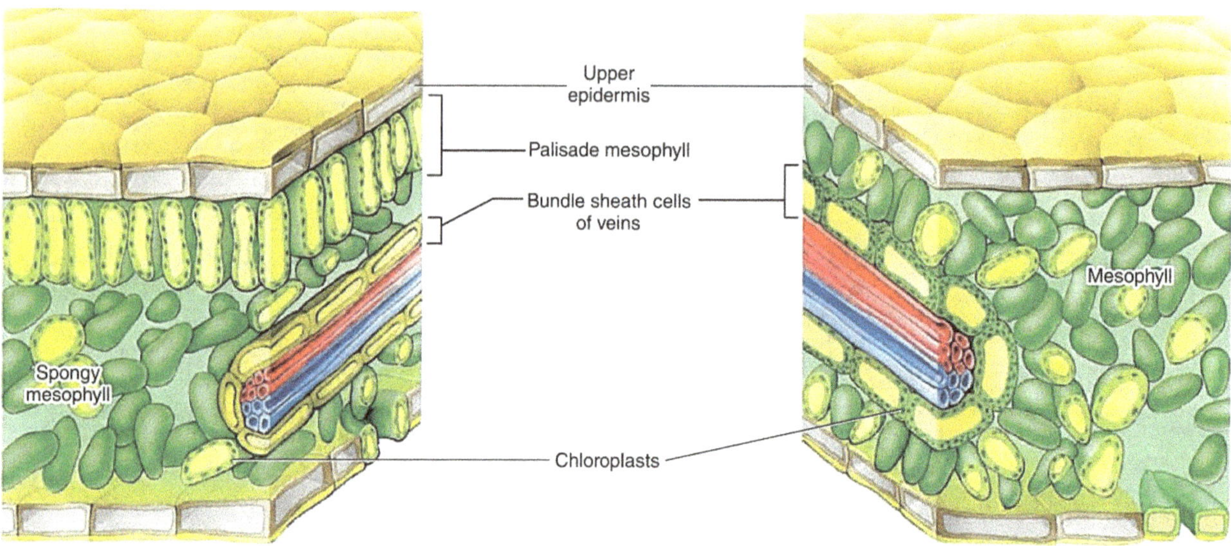

(a) In C_3 plants, the Calvin cycle takes place in the mesophyll cells and the bundle sheath cells are nonphotosynthetic.

(b) In C_4 plants, reactions that fix CO_2 into 4-carbon compounds take place in the mesophyll cells. The 4-carbon compounds are transferred from the mesophyll cells to the photosynthetic bundle sheath cells, where the Calvin cycle takes place.

C$_4$ Plants have evolved to minimize photorespiration

About 1% of plants use the C$_4$ pathway. The **C$_4$ plants** are so named because they preface the Calvin cycle with an alternate mode of carbon fixation that forms a four-carbon compound as its first product, **oxaloacetate**. Among the C$_4$ plants important to agriculture are sugarcane and corn, members of the grass family. When the weather is hot and dry, a C$_4$ plant partially closes its stomata, conserving water but reducing the CO$_2$ concentration in the leaves.

However, sugar is still made because C$_4$ plants use a multistep process that operates even under low CO$_2$ conditions. Photosynthesis begins in mesophyll cells but is completed in **bundle-sheath cells**, cells that are arranged into tightly packed sheaths around the veins of the leaf. In C$_4$ leaves, the loosely arranged mesophyll cells are located between the bundle-sheath cells and the leaf surface, no more than two to three cells away from the bundle-sheath cells. In mesophyll cells, CO$_2$ is incorporated into organic compounds that then move into the bundle sheath cells, where the Calvin cycle takes place.

1. The first step is carried out by an enzyme present only in mesophyll cells called **PEP carboxylase**, which has a much higher affinity for CO$_2$ than does rubisco, and no affinity for O$_2$. This enzyme adds CO$_2$ to **phosphoenolpyruvate (PEP)**, forming the four-carbon product oxaloacetate; it does this even under conditions of lower CO$_2$ concentration and relatively higher O$_2$ concentration.

2. After the CO$_2$ is fixed in the mesophyll cells, the four-carbon products (usually **malate**) are exported to bundle-sheath cells through plasmodesmata.

3. Within the bundle-sheath cells, an enzyme releases CO$_2$ from the four-carbon compounds; the CO$_2$ is re-fixed into organic material by rubisco and the Calvin cycle. The same reaction regenerates pyruvate, which is transported to mesophyll cells. There, ATP is used to convert pyruvate to PEP, which can then accept addition of another CO$_2$, allowing the reaction cycle to continue. This ATP can be thought of, in a sense, as the "price" of concentrating CO$_2$ in the bundle-sheath cells. To generate this extra ATP, bundle-sheath cells carry out cyclic electron flow. In fact, these cells contain PS I but no PS II, so cyclic electron flow is their only photosynthetic mode of generating ATP.

$$\text{malate} + \text{NADP}^+ \longrightarrow \text{pyruvate} + \text{CO}_2 + \text{NADPH}$$

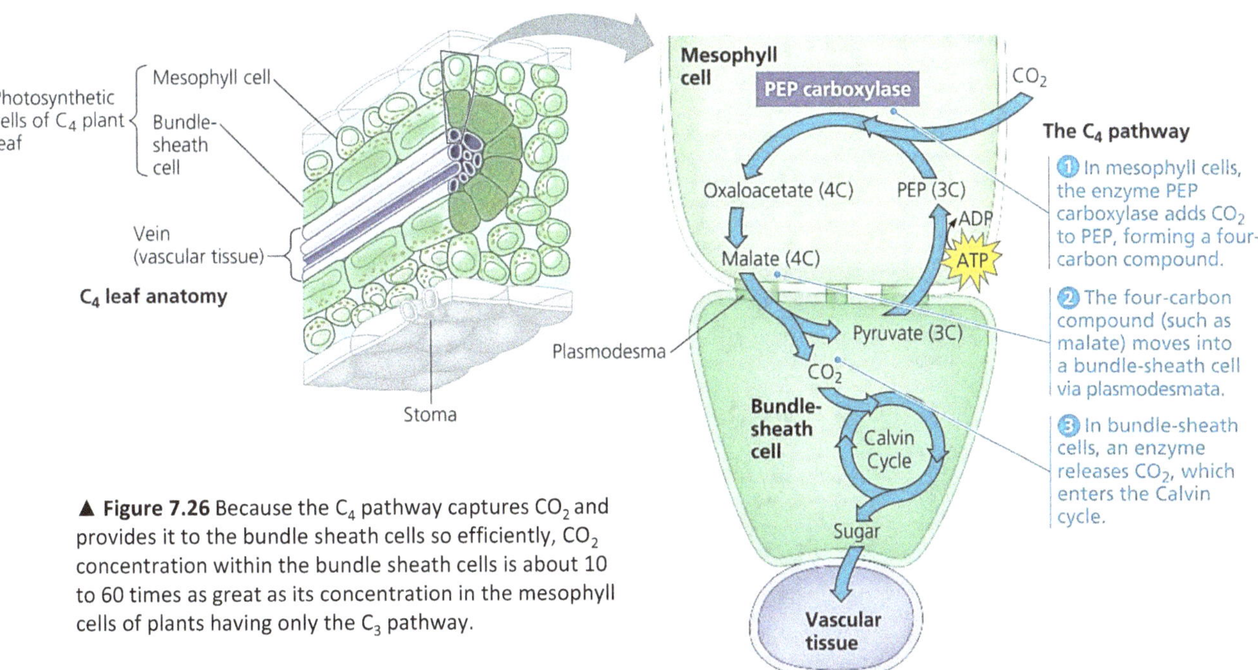

▲ **Figure 7.26** Because the C$_4$ pathway captures CO$_2$ and provides it to the bundle sheath cells so efficiently, CO$_2$ concentration within the bundle sheath cells is about 10 to 60 times as great as its concentration in the mesophyll cells of plants having only the C$_3$ pathway.

In effect, the mesophyll cells of a C_4 plant pump CO_2 into the bundle-sheath cells, keeping the CO_2 concentration in those cells high enough for rubisco to bind CO_2 rather than O_2. The cyclic series of reactions involving PEP carboxylase and the regeneration of PEP can be thought of as an ATP powered pump that concentrates CO_2 where it can be fixed. The C_4 pathway, although it overcomes the problems of photorespiration, does have a cost.

The conversion of pyruvate back to PEP requires breaking two high-energy bonds in ATP. Thus each CO_2 transported into the bundle-sheath cells cost the equivalent of two ATP. To produce a single glucose, this requires 12 additional ATP compared with the Calvin cycle alone. Despite this additional cost, C_4 photosynthesis is advantageous in hot dry climates where photorespiration would remove more than half of the carbon fixed by the usual C_3 pathway alone.

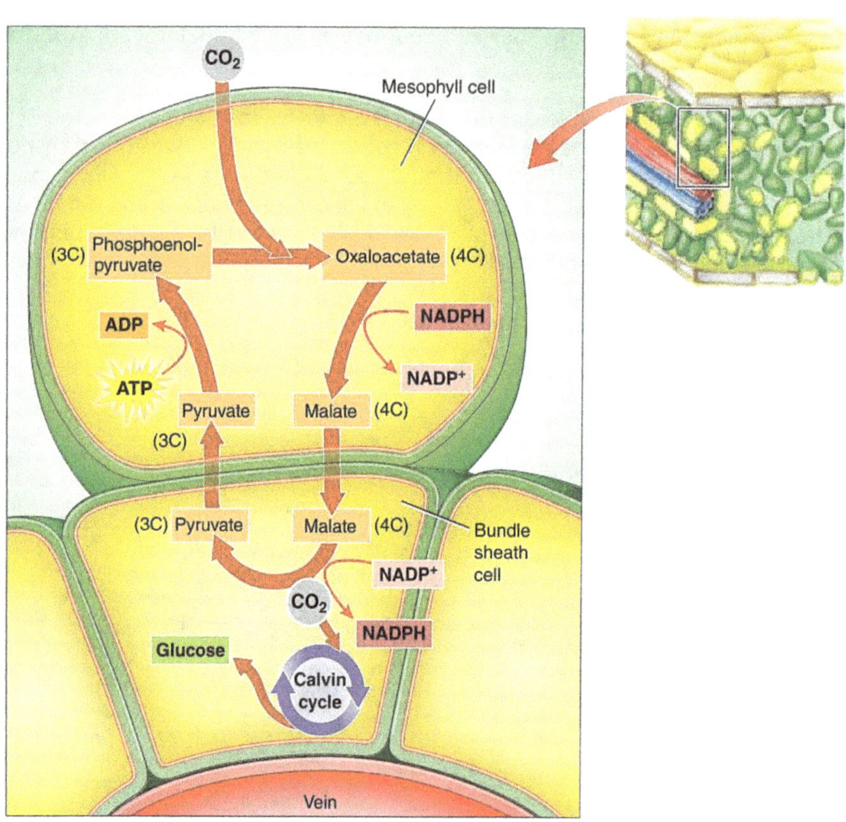

▲ **Figure 7.27** Summary of the C4 pathway

CAM Plants

A second photosynthetic adaptation to arid conditions, **CAM**, has evolved in about 3% to 4% of plant species, including pineapples, many cacti, and other succulent (water-storing) plants, such as aloe and jade plants. These plants open their stomata at night and close them during the day, just the reverse of how other plants behave. Closing stomata during the day helps desert plants conserve water, but it also prevents CO_2 from entering the leaves. During the night, when their stomata are open, these plants take up CO_2 and incorporate it into a variety of organic acids.

This mode of carbon fixation is called **crassulacean acid metabolism**, or **CAM**, after the plant family Crassulaceae, the succulents in which the process was first discovered. The mesophyll cells of CAM plants store the organic acids (malate) they make during the night in their vacuoles until morning, when the stomata close. During the day, when the light reactions can supply ATP and NADPH for the Calvin cycle, CO_2 is released from the organic acids made the night before to become incorporated into sugar in the chloroplasts.

Notice that the CAM pathway is similar to the C_4 pathway in that CO_2 is first incorporated into organic intermediates before it enters the Calvin cycle. The difference is that in C_4 plants, the initial steps of carbon fixation are separated structurally from the Calvin cycle, whereas in CAM plants, the two steps occur within the same cell but at separate times (Keep in mind that CAM, C_4, and C_3 plants all eventually use the Calvin cycle to make sugar from carbon dioxide).

▲ **Figure 7.28** A typical CAM plant. Prickly pear cactus (Opuntia) is a CAM plant. The more than 200 species of Opuntia living today originated in various xeric habitats in North and South America.

The CAM pathway is very similar to the C_4 pathway but with important differences. C_4 plants initially fix CO_2 into 4-carbon organic acids in mesophyll cells. The acids are later decarboxylated to produce CO_2, which is fixed by the C_3 pathway in the bundle sheath cells. In other words, the C_4 and C_3 pathways occur in **different locations** within the leaf of a C_4 plant. In CAM plants the initial fixation of CO_2 occurs at night.

Decarboxylation of malate and subsequent production of sugar from CO_2 by the normal C_3 photosynthetic pathway occur during the day. In other words, the CAM and C_3 pathways occur at **different times** within the same cell of a CAM plant. Although it does not promote rapid growth the way that the C_4 pathway does, the CAM pathway is a very successful adaptation to xeric conditions. CAM plants can exchange gases for photosynthesis and reduce water loss significantly. Plants with CAM photosynthesis survive in deserts where neither C_3 nor C_4 plants can.

	C₃ plant	C₄ plant	CAM plant
Example	 Sycamore	 Corn	 Cactus
Pathway	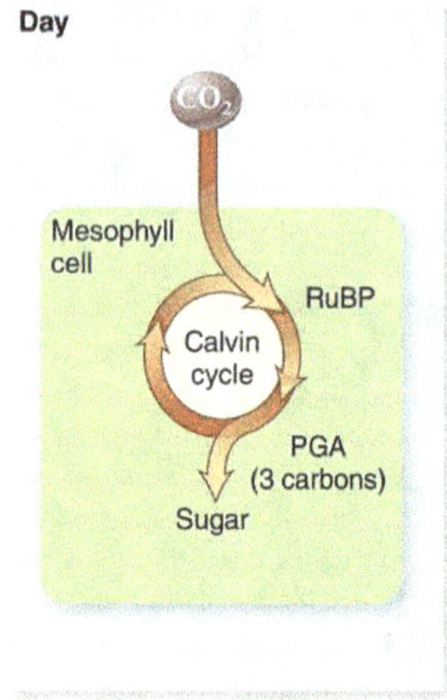 Day	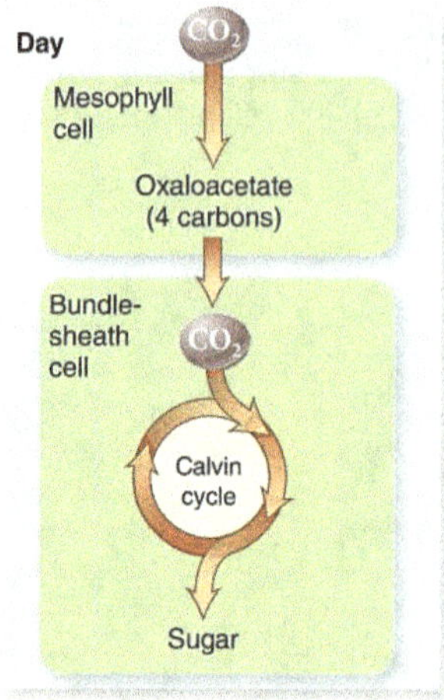 Day	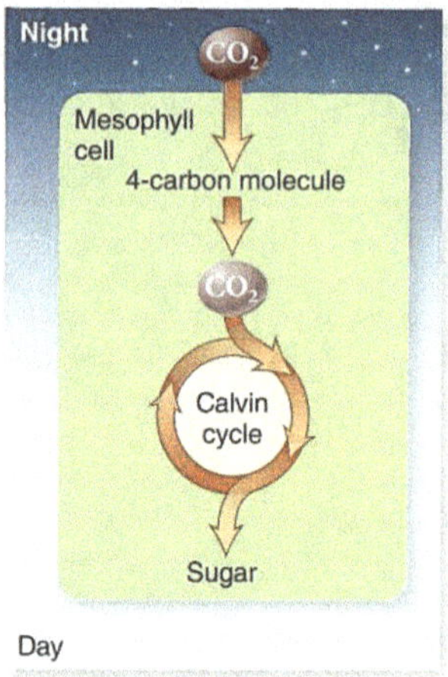 Night / Day
Limitation	Photorespiration	ATP cost	Reduced carbon availability
How plant avoids photorespiration	N/A	Light reactions and carbon reactions occur in separate cells.	CO_2 is absorbed at night; light reactions and carbon reactions occur during the day.
Habitat	Cool, moist	Hot, dry	Hot, dry
% of plant species	95%	1%	3–4%

▲ **Figure 7.29** C₃, C₄, and CAM Pathways Compared. Most plants use the C₃ pathway, which is vulnerable to photorespiration in hot, dry weather. The C₄ and CAM pathways are adaptations that minimize photorespiration.

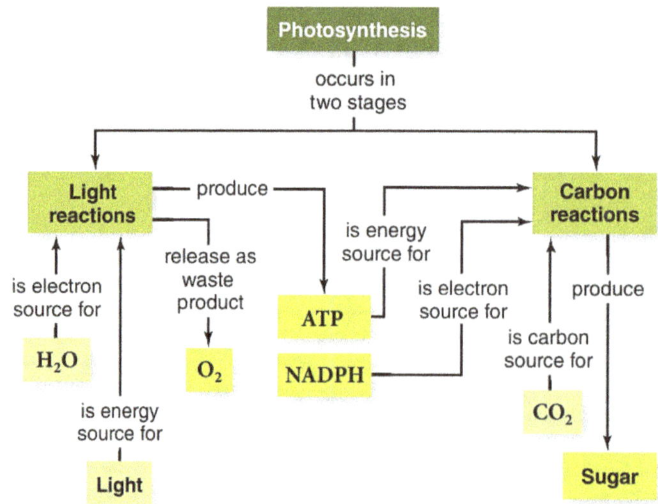

Limiting factors in photosynthesis

Plants need several different factors for photosynthesis to occur. These are:

- the presence of photosynthetic pigments
- a supply of carbon dioxide
- a supply of water
- light energy
- a suitable temperature.

A shortage of any one of these factors reduces the rate of photosynthesis below its maximum possible rate. In practice, the main external factors affecting the rate of photosynthesis are light intensity, light wavelength, temperature and carbon dioxide concentration. A shortage of water usually affects other processes in the plant before it affects photosynthesis. The **limiting factor** for any process is the factor that is in the shortest supply. If you increase the value of the limiting factor, then the rate will increase. Here, if you increase light intensity, then the rate of photosynthesis increases.

For most metabolic reactions, temperature has a large effect, but it has no significant effect on light-dependent reactions. This is because these are driven by energy from light, not the kinetic energy of the reacting molecules. However, the Calvin cycle is affected by temperature, because these are more 'normal' enzyme-controlled reactions. At low light intensities, the limiting factor governing the rate of photosynthesis is the light intensity. It is light that is the factor in shortest supply. You can increase the rate by increasing the light intensity – which is why the 'high light intensity' line is above the 'low light intensity' line.

At high light intensity, the plant has more than enough light, so temperature is the limiting factor. You can increase the rate of photosynthesis by increasing the temperature. This is why the line slopes upwards steeply. But at low light intensity, increasing the temperature does not have much effect on the rate. This is because, at low light intensity, it is light intensity, not temperature, that is the limiting factor. You can only increase the rate substantially by increasing the light intensity.

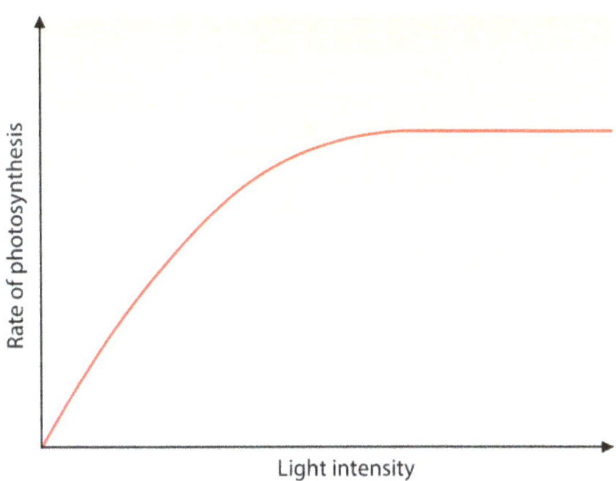

▲ **Figure 7.30** The rate of photosynthesis at different light intensities and constant temperature and carbon dioxide concentration.

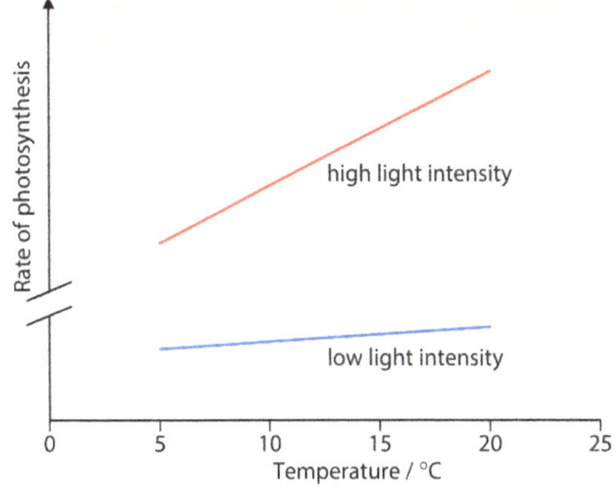

▲ **Figure 7.31** The rate of photosynthesis at two different light intensities and varying temperature.

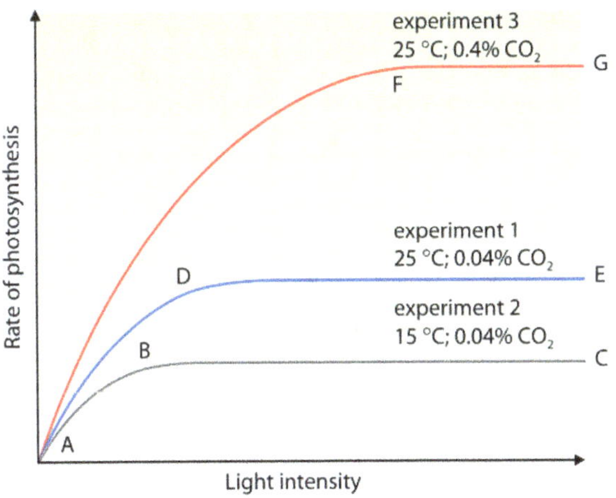

◀ **Figure 7.32** The effects of varying temperature, light intensity and carbon dioxide concentration on the rate of photosynthesis.

KEY CONCEPTS

- Photosynthesis is the conversion of light energy into chemical energy

- Photosynthesis has three stages: absorbing light energy, using this energy to synthesize ATP and NADPH, and using the ATP and NADPH to convert CO_2 to organic molecules. The first two stages consist of light dependent reactions, and the third stage of light-independent reactions.

- Chloroplasts contain internal thylakoid membranes and a fluid matrix called stroma. The photosystems involved in energy capture are found in the thylakoid membranes, and enzymes for assembling organic molecules are in the stroma.

- Early investigations revealed that plants produce O_2 from carbon dioxide and water in the presence of light.

- The light-dependent reactions require light; the light-independent reactions occur in both daylight and darkness. The rate of photosynthesis depends on the amount of light, the CO_2 concentration, and temperature.

- The use of isotopes revealed the individual origins and fates of different molecules in photosynthetic reactions.

- Carbon fixation requires ATP and NADPH, which are products of the light-dependent reactions. As long as these are available, CO_2 is reduced by enzymes in the stroma to form simple sugars.

- Light exists both as a wave and as a particle (photon). Light can remove electrons from some metals by the photoelectric effect, and in photosynthesis, chloroplasts act as photoelectric devices.

- Chlorophyll a is the only pigment that can convert light energy into chemical energy. Chlorophyll b is an accessory pigment that increases the harvest of photons for photosynthesis.

- Carotenoids and other accessory pigments further increase a plant's ability to harvest photons.

- Measurement of O_2 output led to the idea of photosystems—clusters of pigment molecules that channel energy to a reaction center.

- A photosystem is a network of chlorophyll a, accessory pigments, and proteins embedded in the thylakoid membrane. Pigment molecules of the antenna complex harvest photons and feed light energy to the reaction center. The reaction center is composed of two chlorophyll a molecules in a protein matrix that pass an excited electron to an electron acceptor.

- The light reactions can be broken down into four processes: primary photoevent, charge separation, electron transport, and chemiosmosis.

- An excited electron moves along a transport chain and eventually returns to the photosystem. This cyclic process is used to generate a proton gradient. In some bacteria, this can also produce NADPH.

- Photosystem I transfers electrons to $NADP^+$, reducing it to NADPH.

- Photosystem II replaces electrons lost by photosystem I. Electrons lost from photosystem II are replaced by electrons from oxidation of water, which also produces O_2.

- Photosystem II and photosystem I are linked by an electron transport chain; the b_6-f complex in this chain pumps protons into the thylakoid space.

- ATP synthase is a channel enzyme; as protons flow through the channel down their gradient, ADP is phosphorylated producing ATP, similar to the mechanism in mitochondria. Plants can make additional ATP by cyclic photophosphorylation.

- Photosystem II replaces electrons lost by photosystem I. Electrons lost from photosystem II are replaced by electrons from oxidation of water, which also produces O_2.

- Photosystem II and photosystem I are linked by an electron transport chain; the b_6-f complex in this chain pumps protons into the thylakoid space.

- ATP synthase is a channel enzyme; as protons flow through the channel down their gradient, ADP is phosphorylated producing ATP, similar to the mechanism in mitochondria. Plants can make additional ATP by cyclic photophosphorylation.

- Imaging studies suggest that photosystem II is primarily found in the grana, while photosystem I and ATP synthase are found in the stroma lamella.

- The Calvin cycle, also known as C_3 photosynthesis, uses CO_2, ATP, and NADPH to build simple sugars.

- The Calvin cycle occurs in three stages: carbon fixation via the enzyme rubisco's action on RuBP and CO_2; reduction of the resulting 3-carbon PGA to G3P, generating ATP and NADPH; and regeneration of RuBP. Six turns of the cycle fix enough carbon to produce two excess G3Ps used to make one molecule of glucose.

- Rubisco can catalyze the oxidation of RuBP, reversing carbon fixation. Dry, hot conditions tend to increase this reaction. C_4 plants fix carbon by adding CO_2 to a 3-carbon molecule, forming oxaloacetate. Carbon is fixed in one cell by the C_4 pathway, then CO_2 is released in another cell for the Calvin cycle.

- CAM plants use the C_4 pathway during the day when stomata are closed, and the Calvin cycle at night in the same cell.

CHAPTER 8

The Cell Cycle

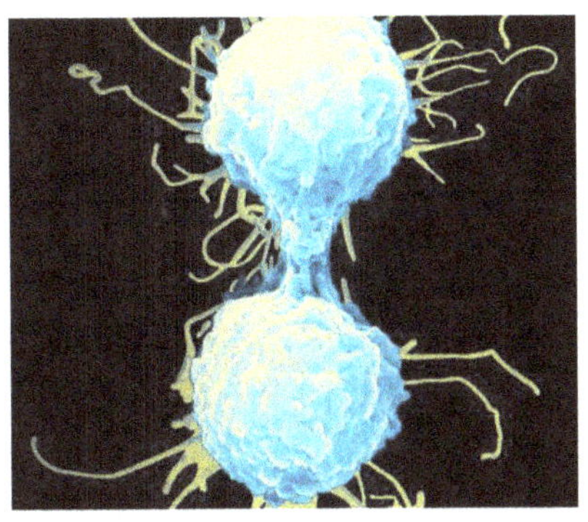

Chapter Contents:

- Cell Division
- Genetic Material
- A chromosome consists of a DNA molecule packed together with proteins
- Distribution of Chromosomes During Eukaryotic Cell Division
- Phases of the Cell Cycle
- Mitosis
- Cytokinesis
- Binary Fission in Bacteria
- The Cell Cycle Control System
- Internal and External Signals at the Checkpoints
- Cancer
- Cell Death Is Part of Life
- Cell Aging
- Cellular Diversity
- Stem Cells
- KEY CONCEPTS

Cell Division

The continuity of life is based on the reproduction of cells, or **cell division**. Cells can only be formed by division of pre-existing cells. Cell division plays several important roles in life. When a prokaryotic cell divides, it is actually reproducing because the process gives rise to a new organism (another cell). The same is true of any unicellular eukaryote, such as the amoeba.

As for multicellular eukaryotes, cell division enables each of these organisms to develop from a single cell—the fertilized egg. All body cells, except the gametes, are called **somatic cells**. In somatic cell division, a cell divides into two identical cells. An important part of somatic cell division is **replication** (duplication) of the **DNA** sequences that make up genes and chromosomes so that the same genetic material can be passed on to the newly formed cells. After somatic cell division, each newly formed cell has the same number of chromosomes as the original cell. Somatic cell division replaces dead or injured cells and adds new ones for tissue growth. For example, skin cells are continually replaced by somatic cell divisions or dividing cells in your bone marrow continuously make new blood cells.

In both prokaryotes and eukaryotes, a crucial function of most cell divisions is the distribution of identical genetic material—**DNA**—to two daughter cells (The exception is meiosis, the special type of eukaryotic cell division that can produce sperm and eggs).

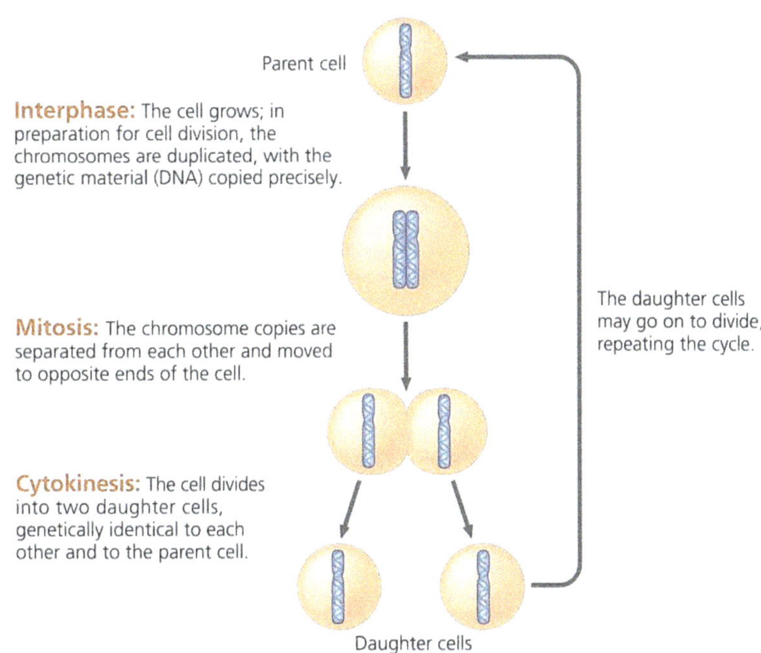

▲ **Figure 8.1** How does one parent cell give rise to two genetically identical daughter cells?

Genetic Material

A cell's **DNA**, its genetic information, is called its **genome**. Although a prokaryotic genome is often a single circular DNA molecule, eukaryotic genomes usually consist of a number of DNA molecules. The overall length of DNA in a eukaryotic cell is enormous. A typical human cell, for example, has about 2 m of DNA—a length about 250,000 times greater than the cell's diameter. Before the cell can divide to form genetically identical daughter cells, all of this DNA must be copied, or replicated, and then the two copies must be separated so that each daughter cell ends up with a complete genome.

The replication and distribution of so much DNA are manageable because the DNA molecules are packaged into structures called **chromosomes** (from the Greek chroma, color, and soma, body), so named because they take up certain dyes used in microscopy. Chromosomes were first observed by the German embryologist Walther Flemming (1843–1905) in 1879, while he was examining the rapidly dividing cells of salamander larvae. Each eukaryotic chromosome consists of one very long, linear DNA molecule associated with many proteins. The DNA molecule carries several hundred to a few thousand **genes**, the units of information that specify an organism's inherited traits, for example, humans have more than 20,000 genes that code for proteins.

Eukaryotic chromosomes consist of **chromatin**, which is a collective term for all of the cell's DNA and its associated proteins. The associated proteins maintain the structure of the chromosome and help control the activity of the genes. These proteins include the many enzymes that help replicate the DNA and transcribe it to a sequence of RNA. Others serve as scaffolds around which DNA entwines, helping to pack the DNA efficiently inside the cell. The chromatin of a chromosome varies in its degree of condensation during the process of cell division.

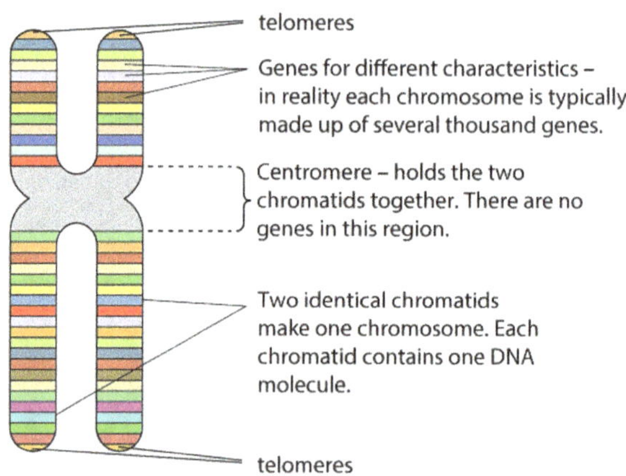

▲ **Figure 8.2** Simplified diagram of the structure of a chromosome.

A chromosome consists of a DNA molecule packed together with proteins

How is DNA packaged into chromosomes, the structures that carry genetic information? The main component of the genome in most bacteria is a double-stranded, circular DNA molecule that is associated with specific proteins. A bacterial chromosome differs from a eukaryotic chromosome in that a eukaryotic chromosome consists of a single linear DNA molecule associated with a large number of proteins.

In E. coli, the chromosomal DNA consists of about 4.6 million nucleotide pairs, representing about 4,400 genes. This is 100 times more DNA than is found in a typical virus, but only about one-thousandth as much DNA as in a human somatic cell. Even so, that is a tremendous amount of DNA to be packaged in such a small container. Stretched out, the DNA of an E. coli cell would measure about a millimeter in length, 1,000 times longer than the region it occupies in the cell. Within a bacterium, however, certain proteins cause the chromosome to coil and "supercoil," densely packing it so that it fills only part of the cell. Unlike the nucleus of a eukaryotic cell, this dense region of DNA in a bacterium, called the **nucleoid**, is not bounded by membrane.

The haploid genome of humans contains about 3.2 billion base pairs and, stretched out, the DNA in a diploid cell would be about 2 meters (1.8 metres) long with 3.2 billion base pairs (bp). A human cell is about 40–60 mm in diameter with a nucleus about 10 mm in diameter. DNA fits into a nucleus because it is packed into a shorter length by histones. Proteins called **histones** are responsible for the main level of DNA packing in interphase chromatin.

Although each histone is small—containing only about 100 amino acids—the total mass of histone in chromatin approximately equals the mass of DNA. More than a fifth of a histone's amino acids are positively charged (lysine or arginine) and therefore bind tightly to the negatively charged DNA. The apparent conservation of histone genes during evolution probably reflects the important role of histones in organizing DNA within cells. The four main types of histones are critical to the next level of DNA packing (A fifth type of histone, called H1, is involved in a further stage of packing).

The role of histones is more than simply structural because their arrangement also affects the activity of the DNA with which they are associated. Histones are increasingly viewed as an important part of the regulation of gene expression, that is, whether genes are turned off or on. Other proteins called **nonhistone proteins** are also associated with DNA; some of these proteins are also important for the structure of chromosomes, whereas others are involved in gene regulation.

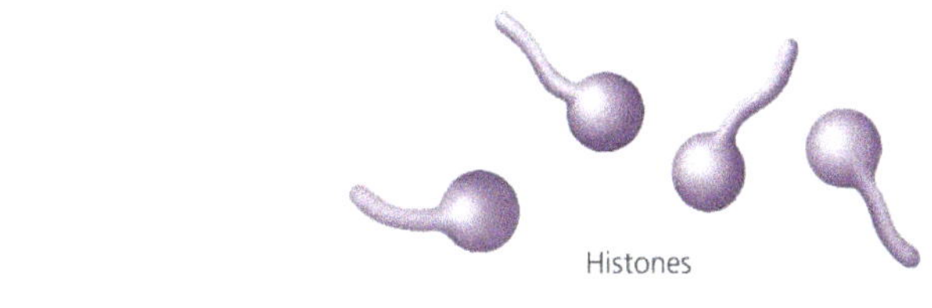

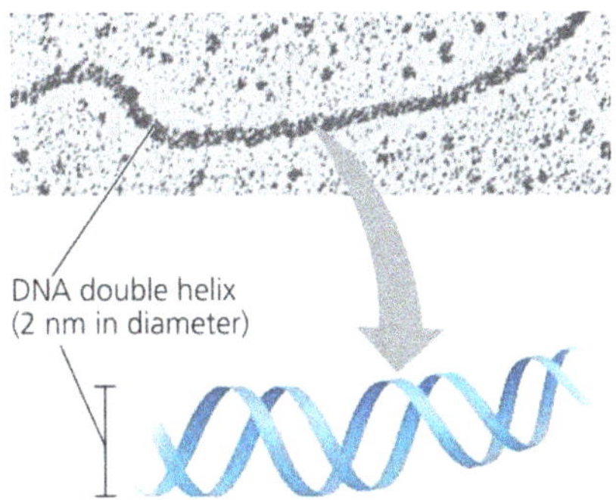

▲ **Figure 8.3** A ribbon model of DNA with sugar-phosphate backbones. The phosphate groups along the backbone contribute a negative charge along the outside of each strand; the double helix alone is 2 nm across.

In electron micrographs, unfolded chromatin is roughly 10 nm in diameter (**the 10-nm chromatin fiber**). Such chromatin resembles beads on a string. Each "bead" is a **nucleosome**, the basic unit of DNA packing; the "string" between beads is called **linker DNA**. The fundamental unit of each nucleosome consists of a beadlike structure with 146 base pairs of DNA wrapped twice around a disc-shaped core of eight histone molecules (two each of four different histone types : **H2A**, **H2B**, **H3**, and **H4**). Each nucleosome also has a larger histone (**H1**) associated with both the wrapped DNA and the surface of the core.

Nucleosomes are very dynamic structures, with H1 loosening and DNA unwrapping at least once every second to allow other proteins, including transcription factors and enzymes, access to the DNA. The amino end of each histone (the histone tail) extends outward from the nucleosome and is involved in regulation of gene expression. In the cell cycle, the histones leave the DNA only briefly during DNA replication. Generally, they do the same during the process of transcription, which requires access to the DNA by transcription proteins. Histones provide a physical means for packing the very long DNA molecules in a compact, orderly way, but they also play an important role in gene regulation.

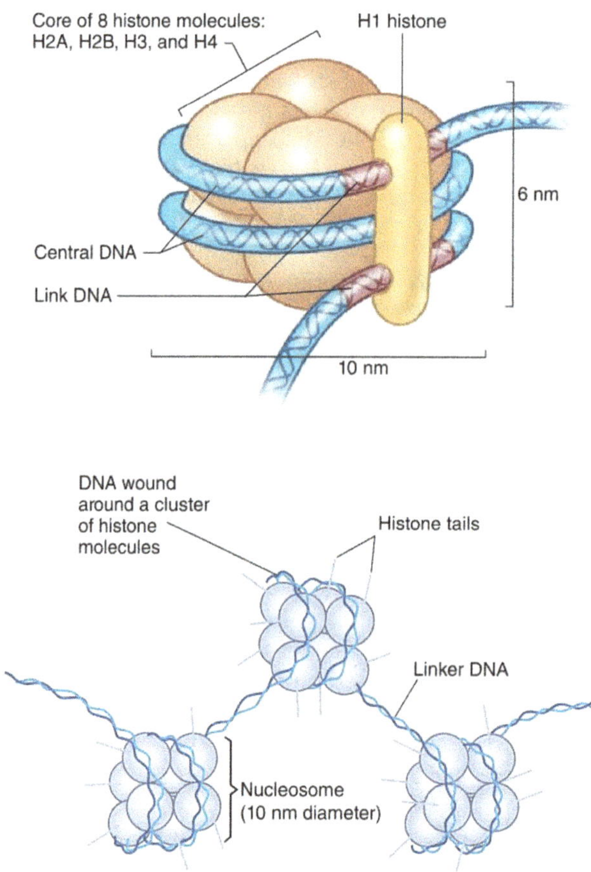

▲ **Figure 8.4** A model for the structure of a nucleosome. Each nucleosome bead contains a set of eight histone molecules, forming a protein core around which the double-stranded DNA winds. The DNA surrounding the histones consists of 146 nucleotide pairs; another segment of DNA, about 60 nucleotide pairs long, links nucleosome beads.

Chromatin is composed of approximately:

- 30% DNA, our genetic material

- 60% globular histone proteins, which package and regulate the DNA

- 10% RNA chains, newly formed or forming

Chromatin undergoes striking changes in how densely it is packed during the course of the cell cycle. In interphase cells stained for light microscopy, the chromatin usually appears as a diffuse mass within the nucleus, with some denser clumps, including in regions of centromeres and telomeres. The less compacted chromatin is called **euchromatin** ("true chromatin"). These active chromatin segments, referred to as **extended chromatin**, are **lightly stained** basophilic areas in the light microscope. The denser-appearing chromatin is called **heterochromatin**. These generally inactive **condensed chromatin** segments are **darker staining** and more easily detected.

DNA in the more open structure of euchromatin is rich with genes, although not all of the genes are transcribed in all cells. Heterochromatin is always more compact than euchromatin, shows little or no transcriptional activity, and includes at least two types of genomic material called **constitutive** and **facultative** heterochromatin. Constitutive heterochromatin is generally similar in all cell types and contains mainly repetitive, gene-poor DNA sequences, including the large chromosomal regions called **centromeres** and **telomeres**, which are located near the middle (most often) and at the ends of chromosomes, respectively. Facultative heterochromatin contains other regions of DNA with genes where transcription is variably inactivated in different cells by epigenetic mechanisms and can undergo reversible transitions from compact, transcriptionally silent states to more open, transcriptionally active conformations.

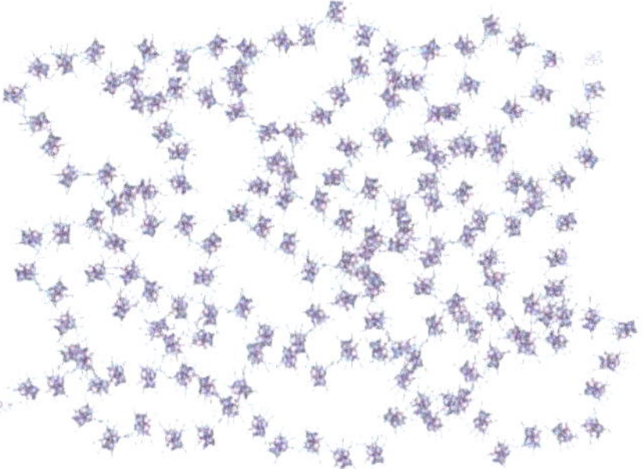

Heterochromatin (densely arranged 10-nm fiber)

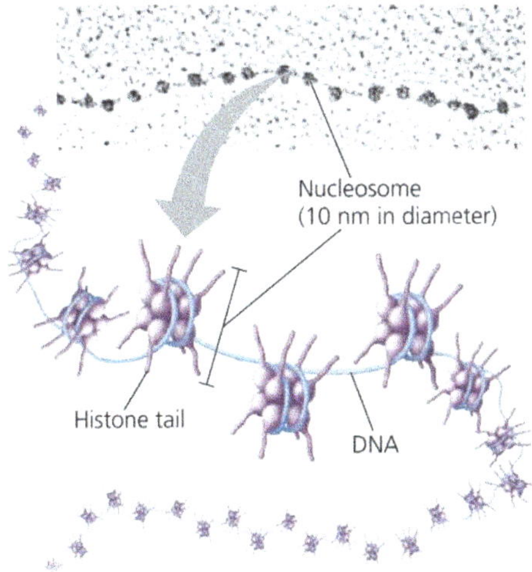

Euchromatin (loosely arranged 10-nm fiber)

The ratio of heterochromatin to euchromatin seen with nuclear staining can provide a rough indicator of a cell's metabolic and biosynthetic activity. Euchromatin predominates in active cells such as large neurons, while heterochromatin is more abundant in cells with little synthetic activity such as circulating lymphocytes. Facultative heterochromatin also occurs in the small, dense "sex chromatin" or **Barr body** which is one of the two large X chromosomes present in human females but not males.

The Barr body remains tightly coiled, while the other X chromosome is uncoiled, transcriptionally active, and not visible. Cells of males have one X chromosome and one Y chromosome; like the other chromosomes, the single X chromosome remains largely euchromatic. Studies show that chromosomal domains with few genes form a layer beneath the nuclear envelope, while domains with many active genes are located deeper in the nucleus.

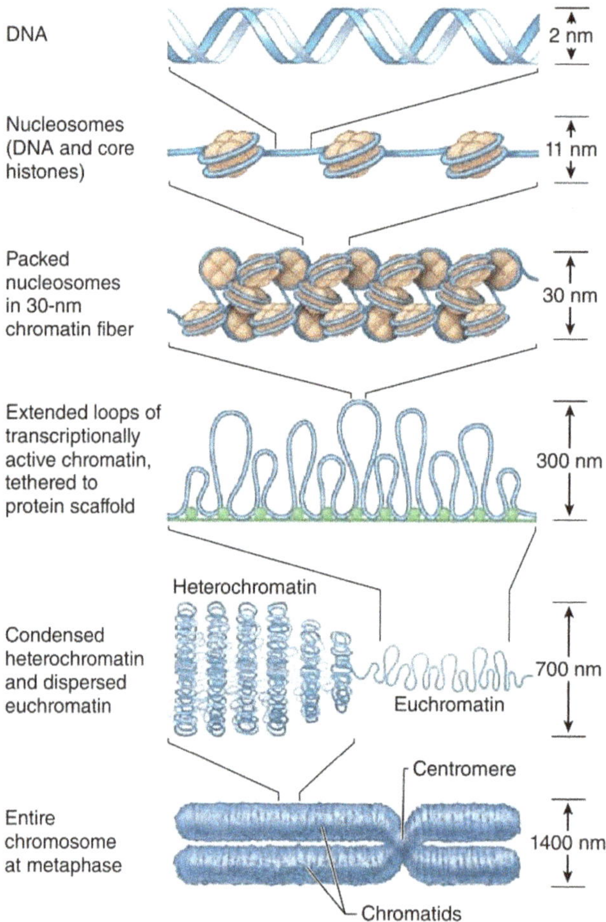

▲ **Figure 8.5** From DNA to chromatin.

For both types of chromatin, the basic organizing unit is the 10-nm fiber—nucleosomes joined by linker DNA. In heterochromatin, this 10-nm fiber is folded and bent back on itself to a much greater degree than in euchromatin, accounting for its denser appearance. Because heterochromatin is so compacted, it is largely inaccessible to the proteins responsible for transcribing the genetic information, a crucial early step in gene expression. So, genes in heterochromatin are generally not expressed. In contrast, the looser packing of euchromatin makes its DNA accessible to those proteins, and the genes present in euchromatin are available for transcription.

The most active body cells have much larger amounts of extended chromatin. The **nucleolus** is a generally spherical, highly basophilic subdomain of nuclei in cells actively engaged in protein synthesis. The intense basophilia of nucleoli is not due to heterochromatin but to the presence of densely concentrated ribosomal RNA (rRNA) that is transcribed, processed, and assembled into ribosomal subunits. Chromosomal regions with the genes for rRNA organize one or more nucleoli in cells requiring intense ribosome production for protein synthesis during growth or secretion.

Ultrastructural analysis of an active nucleolus reveals **fibrillar** and **granular** subregions with different staining characteristics that reflect stages of rRNA maturation (TEM reveals morphologically distinct regions within a nucleolus. Small, light-staining areas are fibrillar centers containing the DNA sequences for the rRNA genes. The darker fibrillar material surrounding the fibrillar centers consists of accumulating rRNA transcripts.

More granular material of the nucleolus contains mainly the large and small ribosomal subunits being assembled from rRNA and ribosomal proteins synthesized in the cytoplasm. Various amounts of heterochromatin are also typically found near the nucleolus, scattered in the euchromatin, and adjacent to the nuclear envelope that separates chromatin from cytoplasm). Molecules of rRNA are processed in the nucleolus and very quickly associate with the ribosomal proteins imported from the cytoplasm via nuclear pores. The newly organized small and large ribosomal subunits are then exported back to the cytoplasm through those same nuclear pores.

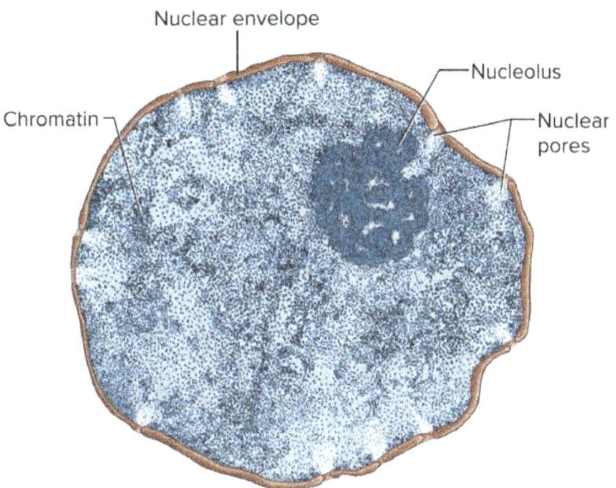

▲ **Figure 8.6** Nucleolus.

Structure: Densely stained filamentous structure within the nucleus. Consists of proteins associated with DNA in regions where information concerning ribosomal proteins is being expressed.

Function: Site of ribosomal RNA synthesis. Assembles RNA and protein components of ribosomal subunits, which then move to the cytoplasm through nuclear pores.

©Don W. Fawcett/Science Source

Although an interphase chromosome lacks an obvious scaffold, there are proteins that further organize the 10-nm fiber into larger compartments and smaller looped domains. One H1 molecule binds both to the nucleosome (at the point where the DNA enters and leaves the core particle) and to the linker DNA. This binding causes the nucleosomes to package into a coiled structure 30 nm in diameter, called the **30-nm chromatin fiber**. One possible model for the 30-nm fiber is the **solenoid** model, with the nucleosomes spiraling helically with about six nucleosomes per turn.

Some of the looped domains appear to be attached to the nuclear lamina, on the inside of the nuclear envelope, and perhaps also to fibers of the nuclear matrix. These attachments may help organize regions of chromatin where genes are active.

The chromatin of each chromosome occupies a specific restricted area within the interphase nucleus, and the chromatin fibers of different chromosomes do not appear to be entangled. As a cell prepares for mitosis, its chromatin becomes organized into loops and coils, eventually condensing into a characteristic number of short, thick metaphase chromosomes that are distinguishable from each other with the light microscope.

Prophase. When mitosis begins, DNA replication has already occurred, so each chromosome consists of two **sister chromatids**. During prophase, the chromatin of each sister chromatid begins to condense. Two related proteins called **condensin II** and **condensin I** play important roles. First, condensin II proteins (red) bind to the 10-nm fiber of DNA and form DNA loops that get larger and larger. The condensin II proteins form a central scaffold from which the loops extend. As the loops grow, the chromosome gets wider and shorter. By the end of prophase, the chromosome is half as long.

▶ **Figure 8.7** Chromatin and chromosome structure. (a) Electron micrograph of chromatin fiber (125,000×). (b) DNA packed in a chromosome. The levels of increasing structural complexity (coiling) are shown in order from the smallest (1–5).

Prometaphase. When prometaphase begins, condensin I proteins (green) bind to DNA outside the central scaffold, making smaller loops out of the larger loops generated by condensin II. The process continues, with more and more loops extending outward, and the chromosome getting denser, shorter, and wider. The scaffold itself also begins to twist into a helix (suggested by the curved gray arrows), allowing even more loops per given length of chromosome (**supercoiling**).

Metaphase. At metaphase, the chromosome is at its most dense, with the most loops per turn, and therefore is at its shortest length. The two sister chromatids are fully condensed.

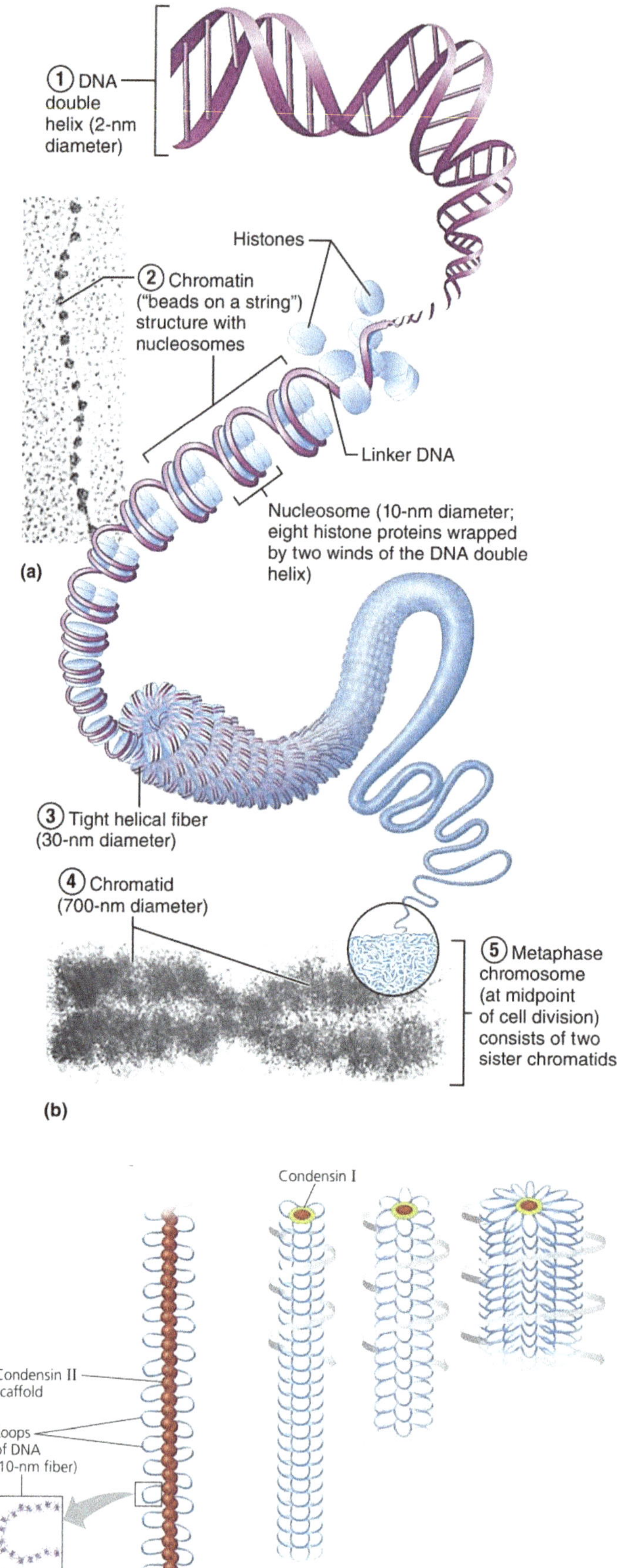

This series of diagrams and transmission electron micrographs depicts a current model for the progressive levels of DNA coiling and folding. The illustration zooms out from a single molecule of DNA to a metaphase chromosome, which is large enough to be seen with a light microscope.

DNA, the double helix

Histones

Nucleosomes, or "beads on a string" (10-nm fiber)

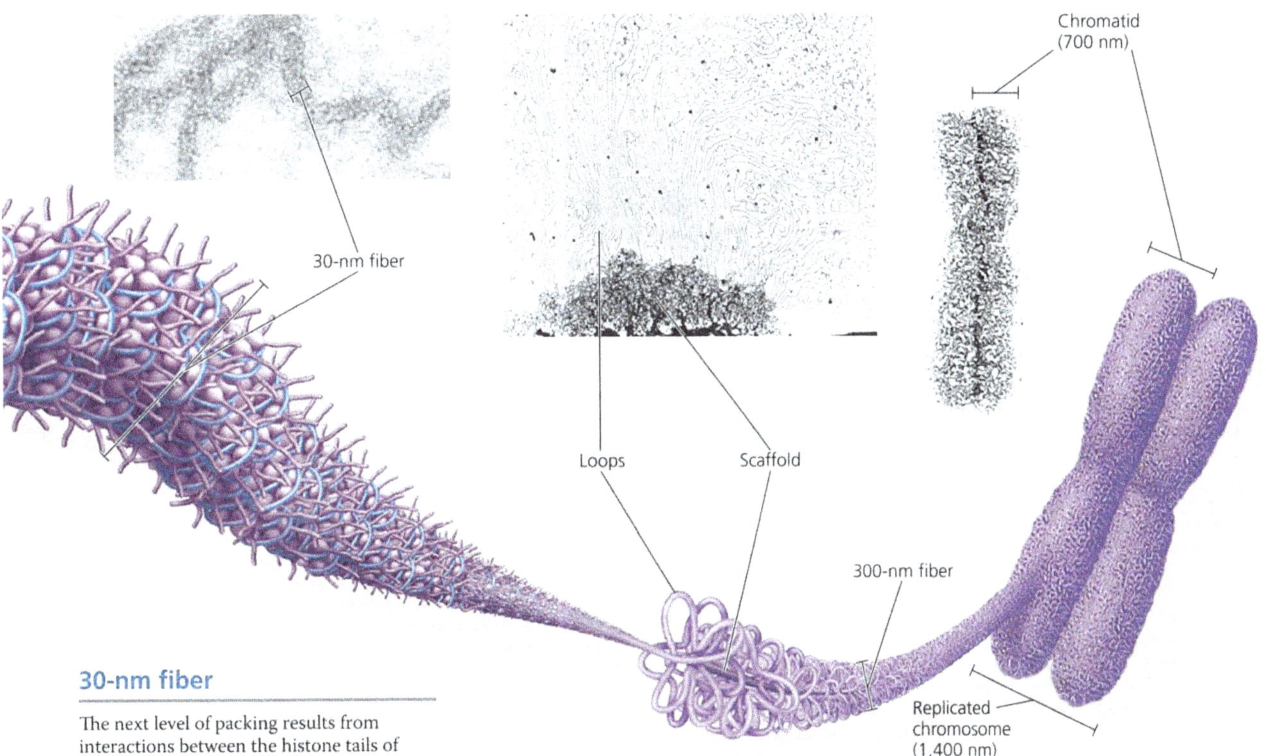

30-nm fiber

The next level of packing results from interactions between the histone tails of one nucleosome and the linker DNA and nucleosomes on either side. A fifth histone, H1, is involved at this level. These interactions cause the extended 10-nm fiber to coil or fold, forming a chromatin fiber roughly 30 nm in thickness, the *30-nm fiber*. Although the 30-nm fiber is quite prevalent in the interphase nucleus, the packing arrangement of nucleosomes in this form of chromatin is still a matter of some debate.

Looped domains (300-nm fiber)

The 30-nm fiber, in turn, forms loops called *looped domains* attached to a chromosome scaffold composed of proteins, thus making up a *300-nm fiber*. The scaffold is rich in one type of topoisomerase, and H1 molecules also appear to be present.

Metaphase chromosome

In a mitotic chromosome, the looped domains themselves coil and fold in a manner not yet fully understood, further compacting all the chromatin to produce the characteristic metaphase chromosome shown in the micrograph above. The width of one chromatid is 700 nm. Particular genes always end up located at the same places in metaphase chromosomes, indicating that the packing steps are highly specific and precise.

Microscopic analysis of chromosomes usually begins with cultured cells arrested in mitotic metaphase by colchicine or other compounds that disrupt microtubules. After processing and staining the cells, the condensed chromosomes of one nucleus are photographed by light microscopy and rearranged digitally to produce a **karyotype** in which stained chromosomes can be analyzed. Note that the 22 pairs of autosomes, as well as the X and Y chromosomes, differ in size, morphology, staining properties, location of the centromere (a constriction found on all chromosomes), the relative length of the two arms on either side of the centromere, and the positions of constricted regions along the arms. A normal karyotype tells us a lot about a body cell.

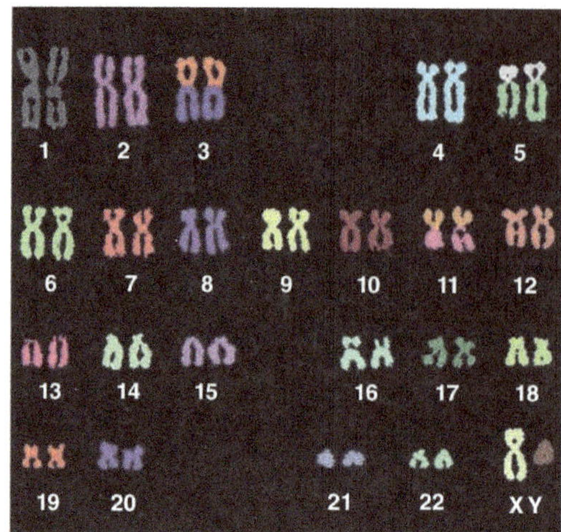

▲ **Figure 8.8** For a diploid human cell, 2 complete sets of chromosomes from a single cell constitute a karyotype of that cell.

By convention, the chromosomes are numbered according to size, with the largest chromosomes having the smallest numbers. For example, human chromosomes 1, 2, and 3 are relatively large, whereas 21 and 22 are the two smallest. This numbering system does not apply to the **sex chromosomes**, which determine the sex of the individual. Sex chromosomes in humans are designated with the letters X and Y; females are XX and males are XY. Y chromosome contains the **SRY** gene that causes testes to develop. The chromosomes that are not sex chromosomes are called **autosomes**. Humans have 22 different autosomes.

Most human cell contains a total of 46 chromosomes. Each cell has two sets because the individual inherited one set from the father and one set from the mother. When the cells of an organism carry two sets of chromosomes, that organism is said to be **diploid**. Geneticists use the letter **n** to represent a set of chromosomes. The number of chromosome sets is called the **ploidy** of a cell or species. Diploid organisms are referred to as **2n** because they have two sets of chromosomes. For example, humans are 2n, where n= 23. Most human cells are **diploid**. An exception is the **gametes**, the sperm and egg cells. Gametes are **haploid**, or 1n, which means they contain one set of chromosomes. For humans, the haploid number is 23 and the diploid number is 46.

Every individual of a given species has a characteristic number of chromosomes in the nuclei of its somatic (body) cells. However, it is not the number of chromosomes that makes each species unique but the information the genes specify. Human cells each have 46 chromosomes, consisting of 23 nearly identical pairs, but humans are not humans merely because we have 46 chromosomes. Some other species—the olive tree, for example—also have 46. Some humans have an abnormal chromosome composition with more or fewer than 46.

Most animal and plant species have between 8 and 50 chromosomes per somatic cell. The number of chromosomes a species has does not indicate the species' complexity or its status within a particular domain or kingdom. A mosquito's cell has 6 chromosomes; grasshoppers, rice plants, and pine trees all have 24; humans have 46; dogs and chickens have 78; a carp has 104. Each of these numbers is even because sexually reproducing organisms inherit one set of chromosomes from each parent.

Human sperm and egg cells, for example, each contain 23 chromosomes; fertilization therefore yields an offspring with 46 chromosomes in every cell. Diploid chromosomes reflect the equal genetic contribution that each parent makes to offspring. We refer to the maternal and paternal chromosomes as being **homologous**, and each one of the pair is termed a homologue.

How similar are homologous chromosomes to each other? The two chromosomes in a homologous pair are nearly identical in size and contain a very similar composition of genetic material. The striking similarity between homologous chromosomes does not apply to the sex chromosomes (for example, X and Y in humans). These chromosomes differ in size and genetic composition. Certain genes found on the X chromosome are not found on the Y chromosome, and vice versa. The X and Y chromosomes are not considered homologous chromosomes, although they do have short regions of homology.

Chromosomes as seen in a karyotype are only present for a brief period during cell division. Prior to replicating, each chromosome is composed of a single DNA molecule that is arranged into the 30-nm fiber. After replication, each chromosome is composed of two identical DNA molecules held together by a complex of proteins called **cohesins,** these chromosomes are said to be **replicated** or **duplicated chromosomes** because the two **sister chromatids** contain the same genes. As the chromosomes become more condensed and arranged about the protein scaffold, they become visible as two strands that are held together at the centromere.

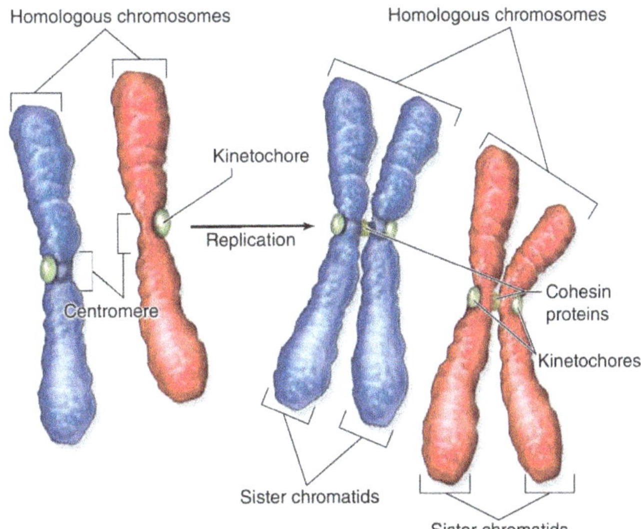

▲ **Figure 8.9** The difference between homologous chromosomes and sister chromatids. Homologous chromosomes are the maternal and paternal copies of the same chromosome. Sister chromatids are the two replicas of a single chromosome held together at their centromeres by cohesin proteins after DNA replication. The kinetochore is composed of proteins found at the centromere that attach to microtubules during mitosis.

Distribution of Chromosomes During Eukaryotic Cell Division

When a cell is not dividing, and even as it replicates its DNA in preparation for cell division, each chromosome is in the form of a long, thin chromatin fiber. After DNA replication, however, the chromosomes condense as a part of cell division: Each chromatin fiber becomes densely coiled and folded, making the chromosomes much shorter and so thick that we can see them with a light microscope. Each duplicated chromosome consists of two sister chromatids, which are joined copies of the original chromosome.

The two chromatids, each containing an identical DNA molecule, are often attached all along their lengths by protein complexes called cohesins; this attachment is known as sister chromatid cohesion. Each sister chromatid has a **centromere**, a region made up of repetitive sequences in the chromosomal DNA where the chromatid is attached most closely to its sister chromatid. This attachment is mediated by proteins that recognize and bind to the centromeric DNA; other bound proteins condense the DNA, giving the duplicated chromosome a narrow "waist."

The position of the centromere is chromosome-specific: it may be central, towards one end or the other, or at or near a chromosome end. The position of a centromere gives the chromosome a particular morphology. The portion of a chromatid to either side of the centromere is referred to as an **arm** of the chromatid. (An unduplicated chromosome has a single centromere, distinguished by the proteins that bind there, and two arms).

Later in the cell division process, the two sister chromatids of each duplicated chromosome separate and move into two new nuclei, one forming at each end of the cell. Once the sister chromatids separate, they are no longer called sister chromatids but are considered individual chromosomes; this is the step that essentially doubles the number of chromosomes during cell division. Thus, each new nucleus receives a collection of chromosomes identical to that of the parent cell.

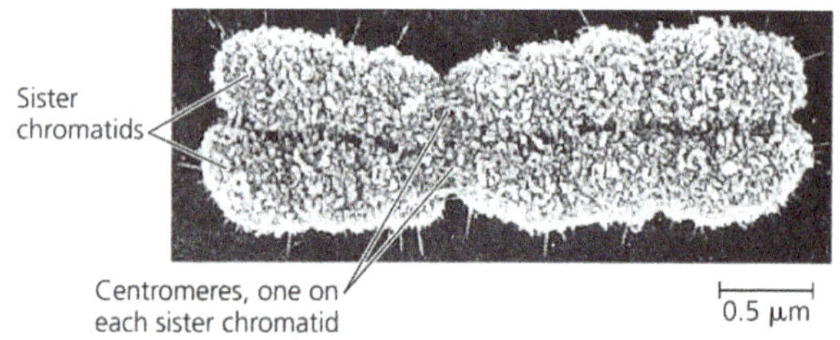

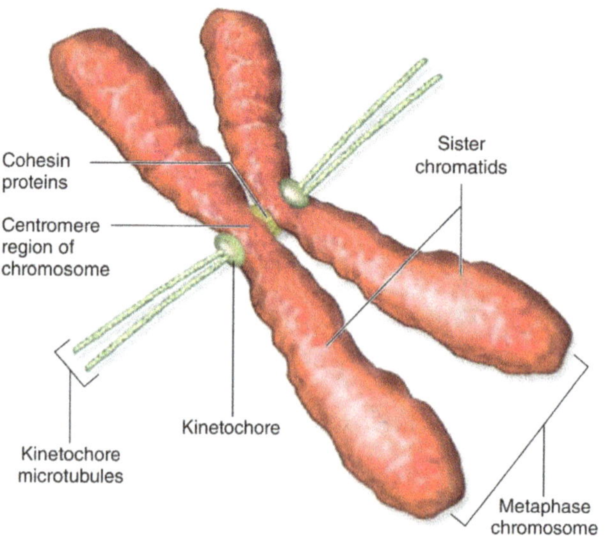

▲ **Figure 8.10** A highly condensed, duplicated human chromosome.

Mitosis, the division of the genetic material in the nucleus, is usually followed immediately by **cytokinesis**, the division of the cytoplasm. One cell has become two, each the genetic equivalent of the parent cell. From a fertilized egg, mitosis and cytokinesis produced the 37 trillion somatic cells that now make up your body (**embryonic development** and **growth**), and the same processes continue to generate new cells to **replace** dead and damaged ones.

Mitotic cell division occurs some 300 million times per minute in your body. The cloning of B and T lymphocytes during the **immune response** is dependent on mitosis too. **Clone** consists of individual organisms, cells, or DNA molecules that are genetically identical to another individual, cell, or DNA molecule from which it was derived.

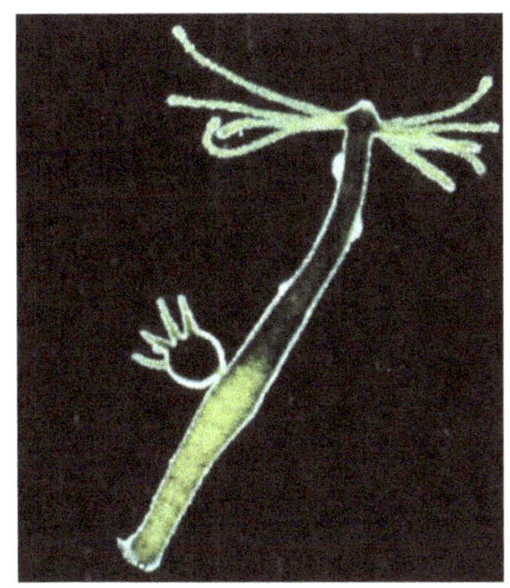

▶ **Figure 8.11** Hydra viridissima with a small new polyp attached, produced by asexual reproduction involving mitosis.

Mitosis also serves as a mechanism of organismal reproduction called **vegetative reproduction** or **asexual reproduction**, which occurs in many kinds of plants and protists and in some animals. In asexual reproduction, daughter cells produced by mitotic cell division grow by further mitosis into complete individuals. For example, asexual reproduction occurs when a single-celled protist such as an amoeba divides by mitosis to produce two separate individuals, or when a **stem cutting** is used to propagate an entire new plant. **Budding** is particularly common in plants. It is most commonly a form of vegetative propagation in which a bud on part of the stem simply grows a new plant. The new plant eventually becomes detached from the parent and lives independently.

The bud may be part of the stem of an overwintering structure such as a bulb or tuber. In contrast, you produce gametes— eggs or sperm—by a variation of cell division called **meiosis**, which yields daughter cells with only one set of chromosomes, half as many chromosomes as the parent cell. Meiosis in humans occurs only in special cells in the ovaries or testes (the gonads). Generating gametes, meiosis reduces the chromosome number from 46 (two sets) to 23 (one set). **Fertilization** fuses two gametes together and returns the chromosome number to 46 (two sets). Mitosis then conserves that number in every somatic cell nucleus of the new human individual.

▶ **Figure 8.12** Chromosome duplication and distribution during cell division.

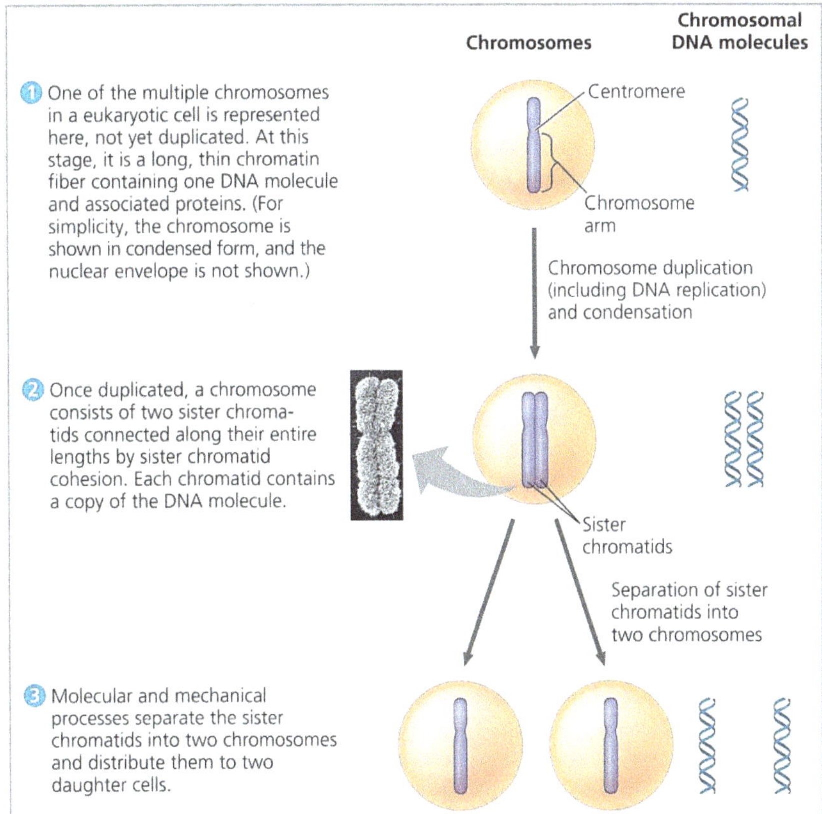

Phases of the Cell Cycle

Mitosis is just one part of the cell cycle, the life of a cell from the time it is first formed during division of a parent cell until its own division into two daughter cells (Biologists use the words daughter or sister in relation to cells, but this is not meant to imply gender). In fact, the **mitotic (M) phase**, which includes both **mitosis** and **cytokinesis**, is usually the shortest part of the cell cycle. The mitotic phase alternates with a much longer stage called **interphase**, which often accounts for about 90% of the cycle.

Interphase can be divided into three phases: the **G₁ phase** ("first gap") which is typically the longest phase, the **S phase** ("synthesis"), and the **G₂ phase** ("second gap"). During all three phases of interphase, actually, a cell grows by producing proteins and cytoplasmic organelles such as mitochondria and endoplasmic reticulum. Duplication of the chromosomes, crucial for eventual division of the cell, occurs entirely during the S phase. Thus, a cell grows (G_1), DNA replicates and histone proteins are synthesized so that the cell can make duplicate copies of its chromosomes (S), grows more as it completes preparations for cell division (G_2), and divides (M).

The daughter cells may then repeat the cycle. A particular human cell might undergo one division in 24 hours. Of this time, the M phase would occupy less than 1 hour, while the S phase might occupy 10–12 hours, or about half the cycle. The rest of the time would be apportioned between the G_1 and G_2 phases. The G_2 phase usually takes 4–6 hours; G_1 is the most variable in length in different types of cells.

Some cells in a multicellular organism divide very infrequently or not at all. These cells spend their time in G_1 doing their job in the organism. Some cells enter a phase called **G₀** which may be temporary or permanent. At any given time, most of the cells in an animal's body are in G_0 phase. Some, such as most cells of nervous tissue, skeletal muscle, and heart muscle lose their ability to divide when they are fully mature or **terminally differentiated** and repairs are made with scar tissue (a fibrous type of connective tissue), so they remain in G_0 phase permanently; others, such as liver cells, can resume G_1 phase in response to factors released during injury.

What factors determine whether or not a cell will divide? First, cell division is controlled by external factors, such as environmental conditions and signaling molecules. Cycling is activated in postmitotic G_0 cells by protein signals from the extracellular environment called **mitogens** or **growth factors** that bind to cell surface receptors and trigger a cascade of kinase signaling in the cells. Second, internal factors affect cell division. These include cell cycle control molecules and checkpoints.

- G_1 phase: 11 hours
- S phase: 8 hours
- G_2 phase: 4 hours
- M phase: 1 hour

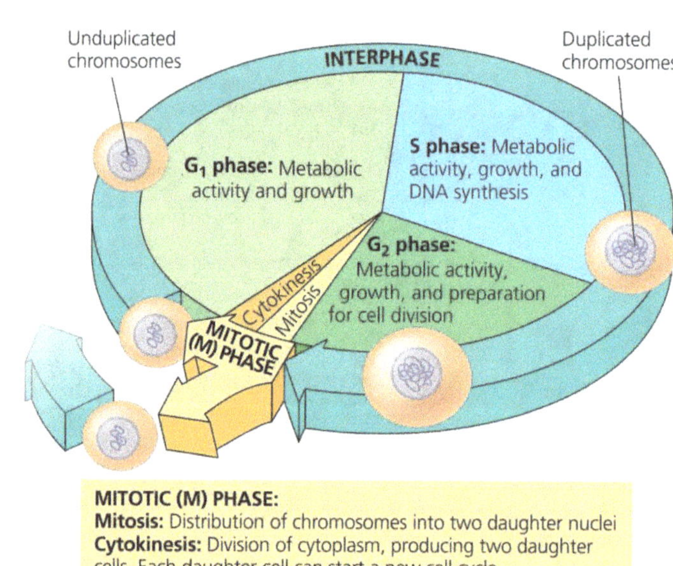

▲ **Figure 8.13** The cell cycle. In a dividing cell, the mitotic (M) phase alternates with interphase, a growth period.

The cell cycle is divided into five phases

The cell cycle is divided into phases based on the key events of genome duplication and segregation.

- **G$_1$ (gap phase 1):** is the primary growth phase of the cell. The term gap phase refers to its filling the gap between cytokinesis and DNA synthesis. For most cells, this is the longest phase.

- **S (synthesis of DNA, the genetic material):** is the phase in which the cell synthesizes a replica of the genome. After the S phase, the sister chromatids appear to share a common centromere, but at the molecular level the DNA of the centromere has actually already replicated, so there are two complete DNA molecules. This means that two chromatids are held together by **cohesin** proteins at the centromere, and each chromatid has its own set of kinetochore proteins. In multicellular animals, most of the cohesins that hold sister chromatids together after replication appear to be replaced by **condensin** as the chromosomes are condensed. This leaves the chromosomes still attached tightly at the centromere, but loosely attached elsewhere.

- **G$_2$ (gap phase 2):** is the second growth phase, and preparation for separation of the newly replicated genome. This phase fills the gap between DNA synthesis and the beginning of mitosis. During this phase microtubules begin to reorganize to form a spindle.

G$_1$, S, and G$_2$ together constitute interphase, the portion of the cell cycle between cell divisions.

- **Mitosis:** is the phase of the cell cycle in which the spindle apparatus assembles, binds to the chromosomes, and moves the sister chromatids apart. Mitosis is the essential step in the separation of the two daughter genomes. It is traditionally subdivided into five stages: prophase, prometaphase, metaphase, anaphase, and telophase.

- **Cytokinesis:** is the phase of the cell cycle when the cytoplasm divides, creating two daughter cells, each is smaller and has less cytoplasm than the mother cell, but is genetically identical to it. In animal cells, the microtubule spindle helps position a contracting ring of actin that constricts like a drawstring to pinch the cell in two. In cells with a cell wall, such as plant cells, a plate forms between the dividing cells.

Mitosis and cytokinesis together are usually referred to collectively as M phase, to distinguish the dividing phase from interphase.

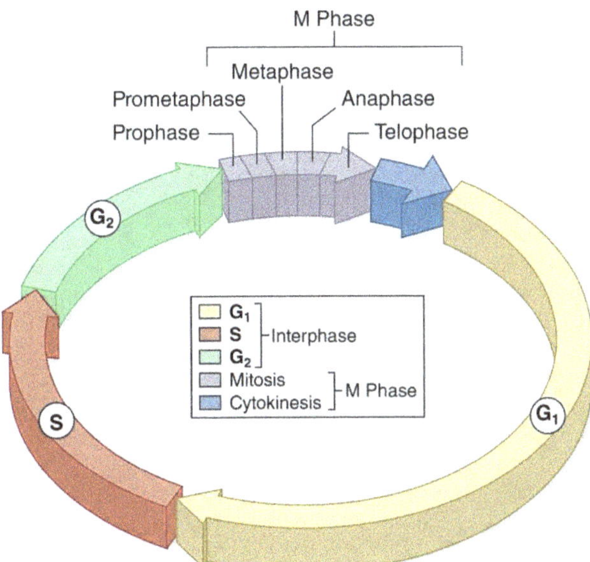

▶ **Figure 8.14** The cell cycle. The cell cycle is depicted as a circle. The first gap phase, G$_1$, involves growth and preparation for DNA synthesis. During S phase, a copy of the genome is synthesized. The second gap phase, G$_2$, prepares the cell for mitosis. During mitosis, replicated chromosomes are partitioned. Cytokinesis divides the cell into two cells with identical genomes.

Mitosis

Mitosis is conventionally broken down into five stages: **prophase**, **prometaphase**, **metaphase**, **anaphase**, and **telophase**. Overlapping with the latter stages of mitosis, cytokinesis completes the mitotic phase. Its duration varies according to cell type, but in human cells it typically lasts about an hour or less. Many of the events of mitosis depend on the **mitotic spindle**, which begins to form in the cytoplasm during prophase. This structure consists of fibers made of microtubules and associated proteins and separates the duplicated chromosomes during anaphase.

The minus ends of these microtubules are at the poles, and the plus ends extend to the cell's midplane. The organization and function of the spindle require the presence of motor proteins and a variety of signaling molecules. While the mitotic spindle assembles, the other microtubules of the cytoskeleton partially disassemble, providing the material used to construct the spindle. The spindle microtubules elongate (polymerize) by incorporating more subunits of the protein tubulin and shorten (depolymerize) by losing subunits.

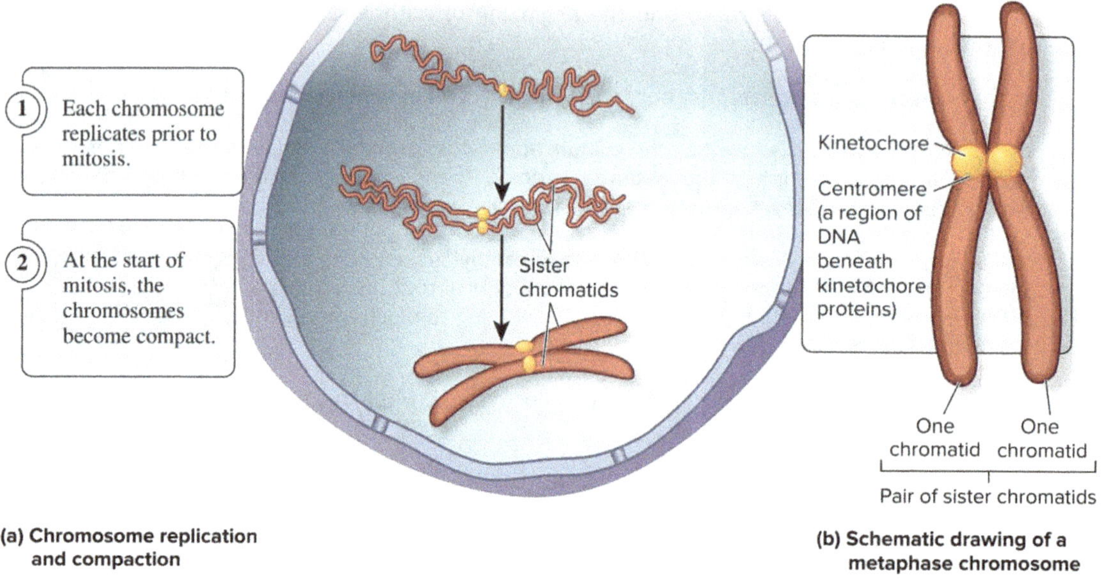

(a) Chromosome replication and compaction

(b) Schematic drawing of a metaphase chromosome

▲ **Figure 8.15** Replication and compaction of chromosomes into pairs of sister chromatids. (a) Chromosomal replication produces a pair of sister chromatids. While the chromosomes are elongated, they are replicated during S phase to produce two copies that are connected and lie parallel to each other. This is a pair of sister chromatids. At the start of mitosis, the sister chromatids condense into more compact structures that are easily seen with a light microscope. (b) A schematic drawing of a metaphase chromosome. This structure has two chromatids that lie side by side. The two chromatids are held together by cohesin proteins (not shown in this drawing). The kinetochore is a group of proteins that are attached to the centromere and play a key role during chromosome sorting.

▶ **Figure 8.16** The mitotic spindle. One end of each microtubule of this animal cell is associated with one of the poles. Astral microtubules (green) radiate in all directions, forming the aster. Kinetochore microtubules (red) connect the kinetochores to the poles, and polar (nonkinetochore) microtubules (blue) overlap at the midplane.

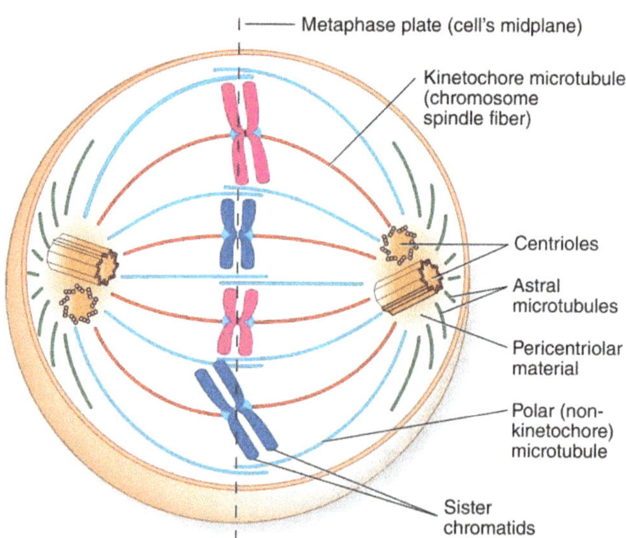

In animal cells, the assembly of spindle microtubules starts at the **centrosome**, a subcellular region containing material that functions throughout the cell cycle to organize the cell's microtubules (it is the main **microtubule organizing center (MTOC)** of animal cells and many protists). Each of the two centrioles is duplicated during the S phase of interphase, yielding two centriole pairs. Animal cells have a pair of centrioles in the middle of each centrosome. The centrioles are surrounded by fibrils that make up the **pericentriolar material**.

The spindle microtubules terminate in the pericentriolar material, but they do not actually touch the centrioles. If the centrioles are destroyed with a laser microbeam, a spindle nevertheless forms during mitosis. In fact, centrioles are not even present in plant cells, which do form mitotic spindles. Current evidence suggests that centrioles organize the pericentriolar material and ensure its duplication when the centrioles duplicate.

During interphase in animal cells, the single centrosome duplicates, forming two centrosomes, which remain near the nucleus. The two centrosomes move apart during prophase and prometaphase of mitosis as spindle microtubules grow out from them. By the end of prometaphase, the two centrosomes, one at each pole of the spindle, are at opposite ends of the cell. An **aster**, a radial array of short microtubules, extends from each centrosome, establishing the two poles of the mitotic spindle.

The spindle includes the centrosomes, the spindle microtubules, and the asters. Angiosperms (flowering plants) and most gymnosperms, such as conifers, lack centrosomes and centrioles. In these organisms, the spindle forms from microtubules that assemble in all directions from multiple MTOCs surrounding the entire nucleus. Then, when the nuclear envelope breaks down at the end of prophase, the spindle moves into the former nuclear region.

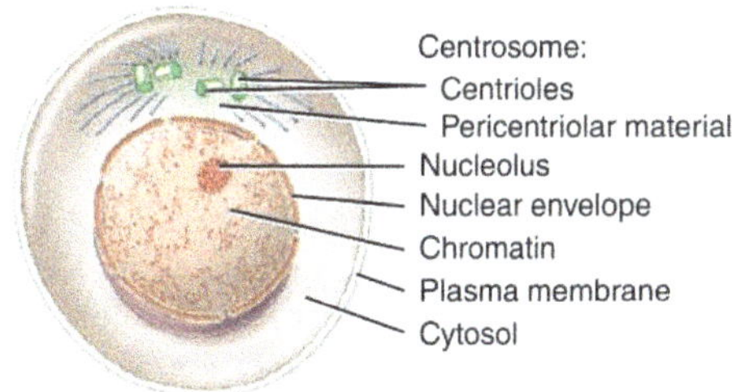

▶ **Figure 8.17** The centrosome and its role in spindle formation. © Cengage Learning 2017

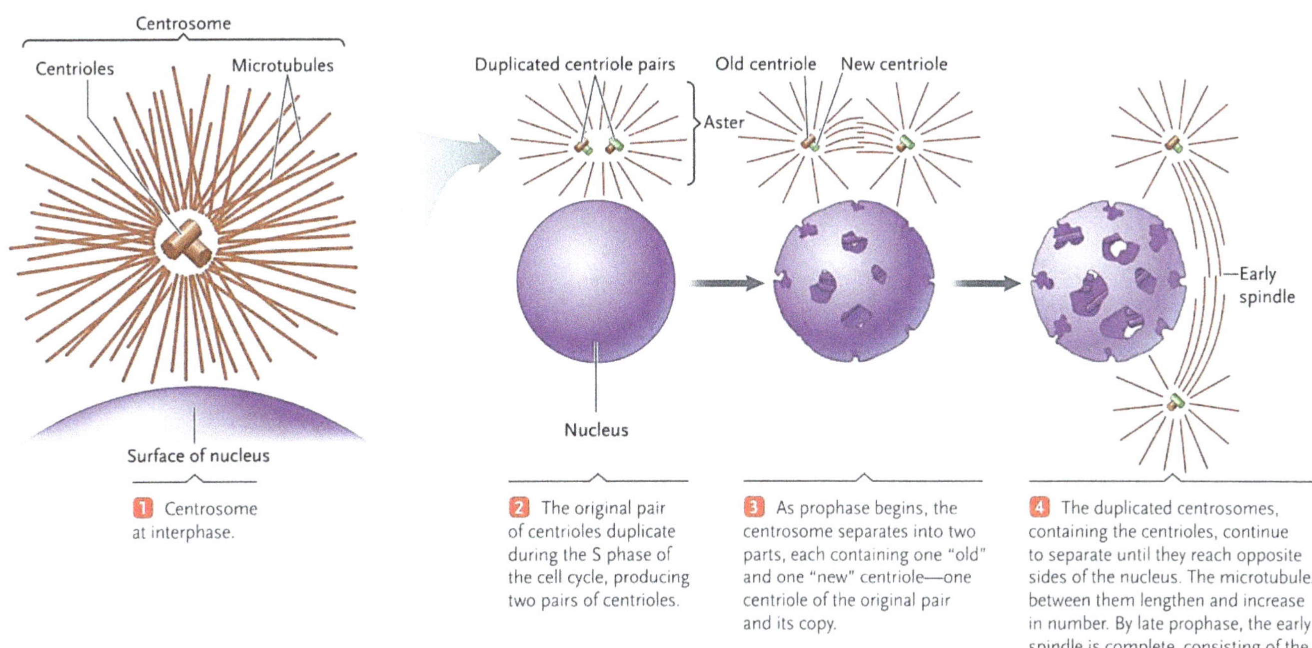

Each of the two sister chromatids of a duplicated chromosome has a **kinetochore**, a structure made up of proteins that have assembled on specific sections of DNA at each centromere. The chromosome's two kinetochores face in opposite directions. During prometaphase the nuclear envelope and associated endoplasmic reticulum break into small pieces, enabling the spindle fibers to reach the chromosomes. Some of the spindle microtubules attach to the kinetochores; these are called **kinetochore microtubules**.

When one of a chromosome's kinetochores is "captured" by microtubules, the chromosome begins to move toward the pole from which those microtubules extend. However, this movement comes to a halt as soon as microtubules from the opposite pole attach to the kinetochore on the other chromatid. What happens next is like a tug-of-war that ends in a draw. The chromosome moves first in one direction and then in the other, back and forth, finally settling midway between the two ends of the cell.

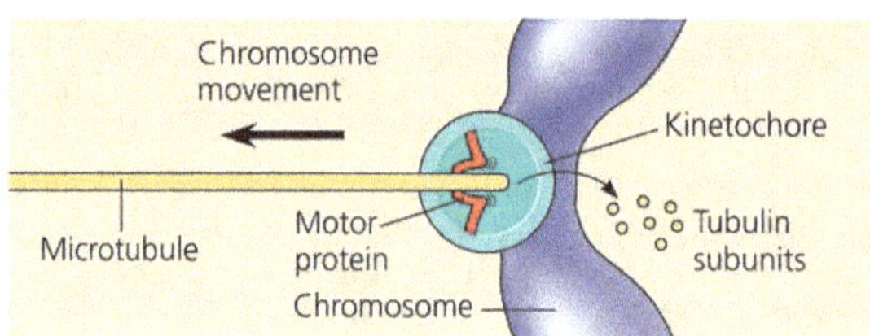

Each chromatid is completely condensed and appears distinct during metaphase. Because individual chromosomes are more obvious at metaphase than at any other time, the **karyotype**, or chromosome composition, is usually checked at this stage for chromosome abnormalities. At metaphase, the centromeres of all the duplicated chromosomes are on a plane midway between the spindle's two poles (A dividing cell can be described as a globe, with an equator that determines the midplane and two opposite poles).

This plane is called the **metaphase plate** or **equatorial plate**, which is an imaginary plate rather than an actual cellular structure. Meanwhile, microtubules that do not attach to kinetochores have been elongating, and by metaphase they overlap and interact with other **nonkinetochore microtubules** (polar microtubules) from the opposite pole of the spindle. By metaphase, the microtubules of the asters have also grown and are in contact with the plasma membrane. The spindle is now complete.

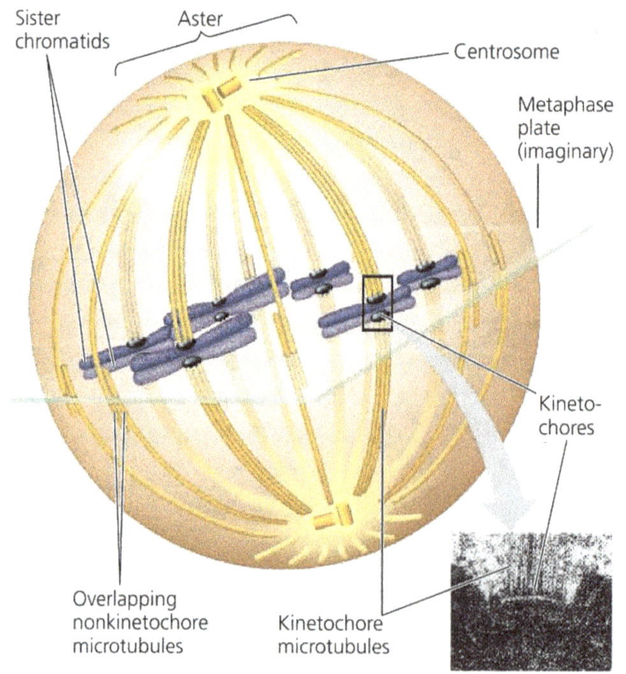

▶ **Figure 8.18** The mitotic spindle at metaphase. The kinetochores of each chromosome's two sister chromatids face in opposite directions. Here, each kinetochore is attached to a cluster of kinetochore microtubules extending from the nearest centrosome. Nonkinetochore microtubules overlap at the metaphase plate.

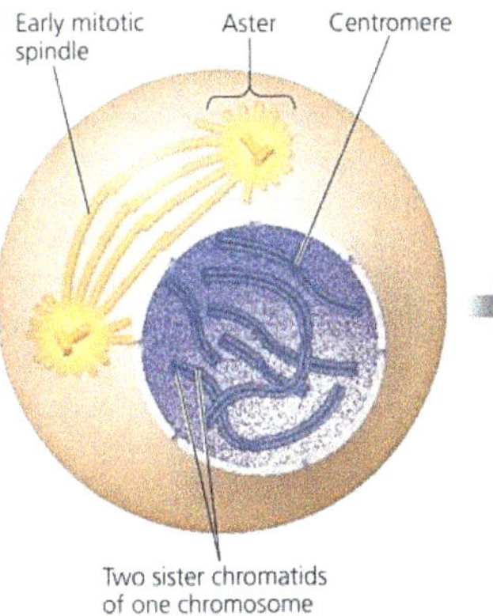

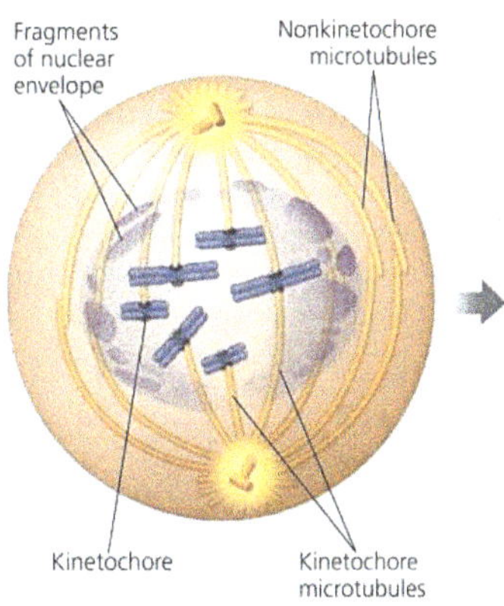

G₂ of Interphase

- A nuclear envelope encloses the nucleus.

- The nucleus contains one or more nucleoli (singular, *nucleolus*).

- Two centrosomes have formed by duplication of a single centrosome. Centrosomes are regions in animal cells that organize the microtubules of the spindle. Each centrosome contains two centrioles.

- Chromosomes, duplicated during S phase, cannot be seen individually because they have not yet condensed.

Prophase

- The chromatin fibers become more tightly coiled, condensing into discrete chromosomes observable with a light microscope.

- The nucleoli disappear.

- Each duplicated chromosome appears as two identical sister chromatids joined at their centromeres and, often, all along their arms by cohesins, resulting in sister chromatid cohesion.

- The mitotic spindle (named for its shape) begins to form. It is composed of the centrosomes and the microtubules that extend from them. The radial arrays of shorter microtubules that extend from the centrosomes are called asters ("stars").

- The centrosomes move away from each other, propelled partly by the lengthening microtubules between them.

Prometaphase

- The nuclear envelope fragments.

- The microtubules extending from each centrosome can now invade the nuclear area.

- The chromosomes have become even more condensed.

- A kinetochore, a specialized protein structure, has now formed at the centromere of each chromatid (thus, two per chromosome).

- Some of the microtubules attach to the kinetochores, becoming "kinetochore microtubules," which jerk the chromosomes back and forth.

- Nonkinetochore microtubules interact with those from the opposite pole of the spindle, lengthening the cell.

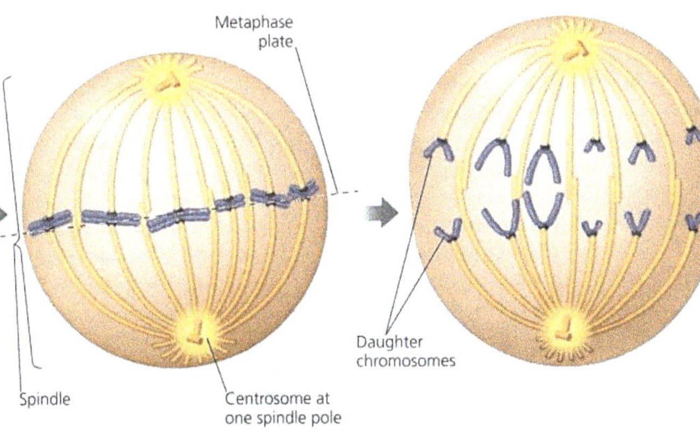

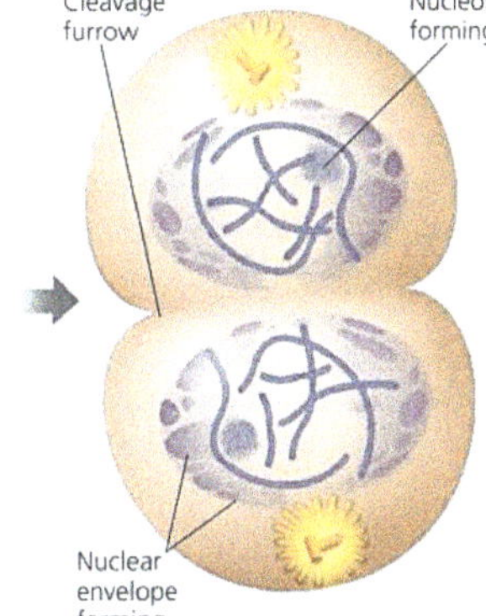

Metaphase

- The centrosomes are now at opposite poles of the cell.

- The chromosomes have all arrived at the *metaphase plate*, a plane that is equidistant between the spindle's two poles. The chromosomes' centromeres lie at the metaphase plate.

- For each chromosome, the kinetochores of the sister chromatids are attached to kinetochore microtubules coming from opposite poles.

Anaphase

- Anaphase is the shortest stage of mitosis, often lasting only a few minutes.

- Anaphase begins when the cohesin proteins are cleaved. This allows the two sister chromatids of each pair to part suddenly. Each chromatid thus becomes an independent chromosome.

- The two new daughter chromosomes begin moving toward opposite ends of the cell as their kinetochore microtubules shorten. Because these microtubules are attached at the centromere region, the centromeres are pulled ahead of the arms, moving at a rate of about 1 μm/min.

- The cell elongates as the nonkinetochore microtubules lengthen.

- By the end of anaphase, the two ends of the cell have identical—and complete—collections of chromosomes.

Telophase

- Two daughter nuclei form in the cell. Nuclear envelopes arise from the fragments of the parent cell's nuclear envelope and other portions of the endomembrane system.

- Nucleoli reappear.

- The chromosomes become less condensed.

- Any remaining spindle microtubules are depolymerized.

- Mitosis, the division of one nucleus into two genetically identical nuclei, is now complete.

Cytokinesis

- The division of the cytoplasm is usually well under way by late telophase, so the two daughter cells appear shortly after the end of mitosis.

- In animal cells, cytokinesis involves the formation of a cleavage furrow, which pinches the cell in two.

The structure of the spindle correlates well with its function during anaphase. Anaphase (the shortest stage of mitosis) begins suddenly when the **cohesins** holding together the sister chromatids of each chromosome are cleaved by an enzyme called **separase**. Once separated, the chromatids become individual chromosomes that move toward opposite ends of the cell.

How do the kinetochore microtubules function in this pole ward movement of chromosomes? Apparently, two mechanisms are in play, both involving motor proteins. Results of a cleverly designed experiment suggested that motor proteins (these motor proteins may work in a similar fashion to the **kinesin** motor) on the kinetochores "walk" the chromosomes along the microtubules, which depolymerize at their kinetochore ends or plus ends that is, closest toward the midplane of the cell after the motor proteins have passed (This is referred to as the "**Pacman**" mechanism). The pulling action is achieved by shortening of the microtubules, both from the pole end and from the kinetochore end.

However, other researchers, working with different cell types or cells from other species, have shown that chromosomes are "**reeled in**" by motor proteins at the spindle poles and that the microtubules depolymerize after they pass by these motor proteins at the poles. Both **walking** and **pulling** mechanisms are used in mitosis, though the relative contributions of the two mechanisms to chromosome movement varies among species and cell types.

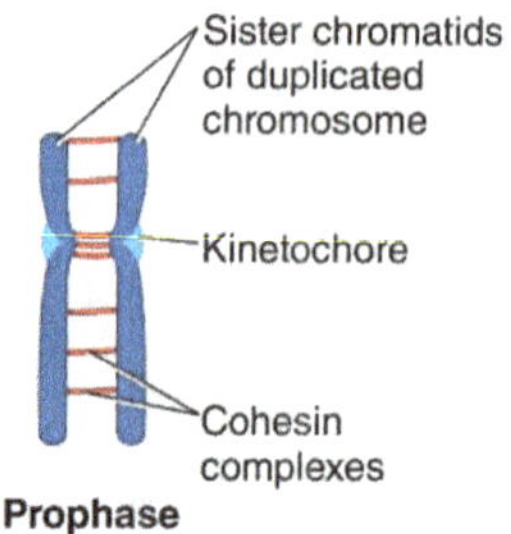

① As mitosis progresses, cohesins dissociate from the duplicated chromosome arms.

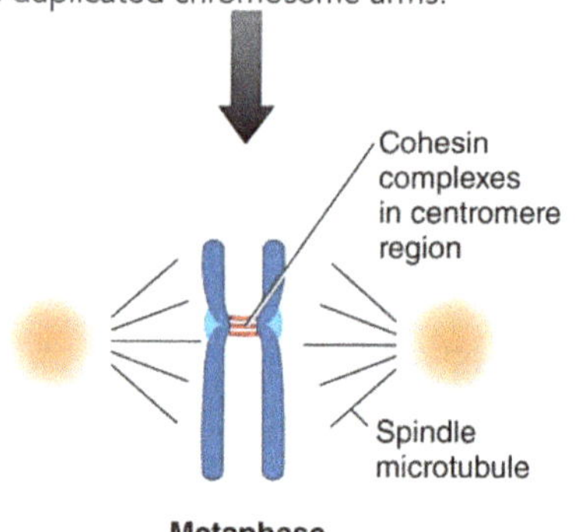

② Cohesins then dissociate at the centromere, allowing the daughter chromosomes to separate during anaphase.

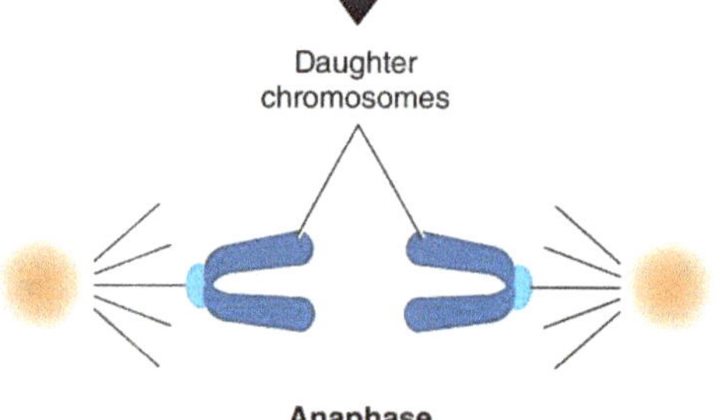

▲ **Figure 8.19** Cohesins. When chromosomes duplicate, sister chromatids are initially linked to one another by protein complexes called cohesins. Cohesin linkages are particularly concentrated in the vicinity of the centromere.

In a dividing animal cell, the nonkinetochore microtubules are responsible for elongating the whole cell during anaphase in two ways:

- Nonkinetochore microtubules from opposite poles overlap each other extensively during metaphase. During anaphase, the region of overlap is reduced as motor proteins attached to the microtubules walk them away from one another, using energy from ATP.

- As the microtubules push apart from each other, their spindle poles are pushed apart, elongating the cell. At the same time, the microtubules lengthen somewhat by the addition of tubulin subunits to their overlapping ends. As a result, the microtubules continue to overlap.

At the end of anaphase, duplicate groups of chromosomes have arrived at opposite ends of the elongated parent cell. Nuclei re-form during telophase. Cytokinesis generally begins during anaphase or telophase, and the spindle eventually disassembles by depolymerization of microtubules. As a result of mitosis division, each daughter nucleus receives exactly the same number and types of chromosomes, and contains the same genetic information, as the parent cell before its chromosomes were duplicated. The equal distribution of daughter chromosomes into each of the two daughter cells that result from cell division is called **chromosome segregation**. The accuracy of chromosome replication and segregation in the mitotic cell cycle creates a group of genetically identical cells—**clones** of the original cell.

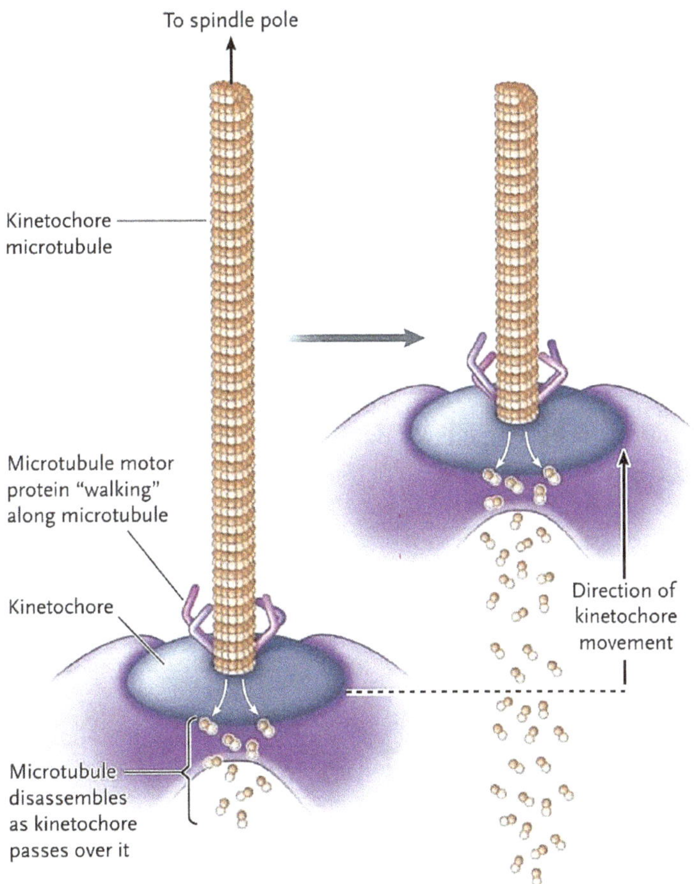

▲ **Figure 8.20** Microtubule motor proteins "walking" the kinetochore of a chromosome along a microtubule. © Cengage Learning 2017

Cytokinesis

In animal cells, cytokinesis occurs by a process known as **cleavage**. The first sign of cleavage is the appearance of a **cleavage furrow**, a shallow groove in the cell surface near the old metaphase plate. On the cytoplasmic side of the furrow is a contractile ring of actin microfilaments associated with molecules of the protein myosin, **actomyosin contractile ring**. The actin microfilaments interact with the myosin molecules, causing the ring to contract.

The contraction of the dividing cell's ring of microfilaments is like the pulling of a drawstring. The cleavage furrow deepens until the parent cell is pinched in two, producing two completely separated cells, each with its own nucleus and its own share of cytosol, organelles, and other subcellular structures. Multinucleated cells form if mitosis is not followed by cytokinesis; this is a normal condition for certain cell types.

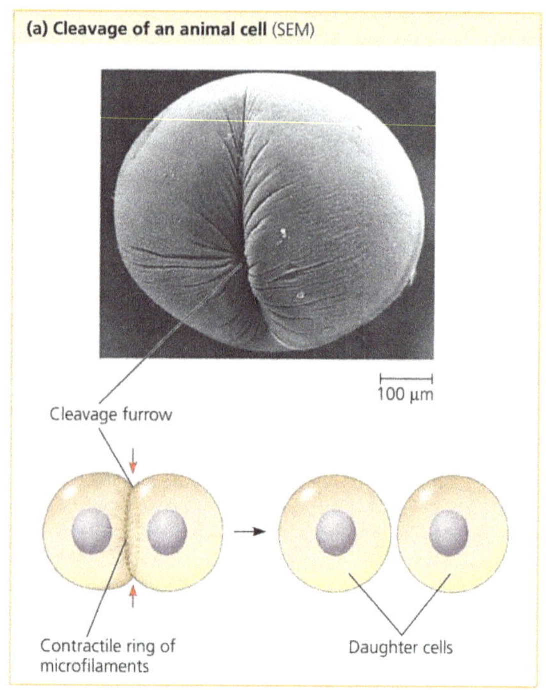

Cytokinesis in plant cells, which have cell walls, is markedly different. There is no cleavage furrow. Instead, during telophase, vesicles derived from the Golgi apparatus move along microtubules to the middle of the cell, where they coalesce, producing a **cell plate**. Cell wall materials carried in the vesicles collect inside the cell plate as it grows. The cell plate enlarges until its surrounding membrane fuses with the plasma membrane along the perimeter of the cell. Two daughter cells result, each with its own plasma membrane. Meanwhile, a new cell wall arising from the contents of the cell plate forms between the daughter cells.

Mitosis provides for the orderly distribution of chromosomes (and of centrioles, if present), but what about the various cytoplasmic organelles? For example, all eukaryotic cells, including plant cells, require mitochondria. Likewise, photosynthetic plant cells cannot carry out photosynthesis without chloroplasts. These organelles contain their own DNA and appear to form by the division of previously existing mitochondria or plastids or their precursors. This **nonmitotic division** process is similar to prokaryotic cell division and generally occurs during interphase. Because many copies of each organelle are present in each cell, organelles are apportioned with the cytoplasm that each daughter cell receives during cytokinesis.

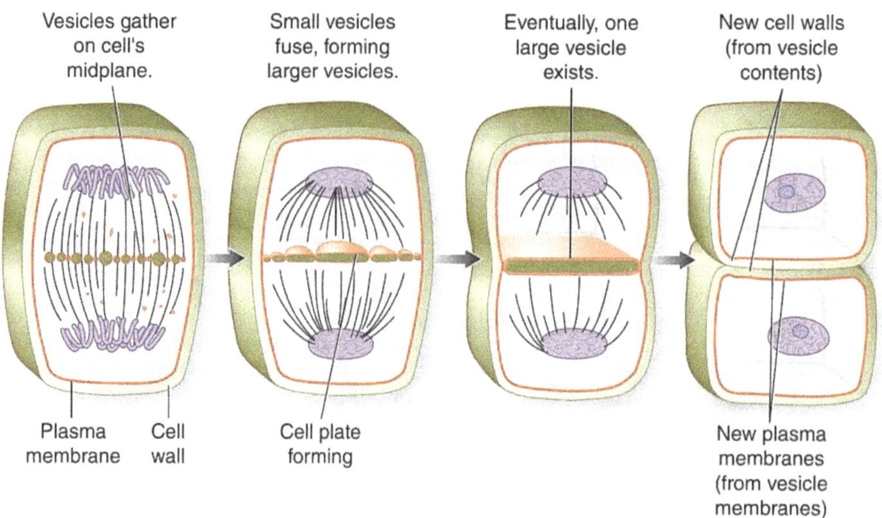

Binary Fission in Bacteria

Prokaryotes (bacteria and archaea) can undergo a type of reproduction in which the cell grows to roughly double its size and then divides to form two cells. The term **binary fission**, meaning "division in half," refers to this process and to the asexual reproduction of single-celled eukaryotes, such as the amoeba. However, the process in eukaryotes involves mitosis, while that in prokaryotes does not. In bacteria, most genes are carried on a single bacterial chromosome that consists of a circular DNA molecule and associated proteins. Inside bacterial cells, the DNA circle is packed and folded into an irregularly shaped mass called the **nucleoid**. The DNA of the nucleoid is suspended directly in the cytoplasm with no surrounding membrane.

Many bacterial cells also contain other DNA molecules, called **plasmids**, in addition to the main chromosome of the nucleoid. Most plasmids are circular, although some are linear. Plasmids have replication origins and are duplicated and distributed to daughter cells together with the bacterial chromosome during cell division. Although bacterial DNA is not organized into nucleosomes, certain positively charged proteins do combine with bacterial DNA. Some of these proteins help organize the DNA into loops, thereby providing some compaction of the molecule. Bacterial DNA also combines with many types of genetic regulatory proteins that have functions similar to those of the nonhistone proteins of eukaryotes.

When prokaryotic cells divide at the maximum rate, DNA replication occupies most of the period between cytoplasmic divisions. As soon as replication is complete, the cytoplasm divides to complete the cell cycle. For example, in E. coli cells, which are capable of dividing every 20 minutes, DNA replication occupies 19 minutes of the entire 20-minute division cycle.

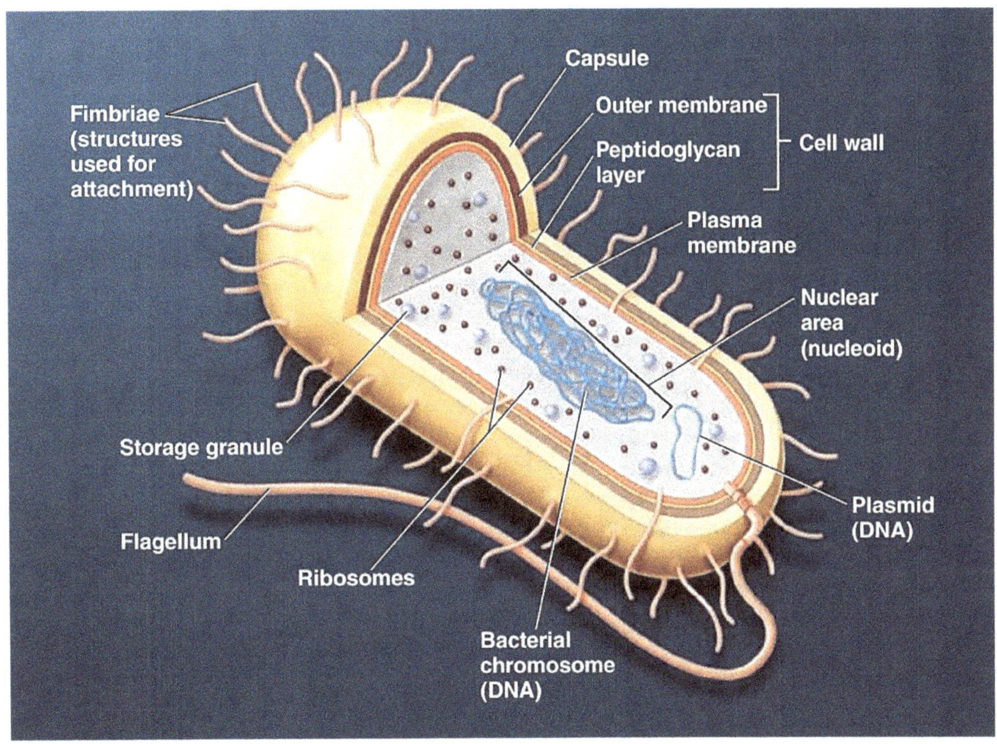

▲ **Figure 8.21** Structure of a prokaryotic cell. This bacillus is a gram-negative bacterium. Prokaryotic cells typically have a nuclear area with a single, circular DNA molecule. They may also have one or more plasmids, small rings of DNA that replicate independently.

In some bacteria, the process of cell division is initiated when the DNA of the bacterial chromosome begins to replicate at a specific place on the chromosome called the **origin of replication**, producing two origins. As the chromosome continues to replicate, one origin moves rapidly toward the opposite end of the cell. While the chromosome is replicating, the cell elongates.

When replication is complete and the bacterium has reached about twice its initial size, proteins cause its plasma membrane to pinch inward, dividing the parent bacterial cell into two daughter cells. In this way, each cell inherits a complete genome. Bacteria don't have visible mitotic spindles or even microtubules, so how bacterial chromosomes move and how their specific location is established and maintained are active areas of research. Several proteins that play important roles have been identified.

The cell's components are partitioned by the growth of new membrane and production of the **septum**. This process, termed **septation**, usually occurs at the midpoint of the cell. It begins with the formation of a ring composed of many copies of the protein **FtsZ**.

The FtsZ protein is found in most prokaryotes, including archaea. It can form filaments and rings, and three-dimensional crystals show a high degree of similarity to eukaryotic tubulin. However, its role in bacterial division is quite different from the role of tubulin in mitosis in eukaryotes. Cytokinesis between the daughter chromosomes is controlled by the **Z ring**, a protein scaffold that holds about 10 different proteins around the cell's midsection.

There the plasma membrane grows inward between the two DNA copies, dividing the cell's cytoplasm in half, and a new transverse cell wall is synthesized between the two cells. Polymerization of one protein resembling eukaryotic actin apparently functions in bacterial chromosome movement during cell division, and another protein that is related to tubulin helps pinch the plasma membrane inward, separating the two bacterial daughter cells.

▶ **Figure 8.22** Bacterial cell division by binary fission. The bacterium E. coli, shown here, has a single, circular chromosome.

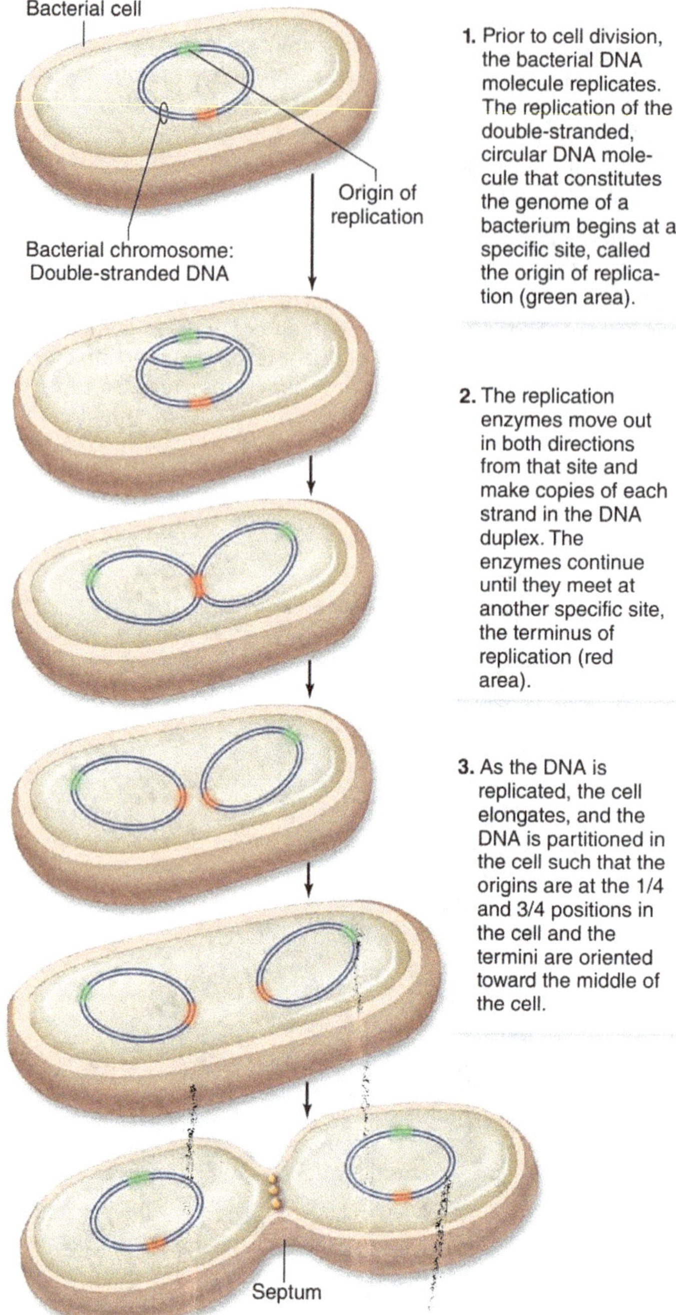

1. Prior to cell division, the bacterial DNA molecule replicates. The replication of the double-stranded, circular DNA molecule that constitutes the genome of a bacterium begins at a specific site, called the origin of replication (green area).

2. The replication enzymes move out in both directions from that site and make copies of each strand in the DNA duplex. The enzymes continue until they meet at another specific site, the terminus of replication (red area).

3. As the DNA is replicated, the cell elongates, and the DNA is partitioned in the cell such that the origins are at the 1/4 and 3/4 positions in the cell and the termini are oriented toward the middle of the cell.

4. Septation then begins, in which new membrane and cell wall material begin to grow and form a septum at approximately the midpoint of the cell. A protein molecule called FtsZ (orange dots) facilitates this process.

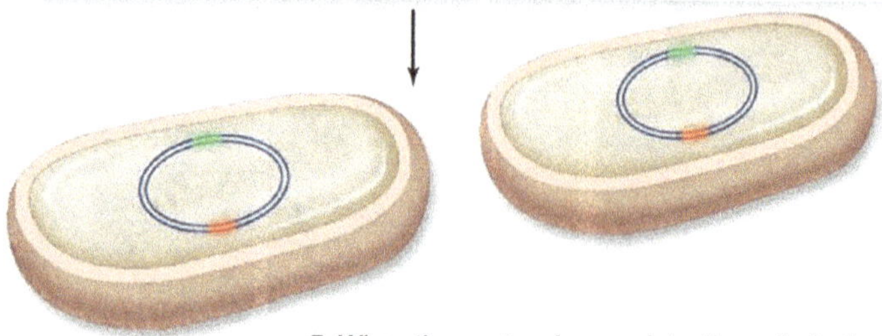

5. When the septum is complete, the cell pinches in two, and two daughter cells are formed, each containing a bacterial DNA molecule.

The Cell Cycle Control System

When conditions are optimal, some prokaryotic cells can divide every 20 minutes. The timing and rate of cell division in different parts of a plant or animal are crucial to normal growth, development, and maintenance. The frequency of cell division varies with the type of cell. Under optimal conditions of nutrition, temperature, and pH, the length of the eukaryotic cell cycle is constant for any given cell type. Under less-favorable conditions, however, the generation time may be longer.

For example, human skin cells divide frequently throughout life, whereas liver cells maintain the ability to divide but keep it in reserve until an appropriate need arises—say, to repair a wound. Some of the most specialized cells, such as fully formed nerve cells and muscle cells, do not divide at all in a mature human. These cell cycle differences result from regulation at the molecular level. The mechanisms of this regulation are of great interest, not only to understand the life cycles of normal cells but also to learn how cancer cells manage to escape the usual controls.

The **cell cycle control system** is a cyclically operating set of molecules in the cell that both triggers and coordinates key events in the cell cycle. The cell cycle is regulated at certain **checkpoints** by both internal and external signals. A checkpoint in the cell cycle is a control point where stop and go-ahead signals can regulate the cycle. Three important checkpoints are found in the G_1, G_2, and M phases as **G_1/S**, **G_2/M**, and **late metaphase** (the **spindle checkpoint**).

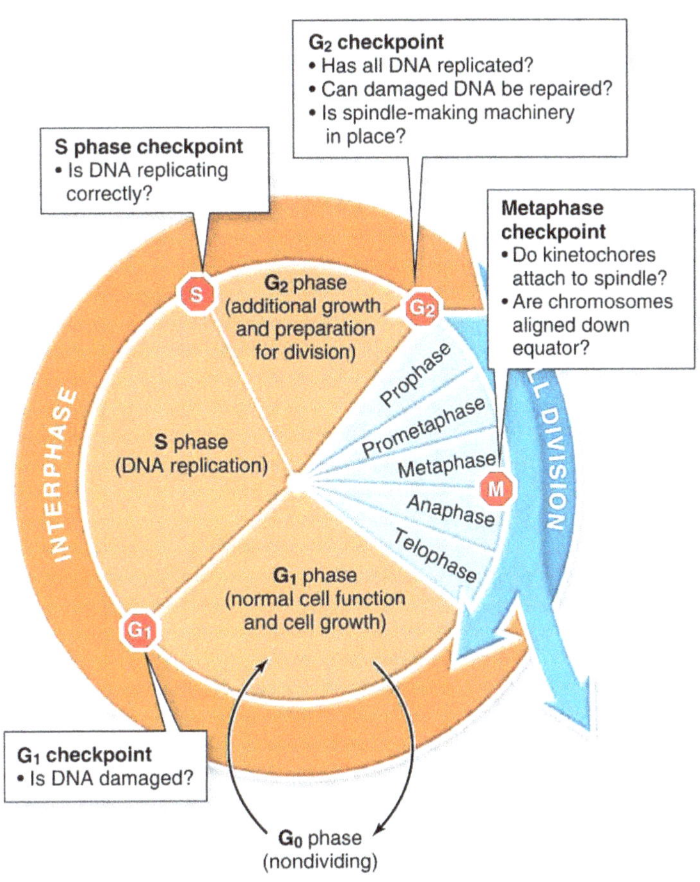

▲ **Figure 8.23** Cell Cycle Control Checkpoints. These checkpoints ensure that a cell completes each stage of the cell cycle correctly before proceeding to the next.

- The **G$_1$/S checkpoint** is the main point in the cell cycle at which the mechanisms governing the cell cycle determine whether the cell will proceed through the rest of the cell cycle and divide. Once it passes this checkpoint, the cell is committed to continue the cell cycle through to cell division in M. For example, the cell cycle arrests (the cell stops proceeding through the cell cycle) at the G$_1$/S checkpoint if the DNA is damaged by radiation or chemicals. If the DNA is damaged, a protein called p53 promotes the expression of genes encoding DNA repair enzymes. Badly damaged DNA prompts p53 to trigger apoptosis, and the cell dies. The G$_1$/S checkpoint is also the primary point at which cells "read" extracellular signals for cell growth and division. Therefore, if a growth factor required for stimulating cell growth is absent, the cells are arrested at this checkpoint (extracellular signals and their effects on the cell cycle are discussed in more detail later).

- Several **S phase checkpoints** ensure that DNA replication occurs properly. If the cell does not have enough nucleotides to complete replication or if a DNA molecule breaks, the cell cycle may pause or stop at this point.

- The **G$_2$/M checkpoint** is at the junction between the G$_2$ and M phases. Passage through this checkpoint commits a cell to mitosis. Cells are arrested at the G$_2$/M checkpoint if DNA was not replicated fully in S, or if the DNA has been damaged by radiation or chemicals or if the spindle-making machinery is not in place. Alternatively, the p53 protein may trigger apoptosis. Complete DNA replication is essential for producing genetically identical daughter cells, highlighting the importance of this checkpoint.

- The **mitotic spindle checkpoint** is within the M phase before metaphase. This checkpoint assesses whether chromosomes are attached properly to the mitotic spindle so that they are aligned correctly at the metaphase plate. The checkpoint is essential for production of genetically identical daughter cells, which depends on separation of daughter chromosomes in anaphase. In turn, that separation depends on the correct alignment of the chromosomes on the spindle in metaphase. Once the cell begins anaphase, it is irreversibly committed to completing M, underlining the importance of the mitotic spindle checkpoint.

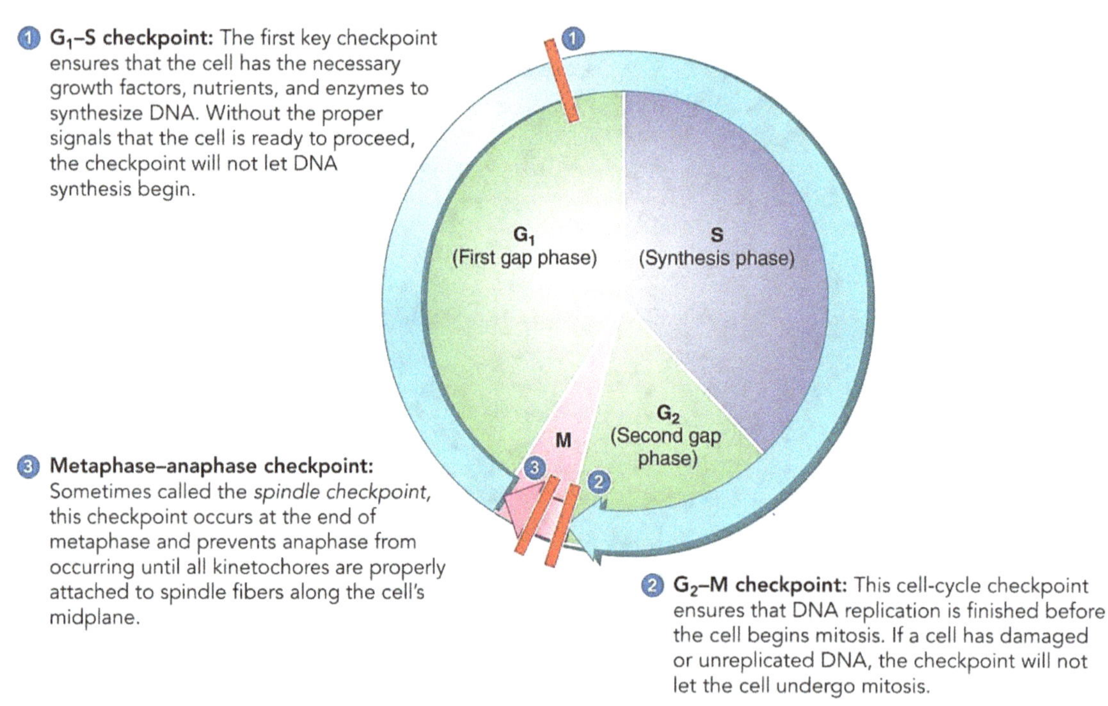

▲ **Figure 8.24** Key checkpoints in the cell cycle. The cell cycle consists of hundreds of sequential events. The red bars show three important checkpoints that determine that previous steps are completed so that the next steps may proceed. Each checkpoint is inactivated after it has performed its function, allowing the cell cycle to continue.

The regulatory molecules are mainly proteins of two types: **protein kinases** and **cyclins**. Protein kinases are enzymes that activate or inactivate other proteins by phosphorylating them (the addition of a phosphate group to the amino acids serine, threonine, and tyrosine in proteins). For example, phosphorylation of the protein p27, known to be a major inhibitor of cell division, is thought to initiate degradation of the protein. As various enzymes are activated or inactivated by phosphorylation, the activities of the cell change. Thus, a decrease in a cell's level of p27 causes a nondividing cell to resume division.

Many of the kinases that drive the cell cycle are actually present at a constant concentration in the growing cell, but much of the time they are in an inactive form. To be active, such a kinase must be attached to a cyclin, a protein that gets its name from its cyclically fluctuating concentration in the cell. Because of this requirement, these kinases are called **cyclin-dependent kinases**, or **Cdks**. Thus, a specific Cdk becomes active when the cell synthesizes the cyclin that binds to it and remains active until that cyclin is degraded.

There are four main types of cyclin in human cells and the levels of these cyclins rise and fall. Unless these cyclins reach a threshold concentration, the cell does not progress to the next stage of the cell cycle. Cyclins therefore control the cell cycle and ensure that cells divide when new cells are needed, but not at other times. For example, cyclin E peaks between the G_1 and S phases of interphase, whereas cyclin B is essentially absent at that time but has its highest concentration between G_2 and mitosis.

Each active Cdk phosphorylates particular target proteins, which play roles in initiating or regulating key events of the cell cycle such as transcription factors that stimulate entry into the next stage of the cell cycle. Regulation of the activity of cyclin–Cdk complexes is integrated with the regulatory events at the cell cycle checkpoints to ensure that daughter cells with damaged DNA or abnormal amounts of DNA are not generated.

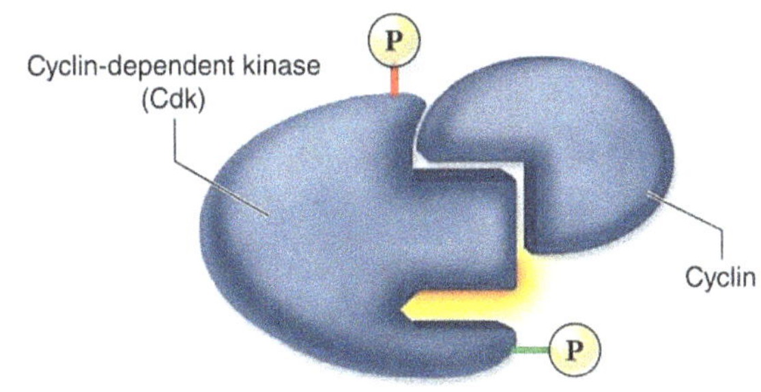

▶ **Figure 8.25** Cdk enzyme forms a complex with cyclin. Cdk is a protein kinase that activates numerous cell proteins by phosphorylating them. Cyclin is a regulatory protein required to activate Cdk. This complex is also called mitosis-promoting factor (MPF). The activity of Cdk is also controlled by the pattern of phosphorylation: Phosphorylation at one site (represented by the red site) inactivates the Cdk, and phosphorylation at another site (represented by the green site) activates the Cdk.

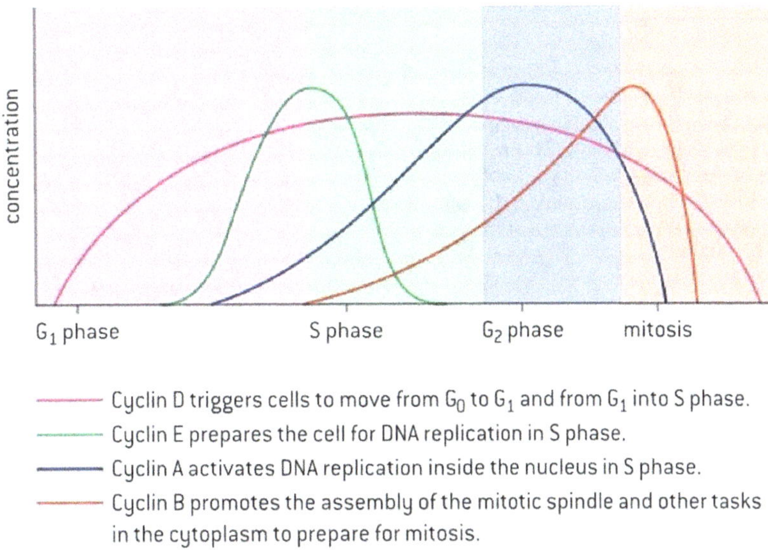

—— Cyclin D triggers cells to move from G_0 to G_1 and from G_1 into S phase.

—— Cyclin E prepares the cell for DNA replication in S phase.

—— Cyclin A activates DNA replication inside the nucleus in S phase.

—— Cyclin B promotes the assembly of the mitotic spindle and other tasks in the cytoplasm to prepare for mitosis.

Three classes of cyclins, each named for the stage of the cell cycle at which they bind and activate Cdks, operate in all eukaryotes: **G₁/S-Cdk**, **S-Cdk** and **M-Cdk** (also called **M phasepromoting factor, MPF**). Each cyclin–Cdk complex phosphorylates a different group of proteins. G1/S cyclin binds to Cdk2 near the end of G1 forming a complex required for the cell to make the transition from G1 to S, and to commit the cell to DNA replication.

S cyclin binds to Cdk2 in the S phase forming a complex required for the initiation of DNA replication and the progression of the cell through S. M cyclin binds to Cdk1 in G2 forming a complex required for the transition from G2 and M, and the progression of the cell through mitosis. In most cells, a fourth class of cyclins, G1 cyclin, binds to two additional Cdks, Cdk4 and Cdk6, before step 1 (the G1/S transition) to form two cyclin–Cdk complexes. These complexes are needed to move the cell through the G1 checkpoint stimulating it then to proceed from G1 to S.

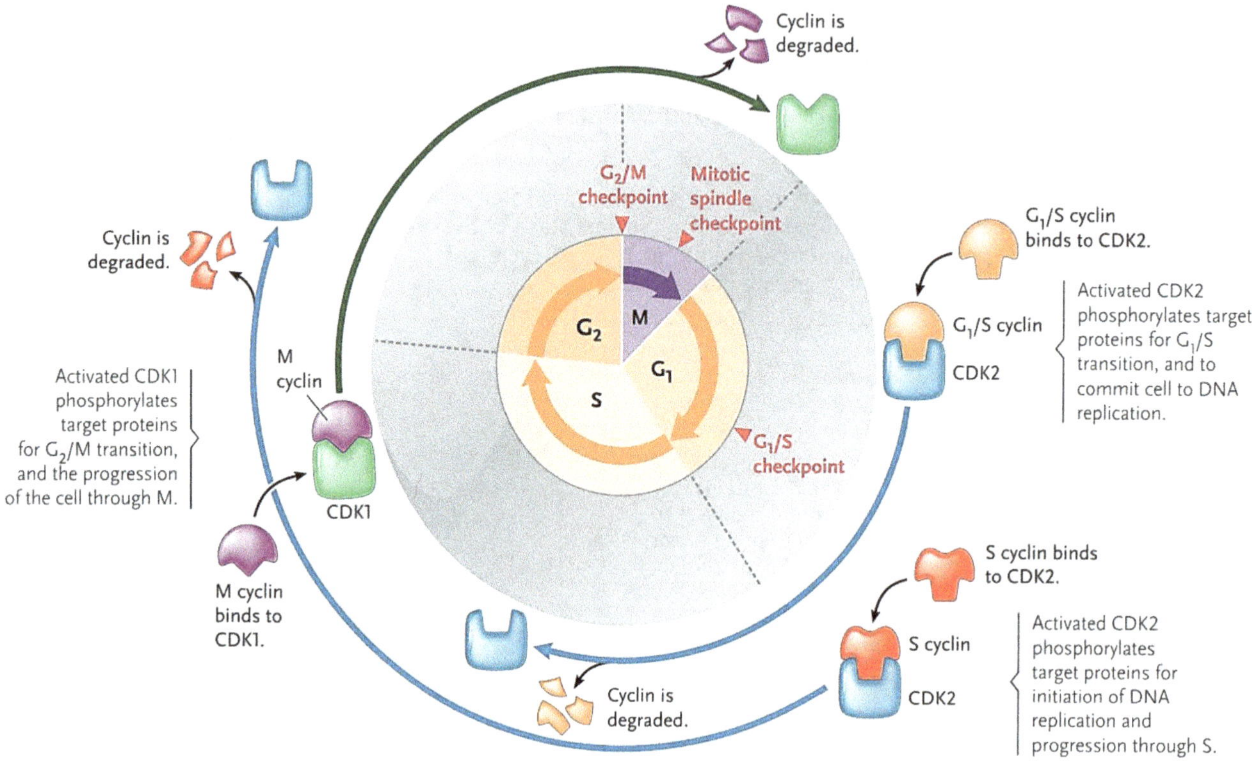

▲ **Figure 8.26** Regulation of the mitotic cell cycle by internal controls. Each cyclin is synthesized and degraded in a regulated way so that it is present only for a particular phase of the cell cycle. During that phase, the Cdk to which it is bound phosphorylates and, thereby, regulates the activity of target proteins in the cell that are involved in initiating or regulating key events of the cell cycle. © Cengage Learning 2017

▶ **Figure 8.27** Checkpoints of the mammalian cell cycle. The more complex mammalian cell cycle is shown. This cycle is still controlled through three main checkpoints. These integrate internal and external signals to control progress through the cycle. These inputs control the state of two different Cdk–cyclin complexes and the anaphase-promoting complex (APC).

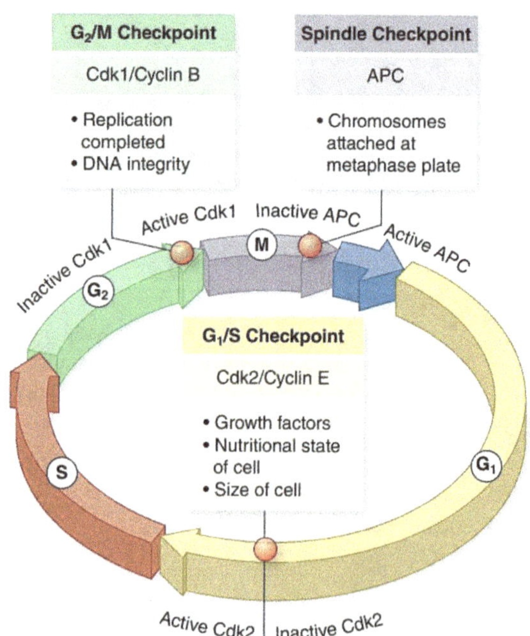

Cycle Phase or Checkpoint	Active Cyclin–CDK Complex	Examples of Target Proteins
Early G_1	Cyclin D–CDK4 or 6	Phosphorylates Rb protein, releasing E2F, a transcription factor that activates genes for many G_1 activities and for cyclin A
Late G_1 /entry of S	Cyclin E–CDK2	Further activation E2F-mediated gene transcription; protein p53; other kinases
Progression through S	Cyclin A–CDK2	DNA polymerase and other proteins for DNA replication
G_2/entry of M	Cyclin A–CDK1	Specific phosphatases and cyclin B
Progression through M	Cyclin B-CDK1	Nuclear lamins; histone H1; chromatin- and centrosome-associated proteins

▲ **Table 8.1** Major cyclin and cyclin-dependent kinase complexes regulating the human cell cycle and important target proteins.

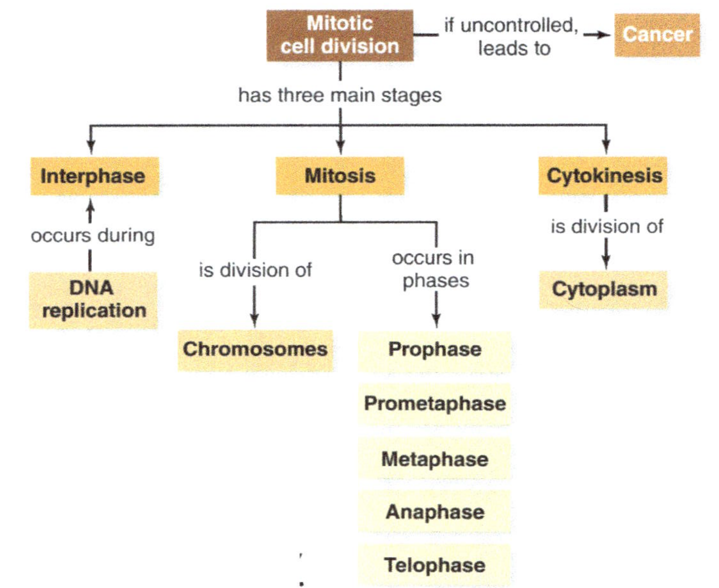

MPF is the cyclin-Cdk complex that was discovered first. MPF as "M-phase-promoting factor", triggers the cell's passage into the M phase, past the G_2 checkpoint. When cyclins that accumulate during G_2 associate with Cdk molecules, the resulting MPF complex is active—it phosphorylates a variety of proteins, initiating mitosis. MPF acts both directly as a kinase and indirectly by activating other kinases. For example, MPF causes phosphorylation of various proteins of the nuclear lamina, which promotes fragmentation of the nuclear envelope during prometaphase of mitosis. There is also evidence that MPF contributes to molecular events required for chromosome condensation and spindle formation during prophase.

M-Cdk also activates another enzyme complex, the **anaphase-promoting complex** (**APC**), toward the end of metaphase. Activated APC degrades an inhibitor of anaphase, and this leads to the separation of sister chromatids by allowing degradation of the cohesins and other proteins that hold the sister chromatids together during metaphase. Later in anaphase, APC directs the degradation of the M cyclin, causing Cdk1 to lose its activity. The loss of Cdk1 activity then allows the mitotic spindle to disassemble, the nuclear envelope to reform around the two clusters of daughter chromosomes in telophase, the cytoplasm then to divide in cytokinesis, the separated chromosomes to become extended again and the cell to exit mitosis.

During anaphase, MPF helps switch itself off by initiating a process that leads to the destruction of its own cyclin. The noncyclin part of MPF, the Cdk, persists in the cell, inactive until it becomes part of MPF again by associating with new cyclin molecules synthesized during the S and G_2 phases of the next round of the cycle. As mentioned, MPF controls the cell's passage through the G_2 checkpoint. Cell behavior at the G_1 checkpoint is also regulated by the activity of cyclin-Cdk protein complexes. Animal cells appear to have at least three Cdk proteins and several different cyclins that operate at this checkpoint.

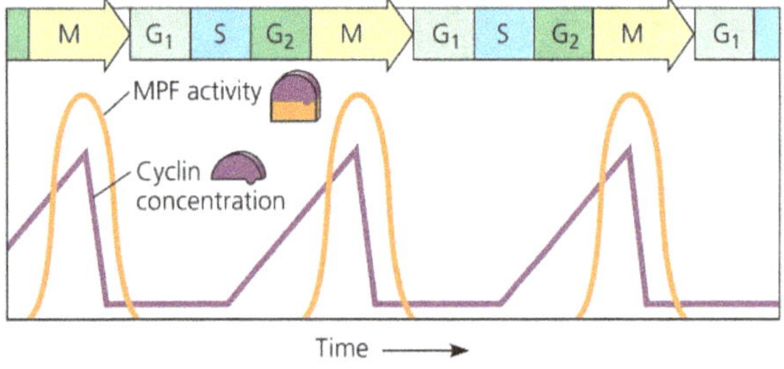

(a) Fluctuation of MPF activity and cyclin concentration during the cell cycle

▶ **Figure 8.28** Molecular control of the cell cycle at the G_2 checkpoint. The steps of the cell cycle are timed by rhythmic fluctuations in the activity of cyclin-dependent kinases (Cdks). Here we focus on a cyclin-Cdk complex in animal cells called MPF, which acts at the G_2 checkpoint as a go-ahead signal, triggering the events of mitosis.

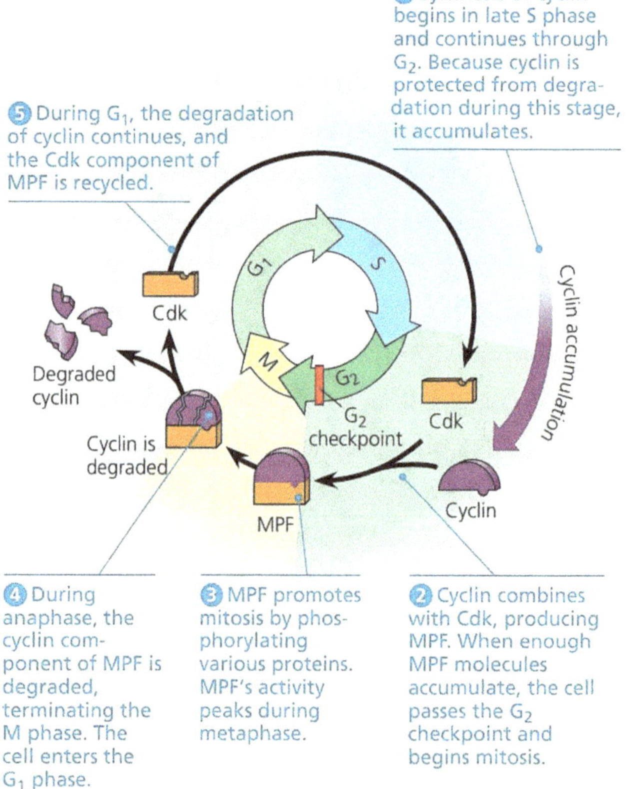

(b) Molecular mechanisms that help regulate the cell cycle

Term	Definition
Chromatin	Collective term for all of the DNA and associated proteins in a cell
Chromosome	A discrete, continuous molecule of DNA wrapped around protein. Eukaryotic cells contain multiple linear chromosomes, whereas bacterial cells each contain one circular chromosome.
Chromatid	One of two identical attached copies that make up a replicated chromosome
Centromere	A small part of a chromosome that attaches sister chromatids to each other
Interphase	Stage of the cell cycle in which chromosomes replicate and the cell grows
G_1 phase	Gap stage of interphase in which the cell grows and carries out its functions
G_0 phase	Gap stage of interphase in which the cell functions but does not divide
G_2 phase	Gap stage of interphase in which the cell produces membrane components and spindle proteins
S phase	Synthesis stage of interphase when DNA replicates
Mitosis	Division of a cell's chromosomes into two identical nuclei
Prophase	Stage of mitosis when chromosomes condense and the spindle begins to form (*pro-* = before)
Prometaphase	Stage of mitosis when the nuclear membrane breaks up and spindle fibers attach to kinetochores
Metaphase	Stage of mitosis when chromosomes are aligned down the center of the cell (*meta-* = middle)
Anaphase	Stage of mitosis when the spindle pulls sister chromatids toward opposite poles of the cell
Telophase	Stage of mitosis when chromosomes arrive at opposite poles and nuclear envelopes form (*telo-* = end)
Cytokinesis	Distribution of cytoplasm to daughter cells following division of a cell's chromosomes
Cleavage furrow	Indentation in cell membrane of an animal cell undergoing cytokinesis
Cell plate	Material that forms the beginnings of the cell wall in a plant cell undergoing cytokinesis
Centrosome	Structure that organizes the microtubules that make up the spindle in animal cells
Spindle	Array of microtubule proteins that move chromosomes during mitosis
Kinetochore	Protein complex to which the spindle fibers attach on a chromosome's centromere

▲ **Table 8.2** Miniglossary of Cell Division Terms.

Internal and External Signals at the Checkpoints

The internal controls that regulate the cell cycle are modified by signaling molecules that originate from outside the dividing cells. In animals, these signaling molecules include the peptide **hormones** and similar proteins called **growth factors**. Animal cells generally have built-in stop signals that halt the cell cycle at checkpoints until overridden by go-ahead signals. Many signals registered at checkpoints come from cellular surveillance mechanisms inside the cell. These signals report whether crucial cellular processes that should have occurred by that point have in fact been completed correctly and thus whether or not the cell cycle should proceed.

Certain factors and signals are important. These include:

- The ratio of cell surface area to cell volume. The amount of nutrients required by a growing cell is directly related to its volume—the greater the volume, the more nutrients are needed. The surface-volume relationships help explain why most cells are microscopic in size.

- Chemical signals such as growth factors and hormones released by other cells.

- The availability of space (how much room there is to grow). Normal cells stop proliferating when they begin touching, a phenomenon known as contact inhibition.

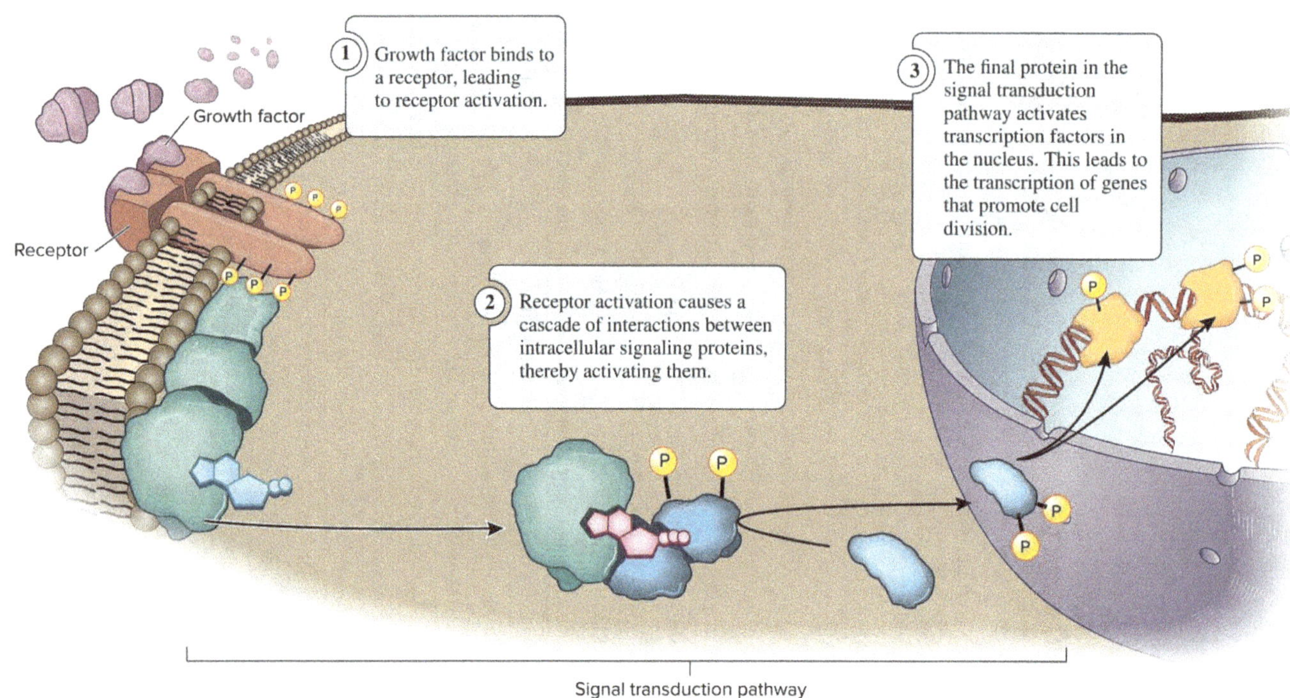

▲ **Figure 8.29** General features of a signal transduction pathway involving a growth factor that promotes cell division.

Checkpoints also register signals from outside the cell. The signals are transmitted within the cell by signal transduction pathways. The external factors bind to receptors at the cell surface, which respond by triggering reactions inside the cell. These reactions often include steps that add inhibiting or stimulating phosphate groups to the cyclin–Cdk complexes, particularly to the Cdks. The reactions triggered by the activated receptor may also directly affect the same proteins regulated by the cyclin–Cdk complexes. The overall effect is to speed, slow, or stop the progress of cell division, depending on the particular hormone or growth factor and the internal pathway that is stimulated.

For many cells, the G_1 checkpoint seems to be the most important. If a cell receives a go-ahead signal at the G_1 checkpoint, it will usually complete the G_1, S, G_2, and M phases and divide. If it does not receive a go-ahead signal at that point, it may exit the cycle, switching into a nondividing state called the G_0 phase. Most cells of the human body are actually in the G_0 phase. As mentioned earlier, mature nerve cells and muscle cells never divide. Other cells, such as liver cells, can be "called back" from the G_0 phase to the cell cycle by external cues, such as growth factors released during injury.

Biologists are currently working out the pathways that link signals originating inside and outside the cell with the responses by cyclin-dependent kinases and other proteins. An example of an internal signal occurs at the third important checkpoint, the M checkpoint. Anaphase, the separation of sister chromatids, does not begin until all the chromosomes are properly attached to the spindle at the metaphase plate. Researchers have learned that as long as some kinetochores are unattached to spindle microtubules, the sister chromatids remain together, delaying anaphase.

Only when the kinetochores of all the chromosomes are properly attached to the spindle does the appropriate regulatory protein complex become activated. Once activated, the complex sets off a chain of molecular events that activates the enzyme separase, which cleaves the cohesins, allowing the sister chromatids to separate. This mechanism ensures that daughter cells do not end up with missing or extra chromosomes. There are checkpoints in addition to those in G_1, G_2, and M. For instance, a checkpoint in S phase stops cells with DNA damage from proceeding in the cell cycle.

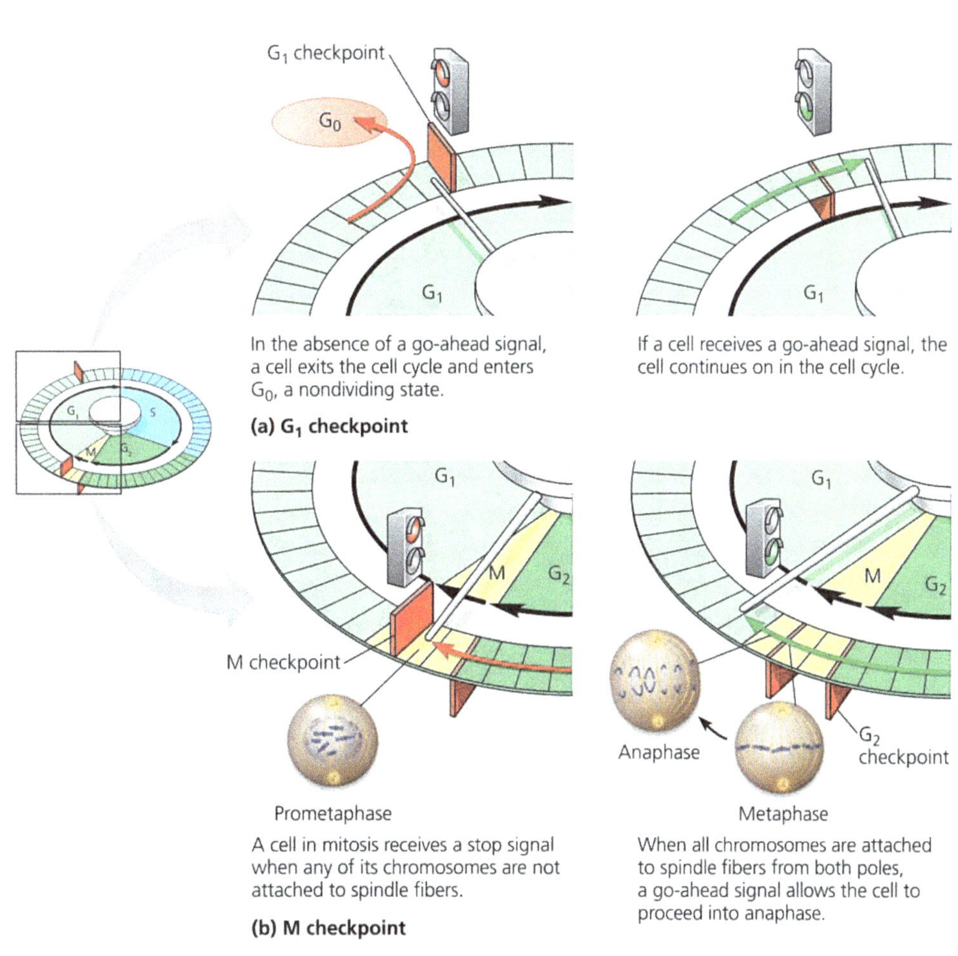

▲ **Figure 8.30** Two important checkpoints.

What about the stop and go-ahead signals themselves—what are the signaling molecules? Studies using animal cells in culture have led to the identification of many external factors, both chemical and physical, that can influence cell division. For example, cells fail to divide if an essential nutrient is lacking in the culture medium. And even if all other conditions are favorable, most types of mammalian cells divide in culture only if the growth medium includes specific **growth factors**. A growth factor is a protein released by certain cells that stimulates other cells to divide. Of the approximately 50 protein growth factors known, some act only on specific types of cells, whereas others act on a broad range of cell types. The cellular selectivity of a particular growth factor depends on which target cells bear its unique receptor.

Some growth factors, such as **platelet-derived growth factor** (PDGF) and **epidermal growth factor** (EGF), affect a broad range of cell types, but others affect only specific types. For example, **nerve growth factor** (NGF) promotes the growth of certain classes of neurons, and **erythropoietin** triggers cell division in red blood cell precursors. Most animal cells need a combination of several different growth factors to overcome the various controls that inhibit cell division. Many types of cancer cells divide in the absence of growth factors. If cells are deprived of appropriate growth factors, they stop at the G_1 checkpoint of the cell cycle. With their growth and division arrested, they remain in this dormant G_0 phase.

The ability to enter G_0 accounts for the incredible diversity seen in the length of the cell cycle in different tissues. Epithelial cells lining the human gut divide more than twice a day, constantly renewing this lining. By contrast, liver cells divide only once every year or two, spending most of their time in the G_0 phase. Mature neurons and muscle cells usually never leave G_0. The effect of an external physical factor on cell division is clearly seen in **density-dependent inhibition**, a phenomenon in which crowded cells stop dividing.

Most animal cells also exhibit **anchorage dependence**. To divide, they must be attached to something, such as the inside of a culture flask or the extracellular matrix of a tissue. Experiments suggest that, like cell density, anchorage is signaled to the cell cycle control system via pathways involving plasma membrane proteins and elements of the cytoskeleton linked to them. Density-dependent inhibition and anchorage dependence appear to function not only in cell culture but also in the body's tissues, checking the growth of cells at some optimal density and location during embryonic development and throughout an organism's life. Cancer cells exhibit neither density-dependent inhibition nor anchorage dependence.

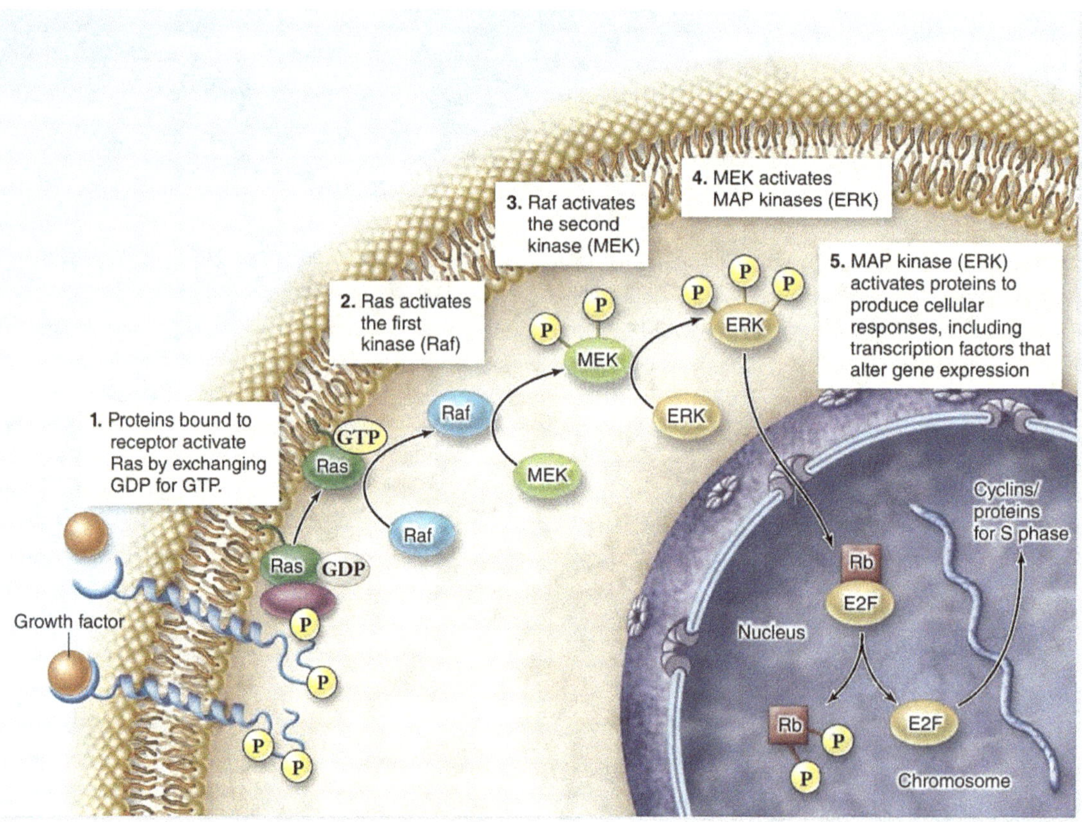

▲ **Figure 8.31** The cell proliferation-signaling pathway. Binding of a growth factor sets in motion a MAP kinase intracellular signaling pathway, which activates nuclear regulatory proteins that trigger cell division. In this example, when the nuclear retinoblastoma protein (Rb) is phosphorylated, another nuclear protein (the transcription factor E_2F) is released and is then able to stimulate the production of cyclin and other proteins necessary for S phase.

Cancer

Cancer is a group of diseases characterized by uncontrolled or abnormal cell proliferation. There are more than 200 different forms of cancer, and the medical profession no longer thinks of cancers as a single disease. When cells in a part of the body divide without control, the excess tissue that develops is called a **tumor** or **neoplasm**. The study of tumors is called **oncology**. Worldwide, cancer is the second leading cause of death in humans, exceeded only by heart disease. For about 10% of cancers, a higher predisposition to develop the disease is an **inherited** or **innate** trait.

Most cancers, though, do not involve genetic changes that are passed from parent to offspring. Rather, cancer is usually an **acquired** condition that typically occurs later in life. Cancer cells do not heed the normal signals that regulate the cell cycle. In culture, they do not stop dividing when growth factors are depleted. A logical hypothesis is that cancer cells do not need growth factors in their culture medium to grow and divide. They may make a required growth factor themselves, or they may have an abnormality in the signaling pathway that conveys the growth factor's signal to the cell cycle control system even in the absence of that factor. Another possibility is an abnormal cell cycle control system. In these scenarios, the underlying basis of the abnormality is almost always a change in one or more genes, for example, a mutation, that alters the function of their protein products, resulting in faulty cell cycle control.

Most mutated cells are affected in some way that results in their early death or their destruction by the body's immune system. Most cells can be replaced, so mutation usually has no harmful effect on the body. Unfortunately, cancer cells manage to escape both cell death and destruction so, although the mutation may originally occur in only one cell, it is passed on to all that cell's descendants. By the time it is detected, a typical tumour usually contains about a billion cancer cells.

Tumors are classified according to their place of origin. **Carcinomas** are cancers of the epithelial tissues, and **adenocarcinomas** are cancers of glandular epithelial cells. Carcinomas include cancer of the skin, breast, liver, pancreas, intestines, lung, prostate, and thyroid. **Sarcomas** are cancers that arise in muscles and connective tissue, such as bone and fibrous connective tissue. **Leukemias** are cancers of the blood, and **lymphomas** are cancers of lymphoid tissue. A **blastoma** is a cancer composed of immature cells. The embryo is formed from three primary germ layers: ectoderm, mesoderm, and endoderm. Each blastoma cell resembles the cells in its original primary germ layer. For example, a nephroblastoma has cells similar to mesoderm cells, because the kidney grows from mesoderm.

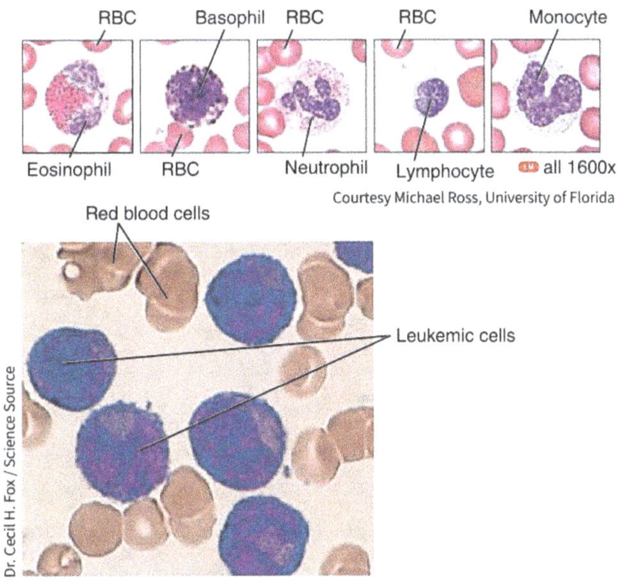

▲ **Figure 8.32** Leukemia cells. Note the odd appearance and higher concentration of these cells, when compared to normal blood cells shown in the inset.

Cancer cell often has altered proteins on its surface, and the body's immune system normally recognizes the cell as "nonself"—an insurgent—and destroys it. However, if the cell evades destruction, it may proliferate and form a tumor, a mass of abnormal cells within otherwise normal tissue. The abnormal cells may remain at the original site if their genetic and cellular changes don't allow them to move to or survive at another site. In that case, the tumor is called a **benign tumor**. Most benign tumors do not cause serious problems (depending on their location) and can be removed by surgery. A **malignant tumor** includes cells whose genetic and cellular changes enable them to spread to new tissues and impair the functions of one or more organs. The disease called **cancer** occurs when the abnormally dividing cells of a malignant neoplasm disrupt body tissues, both physically and metabolically.

An individual with a malignant tumor is said to have cancer. Malignant cells can slip easily into and out of vessels of the circulatory and lymphatic systems. By migrating through these vessels, the cells can establish neoplasms elsewhere in the body. The process in which malignant cells break loose from their home tissue and invade other parts of the body is called **metastasis**. Growing tumors damage surrounding normal tissues by compressing them and interfering with blood supply and nerve function.

Tumors may also break through barriers such as the outer skin, internal cell layers, or the gut wall. The breakthroughs cause bleeding, open the body to infection by microorganisms, and destroy the separation of body compartments necessary for normal functioning. Both compression and breakthroughs can cause pain that, in advanced cases, may become extreme. As tumors increase in mass, the actively growing and dividing cancer cells may deprive normal cells of their required nutrients, leading to generally impaired body functions, muscular weakness, fatigue, and weight loss.

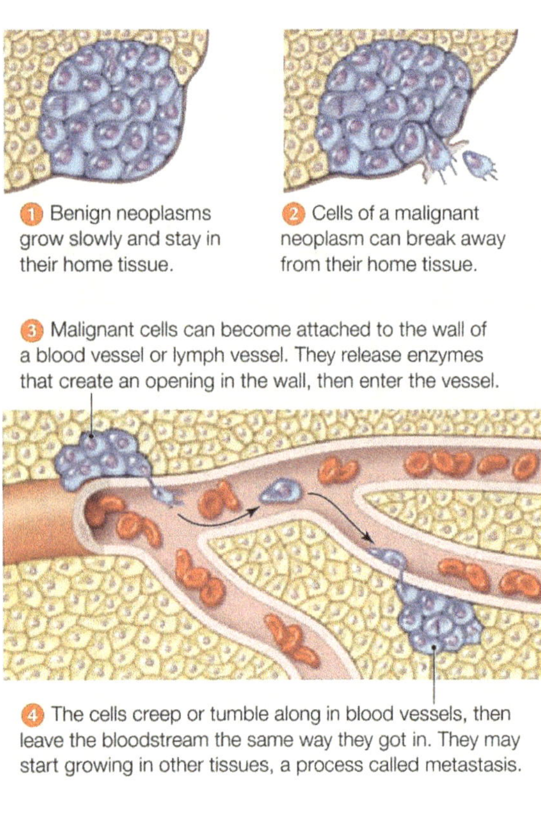

▲ **Figure 8.33** Metastasis. © Cengage Learning.

▶ **Figure 8.34** Benign tumor formation Benign tumors grow in situ, or in place. They push on surrounding tissue rather than infiltrate it, and they do not metastasize, or spread.

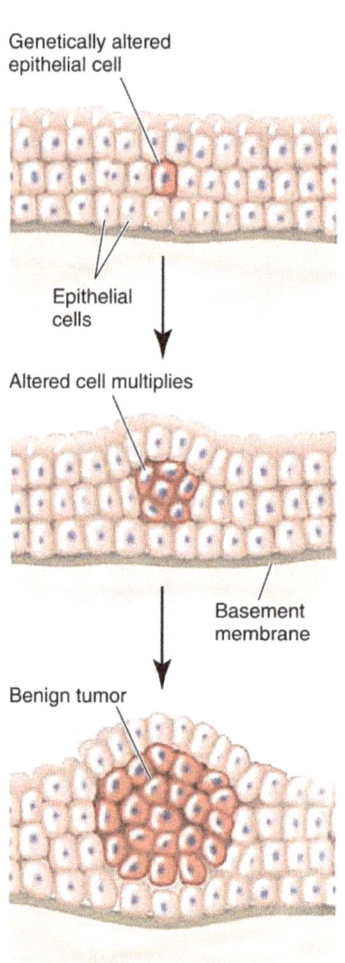

The genes that normally regulate cell growth and division during the cell cycle include genes for growth factors, their receptors, and the intracellular molecules of signaling pathways. Mutations that alter any of these genes in somatic cells can lead to cancer. Research has identified numerous **oncogenes**, genes that can, when introduced into a cell, cause it to become a cancer cell. This identification then led to the discovery of **proto-oncogenes** or **growth-promoting genes**, which are normal cellular genes. The genetic changes that convert proto-oncogenes to oncogenes fall into four main categories: epigenetic changes, translocations, gene amplification, and point mutations.

First, alterations in epigenetic modifications that can lead to abnormal chromatin condensation in a cell are often found in tumor cells. If a mutation in a gene for a chromatin-modifying enzyme leads to loosening of chromatin in a region that is normally not being expressed, a proto-oncogene in that region could be expressed at abnormally high levels. For example, a gene for one such enzyme has been shown to be mutated in 20% of tumor cells analyzed. **Second**, cancer cells are frequently found to contain chromosomes that have broken and rejoined incorrectly, translocating fragments from one chromosome to another.

If a translocated proto-oncogene ends up near an especially active promoter (or other control element), its transcription may increase, making it an oncogene. The **third** main type of genetic change, amplification, increases the number of copies of the proto-oncogene in the cell through repeated gene duplication. **Fourth**, a point mutation either in the promoter or an enhancer that controls a proto-oncogene, could cause an increase in its expression. A point mutation in the coding sequence of the proto-oncogene could change the gene's product to a protein that is more active or more resistant to degradation than the normal protein.

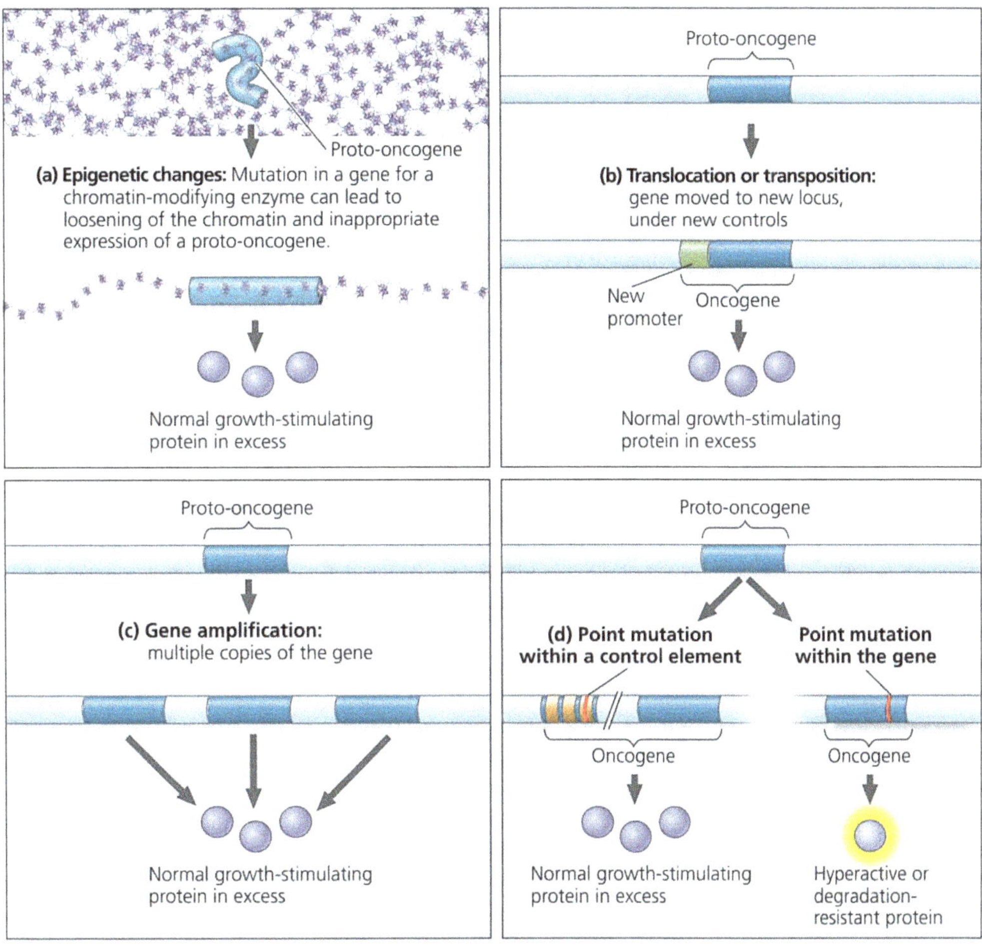

▲ **Figure 8.35** Genetic changes that can turn proto-oncogenes into oncogenes.

In their normal (nonmutated) form, a number of proto-oncogenes code for components of the cyclin/Cdk system that regulates cell division. Others encode proteins that regulate gene expression, form cell surface receptors, or make up elements of the systems controlled by the receptors. A mutation in a proto-oncogene might cause the encoded protein to be abnormally active or expressed at too high a concentration. In that case, the cell cycle will be accelerated, and cancer may develop. Oncogenes cause some cancers of the cervix, bladder, and liver.

The action of proto-oncogenes is often related to signaling by growth factors, and their mutation can lead to loss of growth control in multiple ways. Some protooncogenes encode receptors for growth factors, and others encode proteins involved in signal transduction that act after growth factor receptors. If a receptor for a growth factor becomes mutated such that it is permanently "on," the cell is no longer dependent on the presence of the growth factor for cell division. Only one copy of a proto-oncogene needs to undergo this mutation for uncontrolled division to take place; thus, this change acts like a dominant mutation.

Cyclin D is a proto-oncogene that codes for cyclin directly. When this gene becomes an oncogene, cyclin is readily available all the time. Protein components of cell-signaling pathways function in normal cells but they go wrong with their function in cancer cells. We will focus on the products of two key genes, the **ras proto-oncogene** and the **p53 tumor suppressor gene**. Mutations in ras occur in about 30% of human cancers and mutations in p53 in more than 50%.

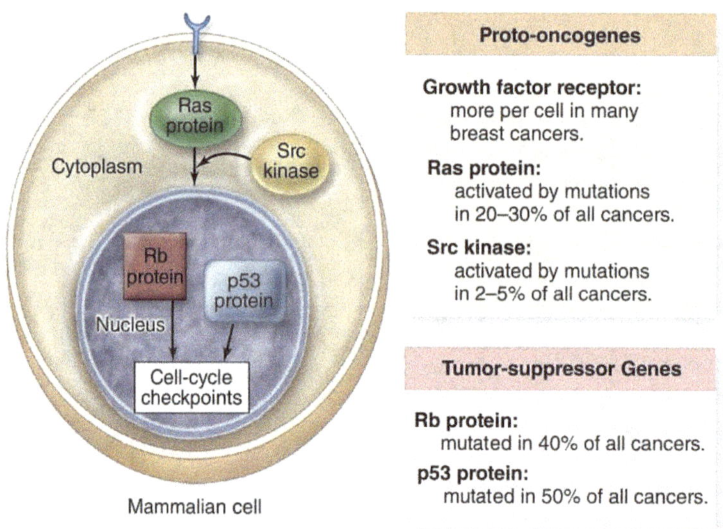

▲ **Figure 8.36** Key proteins associated with human cancers. Mutations in genes encoding key components of the cell division–signaling pathway are responsible for many cancers. Among them are proto-oncogenes encoding growth factor receptors, protein relay switches such as Ras protein, and kinase enzymes such as Src, which act after Ras and growth factor receptors. Mutations that disrupt tumor-suppressor proteins, such as Rb and p53, also foster cancer development.

The Ras protein, encoded by the ras gene (named for rat sarcoma, a connective tissue cancer), is a G protein that relays a signal from a growth factor receptor on the plasma membrane to a cascade of protein kinases. The cellular response at the end of the pathway is the synthesis of a protein that stimulates the cell cycle. Normally, such a pathway will not operate unless triggered by the appropriate growth factor. But certain mutations in the ras gene can lead to production of a hyperactive Ras protein that triggers the kinase cascade even in the absence of growth factor, resulting in increased cell division. In fact, hyperactive versions or excess amounts of any of the pathway's components can have the same outcome: excessive cell division.

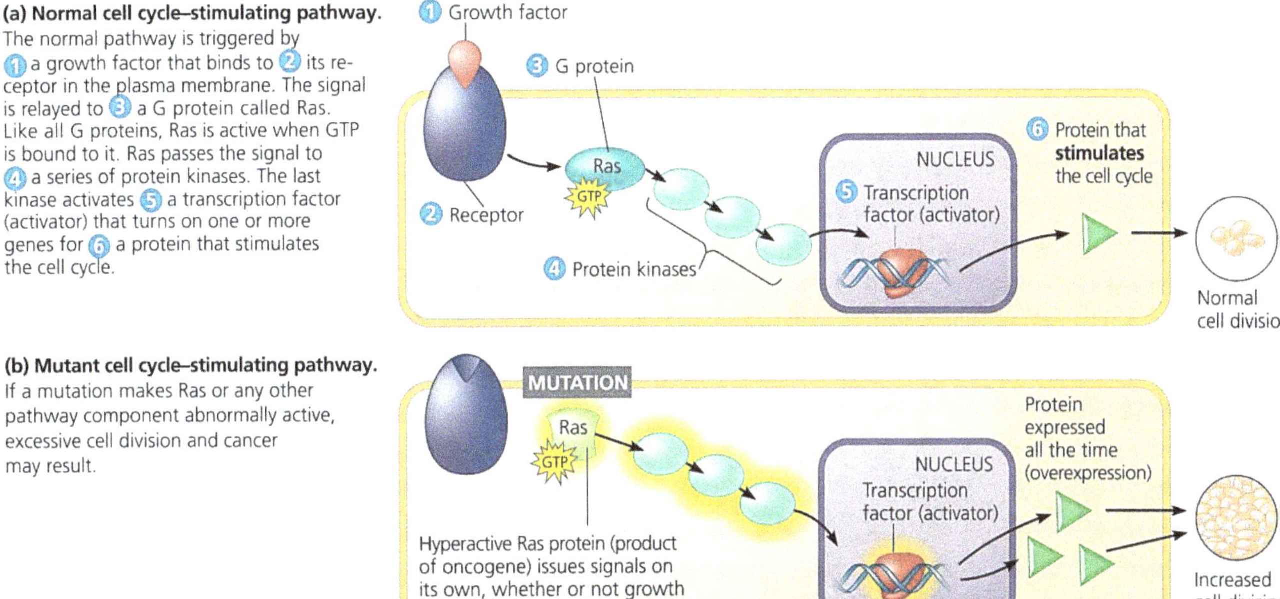

(a) Normal cell cycle–stimulating pathway. The normal pathway is triggered by ① a growth factor that binds to ② its receptor in the plasma membrane. The signal is relayed to ③ a G protein called Ras. Like all G proteins, Ras is active when GTP is bound to it. Ras passes the signal to ④ a series of protein kinases. The last kinase activates ⑤ a transcription factor (activator) that turns on one or more genes for ⑥ a protein that stimulates the cell cycle.

(b) Mutant cell cycle–stimulating pathway. If a mutation makes Ras or any other pathway component abnormally active, excessive cell division and cancer may result.

▲ **Figure 8.37** Normal and mutant cell cycle–stimulating pathway.

After the discovery of proto-oncogenes, a second category of genes related to cancer was identified: the **tumor-suppressor genes** or **growth-stopping genes**, whose normal products inhibit cell division. The protein products of tumor-suppressor genes have various functions. Some repair damaged DNA, which prevents the cell from accumulating cancer-causing mutations. Other tumor-suppressor proteins control adhesion of cells to each other or to the extracellular matrix; proper cell anchorage is crucial in normal tissues and is often absent in cancers. Still other tumor-suppressor proteins are components of cell-signaling pathways that inhibit the cell cycle.

Both copies of a tumor-suppressor gene must lose function for the cancerous phenotype to develop, in contrast to the mutations in proto-oncogenes. Put another way, the proto-oncogenes act in a dominant fashion, and tumor suppressors act in a recessive fashion. Tumor suppressor genes, encode proteins that normally block cancer development; that is, they promote apoptosis or prevent cell division. Inactivating, deleting, or mutating these genes therefore eliminates crucial limits on cell division. The first tumor suppressor identified was the **retinoblastoma susceptibility gene (Rb)**.

The role of the Rb protein in the cell cycle is to integrate signals from growth factors. The Rb protein is called a "pocket protein" because it has binding pockets for other proteins. Its role is therefore to bind important regulatory proteins such as regulatory transcription factor called **E2F** that activates genes required for cell cycle advancement from G1 to S phase, such as cyclins or Cdks.

The binding of Rb to other proteins is controlled by phosphorylation: When it is dephosphorylated, it can bind a variety of regulatory proteins, but loses this capacity when phosphorylated. The action of growth factors results in the phosphorylation of Rb protein by a Cdk. This then brings us full circle, because the phosphorylation of Rb releases previously bound regulatory proteins, resulting in the production of S phase cyclins that are necessary for the cell to pass the G_1/S boundary and begin chromosome replication.

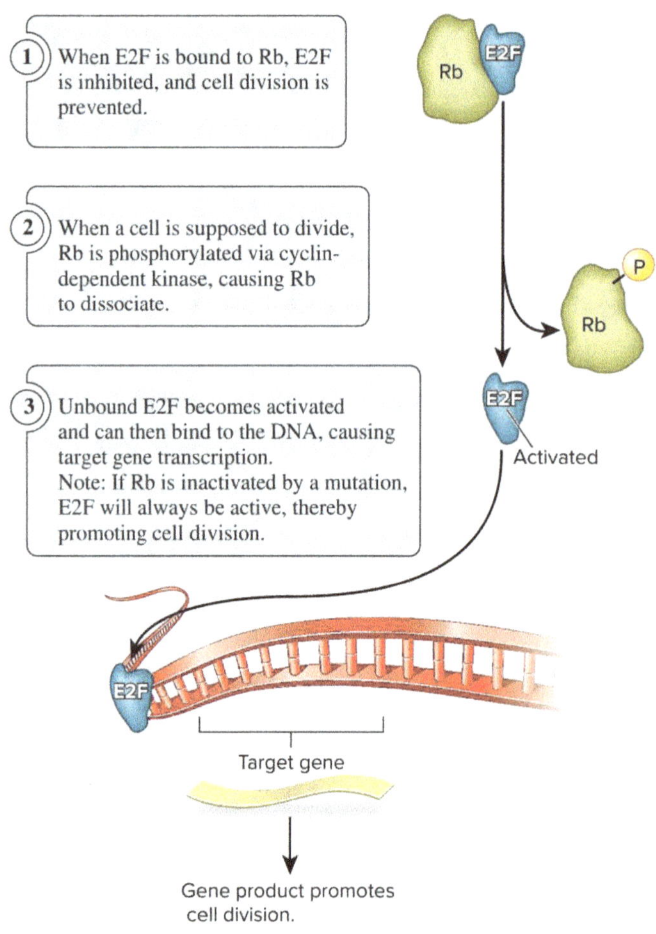

▲ **Figure 8.38** Function of the Rb protein. The Rb protein inhibits the function of E2F, which turns on genes that cause a cell to divide.

One of the most important and studied of the tumor-suppressor genes is named *TP53*, named for the apparent molecular weight of its protein product, **p53** that plays a key role in the G1 checkpoint of cell division and is a specific transcription factor that promotes the synthesis of cell cycle–inhibiting proteins. That is why a mutation that knocks out the p53 gene or in a gene required to activate the p53 protein (for example, a gene called ATM) can lead to excessive cell growth and cancer. The *p53* gene has been called the "guardian angel of the genome." Once the p53 protein is activated by the ATM protein, a protein kinase, after DNA damage, p53 activates several other genes, such as *p21*.

The p21 protein halts the cell cycle by binding to cyclin-dependent kinases, allowing time for the cell to repair the DNA. Researchers recently showed that p53 also activates expression of a group of miRNAs that inhibit the cell cycle. The p53 protein can also turn on genes directly involved in DNA repair. If DNA damage is irreparable, p53 activates "suicide" genes, whose protein products bring about programmed cell death (apoptosis). Thus, p53 acts in several ways to prevent a cell from passing on mutations due to DNA damage. If mutations do accumulate and the cell survives through many divisions—as is more likely if the p53tumor-suppressor gene is defective or missing—cancer may ensue.

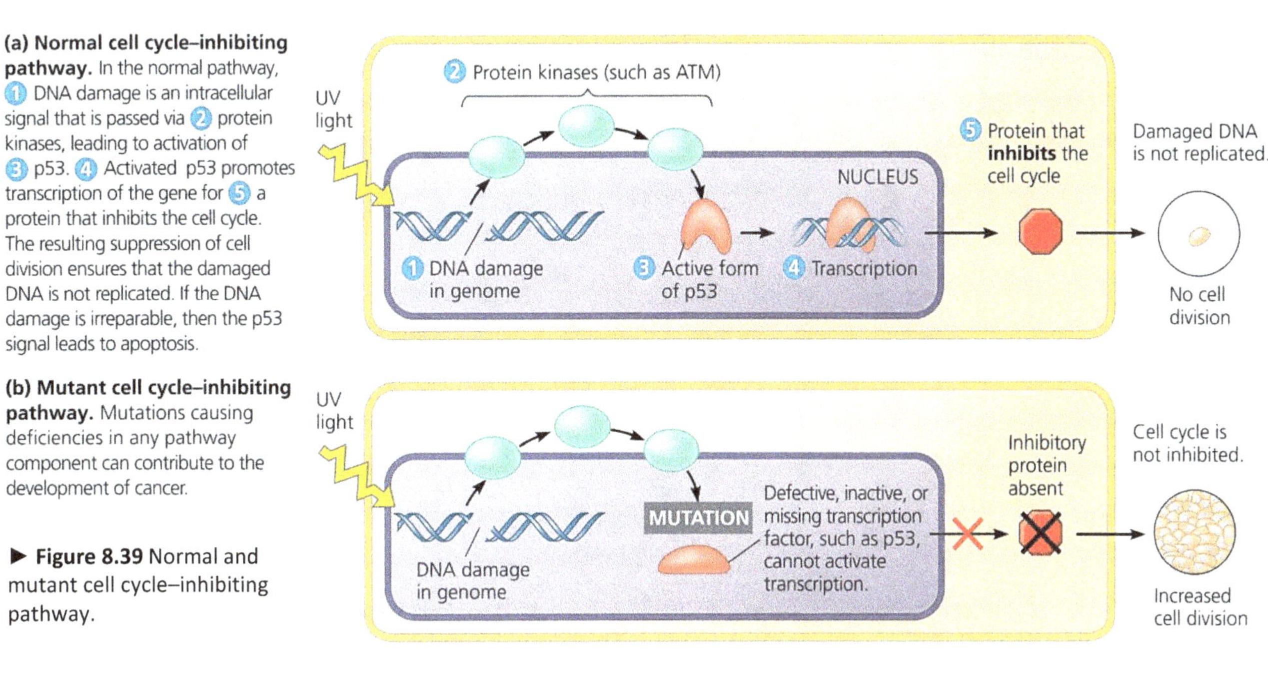

▶ **Figure 8.39** Normal and mutant cell cycle–inhibiting pathway.

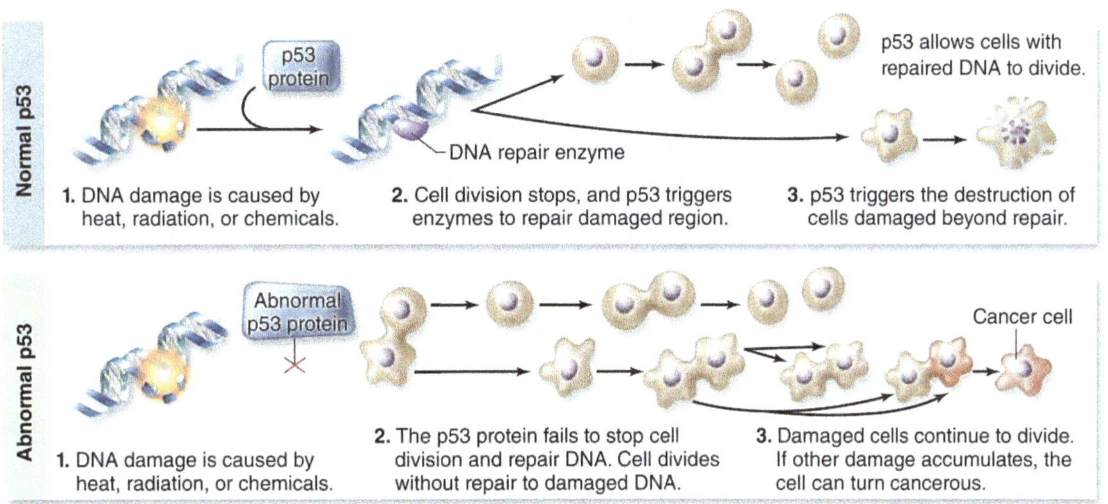

▲ **Figure 8.40** Cell division, cancer, and p53 protein. Normal p53 protein monitors DNA, destroying cells that have irreparable damage to their DNA. Abnormal p53 protein fails to stop cell division and repair DNA. As damaged cells proliferate, cancer develops.

The products of the **BRCA1** and **BRCA2** genes are also tumor suppressors; mutations in these genes give rise to neoplasms, an accumulation of abnormally dividing cells, in the breast, prostate, ovary, and other tissues. BRCA gene products are multifunctional proteins that help repair broken or otherwise damaged DNA. Cells infected with human papillomavirus (HPV) form skin growths called warts because this virus interferes with tumor suppressors.

Some HPV strains are associated with neoplasms that form on the cervix. Mutation of the tumor suppressor gene **Bax** is another example. Its product, the protein Bax, promotes apoptosis. When Bax mutates, Bax protein is not present and apoptosis is less likely to occur. The Bax gene contains a line of eight consecutive G bases in its DNA. When the same base molecules are lined up in this fashion, the gene is more likely to be subject to mutation.

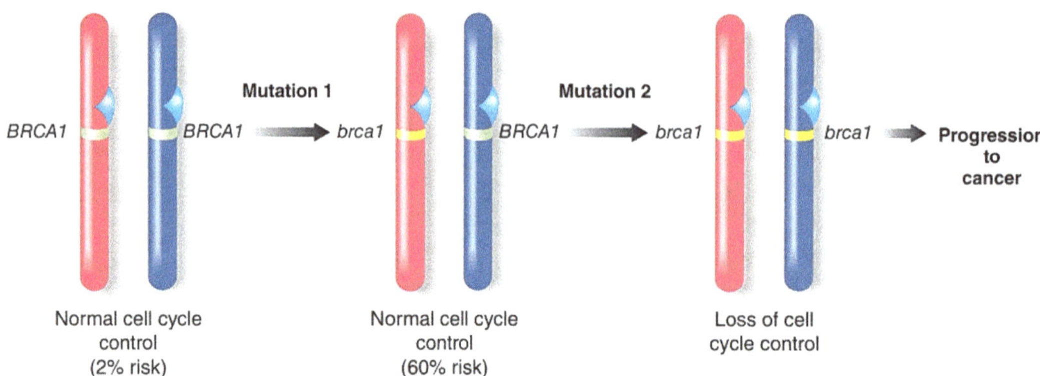

If both the tumor-suppressor genes and proto-oncogenes in the same cell are altered, the result could be cancer.

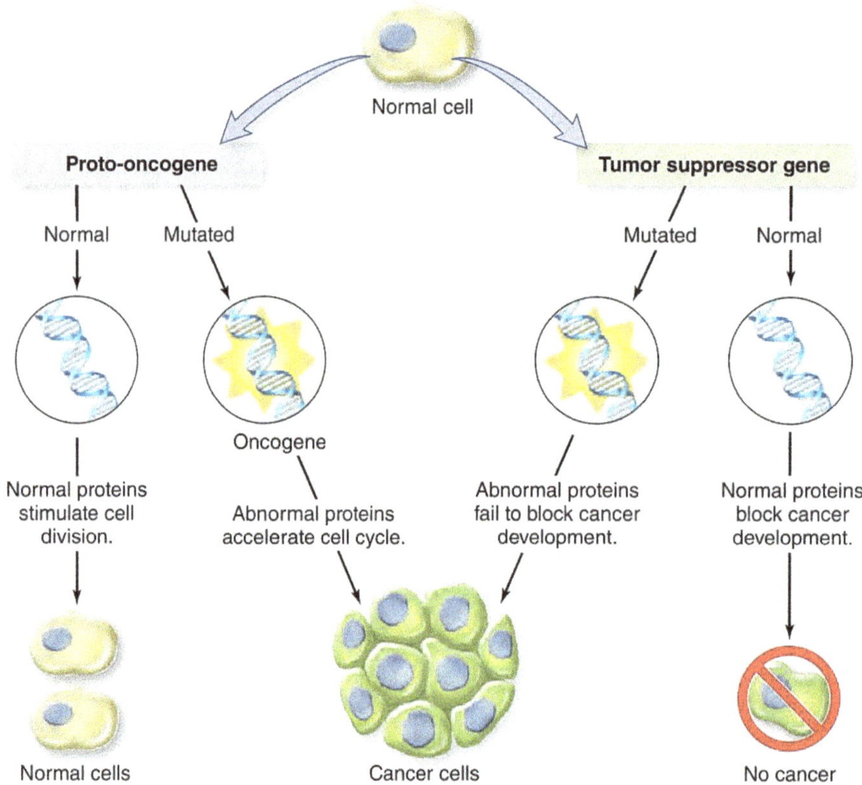

▲ **Figure 8.41** Cancer-Related Genes. Oncogenes and tumor suppressor genes both influence the cell cycle. When proto-oncogenes are mutated, they form oncogenes that accelerate cell division. Tumor suppressor genes encode proteins that normally inhibit cell division, but when the genes are mutated, cancer can develop.

Cancer cells have unique characteristics

First, a cancer cell looks different from a normal cell. Its shape may be different, and it may lose some of the specialized features of its parent cells. Some cancer cells have multiple nuclei. These visible differences allow pathologists to detect cancerous cells by examining tissue under a microscope. Abnormal changes on the cell surface cause cancer cells to lose attachments to neighboring cells and the extracellular matrix, allowing them to spread into nearby tissues. Cancer cells may also secrete signaling molecules that cause blood vessels to grow toward the tumor. A few tumor cells may separate from the original tumor or the **primary tumor**, enter blood vessels and lymph vessels, and travel to other parts of the body. There, they may proliferate and form a new tumor or **secondary cancer mass**. This spread of cancer cells to locations distant from their original site is called **metastasis.**

Second, unlike normal cells, many cancer cells are essentially immortal, ignoring the "clock" that limits normal cells to 50 or so divisions. This cellular clock resides in **telomeres**, the noncoding DNA at the tips of eukaryotic chromosomes. Telomeres consist of hundreds to thousands of repeats of a specific DNA sequence ('**multiple repeat sequences**'). At each cell division, the telomeres lose nucleotides from their ends, so the chromosomes gradually become shorter. After about 50 divisions, the cumulative loss of telomere DNA signals division to cease in a normal cell. Cells that produce an enzyme called **telomerase**, however, can continually add DNA to chromosome tips. Their telomeres stay long, which enables them to divide beyond the 50-or-so division limit. Cancer cells have high levels of telomerase; inactivating this enzyme could therefore have tremendous medical benefits.

Third difference between normal cells and cancer cells lies in growth factors, the chemical signals that stimulate cell division. Normal cells stop dividing once external growth factors are depleted. Many cancer cells, however, divide even in the absence of growth factors.

Fourth, normal cells growing in culture exhibit **contact inhibition**, meaning that they stop dividing when they touch one another in a one-cell-thick layer. Cancer cells lack contact inhibition, so they tend to pile up in culture. In addition, normal cells divide only when attached to a solid surface, a property called anchorage dependence. The observation that cancer cells lack anchorage dependence helps explain how metastasis occurs.

Cancer cells have other unique features, too. For example, a normal cell dies (undergoes apoptosis) when badly damaged, but many cancer cells do not. Cancer cells also send signals that stimulate a process called **angiogenesis**, the development of new blood vessels. Proteins that stimulate angiogenesis in tumors are called **tumor angiogenesis factors** (**TAFs**). The newly sprouted blood vessels boost tumor growth by delivering nutrients and removing wastes. Disrupting angiogenesis is a possible cancer-fighting strategy. The cytoplasm and plasma membrane of malignant cells are altered. The cytoskeleton may be shrunken, disorganized, or both.

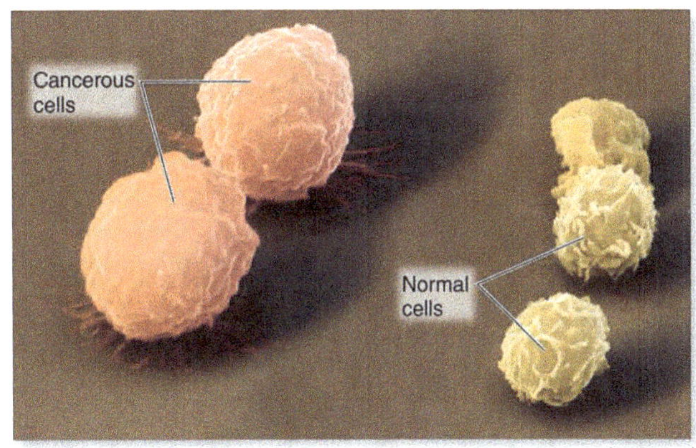

▶ **Figure 8.42** Cancer Cells Are Abnormal. The two cancerous leukemia cells on the left are larger than the normal marrow cells on the right.

Malignant cells typically have an abnormal chromosome number, with some chromosomes present in multiple copies, and others missing or damaged. The balance of metabolism is often shifted, as in an amplified reliance on ATP formation by fermentation rather than aerobic respiration. Cancer cells also lack **differentiation**. A cancer cell is not differentiated, meaning that it has no specified function and therefore can make no contribution to the overall functioning of a particular body part.

Cancer Diagnosis and Treatments

Many cancer cells have antigens on their surfaces that are not found on normal cells of the body. Usually, T cells and NK cells recognize these abnormal antigens in potentially cancerous cells and destroy them. However, cancer cells have mechanisms that allow them to avoid destruction by the immune system. Some types of cancerous cells include mechanisms that actively seek to avoid the body's defenses, while other cancers simply overwhelm the immune system defenses by multiplying more rapidly than they can be killed off. Every tumor starts as one cell gone wild. The cancerous cell must compete with its surrounding cells for nutrients and space. If the cancerous cell has distinct advantages over its neighbors, like ways to avoid cellular apoptosis, the cell will survive and divide and those advantages will be passed on to its descendants.

The cancerous descendants tend to accumulate even more mutations as they divide rapidly and without control, making their progeny even more abnormal. These mutations allow the cells to continually change with each generation. The changes make it difficult for the immune system to identify these cells, and therefore make it harder for the immune system to track down and destroy them. Medical professionals describe the spread of cancer cells as a series of stages.

In one system used to classify colon cancer, for example, a **stage I** (**carcinogenesis**) cancer has started invading tissue layers adjacent to the tumor's origin, but cancerous cells remain confined to the colon. At **stage II**, the tumor has spread to tissues around the colon but has not yet reached nearby lymph nodes (**carcinoma**). **Stage III** (**Angiogenesis**) cancers have spread to organs and lymph nodes near the cancer's origin, and **stage IV** cancers have spread to distant sites. The names and criteria for each stage vary among cancers. In general, however, the lower the stage, the better the prospect for successful treatment.

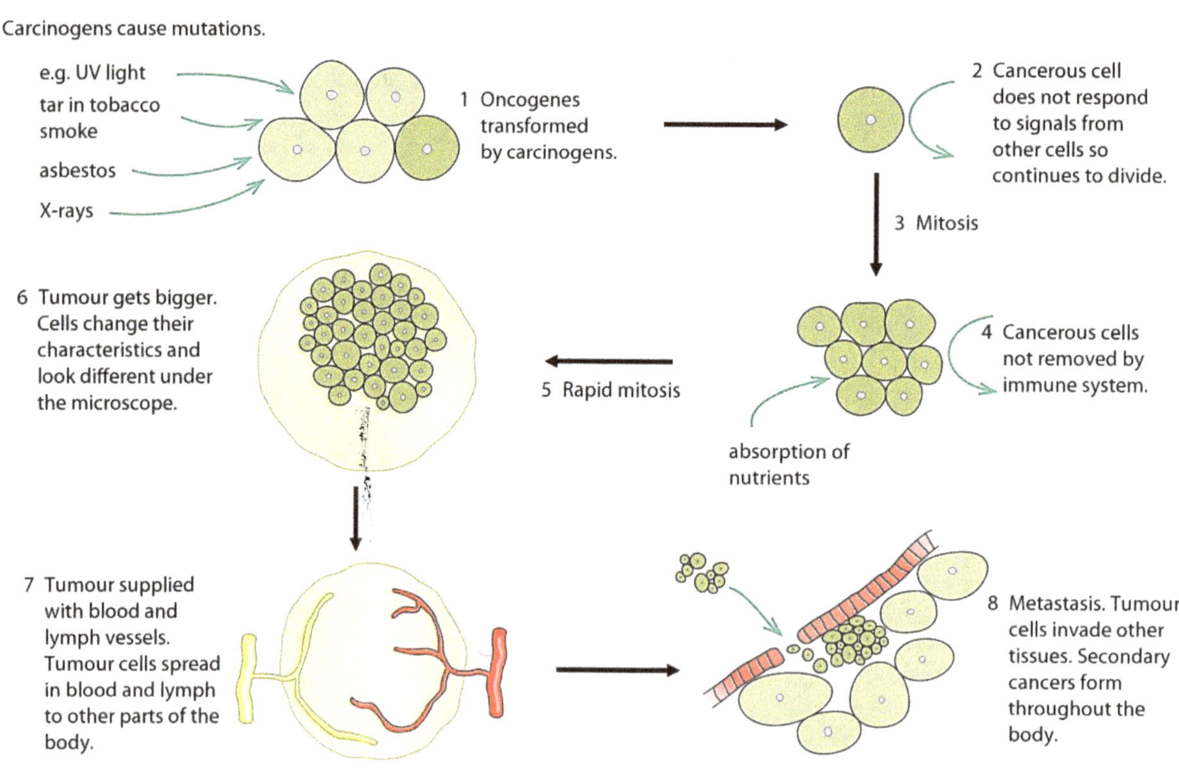

▲ **Figure 8.43** Stages in the development of cancer.

Physicians use many techniques to estimate the stage of a patient's cancer. For example, X-rays, CAT scans, MRIs, PET scans, ultrasound, and other imaging tests are noninvasive ways to detect and measure tumors inside the body. Some neoplasms can be detected early with periodic screening such as Pap tests or dermatology exams. A physician can also use an endoscope to inspect the inside of some organs, such as the esophagus or intestines. The same tool can also collect a **biopsy** sample; pathologists then use microscopes to search the tissue for suspicious cells. Traditional cancer treatments include **surgical tumor removal**, drugs (**chemotherapy**), and **radiation**. Chemotherapy drugs, usually delivered intravenously, are intended to stop cancer cells anywhere in the body from dividing. To treat known or suspected metastatic tumors, chemotherapy is used, in which drugs that are toxic to actively dividing cells are administered through the circulatory system.

Chemotherapeutic drugs interfere with specific steps in the cell cycle. For example, the drug Taxol freezes the mitotic spindle by preventing microtubule depolymerization, which stops actively dividing cells from proceeding past metaphase and leads to their destruction. The side effects of chemotherapy are due to the effects of the drugs on normal cells that divide frequently, due to the function of that cell type in the organism. For example, nausea results from chemotherapy's effects on intestinal cells, hair loss from effects on hair follicle cells, and susceptibility to infection from effects on immune system cells.

Radiation therapy uses directed streams of energy from radioactive isotopes to kill tumor cells in limited areas. A tumor that appears to be localized may be treated with high-energy radiation, which damages DNA in cancer cells much more than DNA in normal cells, apparently because the majority of cancer cells have lost the ability to repair such damage. Another potential treatment for cancer that is currently under development is **virotherapy**, the use of viruses to kill cancer cells. The viruses employed in this strategy are designed so that they specifically target cancer cells without affecting the healthy cells of the body. For example, proteins (such as antibodies) that specifically bind to receptors found only in cancer cells are attached to viruses. Once inside the body, the viruses bind to cancer cells and then infect them. The cancer cells are eventually killed once the viruses cause cellular lysis.

Researchers are also investigating the role of **metastasis regulatory genes** that control the ability of cancer cells to undergo metastasis. Scientists hope to develop therapeutic drugs that can manipulate these genes and, therefore, block metastasis of cancer cells. **Immunotherapy** and **p53 Gene Therapy** are now in clinical trials and are expected to be increasingly used to treat cancer. The success of any cancer treatment depends on many factors, including the type of cancer and the stage in which it is detected. Surgery can cure cancers that have not spread or that have only invaded local lymph nodes. Once cancer metastasizes, however, it becomes difficult to locate and treat all of the tumors. Moreover, DNA replication errors introduce mutations in rapidly dividing cancer cells.

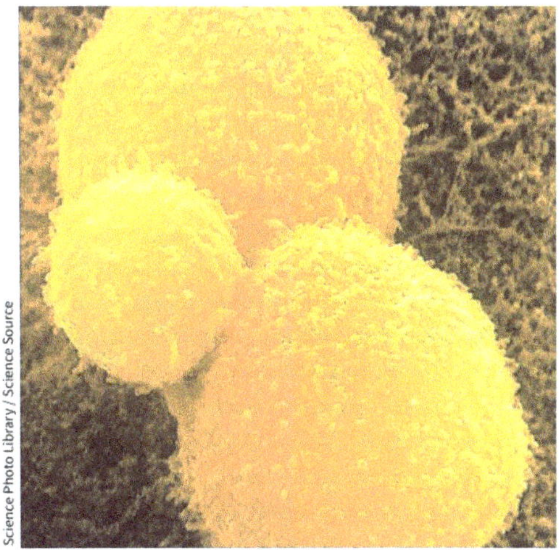

▶ **Figure 8.44** T cell attacking two large tumor cells. T cells and natural killer cells are effective in removing many potentially cancerous cells. However, these cells fail to destroy all tumor cells, so modern medical practices are working to fill in the gaps.

Causes and Prevention of Cancer

- **Heredity:** BRCA1 and BRCA2 are tumor suppressor genes that follow a recessive pattern of inheritance. The RB gene is also a tumor suppressor gene. It takes its name from its association with retinoblastoma, a rare eye cancer that occurs almost exclusively in early childhood. In an affected child, a single mutated copy of the RB gene is inherited. A second mutation, this time to the other copy of the RB gene, causes the cancer to develop. An abnormal RET gene, which predisposes an individual to thyroid cancer, can be passed from parent to child. RET is a protooncogene known to be inherited in an autosomal dominant manner, meaning that only one mutation is needed to increase a predisposition to cancer. The remainder of the mutations necessary for a thyroid cancer to develop are acquired (not inherited).

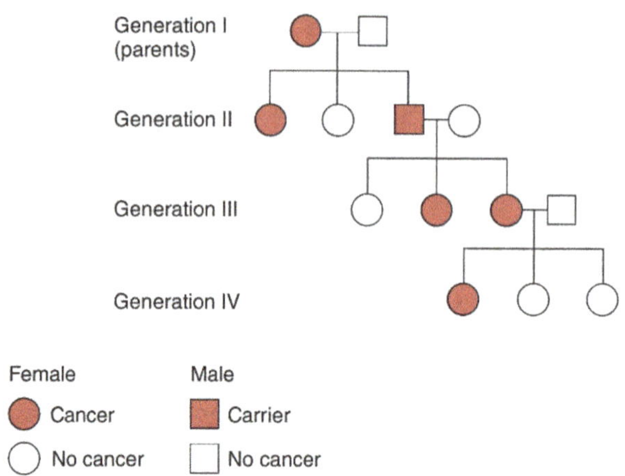

▲ **Figure 8.45** The genetics of cancer One cause of cancer is genetic predispostion. In this pedigree, the female in generation 1 had breast cancer and passed that gene on to her son and one daughter. The son then had three daughters, two who developed breast cancer. Additionally, one of those daughters produced a granddaughter who developed breast cancer.

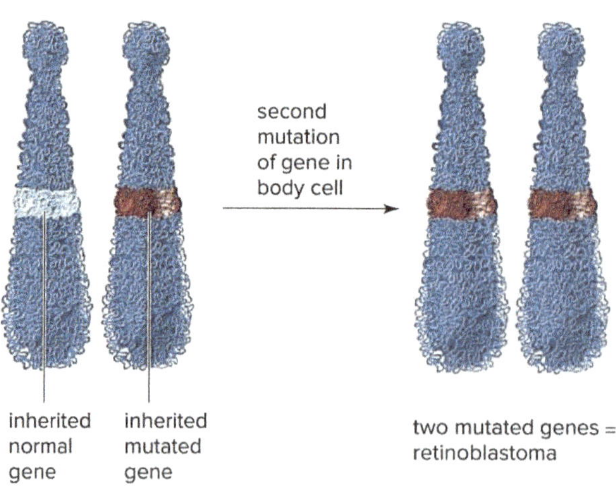

▲ **Figure 8.46** Inheritance of retinoblastoma. A child is at risk for an eye tumor when a mutated copy of the RB gene is inherited, even though a second mutation in the normal copy is required before the tumor develops.

- **Environmental Carcinogens:** Any agent, such as asbestos, that causes cancer is called a **carcinogen** and is described as carcinogenic such as UV radiation, mechanical trauma, certain viral infections caused by oncogenic viruses, chronic inflammation, and many chemicals in tobacco can be **mutagens** which means they damage DNA (cause mutations).

At least four types of DNA viruses—hepatitis B and C viruses, Epstein-Barr virus, and human papillomavirus (HPV)—are believed to cause human cancers. RNA-containing retroviruses, in particular, are known to cause cancers in animals. In humans, the retrovirus HTLV-1 (human T-cell lymphotropic virus, type 1) has been shown to cause hairy cell leukemia. HIV, the virus that causes AIDS, and Kaposi sarcoma–associated herpesvirus (KSHV) are responsible for the development of Kaposi sarcoma and certain lymphomas. This occurs due to the suppression of proper immune system functions.

Viruses can interfere with gene regulation in several ways if they integrate their genetic material into the DNA of a cell. Viral integration may donate an oncogene to the cell, disrupt a tumor-suppressor gene, or convert a protooncogene to an oncogene. Some viruses produce proteins that inactivate p53 and other tumor-suppressor proteins, making the cell more prone to becoming cancerous.

Reducing sun exposure and avoiding tobacco therefore directly reduce cancer risk. Obesity greatly increases the risk of death from cancers of the breast, cervix, uterus, and ovaries in women; obese men have an elevated risk of dying from prostate cancer. High-calorie foods that are rich in animal fats and low in fiber, coupled with a lack of exercise, contribute to high body weight. But scientists remain uncertain why obesity itself is a risk factor for cancer.

Perhaps fat tissue secretes hormones that contribute to metastasis, or maybe obesity reduces immune system function. A healthy lifestyle remains the best way to reduce cancer risk. Recently, there has been a great deal of public concern regarding the presumed danger of nonionizing radiation. This energy form is given off by cell phones, electrical lines, and appliances. Some recent studies have discovered links between these forms of radiation and rare forms of cancer.

All these factors cause mutations changes in DNA that alter the expression of certain genes. However, not all carcinogens, agents that increase the likelihood of developing cancer, do damage because most are eliminated by peroxisomal or lysosomal enzymes or by the immune system. Furthermore, one mutation isn't enough. It takes several genetic changes to transform (convert) a normal cell into a cancerous cell. **Carcinogenesis**, the process by which cancer develops, is a multistep process in which as many as 10 distinct mutations may have to accumulate in a cell before it becomes cancerous.

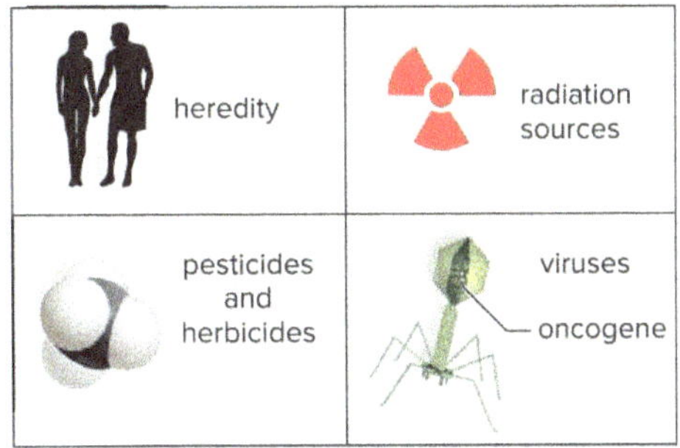

▲ **Figure 8.47** Some sources contributing to development of cancer.

Cell Death Is Part of Life

The development of a multicellular organism requires more than just cell division. Cells also die in predictable ways, carving distinctive structures. Apoptosis is cell death that is a normal part of development. Like cell division, it is a precise, tightly regulated sequence of events. **Apoptosis** is therefore also called "**programmed cell death**." It results in small membrane-enclosed **apoptotic bodies**, which quickly undergo phagocytosis by neighboring cells or cells specialized for debris removal. Apoptosis is different from **necrosis**, which is the "accidental" cell death that follows a cut or bruise. Whereas necrosis is sudden, traumatic, and disorderly, apoptosis results from a precisely coordinated series of events that dismantle a cell. Apoptotic cells do not rupture and release none of their contents, unlike cells that die as a result of injury and undergo necrosis. This difference is highly significant because release of cellular components triggers a local inflammatory reaction and immigration of leukocytes.

A few examples of apoptosis emphasize its significance. In the ovary, apoptosis is the mechanism for both the monthly loss of luteal cells and the removal of excess oocytes and their follicles. Apoptosis was first discovered as programmed cell death in embryos, where it is important in shaping developing organs or body regions, such as the free spaces between embryonic fingers and toes. Apoptosis of excess nerve cells plays an important role in the final development of the central nervous system. Triggered by p53 and other tumor suppressor proteins, apoptosis is the method for eliminating cells whose survival is blocked by lack of nutrients or by damage caused by free radicals or radiation. In all these examples, apoptosis occurs very rapidly, in less time than required for mitosis, and the affected cells are removed without a trace.

During early development, cell death and destruction are normal events. Nature takes few chances. More cells than needed are produced, and excesses are eliminated later in a type of programmed cell death called apoptosis. Apoptosis is particularly common in the developing nervous system. Most organs are well formed and functional long before birth, but the body continues to grow and enlarge by forming new cells throughout childhood and adolescence. Once we reach adult size, cell division is important mainly to replace short-lived cells and repair wounds.

During young adulthood, cell numbers remain fairly constant so that cell division and cell death are in balance and tissue neither overgrows nor shrinks. Both cell division and apoptosis also help protect the organism. For example, cells divide to heal a scraped knee; apoptosis peels away sunburnt skin cells that might otherwise become cancerous. However, local changes in the rate of cell division are common. For example, when a person is anemic, his or her bone marrow undergoes **hyperplasia**, or accelerated growth, to produce red blood cells at a faster rate. If the anemia is remedied, the excessive marrow activity ceases. **Atrophy**, a decrease in size of an organ or body tissue, can result from loss of normal stimulation or from diseases like muscular dystrophy.

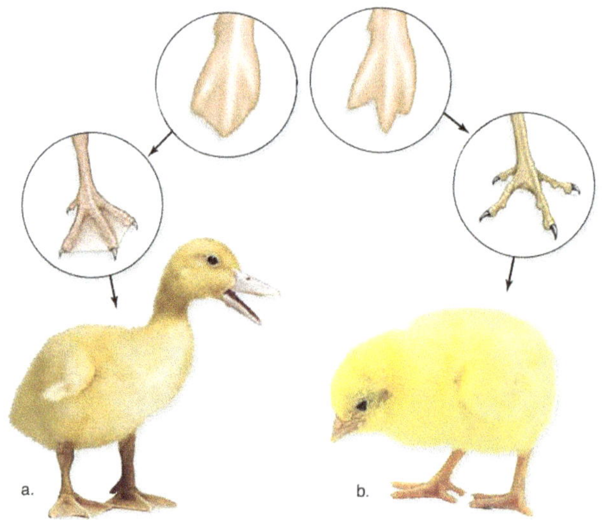

► **Figure 8.48** Apoptosis Carves Toes. The feet of embryonic ducks and chickens have webbing between the digits. (a) A duck's foot retains this webbing. (b) In a developing chicken foot, the toes take shape when the cells between the digits die.

Apoptosis does not use the services of lysosomes as its destructive tool. During the most common pathway of apoptosis:

- **Loss of mitochondrial function and caspase activation:** The mitochondrial membranes become permeable in response to external signals or irreversible internal damage that activate cytoplasmic proteins in the **Bcl-2 family** which regulate the release of death-promoting factors from mitochondria. **Cytochrome c** and other factors leak from the mitochondria into the cytosol, and these factors activate intracellular enzymes called **caspases**. They function as proteases that are sometimes called the "executioners" of the cell. Caspases digest selected cellular proteins such as actin filaments, which are components of the cytoskeleton. The destruction of these microfilaments causes the cell to break into small vesicles that are eventually phagocytized by cells of the immune system.

- **Fragmentation of DNA: Endonucleases** are activated, which cleave DNA between nucleosomes into small fragments.

- **Shrinkage of nuclear and cell volumes:** Destruction of the cytoskeleton and chromatin causes the cell to shrink quickly, producing small structures with dense, darkly stained **pyknotic nuclei.**

- **Cell membrane changes:** The plasma membrane of the shrinking cell undergoes dramatic shape changes, such as "blebbing," as membrane proteins are degraded and lipid mobility increases.

- **Formation and phagocytic removal:** Membrane-bound remnants of cytoplasm and nucleus separate as very small apoptotic bodies. Newly exposed phospholipids on these bodies induce their phagocytosis by neighboring cells or white blood cells.

Apoptosis disposes of unneeded cells; **autophagy** and **proteasomes** dispose of unneeded organelles and proteins. The process called autophagy ("self-eating") sweeps up bits of cytoplasm and excess organelles into double membrane vesicles called **autophagosomes**. They are then delivered to lysosomes for digestion of the contents, which the cell reuses. Autophagy may have evolved as a response to cell starvation and it speeds up in response to several kinds of stress, such as low oxygen, high temperature, or lack of growth factors.

Although autophagy can lead to programmed cell death, it makes a greater contribution to cell survival and provides a fail-safe system against complete self-destruction when such a dire response is not necessary. Autophagy is exactly what the doctor ordered for disposal of large cytoplasmic structures and protein aggregates. But lysosomal enzymes do not have access to soluble proteins that are misfolded, damaged, or unneeded and need to be disposed of. Examples of unneeded proteins include some that are used only in cell division and must be degraded at precise points in the cell cycle, and short-lived transcription factors.

Proteins called **ubiquitins** mark doomed proteins for attack (proteolysis) by attaching to them. The tagged proteins are then hydrolyzed to small peptides by soluble enzymes or by **proteasomes**, giant "waste disposal" complexes composed of protein-digesting enzymes, and the ubiquitin is recycled. Proteasome activity is critical during starvation when these complexes degrade preexisting proteins to provide amino acids for synthesis of new and needed proteins.

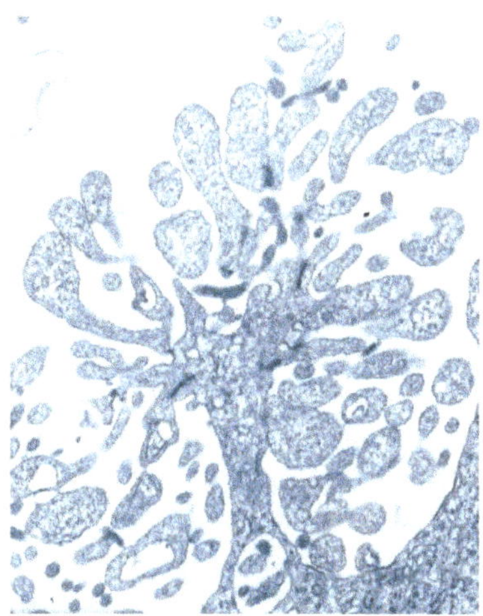

▶ **Figure 8.49** Late apoptosis—apoptotic bodies.

Cell Aging

Aging is a normal process accompanied by a progressive alteration of the body's homeostatic adaptive responses. Cell aging is complicated and has many causes. It produces observable changes in structure and function and increases vulnerability to environmental stress and disease. The exact nature of human aging is still a mystery, but there are several current theories on aging:

- The **mitochondrial theory** places the blame on damage caused by free radicals, resulting in diminished energy production by damaged mitochondria.

- The **immune theory** holds that aging results from a progressive weakening of the immune system; the body loses its ability to fight off pathogens or to heal systemic inflammation, which is also associated with aging and risks for chronic diseases.

- The genetic theory holds that cell aging is "**programmed**" into our genes. Although many millions of new cells normally are produced each minute, several kinds of cells in the body—including skeletal muscle cells and nerve cells—do not divide. Experiments have shown that many other cell types have only a limited capability to divide. Normal cells grown outside the body divide only a certain number of times and then stop. These observations suggest that cessation of mitosis is a normal, genetically programmed event. According to this view, "aging genes" are part of the genetic blueprint at birth. These genes have an important function in normal cells, but their activities slow over time. They bring about aging by slowing down or halting processes vital to life.

- **Telomeres** are nonsensical strings of nucleotides that cap the ends of chromosomes, providing protection. Though telomeres carry no genes, they appear to be vital for chromosomal survival. With each cycle of DNA replication, the telomeres get a bit shorter. Telomerase is an enzyme found in certain specialized cells that has been dubbed the "immortality enzyme" due to its ability to lengthen previously shortened telomeres.

- **Glucose**, the most abundant sugar in the body, plays a role in the aging process. It is haphazardly added to proteins inside and outside cells, forming irreversible cross-links between adjacent protein molecules. With advancing age, more cross-links form, which contributes to the stiffening and loss of elasticity that occur in aging tissues.

- The **wear-and-tear theory** holds that the cumulative effect of assaults, such as environmental toxins, leads to accelerated rates of cell death throughout the body. **Free radicals** produce oxidative damage in lipids, proteins, or nucleic acids. Some effects are wrinkled skin, stiff joints, and hardened arteries. Naturally occurring enzymes in peroxisomes and in the cytosol normally dispose of free radicals. Certain dietary substances, such as vitamin E, vitamin C, beta carotene, zinc, and selenium, are antioxidants that inhibit free radical formation.

- Some theories of aging explain the process at the cellular level, while others concentrate on regulatory mechanisms operating within the entire organism. For example, the immune system may start to attack the body's own cells. This **autoimmune response** might be caused by changes in certain plasma membrane glycoproteins and glycolipids (cell identity markers) that cause antibodies to attach to and mark the cell for destruction. As changes in the proteins on the plasma membrane of cells increase, the autoimmune response intensifies, producing the well-known signs of aging.

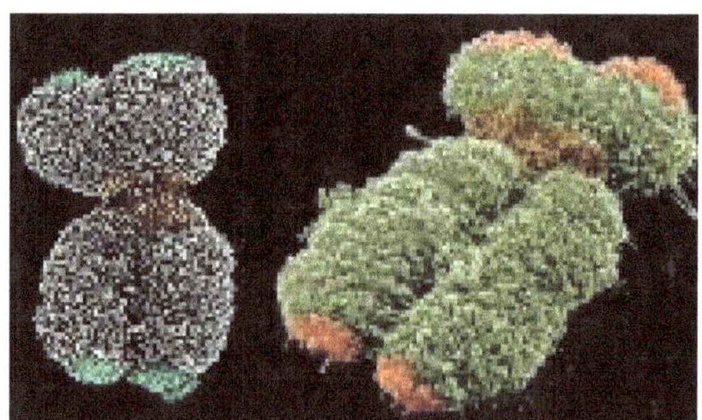

▶ **Figure 8.50** Coloured scanning electron micrographs of human chromosomes showing the location of telomeres at the ends of the chromosomes. Chromatids and centromeres are also clearly visible. Telomeres contain short repeated sequences of DNA. As cells replicate and age, the telomeres gradually get shorter. Stem cells are an exception.

Cellular Diversity

The body of an average human adult is composed of nearly 100 trillion cells that can be classified into about 200 different cell types known as differentiated cells, are organized into diverse and complex structures—such as the eye, hand, and brain— each capable of carrying out many sophisticated activities. Cells vary considerably in size. The shapes of cells also vary considerably, they may be round, oval, flat, cube-shaped, column-shaped, elongated, star-shaped, cylindrical, or disc-shaped. A cell's shape is related to its function in the body. We all begin life as a single cell, the fertilized egg, and all the cells of our body arise from it.

Early in development, cells begin to specialize, some becoming liver cells, some nerve cells, and so on. All our cells carry the same genes, so how can one cell become so different from another? Because genes do not seem to be lost regularly during development (and thus nuclear equivalence is present in different cell types), differences in the molecular composition of cells must be regulated by the activities of different genes. The process of developmental gene regulation is often referred to as **differential gene expression**. The transcription of certain sets of genes is repressed, whereas that of other sets is activated.

Apparently, cells in various regions of the embryo are exposed to different chemical signals that channel them into specific pathways of development. When the embryo consists of just a few cells, the major signals may be nothing more than slight differences in oxygen and carbon dioxide concentrations between the more superficial and the deeper cells. But as development continues, cells release chemicals that influence development of neighboring cells by triggering processes that switch some genes "off" and others "on." Some genes are active in all cells. For example, genes for rRNA and ATP synthesis are "on" in all cells, but genes for synthesizing the enzymes needed to produce thyroxine are "on" only in cells that are going to be part of the thyroid gland. Hence, the story of cell specialization lies in the kinds of proteins made and reflects the activation of different genes in different cell types.

Cell specialization leads to structural variation—different organelles come to predominate in different cells. For example, muscle cells make large amounts of actin and myosin, and their cytoplasm fills with microfilaments. Liver and phagocytic cells produce more lysosomes. The development of specific and distinctive features in cells is called cell **differentiation**. **Somatic cells** are all the cells of the body other than **germ line cells**, which ultimately give rise to a new generation. In animals germ line cells—whose descendants ultimately undergo meiosis and differentiate into **gametes**—are generally set aside early in development.

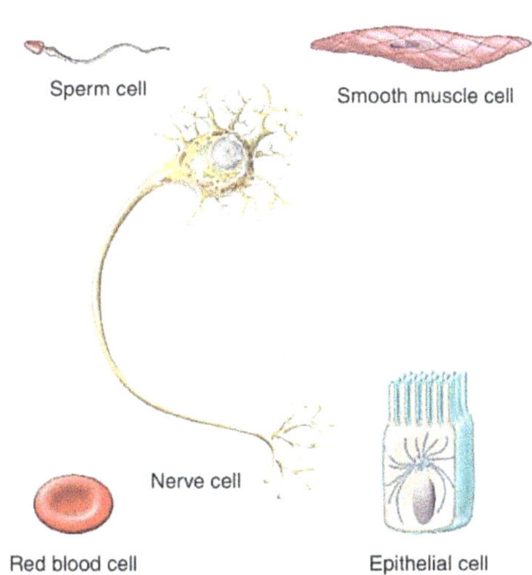

▲ **Figure 8.51** Diverse shapes and sizes of human cells.

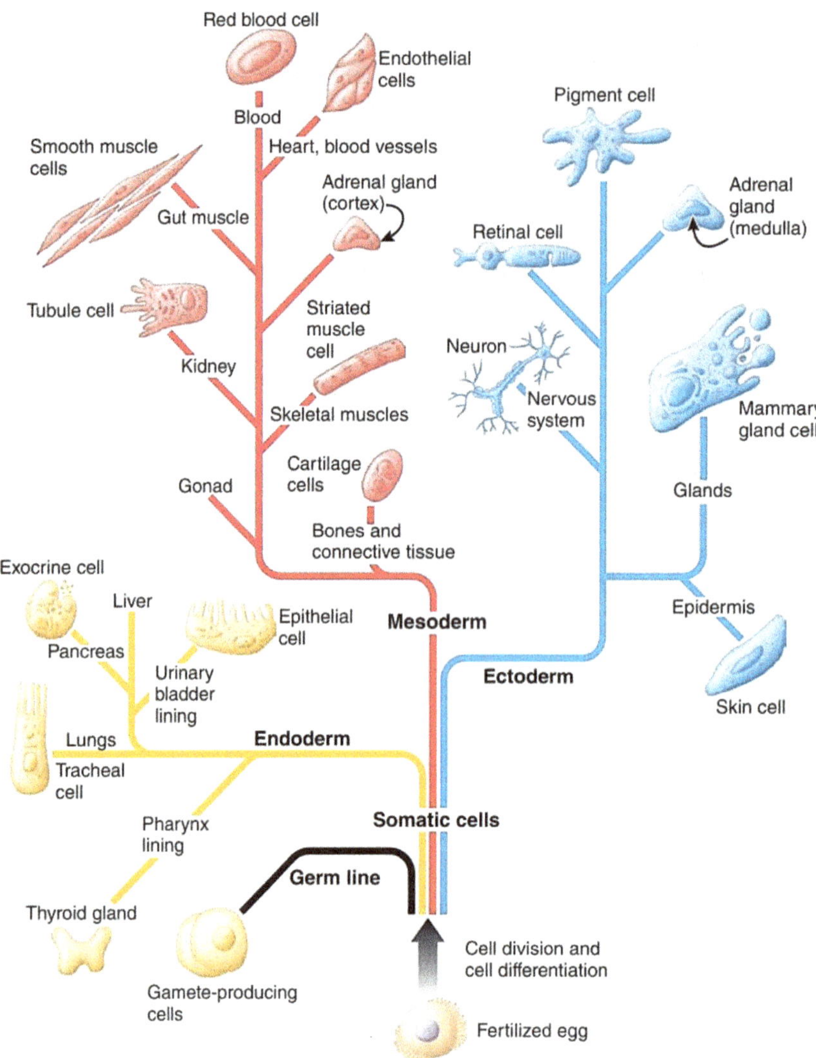

▲ **Figure 8.52** Vertebrate cell lineages. Repeated divisions of the fertilized egg (bottom) result in the establishment of tissues containing groups of specialized cells. Germ line cells (cells that produce the gametes) are set aside early in development. Somatic cells progress along various developmental pathways undergoing a series of commitments that progressively determine their fates.

Stem Cells

Throughout an individual's lifetime, many tissues and organs contain a small population of undifferentiated **stem cells** whose cycling serves to renew the differentiated cells of tissues as needed. A stem cell is a cell that can divide an unlimited number of times (by mitosis). Many stem cells divide infrequently and the divisions are asymmetric; that is, one daughter cell remains as a stem cell, while the other becomes committed to a path that leads to differentiation. Stem cells of many tissues are found in specific locations.

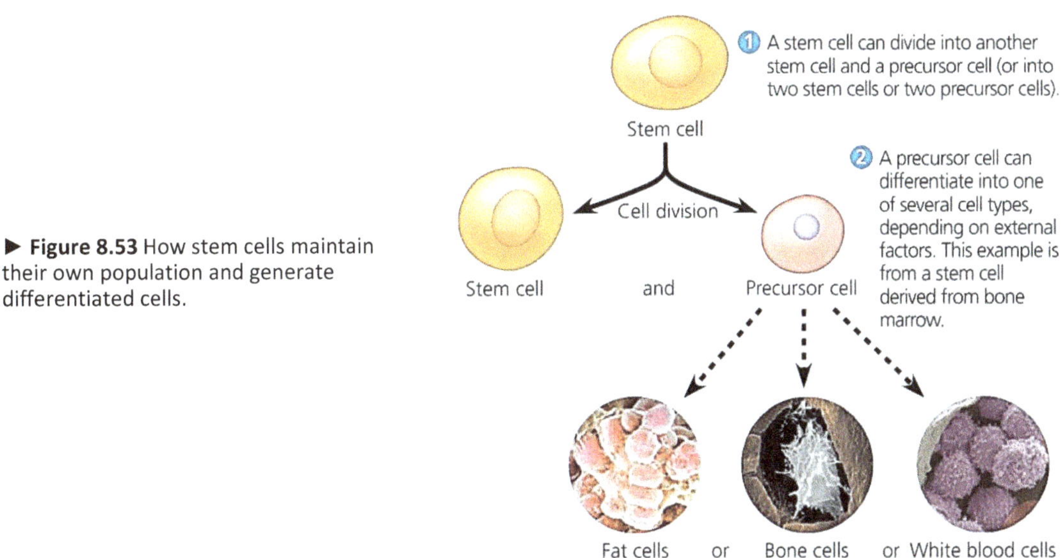

▶ **Figure 8.53** How stem cells maintain their own population and generate differentiated cells.

The power of a stem cell to produce different types of cell is variable and is referred to as its potency. Stem cells that can produce any type of cell are described as **totipotent**. The zygote formed by the fusion of a sperm with an egg at fertilization is totipotent, as are all the cells up to the 16-cell stage of development in humans. After that, some cells become specialized to form the placenta, while others lose this ability but can form all the cells that will lead to the development of the embryo and later the adult. These embryonic stem cells are described as **pluripotent** because they can give rise to many, but not all, of the types of cells in an organism. **Embryonic stem cells (ES cells)**, formed after a zygote has undergone several rounds of cell division to form a 5 or 6 day old **blastocyst**. As tissues, organs and systems develop, cells become more and more specialized. There are more than 200 different types of cell in an adult human body.

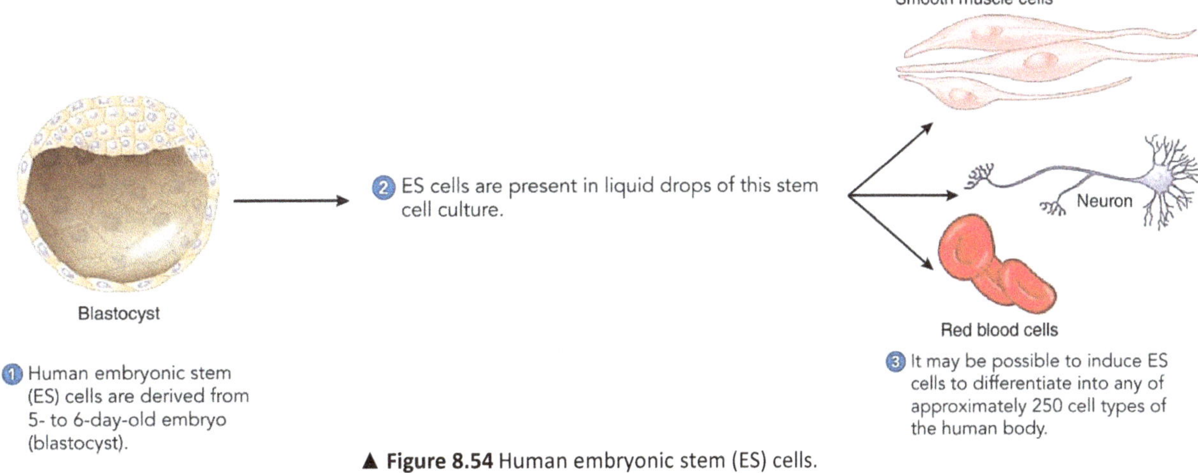

▲ **Figure 8.54** Human embryonic stem (ES) cells.

In the adult, most cells do not divide. However, for growth and repair it is essential that small populations of stem cells remain which can produce new cells. Adult stem cells have already lost some of the potency associated with embryonic stem cells and are no longer pluripotent. They are only able to produce a few types of cell and may be described as **multipotent**.

For example, the stem cells found in bone marrow are of this type. They can replicate any number of times, but can produce only blood cells, such as red blood cells, monocytes, neutrophils and lymphocytes. Mature blood cells have a relatively short lifespan, so the existence of these stem cells is essential. For example, around 250 billion red blood cells and 20 billion white blood cells are lost and must be replaced each day.

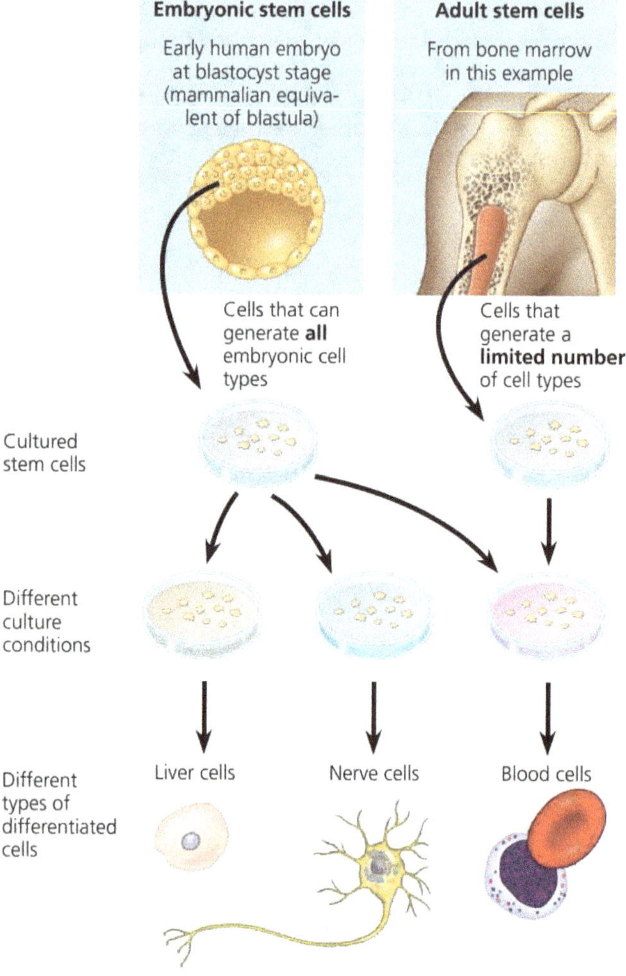

▶ **Figure 8.55** Working with stem cells. Animal stem cells, which can be isolated from early embryos or adult tissues and grown in culture, are self-perpetuating, relatively undifferentiated cells. Embryonic stem cells are easier to grow than adult stem cells and can theoretically give rise to all types of cells in an organism. The range of cell types that can arise from adult stem cells is not yet fully understood.

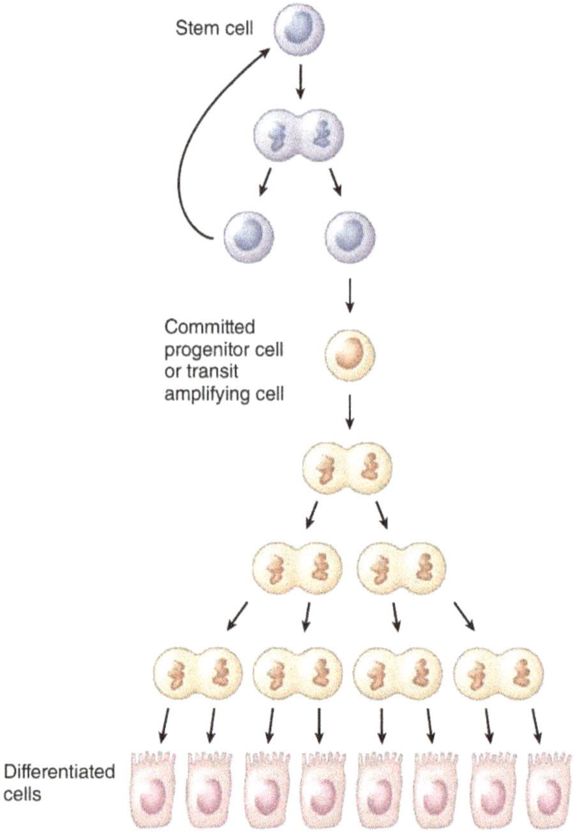

Stem cells are best studied in tissues with rapidly renewing cell populations, including blood cells, skin cells, and cells lining the digestive tract. Most mitotic cells here are not stem cells but the more rapidly dividing progeny of the cells committed to differentiation. They are commonly called **progenitor cells** or transit amplifying cells because they are in transit along the path from the stem cell niche to a differentiated state, while still amplifying by mitosis the number of new cells available for the differentiated tissue. Cells formed by progenitor cells may become terminally differentiated, meaning that renewed cycling cannot occur and the specialized cells exist for only a short time.

In tissues with stable cell populations, such as most connective tissues, smooth muscle, and the cells lining blood vessels, stem cells are not readily apparent and differentiated cells appear to undergo slow and episodic division to maintain tissue integrity.

◀ **Figure 8.56** Diverse shapes and sizes of human cells.

In the adult, stem cells are found throughout the body – for example, in the bone marrow, skin, gut, heart, retina, dental pulp, brain and other sites. Research into stem cells has opened up some exciting medical applications. Stem cell therapy is the introduction of new adult stem cells into damaged tissue to treat disease or injury. Bone marrow transplantation is an example of this therapy that has progressed beyond the experimental stage into routine medical practice. It is used to treat blood and bone marrow diseases, and blood cancers such as leukaemia.

Adult stem cells from bone marrow have long been used in bone marrow transplants as a source of immune system cells in patients whose own immune systems are nonfunctional because of genetic disorders or radiation treatments for cancer. In the future, it is hoped to be able to treat conditions such as diabetes, muscle and nerve damage, and brain disorders such as Parkinson's and Huntington's diseases. Experiments with growing new tissues, or even organs, from isolated stem cells in the laboratory have also been conducted.

Induced Pluripotent Stem (iPS) Cells

In another approach to make stem cells for research and therapy, researchers succeeded in 2007 in learning how to turn back the clock in fully differentiated cells, reprogramming them to act like ES cells. Differentiated cells can be transformed into a type of ES cell by using a modified retrovirus to introduce extra, cloned copies of four "stem cell" master regulatory genes. The "deprogrammed" cells are known as induced pluripotent stem (iPS) cells because, in using this fairly simple laboratory technique to return them to their undifferentiated state, pluripotency has been restored.

There are two major potential uses for human iPS cells. First, cells from patients with diseases have been reprogrammed to become iPS cells, which act as model cells for studying the disease and potential treatments. Human iPS cell lines have already been developed from individuals with type 1 diabetes, Parkinson's disease, Huntington's disease, Down syndrome, and many other diseases. Second, in the field of regenerative medicine, a patient's own cells could be reprogrammed into iPS cells and then used to replace nonfunctional tissues, such as cells of the retina of the eye that have been damaged by a condition called age-related macular degeneration (AMD).

▶ **Figure 8.57** Can a fully differentiated human cell be "deprogrammed" to become a stem cell?

Experiment Shinya Yamanaka and colleagues at Kyoto University, in Japan, used a retroviral vector to introduce four genes into fully differentiated human skin fibroblast cells. The cells were then cultured in a medium that would support growth of stem cells.

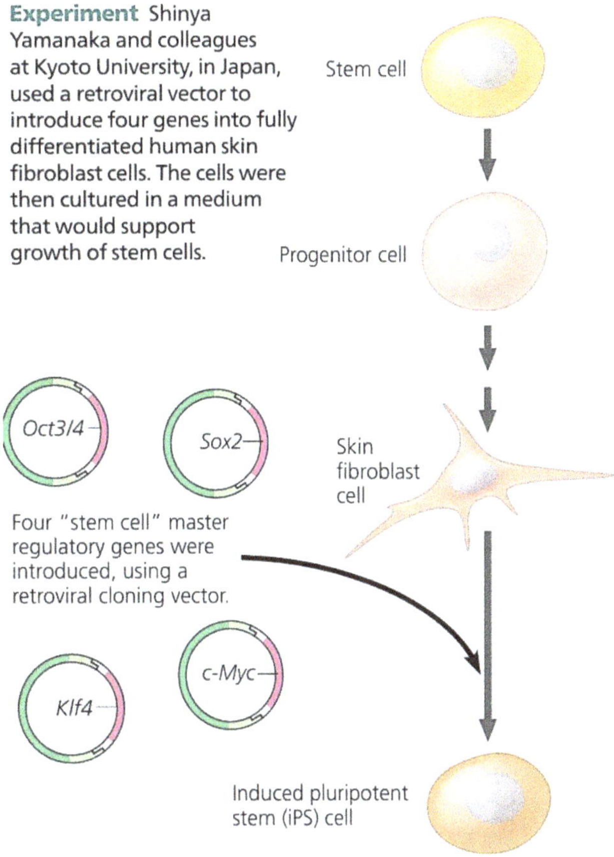

Four "stem cell" master regulatory genes were introduced, using a retroviral cloning vector.

Results Two weeks later, the cells resembled embryonic stem cells in appearance and were actively dividing. Their gene expression patterns, gene methylation patterns, and other characteristics were also consistent with those of embryonic stem cells. The iPS cells were able to differentiate into heart muscle cells, as well as other cell types.

Conclusion The four genes induced differentiated skin cells to become pluripotent stem cells, with characteristics of embryonic stem cells.

Data from K. Takahashi et al., Induction of pluripotent stem cells from adult human fibroblasts by defined factors, *Cell* 131:861–872 (2007).

KEY CONCEPTS

- Prokaryotic cell division is clonal, resulting in two identical cells. Bacterial DNA replication and partitioning of the chromosome are concerted processes.

- DNA replication begins at a specific point, the origin, and proceeds bidirectionally to a specific termination site. Newly replicated chromosomes are segregated to opposite poles at the same time as they are replicated. New cells are separated by septation, which involves insertion of new cell membrane and other cellular materials at the midpoint of the cell. A ring of FtsZ and proteins embedded in the cell membrane expands radially inward, pinching the cell into two new cells.

- The phases of the cell cycle are gap 1 (G_1), synthesis (S), gap 2 (G_2), mitosis, and cytokinesis (C). G_1, S, and G_2 are collectively called interphase, and mitosis and cytokinesis together are called M phase.

- The length of a cell cycle varies with age, cell type, and species. Cells can exit G_1 and enter a nondividing phase called G_0; the G_0 phase can be temporary or permanent.

- G_1, S, and G_2 are the three subphases of interphase. G_1 is the primary growth phase; during S phase, DNA synthesis occurs. G_2 phase occurs after S phase and before mitosis.

- The centromere binds proteins assembled into a disklike structure called a kinetochore where microtubules attach during mitosis. The centromeric DNA is replicated, but the two DNA strands are held together by cohesin proteins.

- In prophase, chromosomes condense, the spindle is formed, and the nuclear envelope disintegrates. In animals cells, centriole pairs separate and migrate to opposite ends of the cell, establishing the axis of nuclear division.

- Chromatids of each chromosome are connected to opposite poles by kinetochore microtubules. They are held at the equator of the cell by the tension of being pulled toward opposite poles.

- At anaphase, cohesin proteins holding sister chromatids together at the centromeres are destroyed, and the chromatids are pulled to opposite poles. This movement is called anaphase A, and the movement of poles farther apart is called anaphase B.

- Telophase reverses the events of prophase and prepares the cell for cytokinesis.

- A contractile ring of actin under the membrane contracts during cytokinesis.

- In plant cells, fusion of vesicles produces a new membrane in the middle of the cell to produce the cell plate.

- Experiments showed that there are positive regulators of mitosis, and that there are proteins produced in synchrony with the cell cycle (cyclins). The positive regulators are cyclin-dependent kinases (Cdks). Cdks are complexes of a kinase and a regulatory molecule called cyclin. They phosphorylate proteins to drive the cell cycle.

- Checkpoints are points at which the cell can assess the accuracy of the process and stop if needed. The G_1/S checkpoint is a commitment to divide; the G_2/M checkpoint ensures DNA integrity; and the spindle checkpoint ensures that all chromosomes are attached to spindle fibers, with bipolar orientation.

- The cycle progresses by the action of Cdks. Yeast have only one CDK enzyme; vertebrates have more than four enzymes. During the G_1 phase, G_1 cyclin combines with Cdc2 kinase to form the Cdk that triggers entry into S phase.

- The anaphase-promoting complex/cyclosome (APC/C) activates a protease that removes cohesins holding the centromeres of sister chromatids together; the result is to trigger anaphase, separating the chromatids and drawing them to opposite poles. The APC/C also triggers destruction of mitotic cyclins to exit mitosis.

- Growth factors, like platelet-derived growth factor (PDGF), stimulate cell division. This acts through a MAP kinase cascade that results in the production of cyclins and activation of Cdks to stimulate cell division in fibroblasts after tissue injury.

- Mutations in proto-oncogenes have dominant, gain-of-function effects leading to cancer. Mutations in tumor-suppressor genes are recessive; loss of function of both copies leads to cancer.

- The gain or loss of chromosomes is usually lethal.

- Chromosomes are composed of chromatin, a complex of DNA, and protein. Heterochromatin is not expressed and euchromatin is expressed. The DNA of a single chromosome is a very long, double-stranded fiber. The DNA is wrapped around a core of eight histones to form a nucleosome, which can be further coiled into a 30-nm fiber in interphase cells. During mitosis, chromosomes are further condensed by arranging coiled 30-nm fibers radially around a protein scaffold.

- Newly replicated chromosomes remain attached at a constricted area called a centromere, consisting of repeated DNA sequences. After replication, a chromosome consists of two sister chromatids held together at the centromere by a complex of proteins called cohesins.

References

- Allott Andrew and Mindorff David, Biology, OXFORD IB DIPLOMA PROGRAMME, OXFORD UNIVERSITY PRESS, 2014.

- Anthony L. Mescher, Junqueira's Basic Histology, 15th Edition, Mc GrawHill, 2018.

- Brooker Robert, Widmaier Eric, Graham Linda, Stiling Peter, Principles of Biology, 4th Edition, McGraw -Hill, 2024.

- Cecie Starr, Christine Evers, Starr Lisa, Biology Today and Tomorrow with Physiology, Broks/Cole, Cengage Learning,4th Edition, 2013.

- Hoefnagels Marielle, Biology, Concepts and Investigations, 3th Edition, McGraw-Hill, 2015.

- Jones Mary, Fosbery Richard, Taylor Dennis, Gregory Jennifer, Biology for Cambridge International AS & A Level, 5th Edition, Cambridge University Press, 2020.

- Kathleen Anne Ireland, Visualizing Human Biology, 5th Edition, Wiley& National Geographic Society, 2018.

- Mader Sylvia, Windelspecht Michael, Human Biology, 15th Edition, McGraw-Hill Education, 2018.

- Marieb Elaine, Hoehn Katja, Human anatomy & Physiology, 10th Edition, Pearson, 2016.

- Raven Peter, Johnson George, Mason Kenneth, Losos Jonathan, Singer Susan, Biology, 11th Edition, McGraw -Hill, 2017.

- Russel Hertz Mcmillan, Biology The Dynamic Science, 4th Edition, Broks/Cole, Cengage Learning, 2017.

- Solomon Eldera, Berg Linda, Martin Charles, Martin Diana, Biology, 11th Edition, Cengage, 2019.

- Tortora Gerard, Derrickson Bryan, Introduction to the human body, 10th Edition, Wiley, 2015.

- Urry Lisa, Cain Michael, Wasserman Steven, Minorsky Peter, Orr Rebecca, Jackson Robert, Reece Jane, Campbell Biology In Focus, Pearson, 2014.

- Urry Lisa, Cain Michael, Wasserman Steven, Minorsky Peter, Orr Rebecca, Campbell Biology, 12th Edition, Pearson, 2021.

- Widmaier Eric, Raff Hershel, Strang Kevin, Shoepe Todd, Vander's Human Physiology, 15th Edition, Mc GrawHill, 2019.